THE GEOLOGY OF DONEGAL:
A Study of Granite Emplacement and Unroofing

a volume in the

REGIONAL GEOLOGY SERIES

edited by

L. U. DE SITTER

Published

GEOLOGY OF THE HIMALAYAS
Augusto Gansser

THE TECTONICS OF THE APPALACHIANS
John Rodgers

THE EAST GREENLAND CALEDONIDES
J. Haller

THE SCANDINAVIAN CALEDONIDES
T. Strand and O. Kulling

THE GEOLOGY OF DONEGAL
W. S. Pitcher and A. R. Berger

THE GEOLOGY OF DONEGAL:
A Study of Granite Emplacement and Unroofing

WALLACE S. PITCHER
Jane Herdman Laboratories of Geology
University of Liverpool

ANTONY R. BERGER
Department of Geology
Memorial University of Newfoundland

With a geological map compiled by
Margaret O. Spencer

A poem is never finished, it is only abandoned
        Auden

WILEY-INTERSCIENCE

a division of John Wiley & Sons, Inc.

New York   London   Sydney   Toronto

Dedicated to the memory of
Professor Herbert Harold Read,
a world-renowned disputant
on the origin of granites.

Copyright © 1972, by John Wiley & Sons, Inc.

All rights reserved. Published simultaneously in Canada.

No part of this book may be reproduced by any means, nor transmitted, nor translated into a machine language without the written permission of the publisher.

Library of Congress Cataloging in Publication Data:
Pitcher, Wallace S.
The Geology of Donegal:
A Study of Granite Emplacement and Unroofing

(Regional geology series)
   1. Intrusions (Geology)—Northern Ireland—Donegal (County) 2. Granite—Northern Ireland—Donegal (County) I. Berger, Antony R., joint author.

II. Title.
QE611.P57   554.16′93   77-39047
ISBN 0-471-69055-4

Printed in the United States of America

10 9 8 7 6 5 4 3 2 1

EDITOR'S NOTE TO REGIONAL GEOLOGY SERIES

The aim of this series on Regional Geology is to add to the available geological literature concise descriptions of large structural units, independent of national boundaries.

It is important that the personal opinion of an author, formed by his work and experience in the structural unit he describes, comes clearly to the attention of the reader. Theorizing about the geological history of a particular kind of structure too often does not take into account the great diversity of the observed phenomena, and then generalizes in an unwarranted way. We aim to give a better basis to these general concepts and thereby stimulate a deeper understanding of the relations between different kinds of structures.

Some of the books describe classical territory where new work has brought new conceptions, others are concerned with hitherto relatively unknown regions, but always the surveys are presented from a fresh aspect.

L. U. DE SITTER

PREFACE

The origin and emplacement of granites has long provided a central debating point in geology. However, with hindsight, it is now easy to see that the granite controversy was bedeviled by overgeneralization, and that now the need is for very detailed studies of particular plutons in different geological settings. Furthermore, an integrated scientific approach is obviously required, though we would submit that a great part of the primary evidence still lies in the field. We refuse to echo Larsen's plaintive cry (1958, p. 30) that 45 years of field work on granites had provided him with no evidence to settle the problem of their origin, or adopt Krauskopf's attitude of hopelessness (1968). The reader of this book must judge how far the data obtained here from both field and laboratory studies have contributed to a local solution of the granite controversy, though we would prefer to regard this work in Donegal as merely a foundation on which others can build.

The detailed exposition of a relatively small portion of the Scottish and Irish Caledonides is justified on many counts. Unlike most of the British Isles, northwest Donegal has been uniformly mapped in recent years. Most of these studies have been published in one form or another, but there is a clear need for a synthesis of this material. Furthermore, modern work in Donegal has an important bearing on contemporary discussions of the whole Caledonian fold belt. Donegal also displays an abundance of geological phenomena that are of more than regional importance, and in this book we shall concentrate on the granites because they provide examples of a wide range of geological behavior that is unsurpassed elsewhere. Indeed, we intend to use these granites as a basis from which to discuss granite phenomena in general; such topics as varying modes of emplacement, internal structures in plutons, and the relationships between granitic plutons and their country rock structures and metamorphism interest us in particular. Throughout this book we have therefore referred widely, even overgenerously, to the relevant literature, attempting a fairly complete coverage to the end of 1970.

Shortly after the Second World War, at the height of one of the recurrent debates on the origin of granite, Dr. (now Professor) Robert Shackleton pointed out to his colleague, Professor H. H. Read, the likely significance of the geology of Donegal in the study of the Caledonides in general and of granites in particular. Together, Read and Shackleton initiated the research program which 24 years later is the subject of this book.

Upwards of 45 workers have been involved, mostly from Imperial College and King's College, London, and from the University of Liverpool. They include M. K. Akaad, M. P. Atherton, A. R. Berger, F. W. Cambray, R. L. Cheesman, P. J. Curtis, W. M. Edmunds, R. W. D. Elwell, M. Fernandes-Davilla, W. J. French, A. R. Gindy, M. G. Ghobrial, A. Hall, J. J. T. Harvey, S. V. P. Iyengar, J. L. Kirwan, E. L. P. Mercy, R. S. Mithal, M. H. Naggar, D. C. Knill, G. L. H. McCall, T. M. S. Obaid, I. C. Pande, W. S. Pitcher, T. C. R. Pulvertaft, H. H. Read, M. J. Rickard, R. M. Shackleton, R. C. Sinha, T. B. Smart, K. K. Srivastava, C. F. Tozer, and G. P. L. Walker. Further contributions have been made by investigators from other schools: J. G. C. Anderson, W. R. Church, G. H. Dury, T. N. George, R. J. Howarth, J. A. E. B. Hubbard, J. Kemp, G. P. Leedal, G. G. Lemon, D. H. Oswald, I. M. Simpson, E. H. T. Whitten, and R. S. R. Wood. Since the first printed work on the geology of Donegal close to 200 memoirs, papers, and abstracts on the area have been published, and 37 Ph.D. and M.Sc. theses have been written.

All these contributions to the geology of Donegal have enabled Margaret Spencer and myself to revise many of the one-inch geological sheets of the County. Sheets Nos. 3, 4, 9, 10, 15, and 23 are complete, and substantial parts of Nos. 16 and 24 have also been redrawn. This particular part of the project has been made possible by the award of an NERC Research Grant, and by the free use of the Ordnance maps granted by the Assistant Director of the Ordnance Survey of Ireland. The topography of Maps 1 and 2 in the back pocket is based on the Ordnance Survey of Ireland by permission of the Government of Ireland (Permit No. 856).

In recent years, Dr. Antony Berger so extended the structural studies that it was clearly appropriate to call upon him to share the authorship of this book; without his contribution this account would be so much the poorer as to be reckoned old-fashioned. Indeed, there are parts of it where this remains so, because no program of field research over so long a period could be continuously updated on all fronts. Further, the geological coverage is so wide that the authors might be excused for not being everywhere authoritative. We have simply done our best in the time available, believing that an accurate account of the remarkable geology of Donegal will usefully add fuel to the active "dialogue" between those engaged in the experimental and observational fields of geological research.

In some part of this book, the general reader may well wish for a simpler statement than is, in fact, provided. But it has been necessary to document many facts for the first time, even of much material unlikely to be published, and, further, some problems have proved so intractable (to the authors) that to present a single model would be more than usually misleading. Indeed, in a few instances the authors have so far differed in opinion that the final agreed version may seem obscure: only the researcher of the future may then fathom the implied discontent of the junior author!

As will be obvious from the names of colleagues mentioned above, much of the work summarized in this book is that of others. Nevertheless, I have had the privilege of seeing it all done, sometimes offering a little guidance, and so might be expected to realize the significance of all the separate parts of the work. However, any compilation of this kind can only be completed by sometimes riding roughshod over the careful work of colleagues, to whom I must offer some kind of apology, taking all the responsibility for the resulting inaccuracies and failings upon myself.

I can only hope that the immense debt of gratitude I owe to Professor H. H. Read is in some part repaid by this compilation. My sincere thanks to my coauthor for bearing with "professorial opinions," to Mrs. M. Spencer, aided by Miss J. Treasure, for a beautiful geological map, to Mr. J. Lynch for his great skill in the preparation of the diagrams, to Mr. D. Smart for equal skill in photography, to Mrs. G. Flinn for the very useful index, to Mr. P. Scott for much general assistance, and to my wife, Stella, for the preliminary editing. My coauthor wishes to acknowledge financial assistance from the Universities of London, Liverpool, and Toronto, and from the Leverhulme Trust.

Thanks also to the following persons for reviewing parts of the manuscript: G. M. Anderson, F. W. Cambray, G. Chinner, M. P. Atherton, W. M. Edmunds, J. Treagus, J. S. Watterson, M. J. Rickard, R. Riddihough, G. Norris, W. M. Schwerdtner, R. James, and M. B. Katz.

Then, on behalf of all those who have worked in Donegal, I gladly take this opportunity to thank the Irish people who have helped in so friendly a way. I must especially mention the MacMonagles of Thorr, the Sweeneys of Dunglow, the Baskins of Ardara, the Breslins of Boyoughter, the McClaffertys of Church Hill, the Colls of Creeslough, the Coyles of Kilmacrenan, the McGinlys of Milford, the Jacksons of Ballybofey, and MacElhinney of Glenveagh Castle.

A final point: on his last visit to Donegal, the late Herbert Read left his socks and boots on a wall for "posterity," and I suspect that in order to finally unravel the geological history of Donegal, posterity will need the magic of these boots, if not the socks!

Liverpool, England   W. S. Pitcher

CONTENTS

INTRODUCTION   1

Introduction   1
Development of Knowledge of Donegal Geology   6
Outline of the Geology of Donegal   9

Part I THE GEOSYNCLINAL PILE AND ITS DEFORMATION

Chapter 1
THE SEDIMENTARY ASSEMBLAGES: STRATA NEAR THE BASE OF THE CAMBRIAN   15

Introduction   15
The Creeslough Succession   18
The Kilmacrenan Succession   26
The Relation between the Creeslough and Kilmacrenan Successions   38
The Sedimentary Facies of the Dalradian of Donegal   39

Chapter 2
THE DOLERITE SILLS: THOLEIITE IN THE EARLY STAGES OF OROGENY   43

Introduction   43
The Quartz-Dolerites   44
The Dolerites of Rosbeg Type   47
Age of the Dolerites   48
The Metamorphism of the Basic Rocks   48
Chemical Composition   49
Discussion   50

Chapter 3
CALEDONIAN REGIONAL STRUCTURE AND METAMORPHISM: A POLYPHASE HISTORY   52

Introduction   52
The Structures of the Creeslough Succession   56
Metamorphic Development of the Creeslough Succession   66
Structure and Metamorphism in the Kilmacrenan Succession Northwest of the Leannan Fault   69
The Structural Relationship between the Creeslough and Kilmacrenan Successions   72

CONTENTS

An Outline of the Structural and Metamorphic Development of the Kilmacrenan Succession Southeast of the Leannan Fault	74
Donegal in Relation to the Caledonian Mobile Belt as a Whole	76
Final Comment	82
A Geological Appendix: the Late Tectonic Lamprophyres	82

Part II THE GRANITES

Chapter 4
THE CALEDONIAN GRANITES AND THEIR ASSOCIATED ROCKS: INTRODUCTORY COMMENTS — 87

Introduction	87
Age Relationships	90
Notes on Nomenclature	91

Chapter 5
THE THORR PLUTON: AN EXAMPLE OF ACTIVE STOPING AND CONTAMINATION — 96

The Outcrop	97
The Envelope	97
The Magma	102
The Reaction between Magma and Country Rocks: an Assimilation Model	111
Chemical Composition: the Need for a Contamination Model	120
The Distribution of the Inclusions: Ghost Stratigraphy and Structure	124
The Origin of the Thorr Rocks	129

Chapter 6
THE FANAD PLUTON: AN UNCOMPLICATED EXAMPLE OF STOPING — 132

Introduction	132
The Outcrop	134
The Contact with the Envelope: an Example of Varying Contact Effect	134
Composition and Texture: Contamination of a Magma	136
The Fabric	139
The Distribution of the Larger Xenoliths: Ghost Stratigraphy Again	139
The Mode of Emplacement: Piecemeal Stoping	139
An Appendix: a Related Dyke Swarm	141

Chapter 7
THE APPINITE SUITE: BASIC ROCKS GENETICALLY ASSOCIATED WITH GRANITE — 143

Introduction	143
General Distribution in Space and Time	148
The General Form of the Intrusions	149
The Intrusive Breccias	156
The Associated Lamprophyres	161
The Origin of the Appinitic Suite	162

Chapter 8
EXAMPLES OF DIAPIRIC INTRUSION — 169

SECTION A. THE ARDARA PLUTON — 169

Introduction — 169
The Components as Magmas — 170
The Contamination Model — 173
The Structure of the Pluton — 174
The Structure of the Envelope — 178
Contact Metamorphism in Relation to the New Structures — 180
The Ardara Pluton as a Diapiric Intrusion — 182

SECTION B. THE TOORIES PLUTON: A FRAGMENT OF ANOTHER DIAPIR — 184

Chapter 9
THE ROSSES CENTERED COMPLEX: AN EXAMPLE OF CAULDRON SUBSIDENCE — 186

Introduction — 186
The Components: Field Relationships — 187
The Components: Composition and Texture — 192
Petrogenesis — 194
Relation of Form of the Contacts to Mode of Emplacement — 198
The Rosses as a Centered Complex — 199

Chapter 10
THE BARNESMORE PLUTON: AN EXAMPLE OF CAULDRON SUBSIDENCE AND THE FORMATION OF IGNEOUS ARCHES — 201

Introduction: Time of Emplacement — 201
The Envelope: a Static Environment — 202
The Major Components — 203
Structural Relationships: Subsidence and Collapse — 204
Later Events: Desilication and Dykes — 205
Mode of Emplacement: Cauldron Subsidence and the Collapse of Roof Arches — 208

Chapter 11
THE MAIN DONEGAL PLUTON: A STUDY IN SYNPLUTONIC DEFORMATION — 210

SECTION A. INTRODUCTION: THE PROBLEM — 211
Preamble — 211
Salient Features and the Pitcher-Read Model — 211

SECTION B. THE INTERNAL FEATURES: A DEFORMATION HISTORY — 215
The Banding: a Unique Structure — 215
The Mineral Alignment — 219
The Dykes: a Chronology of Emplacement — 220
Discussion of Banding and Dykes — 226
The Marginal Zones: Cataclastic Deformation — 227
Petrography: the Importance of Late Mineralogical Changes — 230

SECTION C. THE GRANITE DEFORMATION: ITS GEOMETRY AND PATTERNS OF STRESS AND STRAIN — 234
Introduction — 234

The Mineral Alignment: Kinematic Significance of the L–S Fabric System	234
The Inclusions: Important Parameters for the Analysis of Strain	234
Late Dykes and the Regional Stress Pattern in Central Areas of the Pluton	247
The Deformation in Marginal Areas	249
SECTION D. A DISCUSSION SUMMING UP THE DEFORMATION MODEL	251

Chapter 12
THE ENVELOPE OF THE MAIN DONEGAL PLUTON: A DEBATE ON CHRONOLOGY — 257

Introduction: the Previous Model	258
A Structural Veneer: the Later Events, DMG_2 and DMG_3	260
The Metamorphism of the Metasediments in the Envelope: an Early Event, DMG_1	268
The Metamorphism and Deformation of the Preexisting Igneous Rocks	281
The Special Problem of the Mullions	283
The Correlation of Deformative Events in Envelope and Pluton: the Deformation of the Marginal Sheets	286
The Origin of the Main Donegal Pluton and its Complex Aureole	288

Chapter 13
THE TRAWENAGH BAY PLUTON: AN ENIGMA — 296

Introduction	296
The Margins	297
The Magmas	298
The Relationship between the Trawenagh Bay and Main Donegal Plutons	299
Discussion: Mode and Time of Emplacement	300

Chapter 14
THE CONTROLS OF CONTACT METAMORPHISM — 302

General Statement	302
The Chronology of Reaction and Recrystallization in the Aureoles	303
The Possibility of Changes in Bulk Composition	312
The Effect of Host Rock Composition on Mineral Reactions	314
The Physico-chemical Controls	317
Mineral Paragenesis in Relation to the Physico-chemical Controls: the Importance of Rate Processes	319
Some Further Conclusions	326

Chapter 15
THE FABRIC OF GRANITIC ROCKS: COMMENTS ON GRANITE TECTONICS, THE INTERPRETATION OF TEXTURES, AND ON GHOST STRATIGRAPHY — 328

Preamble: Primary versus Secondary Structures	328
The Rheology of Mobile Granitic Material	329
The Mineral Fabric of Granites: Magmatic Indicators and Mechanisms of Flow	332
Ghost Stratigraphy and Structure in Plutonic Rocks	335

Chapter 16
THE FORM OF THE DONEGAL GRANITES AND THE ORIGIN OF THE MAGMAS — 338

General	338

The Form of the Plutons at Depth 338
The Time of Emplacement Relative to the Regional Thermal Event 341
The Chemical Variation within the Plutons at the Present Level 342
The Bulk Compositions 344
The Source of the Granite Magmas 345
Conclusion: Final Model 350

Part III LATE EVENTS

Chapter 17
FAULTING AND UPLIFT, EROSION, UNROOFING, AND SEDIMENTATION   355

Uplift and Erosion 355
The Wrench Faults 357
The Old Red Sandstone: Intermontane Molasse 363
The Viséan Marine Transgression 366
The Hercynian Mineralization 371

Chapter 18
THE TERTIARY DYKE SWARM: THE CRUST UNDER TENSION   373

BIBLIOGRAPHY I: GENERAL   377

BIBLIOGRAPHY II: THESES ON DONEGAL   411

AUTHOR INDEX   413

SUBJECT INDEX   419

INTRODUCTION

CONTENTS

1. Introduction 1
   (A) Location 1
   (B) Broad Geological Framework 1
   (C) The Donegal Landscape 3
2. Development of Knowledge of Donegal Geology 6
3. Outline of the Geology of Donegal 9

1. Introduction

(A) Location

Donegal forms the northwestern coign of Ireland, its Atlantic coastline indented by many long sea-loughs. The location of our study includes only that part of this long seaboard and its hinterland stretching between Fanad and Donegal Bay, though we shall also touch on the geology of parts of the peninsulas of Inishowen and Slieve League (Fig. I-1). This part of the county, covering some 1250 square miles (3240 square kilometers), is best referred to as "Northwest and Central Donegal," and this indeed is the title of the relevant Memoir of the Geological Survey of Ireland (Hull *et al.*, 1891). Within this area, a varied topography expresses a diverse geological construction, and for the most part, the rocks are splendidly exposed; it is this very factor which has led to the detailed geological investigations that form the subject of this book.

(B) Broad Geological Framework

Donegal is part of the NE–SW trending Caledonian fold belt of Lower Palaeozoic age, and most of its stratigraphic groups and many of its larger structures can be easily recognized in the Highlands of Scotland and in Counties Mayo and Galway to the southwest (Fig. I-1). (See Pitcher and Cheesman, 1954; Read, 1958*b*; Cambray, 1969*b*, for earlier brief reviews of Donegal geology, and Charlesworth, 1963, for a general review of Irish geology.) Most of Donegal is made up of two great groups of rocks. First, there are the metamorphosed and deformed sediments and basic igneous

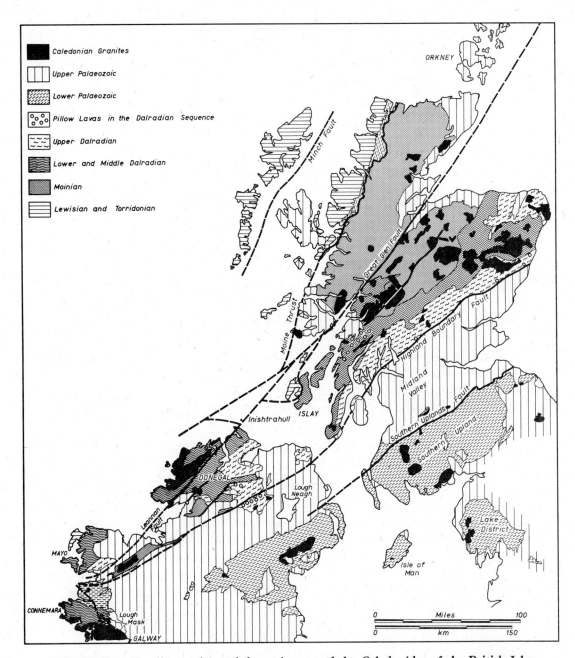

FIGURE I-1. Outline geology of the main part of the Caledonides of the British Isles.

rocks of the Dalradian Series, which pass southward into more highly metamorphosed rocks similar in appearance to the Moinian Series of Scotland. The second major group is formed by the granitic and allied rocks that were emplaced by a wide variety of mechanisms in the latter part of the Caledonian orogeny.

Less important items in terms of area are the overlying cover of Carboniferous Limestone in the basin of southern Donegal, a small outlier of Devonian Old Red Sandstone in eastern Fanad, and a prominent swarm of basaltic dykes of Tertiary age. Despite their smaller outcrop, however, the interpretations of these rocks are topics of some considerable interest. Finally, the records of the Pleistocene, glacial and postglacial, close our history. These may conveniently be discussed together with the physiographic features of Donegal in this introductory chapter.

(C) The Donegal Landscape

The predominantly NE–SW trend of the Dalradian metasediments and of the main mass of granite produces a marked grain in the topography of the northwestern part of the country (see Map 1 in folder). The line of quartzite mountains extending from Errigal northeasterly to Muckish and Crockatee, and the high granitic ridges of the Derryveagh and Glendowan Mountains accentuate this grain, as do the many NE–SW faults. It is interesting to point out here that the most prominent fault grooves, such as those of Glenveagh and Barnesmore Gap, follow dislocations in massive granite along which the net displacement has been relatively minor. On the other hand, the fundamental geological dislocation of Donegal, that of the Leannan Fault, along which there appears to have been upward of 40 kilometers displacement, is marked for the most part by a relatively minor topographic feature.

This physiographic alignment, of course, influences the direction of many important streams, though their detail in granitic terrains is often controlled by the regional joint pattern. In contrast to this, many of the chief river systems drain eastward or westward right across the geological grain, and the main watershed runs roughly north to south not far inland from the west coast (Fig. I-2). Superposition of drainage has undoubtedly occurred. There has been much discussion of the possibility of former land surfaces formed during the Devonian and modified to varying extents during the Carboniferous (Dewey and McKerrow, 1963), the Cretaceous (Dury, 1959), or the mid-Tertiary (George, 1967); but whatever the precise history of events, it appears that the major consequents, such as the Calabber, Swilly, and Finn Rivers, were initiated on this pre-glacial surface.

The highest summits, at just over 670 meters, may represent a high surface widespread in Ireland and Scotland (George, *loc. cit.*, p. 437). There are remnants of another old land surface at 490 meters, more distinct remnants of one at about 245 meters, and a trace of another at 180 meters. But the most obvious peneplain is that developed along the coast as in the Rosses, where there are wide areas at a height of about 60 meters. The rivers are graded to this level and exhibit marked knick points below which there are entrenchments, as for example, on the Gweedore and Clady. This 60 meter surface rises inland to a very distinct break of slope at approximately 137 meters. All these preglacial erosion surfaces are well known in Ireland and elsewhere (*cf.* Miller, 1939; Reffay, 1966); the two lower are thought to be of mid-Tertiary and Pliocene age, respectively.

Many obvious features of the landscape are due to glaciation. From the extremely clear evidence of ice movement provided by striae, *roches moutonnées*, drumlins, and erratic trains, and from the distribution of corries, U-valleys, and moraines, Charlesworth (1924 and 1963, p. 452) has compiled a picture of ice moving radially from a source situated high on the main part of the watershed (Fig. I-3). This center of accumulation was perhaps the most extensive of the gathering grounds for the local ice sheet which covered Ireland and prevented the Scottish ice-cap from advancing farther than the base of the Inishowen peninsula (Hill and Prior, 1968; Colhoun, 1971). At the glacial maximum, the Slieve League Peninsula developed a local ice-cap thick enough to remain independent of this Donegal ice, except in the east where some overriding by the main mass occurred (Dury, 1964).

As Dury has shown, one result of this main phase of glaciation was the breaching of the old watershed by the backward erosion of corrie heads (1957; Fig. I-3). Another result was the scouring of deep U-valleys along fault zones, such as those of Glengesh, Glenveagh, and Barnesmore. The Glendowan and Blue Stack Mountains, which occupied the central position in this ice center, were

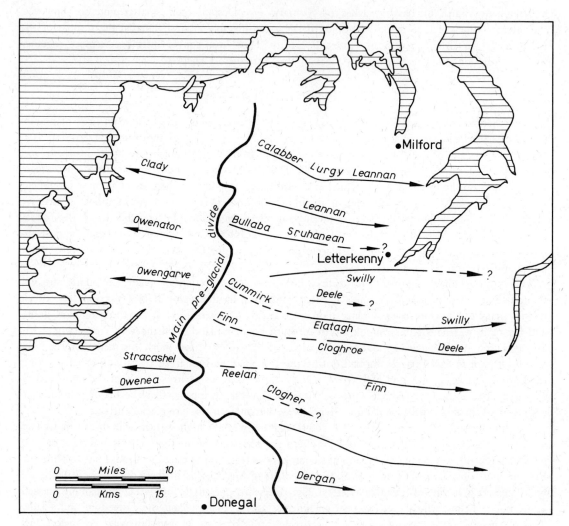

FIGURE I-2. The main watershed of northwestern Ireland showing the consequent river system (after Dury, 1959).

scraped bare of overburden, but the valleys were filled with debris. In these, the withdrawal of the glacier front can be traced back into the corries through a succession of crescent-shaped moraines.

On the coast, true fjords were excavated as exemplified by Mulroy Bay and Lough Swilly. In these and other sheltered positions, such as on Rosguill, Melmore, and Horn Head, where strands join former islands to the mainland, there are remnants of late- and postglacial raised beaches. These rise gradually from sea-level in the south to 5 meters at Bloody Foreland and to 25 meters in eastern Inishowen. Stephens and Synge (1965) reasonably attribute this northeasterly tilt to the great thickness of Scottish ice to the northeast of Inishowen at the time these beaches were formed. For the most part, however, the highly indented coastline with its steep cliffs and headlands is under active erosion. Excellent examples of this are provided by the spectacular cliffs of Malin, Fanad and Horn Heads, Aranmore, Malinmore, and Slieve League, this latter forming one of the highest sea-cliffs in Western Europe.

After the retreat of the ice, forests of pine and birch covered the valleys and lower slopes, and the stumps of these trees can be seen in the basal layers of turf cuttings. At some time during the late

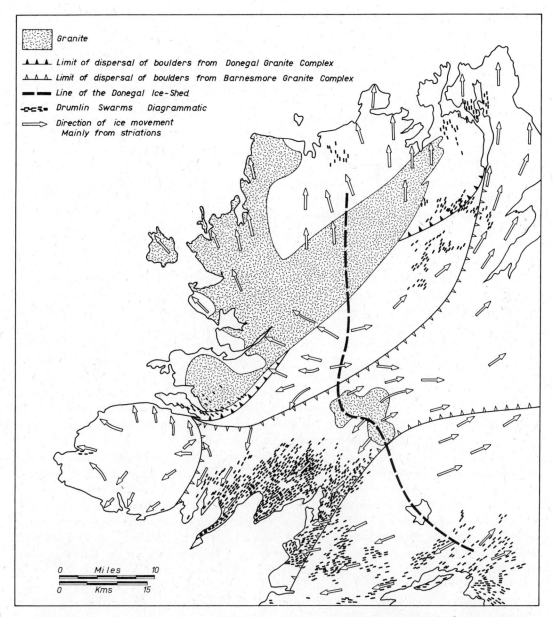

FIGURE I-3. Glacial features of Donegal. Much simplified from Charlesworth (1924).

Boreal period (6000 B.C.), peat began to accumulate, and an irregular increase in rainfall, coupled with the lack of drainage so typical of glaciated areas, continued well into historic time and led to the replacement of these forests by peat (Charlesworth, 1963). Thus, the scene today is one of open, treeless, high moor dotted with innumerable reedy loughs. At high levels, there are square miles of bare rock surfaces, while on lower slopes rocky knolls emerge from the mantle of drift and peat, and only in the valleys do these conceal the bedrock. Recent erosion of the peat bogs in the hill areas, together with the extensive removal of this valuable fuel, has considerably enlarged the area of exposures, so that northwest Donegal is as well exposed as anywhere in the British Isles.

2. Development of Knowledge of Donegal Geology

The first accounts of the rocks of Donegal were published in the early 1800's. Thus Donald Stewart (1800), appointed "Itinerant Mineralogist" by the Royal Dublin Society, James M'Parlan (1802), and Sir Charles Giesecke (1826) all compiled lists of known mineral occurrences in the county. Another pioneer account was that of J. F. Berger, M.D., a Geneva Wernerian, who described the Tertiary dykes of Donegal as part of a larger survey of Northern Ireland (1816).

In 1839, the first geological map of Ireland was published by Griffith, who showed several NE–SW trending belts of metamorphosed sediments in northwest Donegal. These were cut by a large V-shaped area of granite, together with several smaller outlying bodies. Although M'Adam (1834) had earlier described portions of Fanad, Griffith's map appears to have been the first real stimulus to geological research, and marks the beginning in Donegal, as elsewhere in Ireland, of a long succession of geological investigations. One of the most important of these was the work of the Geological Survey of Ireland, which had completed its primary mapping of Ireland on a scale of one inch to the mile by 1887. The Survey maps and accompanying Memoirs for that part of Donegal with which we are concerned here were published in 1888 (Egan *et al.*) and 1891 (Hull *et al.*; Hull, Kilroe and Mitchell), and are fine tributes to the excellent field work of these geologists.

The various metasedimentary rocks of Donegal were early recognized as being similar to those of the Grampian Highlands of Scotland, then considered to be Lower Silurian (*i.e.,* Ordovician) in age (Harkness, 1861). Comparisons were also made at this time with the rocks of eastern Canada. Thus, Sterry Hunt thought that some of the Donegal rocks were lithologically identical to the Laurentian and that others recalled the modified American Palaeozoic (1864, p. 67). Later, Hull (1881, 1882) revived this suggestion and also distinguished Silurian metasediments which he considered lay unconformably on Laurentian gneisses and granites. At this time, Kinahan (1881) criticized these "Archaean geologists" who claimed similarities with the Laurentian; the Donegal rocks were more akin, he argued, to the American Huronian (*i.e.,* Cambro-Silurian). In 1891, however, Kinahan changed his mind and made comparisons with the Precambrian Algonkian, and in 1901 stated finally that the Donegal granitic gneisses were Laurentian and the later granites Huronian.

Sir Archibald Geikie, in 1891, proposed the term "Dalradian" as an "appropriate and useful appellation for the crystalline schists of the north of Ireland and the center and southwest of Scotland" (p. 75). This name is derived from the old kingdom of Dalriada in the north of Ireland, which in ancient times extended its dominion over much of southwest Scotland. Geikie (*ibid.*) commented on the general continuity of these rocks from Scotland to Ireland, and considered that certain limestones, for example, could be traced in both areas for a distance of over 320 kilometers. Modern work has confirmed Geikie's views, and detailed correlations can now be made between the Dalradian metasediments of Donegal and Scotland. These will be discussed in detail later.

In the second half of the 19th century, there was much debate concerning the stratigraphical order of the Dalradian rocks of Donegal and their structural relationships (*e.g.,* Blake, 1862, pp. 294–9; Callaway, 1885, pp. 230–8; Kinahan, 1891; Hull *et al.,* 1891). One of the officers of the Irish Geological Survey, Kinahan (*ibid.*), even went so far as to disclaim all responsibility for his part in the Donegal Memoir, declaring that the editor (Hull) had rewritten it to conform with his own (Hull's) ideas. It is clear that most of these debates stemmed from the absence of satisfactory "way up" criteria, such as current-bedding, washouts, and graded bedding. We may point out here that one of the earliest statements—probably the first—of the importance of sedimentary structures in determining the proper stratigraphic succession of beds was made by Ganly (1857), who observed ripple marks and current-bedding in quartzitic rocks at Carndonagh in northwestern Donegal. Unfortunately, recognition of the significance of such criteria did not come until much later.

The evolution of ideas on the Donegal granites reflects the debate during the last century on the origin of granite itself. It is both instructive and chastening to find that the arguments employed then are very like those used in the last few decades in the discussion of the same perennial topic (*cf.* Read, 1957). It is now clear that the "Donegal Granite" consists of many different granites of different characters and modes of emplacement. This variety must be borne in mind when the early work is read—there is no "Donegal Granite" as the early workers supposed, and their observations and interpretations had only local validity.

In the middle of the 19th century, the Donegal granite debate centered, as always, on the metamorphic versus magmatic origin of the rock. An early statement by Kelly (1853, p. 239) seems to imply an intrusive origin, but the general consensus of opinion among Irish geologists in the 1860's favored a metamorphic genesis(*e.g.*, Scott, 1862, 1864; Blake, 1862; Haughton, 1862, 1864, 1866; Harte, 1867; Westropp, 1867). As Scott put it, the "typical Donegal granite . . . formed from materials existing on the spot without actual fusion . . . by some action insufficient to destroy all bedding or even to convert the whole mass into granite" (1862, p. 289). The "Donegal Granite" was bedded in accordance with the country rocks and was often interstratified with the adjacent sediments. Transitions between granite and country rock could be demonstrated, and the granite was very variable in composition and texture. Trains of sedimentary xenoliths could be traced for miles within the massive granite, and the attitudes of these inclusions agreed with those of the country rocks in general—here we find the modern argument of "ghost stratigraphy" and "ghost structure" being used a century ago. Some portions of the granite were gneissose, with a foliation concordant with that of the adjacent country rocks. Gneissose granitic bodies were recorded in crystalline structureless granite—presumably representing an arrested stage in the transformation. The Irish workers repeatedly emphasized the difference between the "Donegal Granite" and the great Leinster and Mourne Mountain masses, which were not "bedded." Further, as Blake (*loc. cit.*) pointed out, the "Donegal Granite" was unlike other granites in having no lodes associated with it. There was general agreement, therefore, among Irish geologists that the "Donegal Granite" was of metamorphic origin— it was a granitization granite, as we would now say.

Among the native opinion, we may note certain proposals of interest, particularly by the Reverend Samuel Haughton, a perennial contributor to the ever-growing literature on Donegal, and one whose energies were chiefly spent on lengthy comparisons of the chemical composition of the granites of Donegal with those of Leinster, Scotland, and Scandinavia (1862, 1864, 1866, 1870). Haughton made, in passing, the comment that in its central portions the "Donegal Granite" was "perhaps of igneous origin, originally deriving its cleavage planes and gneissose character from the pressure exercised on it by the stratified rock which has been lifted to the north and south to a nearly vertical position" (1862, p. 406).

In 1871, unaware of any of the previous work on Donegal, A. H. Green advanced a metamorphic origin for the "Donegal Granite," and in doing so, brought down upon himself a shower of classic vituperation from Forbes (1872), whose mildest remark was to label the metamorphic origin of granite "a strange hypothesis." Although he had never visited Donegal and was thus unhampered by any firsthand knowledge of the problem, Forbes pointed out that the granite there behaved like any other igneous rock, being sometimes concordant, sometimes discordant—it was certainly displacive and never replacive.

In a major contribution to Donegal geology, Callaway, in 1885, wrote that in Barnesbeg Gap "where I first touched the granite" it was "perfectly clear that I was on the margin of an intrusive mass" (p. 223). Callaway found granite veins penetrating the country rock schists along their foliation planes, and argued that the gneissose structure of the marginal parts of the granite was due to pressure after consolidation of the magma. Inclusions of sediments in the granite had perfectly sharp margins, and there was no correlation between the composition of the granite and its immediate inclusions, nor were there transitions between the granite and its country rocks. Although Callaway's work was not wholly accepted by his contemporaries, Hull (1886) and Kinahan (1888) both began to separate out different events in the granite story; intrusive granites were now distinguished from earlier and later metamorphic granites and gneisses.

The intrusive nature of the "Donegal Granite" was confirmed in the Memoir on Northwest and Central Donegal (Hull *et al.*, 1891). The foliated granite of Glenveagh was traced "with uninterrupted passage" into the unfoliated granite of Dunglow and Trawenagh Bay, which there truncated the country rock sediments (*ibid.*, p. 17). Hull stated "that the intrusive granite of the north coast, as well as that of Barnesmore, is all of one age—an age indeterminate, indeed, but later than that to which the chief metamorphism belongs" (*ibid.*, p. 18). Both Hull (*ibid.*) and Kilroe (in Hull, Kilroe and Mitchell, 1891, pp. 28–30) emphasized that the parallelism of the inclusions of country rock within the granite and the granite foliation itself were the result of later shearing. Kilroe concluded that "the intrusion of the granite marks an important epoch in the geological history of the Donegal area, having occurred, it would seem, between earlier and later periods of metamorphism (*ibid.*).

Early in this present century, much of the earlier work received the comments of Grenville Cole who, between 1900 and 1906, published a series of papers written in a somewhat romantic style. In 1902, he attributed the foliation of the granite to flow during emplacement and "emphasized to some extent by subsequent pressure" (p. 204). The various structures seen in the contact zones were due to "that bodily intermingling and incorporation of two dissimilar masses which results in the formation of composite gneisses" (*ibid.*)—in modern terms they were the products of migmatization of some kind. Cole later considered that the mixed character of the granitic gneisses was due to the incorporation in a rising granitic "dome," particularly in the marginal portions, of "previously foliated sedimentary and igneous material" (1906a, p. 80). The granite penetrated the country rocks along structural planes, such as bedding and schistosity, and its foliation was now interpreted as a relic of the preexisting sediments; this was the "extreme case of *lit-par-lit* injection" (1906b, p. 348).

Finally, in this review of the early work on the "Donegal Granite," we may cite two short contributions by Andrew (1928a and 1928b), in which he supported the *lit-par-lit* mechanism proposed by Cole and attributed the marginal gneissosity of the same granite to flow of partially crystallized magma, and not to postconsolidational shear, as the Survey Memoirs had stated.

In the years immediately following the Second World War, Donegal once again became the subject of many geological studies. At this time, the widespread and often acrimonious controversy about the origin of granite—magmatic or transformist—was once more at the center of geological discussion (*e.g., Report of the 18th International Geological Congress, London,* 1948, Part III, published 1950; Gilluly, 1948; Read, 1957). One of the chief protagonists, H. H. Read, then Professor at Imperial College, London, felt that a detailed study of some large, well exposed granite body would prove instructive. The granites of Donegal seemed to fit this description, and the complete divergence of opinion revealed in the earlier accounts suggested that a wide variety of granitic phenomena might be found here. Accordingly, a team attack was begun by students from Imperial College, first under the direction of Read, and later under the joint leadership of Read and Pitcher, the senior author of this book and one of the original research team (see Read, 1955, 1958b). Subsequent contributions have also come from other British universities.

Individual areas in the granitic terrain and the Dalradian metasediments have been mapped, chiefly on a scale of 6 inches to the mile ($1'' = 880'$), by Gindy (1953a), Mithal (1952T*), Pitcher (1953a and 1953b), Shackleton (in Pitcher and Cheesman, 1954), McCall (1954), Pande (1954T), Iyengar (with Pitcher and Read, 1954), Tozer (1955a), Cheesman (1951T and 1956T), Akaad (1956a and 1956b), Whitten (1957b), Pitcher and Read (1959), D. C. and J. L. Knill (1961), Pulvertaft (1961), Rickard (1962), Cambray (1964T), Kemp (1966T), and Wood (1970T). Figure I-4 shows the location of these areas. Detailed problems of structure, mineralogy, and geochemistry have

---

* The letter "T" after a reference denotes a thesis listed in a separate bibliography at the end of this book.

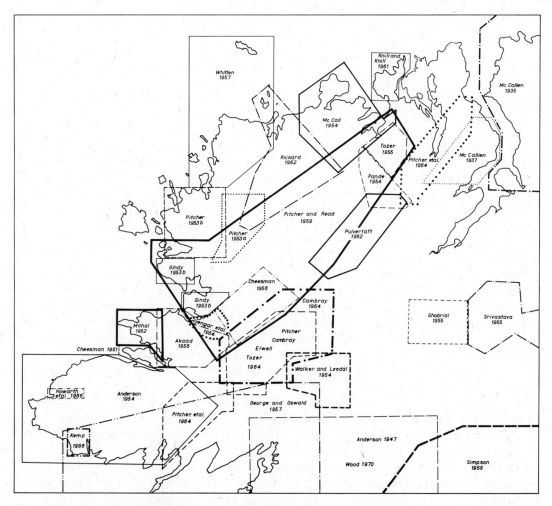

FIGURE I-4. Individual areas investigated during the latest phase of geological research in Donegal. Unpublished theses are referred to in the cases of Cheesman, Mithal, Cambray, Kemp, Ghobrial, Srivastava, and Wood.

been studied by Pitcher and Sinha (1958), Mercy (1960a and 1960b), French (1966), Hall (1966a,c; 1967c), Berger (1967T), and Naggar (1968T), among others. The work of these and other research workers forms the basis for this book, in which we give, first, a general outline of the geological development of northwest and central Donegal. Subsequent chapters will enlarge upon each of the main geological events, particularly, as we have already indicated, on those relating to the evolution of the plutonic rocks.

3. Outline of the Geology of Donegal

We give here a summary of the geological history of northwest and central Donegal in order that the relationships between the various events may be available during the detailed individual discussions that follow in later chapters (see Map 2 in folder). Although this book is concerned chiefly with that area, northwest of the Leannan Fault, which contains the main complex of granites and their envelopes, we shall also consider certain aspects of the stratigraphy southeast of this line, and in addition, discuss the outlying granite of Barnesmore.

(i) Our record starts with the deposition of great thicknesses of *Dalradian* and *Moinian* sediments. By using various sedimentary and structural criteria, the Imperial College research team erected five detailed Dalradian successions, but these have recently been consolidated into two great divisions by Pitcher and Shackleton (1966).

In northwest Donegal, the sediments are grouped into the *Creeslough Succession,* which includes the former Creeslough, Maas, and Fintown Successions. This is a sequence of nine formations and groups totaling up to 4900 meters of strata, which closely resembles the Ballachulish Succession of the Scottish Dalradian. To the southeast, the Creeslough rocks are overlain structurally by the *Kilmacrenan Succession,* a fairly simple sequence of eight members which may reach a thickness of some 3000 meters. This contains a prominent tillite and several other significant formations that enable a clear correlation to be made with the Islay Succession of Scotland. The true stratigraphic relationship between the Creeslough and Kilmacrenan Successions is not certain, for wherever they adjoin one another in the British Isles, they are separated by thrusting or faulting.

On the southeast side of the Leannan Fault, in southwest Donegal, there is a group of metasediments distinct in metamorphic and structural style, and formerly termed the Central Donegal Succession. However, since it also contains a tillite and passes northeastward into rocks clearly belonging to the Kilmacrenan Succession, this central Donegal group is now considered to represent no more than a different facies of the former sequence. In the south, around Lough Derg, the Central Donegal Facies is bounded by a group of psammitic schists and gneisses very similar in appearance to the typical Moine metasediments of Sutherland and Inverness in Scotland (Anderson, 1947; Powell, 1965; Church, 1969).

The age of the Dalradian sediments of the British Isles has been the subject of much discussion in recent years. In western Ireland, the Dalradian Connemara Schists are unconformably overlain by Arenig sediments and volcanics (Dewey *et al.,* 1970), and in Perthshire a limestone in the uppermost Dalradian contains a late Lower Cambrian fauna (Pringle, 1940; Stone, 1957; Harris, 1969)—the only diagnostic fossils found anywhere in the Dalradian. On the other hand, the prominent tillite in the Lower Dalradian succession is clearly the stratigraphic equivalent of similar beds in the late Precambrian of Norway, Spitzbergen, and Greenland (Kilburn *et al.,* 1965, Spencer, 1971). Accordingly, the Dalradian sedimentary sequence can be considered to range from late Precambrian to early Cambrian (see Leggo and Pidgeon, 1970).

(ii) After the deposition of all the Dalradian sediments, and before the onset of any regional metamorphism or deformation, a great volume of basic magma was intruded as sills of varying thickness. Two compositional types of tholeiite are represented; one of lesser extent shows the effect of early autometamorphism, while the other, of much wider extent, is even now represented by ophitic dolerites in parts of sills which have resisted regional metamorphism. The larger of these intrusions produced thermal aureoles in the adjacent sediments, and these have in places escaped obliteration during the later events.

(iii) These intrusions were followed by a series of tectonic and metamorphic events. For the most part, folding took place on Caledonoid (NE–SW) axes, but in western Donegal, the structural trend swings E–W to NW–SE, giving rise to a significant pattern also marked throughout western Ireland. During these fold movements, the pelitic rocks advanced in metamorphic grade to biotite-muscovite-chlorite schists, which locally contain garnet, and at the same time, the basic sills were converted to amphibole-chlorite-quartz-oligoclase metadolerites. Apart from considerable changes near the granites, the grade of metamorphism northwest of the Leannan Fault is low and fairly constant in facies, but to the southeast, the rocks of central Donegal are considerably more metamorphosed, and a progressive metamorphism, which reaches kyanite grade, is recorded. It is not possible, however, to draw simple isogradic lines across Donegal because of the several phases of metamorphism involved, both prograde and retrograde.

The age of the folding and metamorphism of the Dalradian presents many problems, as a glance at the recent literature will show (*e.g.,* Brown *et al.,* 1965; Sabine and Watson, 1965; Leggo *et al.,* 1966; Harper, 1967; Bell, 1968; Dewey, 1969; Rast and Crimes, 1969; Leggo *et al.,* 1969; Brown and Miller, 1969a,b; Fitch *et al.,* 1969). Most of the isotopic ages range between 530 and 400 million years, and although the interpretation of such dates is open to question, there is much evidence that the younger ages (largely by K-Ar) represent merely stages in cooling and uplift (see especially Harper, *loc. cit.,* and Bell, *loc. cit.*), and that it is the older (mostly by Rb/Sr) that more closely date the metamorphic activity. Despite these problems in isotopic dating, it is clear in the field that, in Connemara, all the recognizable tectonic and metamorphic events affecting the Dalradian took place prior to the deposition of Arenig sediments (Dewey *et al.,* 1970). We are not impressed by arguments that all the structural and metamorphic changes in the Dalradian in one part of the Caledonides post-date all those in another, and we conclude that the major folding and metamorphism of the Dalradian, that is, the *Caledonian I* of Read (1961) and the Grampian of Rast and Crimes (1969), is post-Lower Cambrian and pre-Ordovician. We shall return in Chapter 4 to a discussion of this chronology and of certain later, less important structural and metamorphic events. However, for the present, we follow Read (*loc. cit.*) in grouping these later, end-Silurian phases under the heading *Caledonian II.*

(iv) Both during and after the latest stage of the deformation history, a varied suite of minor *lamprophyric intrusions* were emplaced into the metamorphic complex over a very wide area.

(v) Following the major structural and metamorphic events in Donegal, there was a major episode of *granite emplacement,* producing relatively high-level plutons, though some have features typical of more deep-seated bodies. Recent Rb/Sr work by Leggo (1969) has given a date of 470 ± 1 million years (recalculated by Leggo and Pidgeon, 1970, as 498 ± 5 million years) for the Donegal Granite Complex.

In all, eight different granite units have been distinguished, exhibiting a variety of mechanisms of emplacement and producing correspondingly different contact effects on their host rocks. The Thorr and Fanad Plutons were emplaced by reactive stoping, those of Ardara and Toories by diapiric action, the Rosses and Barnesmore by cauldron subsidence, Trawenagh Bay possibly by piecemeal stoping, and the Main Donegal Pluton by a process of magmatic wedging greatly complicated by an intense deformation which affected both the Granite and its envelope prior to its complete consolidation.

With some of these granites, there are connected suites of dykes of varying character. These include the appinitic lamprophyres and felsites which are especially concentrated around the Ardara Pluton, the porphyritic microgranites radiating from the Rosses Ring Complex, the microgranites centering on the Barnesmore Granite, and a substantial swarm of acid dykes apparently attached to the Fanad Granite. In addition, there are several distinct suites of pegmatites, microgranites, and felsites associated with the Main Donegal Granite.

(vi) After the period of granite emplacement, Donegal, in common with the rest of the Caledonides, was affected by a great system of *transcurrent faults,* of which the most powerful member, the Leannan Fault, is probably a splay of the Great Glen fault of Scotland (Pitcher *et al.,* 1964; Pitcher, 1969). The main movements on these faults were pre-Carboniferous in age, but many were reactivated at later dates.

(vii) The Dalradian metasediments were overlain by *Devonian rocks* of the Old Red Sandstone facies, though in Donegal these are found only in a small area in eastern Fanad, where they are affected by posthumous movements on the Leannan Fault. During early *Carboniferous* times, southern Donegal, at least, was gradually submerged beneath the transgressive seas of the Viséan, and thick deposits of highly fossiliferous limestones, sandstones, and shales accumulated in large basins.

(viii) *Basaltic dykes of Tertiary age* are well represented throughout Donegal, and are especially numerous in two swarms of northwesterly trend. One of these traverses the Main Donegal Granite and the other Barnesmore mass.

In the succeeding chapters of this book, we shall devote attention first to the stratigraphy, structure, and metamorphism of the Dalradian rocks (Part I), and second to the Donegal granites, their emplacement, and contact effects (Part II). Later chapters will deal in less detail with the fault pattern, Devonian and Carboniferous sedimentation, and the Tertiary dykes (Part III).

PART I

THE GEOSYNCLINAL PILE AND ITS DEFORMATION

CHAPTER 1

THE SEDIMENTARY ASSEMBLAGES: STRATA NEAR THE BASE OF THE CAMBRIAN

CONTENTS

1. Introduction 15
    (A) Preamble 15
    (B) The Determination of the Sequence and Stratigraphic Order 16
    (C) Stratigraphic Nomenclature 16
    (D) The Sedimentary Sequences 16
2. The Creeslough Succession 18
    (A) Distribution 18
    (B) The Sequence 18
    (C) Comparison with Scotland 25
3. The Kilmacrenan Succession 26
    (A) Previous Nomenclature 26
    (B) Sequence and Distribution 26
    (C) The Northwest Donegal Facies 27
    (D) The Central Donegal Facies 33
    (E) Comparison of the Two Facies of the Kilmacrenan Succession 35
    (F) The Southern Limit of the Kilmacrenan Succession 35
    (G) Comparison with Scotland 36
4. The Relation between the Creeslough and Kilmacrenan Successions 38
5. The Sedimentary Facies of the Dalradian of Donegal 39

1. Introduction

(A) Preamble

In this chapter, we shall deal with the primary features of the great pile of miogeosynclinal rocks which make up the Dalradian (Eocambrian–Cambrian) of Donegal. Despite the long and complicated tectonic and metamorphic history, the original condition of these sediments has rarely been obliterated, and we are able to construct the detailed stratigraphic sequences which are necessary in any structural analysis of metamorphic rocks, as Billings (1950) and Read (1958a) have emphasized.

(B) The Determination of Sequence and Stratigraphic Order

The recognition of the original lithology of these unfossiliferous metasediments allows the stratigraphic order to be deduced in the normal way from sedimentary structures, such as cross-lamination, graded bedding, washouts, load and slump structures, mud cracks, and the like. Transitional passages from one formation to another can usually be demonstrated, and the persistence of associations of beds along their strike for distances of tens and even hundreds of miles convinces us that we are dealing with reasonably continuous sequences.

Nevertheless, we are aware of the difficulties in interpreting sedimentary structures in deformed rocks (see Bowes and Jones, 1958), and there are many examples where the shearing out of one limb of a tight fold in quartzite produces a structure simulating deformed cross-bedding. Further, it is well known that current ripples can resemble tectonic ripples, though these are relatively easily distinguished, using criteria which have been outlined by Spry (1963c). Other possibilities of confusion exist, such as the cases where metamorphism has led, by the differential growth of minerals, to an apparent reversal in the grain size relationships in graded beds, as early described by Read in Banffshire (1936). All in all, however, we are confident that the way up determinations which have led to the stratigraphic sequences discussed below are based on real sedimentary structures and not on those simulated by deformation. Indeed, in some competent beds, deformation is at a minimum, and a particularly good example is provided by the contact aureoles of metadolerite sills, where the static hornfelses have resisted the later regional deformations and have thus preserved their primary sedimentary features more or less intact.

In deformed rock sequences, it is generally impossible to deduce the original thickness, and there is even considerable difficulty in estimating the "unfolded" thickness, especially as the effects of the deformation will be dependent on lithology. Nevertheless, we consider it worthwhile to provide even crude estimates.

(C) Stratigraphic Nomenclature

In the revision of the stratigraphy of the Dalradian rocks which follows, we take the opportunity of bringing the nomenclature into line with the British Code of Stratigraphic Nomenclature (George et al., 1969). We continue to use, however, the informal term "succession" instead of "group," because this usage is so well established in the literature of the Caledonides.

We shall also make use of the well-known fact that certain lithological types tend to occur together and, in consequence, the use of such sedimentary associations (Knill, 1959a) or affiliations (Roberts, 1966a) results in a wider area of confidence in long distance correlation; this is due to the greater lateral extent of rock types belonging to a given sedimentary association compared with the area of deposition of any one member or rock type.

(D) The Sedimentary Sequences

We have already mentioned that the Imperial College research team erected five more or less separate "successions" from the Dalradian rocks of northwest and central Donegal. The "Creeslough Succession" comprised those sediments north and northwest of the Main Donegal Granite and terminating northeastward in northern Fanad (McCall, 1954; Knill and Knill, 1961; Rickard, 1962).*

---

* See Fig. I-4 for location of individual map areas.

The "Maas Succession" (Gindy, 1953a; Mithal, 1952T; Iyengar et al., 1954; Akaad, 1956a) was located west of the Donegal Granite Complex on the north and south side of Gweebarra Bay, while the "Fintown Succession" (Pande, 1954T; Tozer, 1955T; Cheesman, 1956T; Akaad, loc. cit.; Pitcher and Read, 1959; Pulvertaft, 1961; Cambray, 1969a,b) was composed of a narrow strip of rocks lying between the Main Granite to the northwest and an important thrust—the Knockateen Slide—to the southeast. This slide itself formed the northwestern boundary of the "Kilmacrenan Succession," which extended southwestward into Slieve League and northeastward into Fanad and Inishowen (Callaway, 1885; McCallien, 1935; 1937; Anderson, 1953; Pitcher and Read, loc. cit.; Pulvertaft, loc. cit.; Pitcher et al., 1964; Cambray, 1969a). Lastly, on the southeast side of the Leannan Fault, around the Blue Stack Mountains, a series of highly deformed and metamorphosed sediments was grouped as the "Central Donegal Succession" (Pitcher et al., loc. cit.).

It has long been apparent, however, that the Maas, Fintown, and Creeslough "Successions" have certain common characteristics and should be connectable through the "ghost stratigraphy" of the Thorr Granodiorite and the Main Granite. Pitcher and Shackleton (1966) have recently made just such a correlation and, in doing so, have greatly simplified the stratigraphy of northwest Donegal (Fig. 1-1 and Map 2 in folder). It is clear now that these three groupings comprise one great sequence, termed here the *Creeslough Succession,* after the region in which all its members outcrop. This is clearly the equivalent of the Ballachulish Succession of Argyll (McCallien, 1937; Pitcher and Shackleton, loc. cit.).

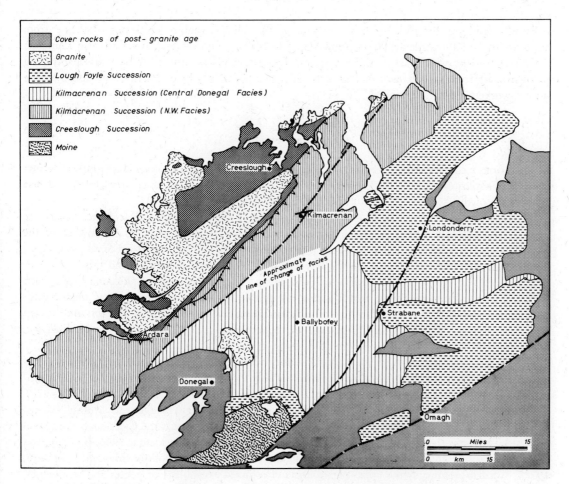

FIGURE 1-1. The main sedimentary units of northwestern Ireland.

Recent mapping in the Ballybofey area has also confirmed earlier suspicions that the "Central Donegal Succession" could be directly connected with the rocks of the Kilmacrenan sequence; there is no longer any reason for separating them, except as two different facies of the one *Kilmacrenan Succession*. In the Inishowen peninsula, this succession passes upward from well-sorted argillite and arenite sequences to massive greywacke units. It was this great change in facies which was used by Callaway (1885) to define the base of his "Lough Foyle Series," and we shall follow his example. Conveniently, the base of this *Lough Foyle Succession* coincides with the junction between the Middle and Upper Dalradian as defined in Scotland by Rast (1963). As we shall see later, the two latter successions are easily correlated with the Islay and Loch Awe successions of southwestern Scotland (McCallien, 1935; Johnstone, 1966).

Since the Lough Foyle rocks do not appear within the area covered by our map, we shall confine our attention in this discussion to the Creeslough and Kilmacrenan sequences, and open the account with the original characters of the Creeslough rocks.

2. The Creeslough Succession

(A) Distribution

Broadly speaking, the rocks of this succession occupy all the country northwest of a line from northeastern Fanad to Ardara (Fig. 1-1 and Map 1 in folder), and representatives are also found on the islands off the western and northern coasts. The ground is for the most part reasonably well exposed apart from certain drift and sand covered areas, and has been described in detail by the various workers already mentioned.

(B) The Sequence

The various members of the Creeslough Succession are listed in Table 1-1, together with their former names and estimates of their thicknesses, and Fig. 1-2 gives a diagrammatic representation of the stratigraphy of this sequence in two of the main areas in which it outcrops.

Upward transitions can usually be established, even though there is a considerable degree of tectonic thinning and even sliding between members of markedly differing competencies. Northwest of the Main Donegal Granite, for example, there is a tectonic break—the *Horn Head Slide*—above the massive Ards Quartzite which cuts out the entire Sessiagh–Clonmass Formation and the Lower Falcarragh Pelites in one area (Rickard, 1962, and see p. 56). Other stratigraphic units have been thickened by the complex folding, which is also responsible for the repetition of many units. Nevertheless, the maxima given for the thickness of each formation and the total of 5000 meters are thought to be fair estimates. The individual formations of the Creeslough Succession will now be discussed, beginning with the *Creeslough Formation,* the lowest exposed.

This thick assemblage of diversified rocks, dominantly calcareous, forms a wide belt running northeastward from Dunlewy to Mulroy Bay. For much of this distance, it lies within the highly deformed envelope of the Main Donegal Granite, so that it is difficult to be sure of the continuity of the succession within the formation. However, the presence between Lackagh Bridge and Mulroy Bay of downfolded portions of the Ards Pelites and Ards Quartzites indicates that the Creeslough Formation here, at least, forms the core of a considerable anticline (Pitcher and Shackleton, 1966, p. 149), and detailed mapping in the Dunlewy area has established several major repetitions of lithologies (Rickard, 1971). What is presumably the upper part consists of a mixed assemblage of pelites and semipelites, some distinctly calcareous and with intercalations of thin limestone. Southeast of these rocks, what we take to be the lower part of the formation is formed by thinly striped calc-pelites with a thin, persistent, wholly pelitic band, and three thick limestones: the Duntally, Brockagh, and Creevagh* limestone members of McCall (1954), of which the first is estimated to reach a thickness of 120 meters.

The Creeslough Formation grades upward by intercalation into dark pelites, and in the transition zone near Altan Lough, marbles and impure limestones form a fairly distinct horizon, named the *Altan Limestone* by Rickard. Above this, the *Ards Pelites* make up a thick formation of pyritous graphitic pelites, occurring chiefly in a long belt extending from Dunlewy northeastward to Fanad. They also occupy the cores of two anticlinal folds in the Crohy Hills, where Gindy (1953a) included them in his "Crohy Semipelitic Group."

Though the greater part of the formation is of a uniformly black pelite, interbanding of grey semi-pelite and black pelite is common in the upper part, especially in the Crohy Hills. Impure calcareous bands and occasional fine-grained cherty bands are also present, as are infrequent dolomitic limestones, and toward the top of the formation, a conspicuous greyish micaceous limestone is always present. Graded bedding with a concentration of carbonaceous material in the finer parts is frequently seen, and there are load casts, washouts, and sedimentary dykes in places, as for example, in the excellent shore exposure at Ards Priory, north of Creeslough (Fig. 1-3a). The whole assemblage of lithologies and associated structures forms a fine example of a typical black slate association.

The banded pelites of the upper part of the formation pass upward into the Ards Quartzite by the incoming of impure quartzite beds and the reduction of the pelitic material to thin partings. This important transitional member averages 30–60 meters in thickness, but may reach as much as 90 meters in the Crohy Hills.

The outcrop of the *Ards Quartzite*** forms the belt of hill country stretching southward from Bloody Foreland to Dunlewy and thence northeastward through the Muckish–Errigal range of mountains to Fanad. Aranmore Island, Horn Head, and the Crohy Hills are also largely composed of this horizon, and minor developments are found in the Meenalargan Hills, at Cor near the Gweebarra estuary, and at Lackagh Bridge and Glenieraragh on either side of the northeastern portion of the Main Donegal Granite.

Over most of its outcrop, this thick formation is a fairly homogeneous, feldspathic sandstone of varying grain size and grade. Pebble beds and lenses are common, with individual pebbles reaching a maximum length of one inch. These consist for the most part of quartz, but feldspar (chiefly microcline) and quartzite itself are also represented. Bedding is sometimes marked by lines of dark heavy minerals, often rich in zircon, or by feldspathic streaks and thin semipelitic laminae. In the western outcrops, pelitic intercalations up to $1\frac{1}{2}$ or 3 meters thick are also present. Cross-bedding is well developed, sometimes on a large scale, with laminae 2 meters long and $1\frac{1}{4}$ meters high; graded bedding, ripple marks, washouts, and small-scale slump structures are common.

Above this relatively uniform quartzite is a very rapid change to a varied assemblage of pelitic, quartzitic, and calcareous horizons which form the most diversified stratigraphic unit in Donegal. Detailed descriptions of this *Sessiagh–Clonmass Formation* in its several outcrop areas have been given by Iyengar et al., (1954), McCall (1954), Pulvertaft (1961), Knill and Knill (1961), and Rickard (1962), and need not be repeated here. It is sufficient that the diagram (Fig. 1-2) shows, in a very schematic way, the major facies variations within this formation—between Creeslough and Maas—together with the stratigraphic terminology in use before the final overall correlations were made.

---

* The name Creevagh Limestone is evidently a misnomer, as Creevagh townland includes the Duntally horizon.
** The name *Ards Quartzite* was introduced by McCall (1954), but Rickard (1962) preferred the term "Errigal Quartzite" for this formation. However, since this latter name was used by the Irish Geological Survey in 1891 for all the Donegal quartzites, which we now allocate to several horizons, it seems best to retain McCall's term.

TABLE 1-1. Local Sequences of the Creeslough Succession in Donegal*

Standard Creeslough Succession	Loughros and Rosbeg (Akaad, 1956a; Pitcher and Shackleton, 1966)	Fintown "Succession" (Akaad, 1956; Pitcher and Read, 1959; Pulvertaft, 1961; Pitcher and Shackleton, 1966)	Maas "Succession" (Iyengar et al., 1954; Akaad, 1956)
	—Knockateen Slide—		
Loughros Formation	Semipelites of Loughros Quartzite of Loughros	Fintown Siliceous Flags	
Upper Falcarragh Pelites	Rosbeg Semipelites	Glenties Series (in part) (Glendowan Pelites) Losset–Lough Greenan Pelites and Semipelites	*top not seen* Cleengort Pelites (Clooney Pelites)
Falcarragh Limestone		Lough Salt Limestone (Glenaboghill Limestone) (Claggan Lough Limestone)	Portnoo Limestone
Lower Falcarragh Pelites		Claggan Lough Pelites	Maas Semipelites
Sessiagh–Clonmass Formation		Staghall Group	Mulnamin Calc-Silicate Group Mulnamin Siliceous Flags
Ards Quartzite		Glenieraragh Siliceous Flags	Cor Quartzite
		—base not seen—	
Ards Pelites			
Creeslough Formation			

* Alternative names in parentheses.

In the type area northwest of Creeslough, the Formation consists of five members, two flaggy quartzite horizons and three dolomitic limestone units as follows:

Port Dolomitic Limestone	Blue schistose limestone, minor calc-flags, and pelites.
Sessiagh Banded Quartzites	Flaggy quartzite with micaceous bands.
Marble Hill Dolomitic Limestone	Dolomitic limestone passing up into calcareous phyllites.
Clonmass Banded Quartzites	Flaggy quartzite passing up into pelitic schist.
Clonmass Dolomitic Limestone	Dolomitic limestone and pelitic and calcareous schist.

Most of these horizons can be followed with little difficulty from Rosguill to Dunlewy, though there are many lithological variations. For example, the Clonmass Limestone Member on Breaghy Head and

Lettermacaward and Crohy (Gindy, 1953a; Pitcher and Shackleton, 1966)	Creeslough-Errigal (McCall, 1954; Rickard, 1962)	Rosguill (Knill and Knill, 1961)	Ballachulish (Argyllshire)	Thickness Range of Formations (Ireland)
	—top not seen—			
	Inishboffin Banded Semipelites	Melmore Mixed Group [?] Boyeeghter Quartzites Meencoolagh Banded Group		155 meters
Toome Lough Semipelitic–Pelitic Group	Upper Falcarragh Pelites			255–1230 meters
Lettermacaward Limestone	Falcarragh Limestone	Falcarragh Pelitic Schists	Lismore Limestone and Cuil Bay Slates	155–320 meters
Lettermacaward Semipellite–Pelite Group	Lower Falcarragh Pelites		Appin Phyllites	255–615 meters
Lettermacaward Alternating Group (Lough Pollrory Alternating Group)	Port Limestone Sessiagh Banded Quartzites Marble Hill Limestone Clonmass Banded Quartzites Clonmass Limestone	Sessiagh Quartzites Marble Hill Limestone Clonmass Banded Quartzites Mevagh Mixed Group	Appin Limestone	370–770 meters
Cor Quartzite (Crohy Quartzite)	Errigal Quartzite (Lackagh Bridge Quartzite) (Ards Quartzite)	Ards Quartzite	Appin Quartzite Appin Striped Transition Series	310–615 meters
Crohy Pelites and Semipelites base not seen	Ards Black Pelites	Ards Black Pelitic Schists	Ballachulish Slates	255–460 meters
	Altan Limestone Creeslough Group —base not seen—	Creeslough Group	Ballachulish Limestone Leven Schists (in part)	310–925 meters

in Rosguill (here termed the "Mevagh Mixed Group" by the Knills), though dominantly calcareous, includes a variety of calc-pelites, pelites, semipelites, and psammites; and lateral facies changes can be traced in many localities (Knill and Knill, *loc. cit.*, p. 275). Lithological variations on a larger scale can be seen in the highest member of the formation, which consists of a bluish dolomite with cream and buff dolomitic flags and calc-pelites at Port Lough, but further to the southwest, is represented by limestone and calc-pelite bands intercalated in the uppermost part of the Sessiagh Quartzite.

The major change comes between the outcrops north and northwest of the Main Donegal Granite and those in the Maas and Fintown–Church Hill areas. Here the Sessiagh–Clonmass Formation is much more calcareous, and a relatively thin flaggy member is followed upward by a thick sequence of what were formerly interbedded limestones, dolomitic marls, and sandy shales. This is, however,

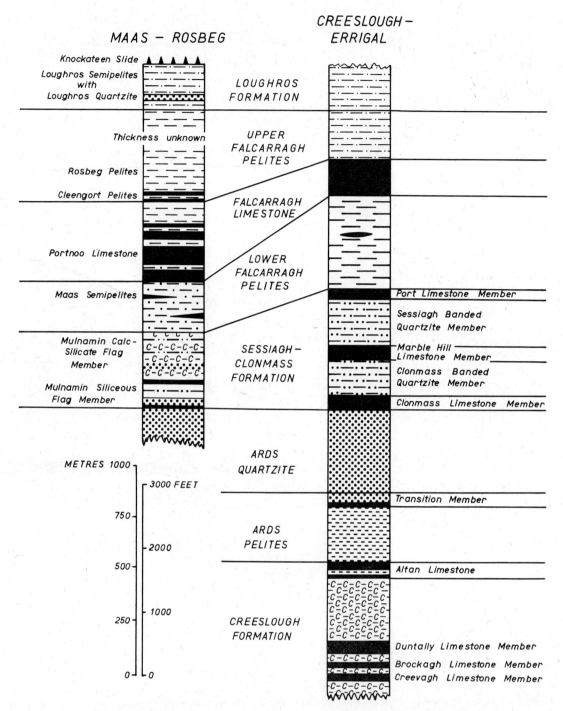

FIGURE 1-2. The two facies variants of the Creeslough Succession.

FIGURE 1-3. (a) Inverted section of load-casts in the upper Ards Pelites. Cove 1000 meters northwest of Ards House, Ards. (b) Junction between Ards Quartzite (bottom) and the Falcarragh Pelite (top) marking the Horn Head Slide, Shore at Ardsbeg, west of Falcarragh. (c) Sun cracks in a sandstone of the Clonmass-Sessiagh Formation, Maas townland. (d) One of the upper tillite horizons in the Port Askaig Tillite showing granite clasts in a siltstone matrix Carrowkeel, south-east of Kindrum, Fanad.

everywhere succeeded by what are clearly the Lower Falcarragh Pelites, and the correlation with the Sessiagh–Clonmass Formation of north Donegal is secure (Pitcher and Shackleton, 1966).

It is therefore clear that the rapid and marked changes in conditions of sedimentation account for many of the lateral variations in the Sessiagh–Clonmass Formation; nevertheless, tectonic thinning also played an important part in determining the stratigraphic sequence, particularly at the base of the Formation. Thus, McCall showed that the junction between the Ards Quartzite and the Sessiagh–Clonmass Formation is marked by the great Horn Head Slide. Though this often runs parallel to the stratigraphy, it cuts out the entire Sessiagh–Clonmass Formation around Dunfanaghy, so that on Horn Head the Lower Falcarragh Pelites make direct contact with the Ards Quartzite. A similar situation exists in the Dunlewy area where individual members of the Sessiagh–Clonmass are pinched and eventually cut out entirely; indeed, if Rickard's (1962) extension of the Slide to Inishdooey is correct, some 1500 meters of sediments are missing here between the Ards Quartzite and the Loughros Formation. At Ardsbeg, near Gortahork, the slide plane is very well exposed (Fig. 1-3b), and apart from local mylonitization and pinch-and-swell structure, there is little reason for suspecting from this outcrop alone that a thousand meters or so of strata are missing. On the other hand, in none of these horizons is there any local change of facies near to the slide position, such as might be expected if a major onlap is involved. Because of these difficulties, we cannot be certain that some beds are not also missing along most of the junction between the Ards Quartzite and the Sessiagh–Clonmass Formation, even though that portion from Errigal to Muckish seems to represent a normal stratigraphic boundary. Indeed, to the northeast and at this level, in Rosguill, the "normal" Clonmass Limestone Member (75 meters) is replaced by the much thicker (300 meters) "Mevagh Mixed Group" (Knill and Knill, 1961, p. 275).

Furthermore, nearby in Fanad, roughly 300 meters of pelites and semipelites follow in apparently natural succession above the Ards Quartzite, giving a possible total of 600 meters of sediment above the Ards Quartzite and below the top of the Clonmass Limestone. It is impossible to determine how much of such thickness variation is tectonic and how much primary; we think the hiatus at the base of the Sessiagh–Clonmass Formation is of tectonic origin but are certain that the major diversities within the Formation represent original sedimentary facies changes.

As might well be expected in such lithologies, sedimentary structures are preserved; thus graded bedding is well preserved in the hornfelses of the Clonmass metadolorite sill, where such grading is metamorphically reversed (McCall, 1954, p. 162), and mud cracks have been noted in the banded quartzite members of both the Creeslough-Dunlewy and the Maas areas (Fig. 1-3c). Trace fossils have also been identified (Wolfe, 1969).

A thick, monotonous, pelitic formation everywhere blankets the variable horizon we have just described. This widespread formation, the *Lower Falcarragh Pelites,* is well developed in several areas in western and northwestern Donegal. In northwest Donegal, north of the Errigal–Muckish range, it consists of up to 600 meters of grey and black striped carbonaceous pelites, with occasional semipelitic and calcareous layers with abundant pyrite. Thin limestone bands are occasionally present and become more abundant toward the west, possibly indicating a slight facies change in that direction.

Around the Gweebarra Estuary in the Maas district, the Lower Falcarragh Pelites consist of about 260 meters of homogeneous purplish semipelites with two or three minor impure calcareous horizons, and further to the northeast, in the Gartan Lough area, this same formation is represented by thin pelites and semipelites lying below a thick limestone, formerly known as the "Claggan Lough Limestone" (Pulvertaft, 1961, p. 257), which we now correlate with the Falcarragh Limestone of the northwest.

We follow Rickard (1962) in using this important and widespread *Falcarragh Limestone* to separate the pelites of Falcarragh into two formations. Over most of its outcrop in northwest Donegal, this is a grey or blue marble with partings of pelite and semipelite, but in Rosguill a distinct limestone unit seems to be missing from a great thickness of pelites, though it may be represented in eastern Rosguill by three lesser limestones with interbedded quartzites (Knill and Knill, 1961).

Around the Gweebarra Estuary, the Falcarragh Limestone (formerly termed here the "Portnoo Limestone") is very well exposed, particularly at Gweebarra Bridge and Portnoo. Again, the massive grey marble, here some 225 meters thick, has psammitic and semipelitic intercalations which become important toward the top of the formation, where they provide a passage to the overlying pelites (Akaad, 1956b). Along the Fintown zone and further northeast along the southeastern margin of the Main Donegal Granite, this same limestone is folded on a major scale and appears as several distinct horizons, especially well shown at Claggan Lough and Lough Salt (Pitcher and Shackleton, 1966).

There is everywhere a clear gradation up into the *Upper Falcarragh Pelites.* In its type locality around Falcarragh, this formation consists of pelites, characteristically striped by lighter-colored semipelitic bands, the individual layers varying from a few millimeters to 10 centimeters in thickness (Rickard, 1962, p. 214); there are wisps and discontinuous layers of pale sand and silt with minute load casts, graded bedding, lamination, and ripple-marks. Their mineralogy here is similar to that of the Lower Falcarragh Pelites, but carbonaceous matter is less abundant. Around the Gweebarra Estuary, these Upper Falcarragh Pelites are again well exposed, and near the Gweebarra Bridge they consist of dark-grey massive pelites (the "Cleengort Pelites") which pass upward into rocks of a more semipelitic character and a marked band of limestone. In the Portnoo area, this formation (the "Clooney Pelites") is again more pelitic in the lower portions and more semipelitic near the top; and in these semipelites, rhythmic banding and graded bedding is fairly common.

Such rocks extend into the northern part of the Loughros Peninsula (Mithal, 1952T). Here, the upper part of the Upper Falcarragh Pelites (the "Rosbeg Semipelites," see Pitcher and Shackleton, 1966) has the same lithology as at Falcarragh, but the great width of outcrop suggests that a very great thickness of strata may be present (Akaad, 1956a, p. 267). If so, the Upper Falcarragh Formation is much thicker here than in northwestern Donegal.

Rocks of this lithology also occur along southeastern margin of the Main Donegal Granite, from Fintown to Lough Greenan, as two separated horizons which are now known to be but a folded repetition of the one formation (Pitcher and Shackleton, 1966).

Rocks of the very highest division of the Creeslough Succession, the *Loughros Formation,* occur in four widely separated areas in Donegal, in each of which they occur in natural succession to the Upper Falcarragh Pelites. Thus, in southern Loughros, the Upper Falcarragh Pelites are overlain by a composite formation of quartzite and banded semipelites which extends farther to the northeast, to Glenties and Fintown. Much farther north, on the island of Inishboffin, Rickard (1962) has described a similar series of banded rocks lying above the Upper Falcarragh Pelites, and in northern Rosguill, the Knills (1961, p. 276) found inclusions of similar lithology in granitic rocks, though here calc-flags, limestones, and semipelites occur together with the siliceous flags and coarse quartzites. At this stratigraphic level, these workers also distinguished a group of pink or white feldspathic quartzites, often with pebbly bands and a variable amount of semipelitic material. These were referred to locally as the "Boyeeghter Quartzites," and we provisionally allocate them to the Loughros Formation.

We have now seen that there are, from place to place, many variations in lithology and in sequence in the rocks of the Creeslough Succession. Although there are certainly important tectonic complications here, we must again emphasize our conviction that lateral changes in sedimentary facies have played an essential part in determining these local successions. Despite these variations, the major lithostratigraphical divisions are sufficiently distinct to enable us to make useful comparisons with similar strata in the southwest highlands of Scotland.

(C) Comparison with Scotland

The correlation of the Dalradian succession in northwestern Donegal with that in Argyllshire (see Table 1-1), first suggested by McCallien (1937, Plate III), is very easy because many of the formations of Ballachulish and Creeslough are remarkably similar in lithology (Pitcher and Shackleton, 1966). There are only two contrasts worthy of special comment. First, the Appin Limestone of Argyll is much thinner than its equivalent in Donegal (130 meters compared with up to 900 meters), mainly as a result of an increase in the psammitic element in the latter area. Secondly, the Creeslough Formation is much more a calcareous pelite in lithology than the equivalent Ballachulish Limestone-Leven Schists of Argyll, both in having a greater preponderance of calc-schists and in the development of thick limestones. We shall comment further on this extension of the Argyllshire sequence after we have dealt with the overlying succession.

3. The Kilmacrenan Succession

(A) Previous Nomenclature

The succession is best exposed on the northeast coast of Inishowen, where the stratigraphic relations were first fully described by Callaway (1885) and later more exactly defined by McCallien (1935, and see Table 1-2). As noted previously, it passes up into the Lough Foyle Succession.

TABLE 1-2. McCallien's Successions

Inishowen (1935)		Rathmullan (1937)	This Book
Lough Foyle Series (of Callaway)	Greencastle Green Beds Inishowen Grits and Phyllites Cloghan Green Beds Fahan Slate–Grit Group Inch Island Limestone Group Culdaff Limestone	Killygarvan Limestone	Lough Foyle Succession
	Crana Quartzite	Upper Quartzite Killygarvan Flags Lower Quartzite	
Kilmacrenan Series (of Callaway)	Stragill Group Linsfort Black Schists Glengad Schists Lag Limestone	Milford Schists	Kilmacrenan Succession
	Malin Head Quartzite	Knockalla Quartzite	

Unfortunately, the lowermost members, including the important tillite, are not reached in this fine Inishowen coast section, but we can safely continue the succession downward in the adjacent peninsula of Fanad, even though these exposures lie on the other side of the Leannan Fault. We then find it convenient to modify slightly McCallien's standard succession at three points as follows:

(1) We would formally name an important dolomitic limestone, lying between the "Malin Head Quartzite" and the "Glengad Schists" (the Lag Limestone of McCallien, 1935, p. 411), the *Cranford Limestone,* after a locality on Mulroy Bay where this horizon is best developed.

(2) We regard the "Linsfort Black Schists" and the "Stragill Group" as local subdivisions of the "Glengad Schists," and unite all three pelitic members sandwiched between the two main quartzites as the *Termon Formation* (cf. McCallien, 1935, p. 415).

(3) Finally, we consider that the slide at the base of the Lower Crana Quartzite (McCallien, 1937, p. 56) is of only local importance. The thinly striped pelitic lithologies of his uppermost Milford Schists are identical with those of the interbeds within the lower part of the quartzite; indeed, a transition series is clearly present wherever faulting or thrusting are absent, as in the townlands of Boheolan and Lough Keel, southeast and northeast of Kilmacrenan, respectively.

(B) Sequence and Distribution

It is easy to follow the Kilmacrenan Succession throughout northwest and central Donegal (Fig. 1-1 and Map 1 in folder), and in Table 1-3 we advance a correlation of all the local sequences, together with a suggested standard nomenclature for the county as a whole. This is based on several marker horizons—two main quartzites, a tillite, and the graphitic pelites—so that there can be little doubt of its validity. Nevertheless, the correlation reveals important changes of facies—so much so, that there is a need to distinguish the two most contrasted sequences as the *Northwest Donegal Facies* and the *Central Donegal Facies*.

There is a complication here in that these two facies in their most contrasted lithologies are brought into juxtaposition by the considerable strike-slip movement on the Leannan Fault, which, it may be remembered, is a major splay of the Great Glen Fault of Scotland. The overall relationships are best appreciated from the greatly simplified diagram of Fig. 1-1, from which it will be seen that the Northwest Donegal Facies runs in a straight line parallel to the Leannan Fault for some 113 kilometers from coast to coast, maintaining its detailed stratigraphic relationships almost unimpaired. In contrast, the Central Donegal Facies is disposed in the form of large complex fold structures askew to the fault, and shows considerable changes in facies along its strike.

(C) The Northwest Donegal Facies

The stratigraphic order within this part of the Kilmacrenan Succession is easily determined from the abundant sedimentary structures, and Fig. 1-4 shows in a schematic way the lithology and sequence in three major areas.

In his description of the Slieve League peninsula, Anderson (1954) grouped together a series of pelitic and semipelitic schists, quartzites, calc-pelites, and limestones, all of which lay below the Slieve Tooey Quartzite, as the lowest of his stratigraphic units—the "Killybegs Group." This thick horizon he traced eastward into central Donegal, where it appeared to pass down conformably into the "Lough Derg Psammitic Group," which he had earlier correlated with the Moine Series of Scotland (1947). Recent mapping around Carrick (Kemp, 1966T) and Glencolumbkille (Howarth *et al.*, 1966; Howarth, 1971) has confirmed this general way up, though the detailed stratigraphy and structure here are clearly more complex than Anderson recognized. However, in the eastern portions of the Slieve League promontory, it is also clear that the "Killybegs Group" lies stratigraphically above the Slieve Tooey Quartzite (Pitcher *et al.*, 1964, Fig. 5), and in the Ballybofey area, as we shall show later, the succession also youngs southward toward Lough Derg. Clearly, the "Killybegs Group" as originally defined must be abandoned, and we shall take the *Glencolumbkille Pelites* (Kemp; Howarth *et al., loc. cit.*) as the base of the Kilmacrenan Succession.

This formation, another black slate assemblage, consists of dull black, graphitic pelites with horizons in which thin psammitic layers alternate with darker pelitic bands and thin graphitic limestones. There is clearly no break between these strata and the overlying *Glencolumbkille Limestone*, which consists of a light grey calcite marble below and a cream or grey dolomite above, nearly everywhere separated by a striped grey pelitic member.

The *Port Askaig Tillite* which follows has been rather fully described, as befits its importance in Eocambrian stratigraphy (Kilburn *et al.*, 1965; Howarth *et al.*, 1966; Howarth, 1970, 1971; Spencer, 1971). Here we break the tradition of using local place names for the nomenclature of the Donegal Dalradian, since this widespread marker horizon was first described at Port Askaig on the Island of Islay off the coast of Argyllshire (MacCulloch, 1819).

The Tillite is particularly well displayed in Donegal, where it thickens northeastward, measuring 60–120 meters in Glencolumbkille, some 525 meters in Fanad, and 900 meters in Islay. There is always a succession of separate tillites, of which 13 have been recognized in Glencolumbkille and 12 in Fanad; this contrasts with 13 in Islay and 38 in the Garvellachs, even farther to the northeast (Spencer, *loc. cit.*). The top of the formation is conveniently marked by a thin pelite in Glencolumbkille, which is probably the equivalent of a thin-bedded dolomite-quartzite member in Fanad.

These boulder beds are mixtites characterized by the presence of randomly distributed, unsorted

TABLE 1-3. Local Sequences of the Kilmacrenan (Islay) Succession in Donegal*

Standard Succession	Glencolumbkille	Slieve League	Central Donegal	Glenties–Fintown
	(Anderson, 1954; Howarth et al., 1966)	(Kemp, 1966T)	(Pitcher et al., 1964)	(Pitcher et al., 1964; Cambray, 1969a)
				fault
Crana Quartzite				Kilmacrenan Grits
				Boheolan Quartzite
		top not seen		
Termon Formation		Teelin Point Formation	Lough Eske Formation (Swilly Chloritic Schist)	Upper Termon Schists
				Knockletteragh Grits
				Lower Termon Schists
				Cranford Limestone
Slieve League Formation		Teelin Schist	Boultypatrick Grit	
		Carrigan Head Formation	Croaghubbrid Dark Schists	
		Crockunna Schist		
	top not seen			
Slieve Tooey Quartzite	Slieve Tooey Quartzite	Slieve Tooey Quartzite	Gaugin (Silver Hill) Quartzite	Slieve Tooey (Aghla) Quartzite
Port Askaig Tillite	Kiltyfanned Schist Boulder Bed	Dolomitic Beds Boulder Beds	Boulder Bed	Boulder Bed
				Knockateen Thrust
Glencolumbkille Limestone	Glencolumbkille Dolomite	Limestone Formation ⎰Dolomite ⎱Striped Schist ⎱Limestone	Owengarve Formation	
	Glen Head Schist			
	Skelpoonagh Bay Limestone			
Glencolumbkille Pelites	Glencolumbkille Schist—Killybegs (in part)	Glencolumbkille Schists		Fintown Dark Schists
		base not seen		*Knockateen Thrust*

* Alternative names in parentheses.

clasts, ranging in size from granules to boulders, exceptionally more than 1 meter in diameter, of igneous and sedimentary rocks in a finer unbedded matrix (Fig. 1-3d). The matrix of the tillites is dolomitic at the base, becoming increasingly psammitic toward the top, and in sympathy with this change, the psammitic interbeds everywhere thicken upward. Within them, large-scale cross-stratification and current bedding are not uncommon, and, taken together with the evidence of local erosion and winnowing at the top of tillite horizons, strongly suggest shallow water conditions.

The clast compositions everywhere fall into two contrasted groups, the extrabasinal igneous and metamorphic fragments and the intrabasinal sedimentary rocks. The former includes subrounded

Fintown–Kilmacrenan	Fanad	Inishowen and Rathmullan	Islay–Loch Awe (Argyllshire)	Thickness Range (Ireland)
(Pitcher and Read, 1959; Pulvertaft, 1961)	(Shackleton, in Pitcher and Cheesman, 1954; Pitcher et al., 1964; Kilburn et al., 1965)	(McCallien, 1935, 1937)	(Knill, 1963)	
fault	*top not seen*			
Kilmacrenan Grits	Kilmacrenan Grits Boheolan Quartzite	Upper Crana Quartzite* Killygarvan Flags Lower Crana Quartzite	Crinan Grits	1525 meters
Gorteen Lough Striped Schists Termon Green Schists	Termon Group	Milford Schists { Stragill Group, Linsfort Black Schists, Glengad Schist }	Ardrishaig–Craignish Phyllites	675–2680 meters
Cranford Dolomitic Limestone (Gartan Bridge Limestone)	Cranford Limestone	Lag Limestone	Shuna Limestone	
			Easdale Slates Scarba Transition Group Scarba Conglomerate Group Jura Slates	1000–1780 meters
Slieve Tooey (Knockateen) Quartzite	Fanad Quartzite	Malin Head (Knockalla) Quartzite	Islay Quartzite	370–2300 meters
Boulder Bed	Dolomite Boulder Bed		Port Askaig Tillite	60–525 meters
Knockateen Thrust				
	Fanad Dolomitic Limestone		Islay Limestone	90–155 meters
Lough Finn Dark Schists and Limestones	Rosnakill Pelitic Schist		Mull of Oa Phyllites	125–310 meters
Knockateen Thrust	*base not seen*			

* For upward continuation, see Table 1-2.

clasts of vein quartz, quartzite, quartz conglomerate, gneiss, and granite, most of which have presumably traveled far; the latter group consists largely of dolomites which are usually angular and apparently derived locally from the Glencolumbkille Limestone and other underlying calcareous horizons, and perhaps some quartzite. However, only three rock types are really common as clasts: quartzite, granite, and dolomite. In the lower tillites, dolomite clasts are by far the most abundant, while in the upper members, the proportion changes in favor of granite and quartzite.

Very full details of the Tillite are provided in the works already cited. The general conclusion is that it represents a subaqueous till as, indeed, was suggested long ago by Thomson (1877) and

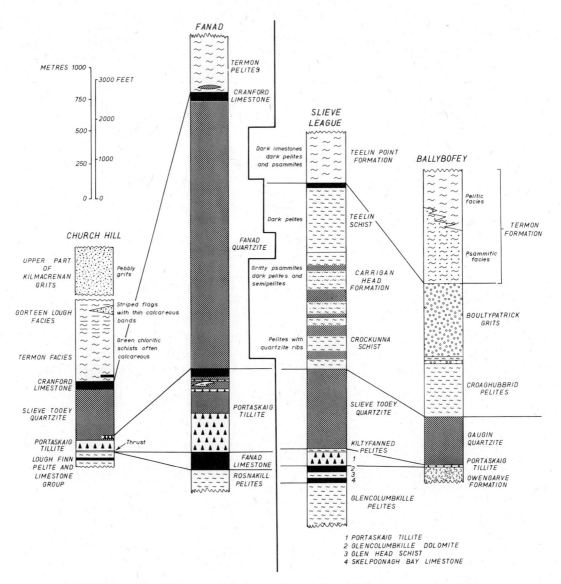

FIGURE 1-4. The several facies variants of the Kilmacrenan Succession.

M'Henry (in Hull, Kinahan et al., 1891, p. 50). The possibility that submarine mudflows were largely responsible (cf. Schermerhorn and Stanton, 1963) is rejected because of the clear evidence of rafting of the extrabasinal clasts and of shallow water, even emergent conditions. This view is confirmed by the presence of thin-bedded sand seams in individual tills, such as are found in modern analogues, by the common association with dolomites, which suggests a much more definite environmental control over sedimentation than subaqueous mud flows or slide-breccias would imply, and finally, by the extent of the formation over so wide an area and with so similar a basic stratigraphy (see Spencer, 1971). The importance of this and other similar tillites throughout the world in indicating a late Precambrian Varangian glaciation has been discussed at length in the literature (e.g., Harland, 1964a,b) and need not be further dealt with here.

As far as local stratigraphy is concerned, this important and constant horizon, lying as it does at the base of a great quartzite which we are to examine next, provides an ideal marker horizon and one which will help to solve the problem of correlation met with in central Donegal. On a more general basis, a case might well be made out for taking the Port Askaig horizon as the base of the Phanerozoic (*cf.* Spjeldnaes, 1959).

The overlying *Slieve Tooey Quartzite* is, in terms of original lithology, dominantly a quartz arenite with extensive beds of feldspathic arenite and occasional calcareous horizons, especially in higher levels. Sorting is usually good, but graded pebbly layers are quite common. Bedding in the quartzite is often marked by darker heavy-mineral-rich layers; pelites and semipelites are present as partings and, occasionally, as more substantial beds. There is ripple bedding and cross-stratification, the latter being of the planar type and usually with single sets which range up to almost 2 meters and are homogeneous in lithology. This formation is clearly the thickest in the succession, but because of structural complications, the actual measurements are difficult to make. It is more than 760 meters thick in Glencolumbkille, 2280 meters on Aghla Mountain, and probably of the same order of thickness in northeast Donegal.

Northeast of Glenties, there is a very sharp upward change of facies into the calcareous horizon which follows, the *Cranford Limestone,* which we define as the lowest member of the overlying pelitic unit, the Termon Formation. Though to some extent this calcareous facies is foreshadowed by occasional thin calcareous beds in the upper part of the Quartzite, this level represents one of the most marked changes in lithology in the succession. In fact, near Cranford, there are exposures of a quartz conglomerate at the junction of the two formations, and southeast of the Quartzite on Aghla, a dolomite breccia is present (Cambray, 1969*a*). Thus, it seems possible, as we shall note later, that this level represents a period of nondeposition and concomitant erosion.

At Cranford, 240 meters of calcareous beds consist of a lower, yellow-weathering dolomite with some bands of coarse grit and an upper grey limestone with interbeds of pelite. Such a full sequence is rarely seen elsewhere; in the southwest, near Aghla, a thin dolomite underlies some 135 meters of limestone, and at Lag in Inishowen, only calcite limestones have been recorded.

Despite local attenuation, mainly we think due to deformation, this calcareous member is present over much of the 93 kilometers of outcrop from Inishowen to the Glen of Glenties. Southwestward, however, beyond this last locality, the overlying pelites follow directly on the quartzite, and as there is here a transitional interbanding of these two lithologies, it seems that the limestone is really absent and that we are dealing with a true facies change.

Even farther southwestward, in the southern part of Slieve League peninsula, the Slieve Tooey Quartzite is overlain by a thick pelitic-psammitic succession. Kemp (1966T) has provided details of this local variation which we summarize here:

Teelin Point Formation	105 meters	dark limestone, dark pelites, and psammites
Teelin Schist	480 meters	dark pelites
Carrigan Head Formation	300–600 meters	gritty psammites, dark pelites, and semipelites
Crockunna Schists	195–600 meters	pelites with quartzite ribs

Though the exact way in which this sequence can be correlated with the Cranford member and the dark pelites immediately above it cannot be determined, we are inclined to believe that the "Teelin Point Formation" can be regarded as the time equivalent of the Cranford Limestone. Thus it is possible that the underlying sediments, which we group together as the *Slieve League Formation,* represent material accumulating during the period of nondeposition in the areas father northwest already mentioned.

In most of northern Donegal, however, there is certainly no doubt that the Cranford Limestone Member passes by rapid gradation into the overlying pelites which form the greater part of the *Termon Formation.* These pelites, like the quartzite not far beneath them, form one of the thickest and most continuous divisions of this succession, not only in Ireland, but also in southwest Scotland; on the Atlantic coast of Inishowen, it must be very nearly 2,440 meters.

This thick, dominantly pelitic formation can be subdivided into various members in the several areas where it can be mapped in detail (Table 1-3). Throughout much of northern Donegal, the lower pelites are often dark and graphitic, and are interbedded with occasional thin dolomites and lenses of arenite. Upward, these dark pelites become greenish and calcareous and interbanded with thin beds of a fine-grained greenish psammite, formerly termed the Boheolan Quartzite (Pitcher et al., 1964). Great lenses of pale, often calcareous, psammite appear in a central position in the Termon Formation and are especially well shown in the Cranford area. Near Glenties, their place is taken by pebbly grits, the "Knockletteragh Grits," which occur, again as lenses, in a somewhat lower stratigraphic position along a considerable part of the outcrop of the formation.

There is a return to the green, calcareous pelite at higher horizons in this formation, and here we meet a particularly interesting phenomenon. The Termon pelites at this level are intruded in places by great dolerite sills which have contact-hardened the country rocks on their flanks, and so protected them from the subsequent deformations. The original sedimentary character is thus preserved, so that we can see that the Termon Formation here has been produced from an association of remarkably evenly bedded and thinly bedded siltstones, sometimes with ripple-marks and thin calcareous bands. Indeed, they contrast so strongly with the highly contorted schists, that Pulvertaft (1961) distinguished them as the "Gorteen Lough Striped Flags and Pelites." This thin banding is characteristic of both this and certain lower members of the succeeding formation, the Crana Quartzite, into which there is a clear gradation; the siltstones or their equivalent phyllites become dark upward at the same time that thin interbeds of platy psammite appear.

The *Crana Quartzite,* which is at least 1,525 meters thick in Inishowen, represents, as a turbidite sequence, a completely different arenaceous facies from the Slieve Tooey Quartzite, and foreshadows the greywacke facies of the overlying strata. Southeast of the Leannan Fault, it was properly divided by McCallien (1937) into three members: a lower fine grained division, the "Lower Crana Quartzite," in which thick beds of fine psammite, occasionally graded, are interbedded by striped semipelites; a middle division, the "Killygarvan Flags," consisting of striped psammites and some thin limestones; and an upper division of graded pebbly grits, the "Upper Crana Quartzite." Northwest of the Leannan Fault, these lower divisions are commonly cut out by faulting, and in fact, the Upper Crana Quartzite (as the Kilmacrenan Grit) forms a long fault strip (Pulvertaft, 1961, p. 263) within the Leannan Fault complex.

Throughout the Upper Crana Quartzite, cross-stratification is very subordinate to grading and is lacking altogether in the Lower Quartzite. This grading is spectacular, bedding units varying in thickness from 0.3–1 meter with basal layers of pebbles averaging 10–12 centimeters. Quartz (both single quartz crystals and vein quartz) is dominant, orthoclase and potash feldspar is usually less than 30%, and the grains as a whole are subrounded and set in a scanty micaceous matrix.

In a way reminiscent of the Slieve Tooey–Cranford junction, there is in Inishowen a sharp change upward from the Crana Quartzite to a thick limestone interbedded with dark pelites, the Culdaff Limestone of McCallien (1935), which is, in turn, gradationally followed by a great series of chloritic slates and greywackes, distinct in lithology from the horizons already described. We use this pronounced change of facies to mark the top of the Kilmacrenan Succession, though Callaway (1885), appreciating that the change from well-sorted arenite to graded wacke was situated lower in the sequence, placed the base of the Lough Foyle Succession at the bottom of the Crana Quartzite (see Table 1-2). However, we think that the base of the Culdaff Limestone is more certain, and, as we mentioned earlier, it is conveniently the junction between the Upper and Middle Dalradian as defined in Scotland by Rast (1963).

THE SEDIMENTARY ASSEMBLAGES

(D) The Central Donegal Facies

On the southeastern side of the Leannan Fault (Fig. 1-1 and Map 2 in folder), the Kilmacrenan Succession begins to change facies southward, finally becoming so different that Pitcher and others (1964) considered that the central Donegal rocks represented a succession quite distinct from any exposed elsewhere in Donegal. We approach this problem afresh by first tabling this new succession in its true stratigraphic order, incidentally making some modification of previous work, and then discussing the correlation of this facies with the northwestern facies of the Kilmacrenan Succession. In this, we have drawn heavily on the unpublished work of Elwell, Tozer, and Wood.

In their main outcrop in mid-Donegal, these metasediments become progressively more strongly deformed to the southwest, with the result that original structures have been obliterated, lithologies disguised, and estimates of thickness rendered of little value. Nevertheless, a true succession can be worked out and correlated with the standard succession as is demonstrated in Table 1-4. In what follows, we use a separate nomenclature in order to draw attention to the considerable contrast in sedimentary facies between the northwest and central Donegal variants of the Kilmacrenan Succession.

TABLE 1-4. Dalradian Successions in Central Donegal, Tyrone, and Inishowen

Inishowen and Rathmullan (McCallien, 1933, 1937)	Western Tyrone and Southern Donegal (Pitcher, Shackleton, and Wood, 1971)		Central Donegal (this book)	Standard Kilmacrenan Succession (this book)
Greencastle Green Beds				
Inishowen Head Grits and Slates	Croaghgarrow Formation	Oughtradeen Group (Anderson, 1948)		
Cloghan Green Beds	Shanaghy Green Beds			
Fahan Group	Mullyfa Formation			
Inch and Culdaff Limestones	Aghyaran Formation		Convoy Formation	Culdaff Limestone
Upper Crana Quartzite				Upper Crana Quartzite
Killygarvan Flags				
Lower Crana Quartzite	Killeter Quartzite		Killeter Quartzite	Lower Crana Quartzite
Milford Schists	Lough Eske Formation	Calcareous Group (Anderson, 1954)	Termon/Lough Eske Formation	Termon Formation
Lag Limestone			Boultypatrick Grit / Croaghubbrid Pelites	Slieve League Formation
Malin Head Quartzite			Gaugin Qaurtzite	Slieve Tooey Quartzite
			Boulder Bed	Port Askaig Tillite
			Owengarve Formation	Glencolumbkille Limestone / Glencolumbkille Pelites

The stratigraphically lowermost *Owengarve Formation* only appears in the narrow core of a major isocline which curves around the Gaugin Mountain mass (see Fig. 1-5). It is here seen to consist of a banded association of calcite limestone, calc-silicates, fine-grained psammites, and some pelites. The overlying *Gaugin Quartzite* (equal to the Silver Hill Quartzite of Pitcher et al., 1964) is a quartz-arenite in which parallel bedding and cross-lamination can occasionally be recognized. Associated with it is a psammite boulder-bed which forms limited exposures southeast of Gaugin Mountain in the nearby locality of Priesttown, there are large erratic blocks of both dolomitic and psammitic tillite with the characteristic association of dolomitic and granitic clasts. Despite the paucity of exposure, we feel certain that we are dealing with a true representative of the *Port Askaig Tillite*.

The quartzite passes upward through a thin, pale-colored pelite, which forms a good marker horizon, into a monotonous formation of dark pelites, the *Croaghubbrid Pelites*. Then, by the incoming of grit bands, there is an extended transition upward into a distinctive psammitic horizon, the *Boultypatrick Grit*; we draw the lower boundary of this formation at the first important appearance of the psammatic bands. Upward, the dark pelitic interbeds gradually disappear, leaving a coarse grit with even coarser pebbly lenses. There is then a fairly rapid upward change to a finer-grained psammite, which becomes increasingly interbedded with greenish pelites; these and the fine psammite we assign to the *Termon Formation*.

The key to these correlations lies in showing that the outcrop of the Gaugin Quartzite in the area due west of Ballybofey follows the form of a major overturned fold (Fig. 1-5). On the northern right way up limb of this Ballybofey fold, the Boultypatrick Grit passes up into a formation where psammite is subordinate to pelite, while on the southern limb, this same formation is dominantly psammitic. Then, around the closure of this fold, it is easy to demonstrate a transition in which the

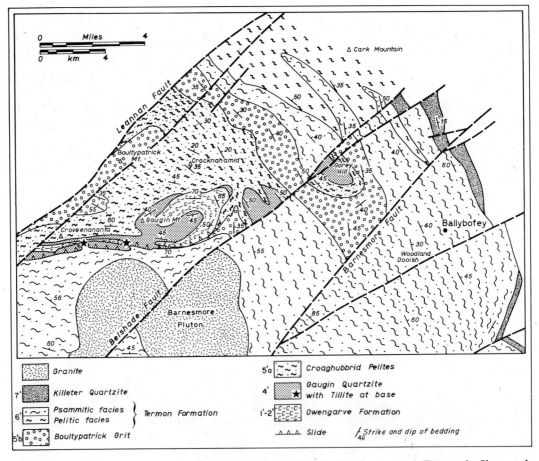

FIGURE 1-5. The geology of the area west of Ballybofey in central Donegal. Shows the outcrop of the polyphase Ballybofey fold and the lateral change of facies in the Termon Formation. After Pitcher, Shackleton and Wood (1971).

pelitic element becomes increasingly reduced southward by interbanding by psammite and eventually passes into what Pitcher and others (1964) defined as the Lough Eske Group (Table 1-4). The latter are therefore the facies equivalent of the Termon Pelites, and possibly even a large part of the Boultypatrick Grit. The final key in the correlation of the two facies is the demonstration by Pitcher, Shackleton, and Wood (1971) that the overlying *Killeter Quartzite* can be followed right around the closure of the Ballybofey fold, eventually forming part of the inverted succession on the lower limb, as exposed north of Lough Derg (Map 2 in folder).

This clean, massive, white quartzite, despite its greatly varying thickness (0–1000+ meters), forms a convenient marker horizon and also the beginning of a very considerable upward change in facies, for it passes by transition into a thick calcareous series, the *Convoy Formation*. Such rocks do not, however, enter our map area due to the limitation of their outcrop by the southward plunge of yet another major Synclinal Fold (unnamed) (Map 2 in folder). This duplicates the pelitic facies of the Termon Formation around Letterkenny, and so extends its outcrop almost as far as Kilmacrenan. Here again, the Termon is folded around a synclinal core of Crana Quartzite, whence it can be followed in continuous outcrop into the area of the type succession in Inishowen (see Map 2 in folder).

Thus we have established a correlation between both limbs of the important Ballybofey fold, and on a more regional basis, with the Northwest Donegal Facies (Table 1-4).

(E) Comparison of the Two Facies of the Kilmacrenan Succession

The correlation between the two areas calls for comment on the nature of the facies changes involved. The lower part of the succession in central Donegal is not unlike its equivalent in northwest Donegal; thus, both calcareous and psammitic facies of the tillite are present and lie sandwiched between a calcareous horizon and a quartzite. The Glencolumbkille Limestone, which normally occupies the lower position in the Northwest Facies, is, however, only represented in central Donegal by the calcareous schists of the Owengarve Formation.

It is among those rocks lying between the Slieve Tooey (Gaugin) and Crana Quartzites that the greatest differences exist, though even here, the changes in one facies are foreshadowed in the other. Thus, the fine-grained psammitic intercalations in the type areas are very similar in lithology to the intercalations in the Termon Formation; further, the southwesterly replacement in the Northwestern Facies of the Cranford Limestone by a thick sequence of dark pelites and gritty psammites represents an approach to the situation in central Donegal. Thus, a most important point emerges—that the horizon immediately above the Slieve Tooey Quartzite in Donegal is the locus for great facies changes, and it seems, as noted previously, that an extensive area of "shallows" existed in the north. In fact, the base of the Cranford Limestone appears to represent a major break in sedimentation—we return to this later.

(F) The Southern Limit of the Kilmacrenan Succession

So far, we have only considered in any detail two of the great successions which make up the Dalradian of Donegal. The information is not yet available to enable us to describe the whole outcrop of the greywacke-argillite lithologies which make up the conformably overlying *Lough Foyle Succession* (Table 1-5). However, we do need to say a little more about the rocks lying at the boundary, in order to discuss the wider correlations.

As we have seen, the Killeter Quartzite abruptly ends the monotony of the Termon (Lough Eske) Formation and passes up by gradation into the calcareous Convoy Formation. The latter, because of its considerable thickness (up to 1000 meters) and its disposition in the limbs of a highly inclined, overturned fold, occupies a large part of the Raphoe–Convoy area of eastern Donegal whence, like all the underlying divisions, it can be traced into Tyrone (as the Aghyaran Formation) and around the nose of the Ballybofey fold (Map 2 in folder). In this Formation, thick beds of limestone or dolomite are intercalated with pelites, now albite-muscovite-schists; and this assemblage passes up into a very thick formation of quartzofeldspathic semipelites with some gritty or pebbly horizons—the Mullyfa Formation, according to Wood (1970T)—which has a distinctly Upper Dalradian aspect, *i.e.,* it is of greywacke type.

The problem is then the correlation with standard succession in Inishowen, for it is clear that a considerable change of facies is involved at this level. Thus, in Inishowen, the thick Crana Quartzite with its dominance of graded grits and subordinate dark pelites is topped by a series of dark limestones and phyllites, the Culdaff Limestone. Around Raphoe, at this general level, there is less quartzite and more of the calcareous and pelitic lithologies, and yet farther southwest, in Tyrone, it is the pelites which finally predominate (see Map 2 in folder).

Thus, it is difficult to make precise correlations: any boundaries are bound to be diachronous. Nevertheless, it seems likely that the pair Crana Quartzite–Culdaff Limestone are generally time equivalent to the Killeter Quartzite and Convoy–Aghyaran Formation taken together; though it may be that the Killeter could be directly equated with the Lower Crana (Table 1-4). Thus, the accepted lower limit of the Upper Dalradian (the base of the Culdaff Limestone) is not easy to define in central Donegal; *for convenience,* we place it beneath the uppermost limestone member of the Convoy Formation (shown separately on Map 2 in folder).

The probable correlations are presented in Table 1-5, from which it can be seen that the basic stratigraphic relations of most of Donegal present no considerable problem up to the borders of the County, where the Upper Dalradian with its undifferentiated semipelites, psammites, green beds, and pillow lavas (McCallien, 1936; Hartley, 1938) takes over the outcrop.

In southern Donegal, however, it has long been puzzling that the Dalradian rocks which we now recognize as psammitic facies of the Termon Formation should young toward the Lough Derg Psammites which Anderson (1947) correlated with the pre-Dalradian Moine Series of Scotland (see also Powell, 1965; Church, 1969). We now know from the work of Pitcher and others (1971) that this is indeed the case and that the Termon Formation gives way southward to representatives of the Crana Quartzite and to lower members of the Lough Foyle Succession. Thus it is certain that these inverted Middle and Upper Dalradian structurally overlie the Lough Derg (Moine) rocks along a major tectonic break, the Lough Derg (Ballykillowen Hill) Slide (see Map 2).

(G) Comparison with Scotland

There is a very strong similarity between the lower part of the Kilmacrenan and Islay Successions in Donegal and southwestern Scotland, respectively (Table 1-5).

In particular, the Tillite and the underlying calcareous rocks below it clearly persist throughout the Dalradian from Connemara to Argyll (Kilburn *et al.,* 1965). There are local changes, as has been so well demonstrated by Spencer (1971) and Howarth (1971), but the essential stratigraphy is everywhere maintained. There is always a dolomite at the base (with the apparent exception of central Donegal), followed by a sequence of dolomitic tillites which grade up into psammitic tillites with increasingly important sandstone interbeds. The latter, however, thicken toward Argyll, where there is also a much greater development of the lower dolomitic tillites; these are changes which coincide with a great thickening of the overlying blanket of quartzite (see Spencer, *loc. cit.*). Further, the thin pelite marking the top of the Tillite in Glencolumbkille—and even in localities in Connemara—gradually changes northeastward into the sandy algal-bearing dolomites of Islay.

In Argyll, the quartzite formation reaches the astonishing thickness of 5 kilometers, and if we accept Knill's (1963) conjecture that it also thins southeastward, we begin to get a picture of a huge lens of quartzite. In fact, it seems that the vast spread of the Slieve Tooey-Islay Quartzite throughout Scotland and Ireland is in the form of great lens-like accumulations of sand strung out along one horizon.

THE SEDIMENTARY ASSEMBLAGES

TABLE 1-5. Correlation of Kilmacrenan (Donegal) and Islay–Loch Awe Successions

Dominant Lithologies			Donegal (McCallien, 1935, and this book)	Argyll (Knill, 1963)
Wackes and chloritic argillites; calcareous horizons rare, slump structures and slide breccias common	Upper Dalradian	LOUGH FOYLE SUCCESSION	Greencastle Green Beds Inishowen Head Grits and Phyllites	Loch Avich Grits
			Cloghan Green Beds Fahan Slate–Grit Group	Tayvallich Lavas
Impure limestone transitional to higher beds			Inch Limestone Group Culdaff Limestone	Tayvallich Limestone
			Crana Quartzite	Crinan Grits
Quartzose and feldspathic arenites and argillites, calcareous horizons throughout; laminated bedding common	Middle Dalradian	KILMACRENAN SUCCESSION	Termon Formation (Cranford Limestone)	Ardrishaig–Craignish Phyllites (Shuna Limestone)
			Slieve League Formation	Easdale Slates Scarba Transition Group Scarba Conglomerate Group Jura Slates
			Slieve Tooey Quartzite	Islay Quartzite
			Port Askaig Tillite	Port Askaig Tillite
Port Askaig Limestone and black pelite	Lower Dalradian		Glencolumbkille Limestone	Islay Limestone
			Glencolumbkille Pelites	Mull of Oa Phyllites
				Maol an Fhithich Quartzite

As regards the overlying strata, we have shown in Donegal that the Cranford Limestone gives way southwestwards to a thick, dark pelite-psammite sequence, the Slieve League Formation. A similar change is apparent when this horizon is followed northeastward into Argyll where the thin band of Jura Slates, the slide breccias of the Scarba Conglomerate, the variable lithologies of the Scarba Transition Group, the graphitic Easdale Slates, and the Shuna Limestone all appear between the Islay Quartzite and the overlying pale pelites—the Ardrishaig–Craignish Phyllites (Knill, 1963), rocks which have much the same mixed lithology as the Termon Formation. Assembled in the form of the sketch-section of Fig. 1-6, these relationships clearly suggest the existence of a zone of shallows flanked by subsiding troughs in which dark carbonaceous shales were being more continuously deposited. According to Roberts (1966a) and Knill (1959, 1963), it is the southeasterly trough which is of greatest importance in the control of sedimentation throughout the whole of the southeast belt of the Dalradian.

The overlying thick group of pale colored pelites (*i.e.*, the Termon–Ardrishaig–Craignish Formation), with its thin psammitic lenses and occasional dark schists and limestones, is remarkably continuous. Yet even in this formation, as we have seen in central Donegal, there is a very considerable change of facies, shown by the southwesterly increase in the psammitic element.

The passage upward into coarser arenites (Crana–Crinan Quartzites) characterized by large-scale graded beds is again a widespread feature, though, as we have seen, these pass laterally *southwestward* into a mixed calcareous facies. Taken together with a similar finding (Knill, 1963) in Argyll, where the Crinan Grits are laterally equivalent to calcareous sediments lying to the *northwest,* it would seem that the grits were deposited in a fairly narrow NE–SW trending trough.

We finally arrive then at the clear possibility of correlating, with some precision, the Kilmacrenan–Islay Succession from Connemara to Argyll, a distance of some 402 kilometers. There is, in fact, little difficulty in extending the correlation to the Banffshire coast of northeast Scotland (see Johnson, 1965), though we will not deal here with the details of the considerable facies changes that are revealed.

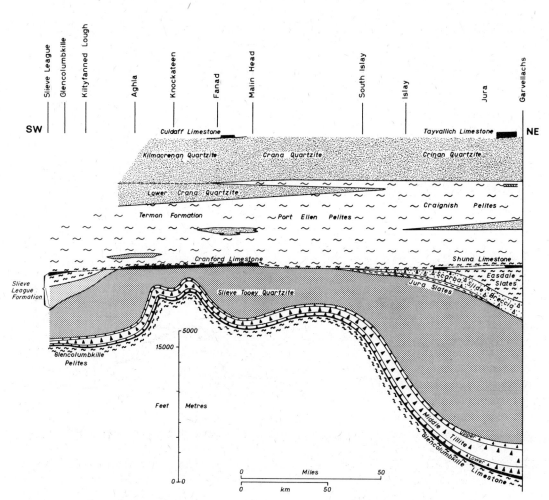

FIGURE 1-6. Schematic section along the regional trend showing the lateral changes in the northwest facies of the Kilmacrenan succession.

4. The Relation between the Creeslough and Kilmacrenan Successions

The relation between the Ballachulish and Islay Successions in Argyll has always been the subject of considerable controversy (Johnstone, 1966; Rast, 1963). There is a general similarity of lithology, and both can apparently be shown to pass directly downward into psammitic lithologies, usually referred to as the Moine Series (*loc. cit.*; Watson, 1963; Anderson, 1956). A widely held explanation of the relationship of the two successions is that they are but facies variants of a single formation (Anderson, 1948, 1965; Johnstone, *loc. cit.,* Voll, 1964), and correlations have been made on this basis. On the other hand, the lack of detailed correlation between the two successions (Rast and Litherland, 1970)—especially the absence of the tillite in the Ballachulish rocks—suggests that they originated at different times and probably even in quite different areas, and that they are now brought into juxtaposition on a powerful dislocation, the Iltay (Islay–Loch Tay) Boundary Slide.

In Donegal, the equivalent successions, the Creeslough and Kilmacrenan, are in such close relationship to one another that one might suppose that a definitive answer to the problem of their relationship could be easily obtained. Certainly, there is a dislocation—the Knockateen Slide—at the base of the Slieve Tooey Quartzite, but below this, there are recognizable slivers of the Glencolumbkille Pelite, such as at Mossfield, near Gartan Lough, and at Fintown. Assuming that these slices do not actually belong to the uppermost Creeslough Succession, their presence requires that we postulate a second dislocation separating these Glencolumbkille rocks from the Loughros Formation, the highest member of the Creeslough Succession.

From the way in which horizons at this level are locally cut out—and not all of the relationships can be explained by the later faulting—this proposition seems sound enough; the Knockateen Slide is, not surprisingly, a compound dislocation. Nevertheless, it is clear that for the 56 kilometers from Loughros Bay northeastward to Glen (see Map 1 in folder), the highest members of the Ballachulish Succession are in close juxtaposition to low-placed members of the Islay Succession, so that it would be reasonable to propose that the one succession followed conformably on the other, and that the slide was only of relatively minor importance in separating the two. What is not in agreement with this proposition, however, is that the Slieve Tooey Quartzite northeast of the village of Glen transgresses sharply, by a combination of faulting and thrusting, across the Creeslough Succession until in northern Fanad, it and its underlying tillite and limestone formations make contact with the Ards Pelites and the Ards Quartzite, both well down in the Creeslough sequence. There is also the complication that the major folds on either side of the Knockateen Slide face in opposite directions (see Chapter 3, Section 5).

The problem of the relationship of the successions cannot, it seems, be finally solved in Donegal. A slide (the Knockateen) equivalent to the Iltay Boundary Slide of Argyll separates the two (as it does in Mayo and Galway: see Phillips *et al.,* 1969), but it is a matter of opinion whether the movement involved is considerable enough to bring two different facies together. There are no lateral facies changes which would lead one to suppose a transition of one into the other in Donegal, and further, on the basis of our better understanding of the role of the tillites, we would find it difficult to account for the absence of such rocks in the Creeslough–Ballachulish sequence, if these are indeed but facies variants of those of Kilmacrenan and Islay. Our interim conclusion is, in fact, that the former lie naturally under the latter, as Rast and Litherland (1970) have recently claimed for the Scottish situation.

5. The Sedimentary Facies of the Dalradian of Donegal

The sediments of the two lowermost successions of Donegal represent a gradual change in facies, from a combination of current-bedded quartzite and black slate affiliations upward into a graded quartzite affiliation, which, in turn, heralds the establishment of a greywacke (turbidite) facies. In general, lowermost formations consist of well-differentiated and well-sorted rocks, and the presence of clean, coarse-grained feldspathic sandstones is especially characteristic. Individual beds often run for kilometers with little change in facies. The sedimentary structures, such as large-scale cross-lamination, very coarse grading in pebble bands, and mud cracks suggest that some units were deposited under conditions of near emergence. Added to this, is the evidence for winnowing at the top of some individual beds of tillite and of local areas of shallows. Indeed, we would claim a relatively shallow-water origin for all the lithological types, not excluding the black shales for, over a considerable thickness, even these are rhythmically interbanded with sandy beds, as at the base of the most typically shallow-water deposit of the whole succession, the Ards Quartzite. The presence of rather pure limestones and dolomites, either alternating with quartzitic flags or interbedding with tillites, confirms this general claim.

There are, however, important changes in facies and thickness which fit in with the concept of differential subsidence; there were clearly troughs and basins in which sedimentation was more continuous than on the intervening swells, where conditions were often those of near emergence (*cf.* Fig. 1-6). On the margins of these areas of shallows, as for example, in Argyll, calcareous rocks, often dolomitic and with stromatolite "reefs" (Knill, 1963, p. 101), give place to dark shales and slide breccias. We have demonstrated this situation for the rocks in the middle part of the Kilmacrenan–Islay succession, and Knill (*ibid.*) and Roberts (1966*a*) have provided elegant studies of similar facies changes in the higher levels of the same succession. Nevertheless, we repeat our contention that the actual depth of water was never very great, and though differential subsidence was clearly operative, it is still a remarkable fact, though almost a commonplace of geology, that deposition and subsidence kept pace over a long period and over an immense area.

Information concerning directions of current flow and of provenance of the material is, as yet, very confusing, probably because there is so little of it. The only determinations for the Creeslough Succession are those by Knill and Knill from quartzites on Rosguill, in which primary structures indicate sedimentary transport from the west and north (1961, p. 277). In the Slieve Tooey Quartzite at Glencolumbkille, the direction is consistently from the south (Howarth *et al.*, 1966, p. 140), but Howarth (1971) records an almost radial paleocurrent pattern in the underlying Tillite, and Knill found evidence of currents from the northeast in the Crana Quartzite in Inishowen and from the west and northwest in higher levels here (1963, p. 116). The latter, at least, are at variance with transport directions in Islay and Jura to the northwest (*ibid*), and it is clear that more work is needed along these lines. Now that the stratigraphy and structure are known over considerable distances, it would be possible to make many such studies.

It is worth while making some special reference to the Slieve Tooey–Islay Quartzite—the Central Highland Quartzite of Perthshire—if only because it must be the most continuous and important marine sandstone horizon in the British Isles. The presence of abundant large scale cross-laminae, worm casts and mud cracks, coupled with the great thickness, homogeneity of lithology, and excellence of sorting, indicates rapid shallow-water deposition of a much-worked sand produced by extensive erosion of a granitic and metamorphic land area, which, incidentally, may have been the source of the extra-basinal clasts in the Tillite.

Following Knill, it is tempting to regard the Jura occurrence as representing the deposits of a delta (1963, p. 107), but this is probably too simple a view of the mode of origin of what we conceive to be—taking the entire outcrop into account—a series of immense sand lenses; it would be most difficult to believe that these were all attached to "Mississippian" river systems. But if differential subsidence provides the place of deposition, there is still the problem of the origin of this vast accumulation of sand. Perhaps it represents, as suggested by Shackleton, the outwash deposits from the final melting of the ice after the great Varangian Glaciation (see also Howarth, 1971).

In more general terms, thick and extensive limestone-bearing sequences of shallow-water sediments are typical of what has come to be known as the miogeosynclinal environment. The identification of the lower successions of the Dalradian as belonging to this environment implies that the members were deposited in a sedimentary trough marginal to a continental mass, and a northeasterly trending shoreline is suggested by the remarkable persistence of the several Dalradian successions, each with a characteristic lithology, from northeastern Scotland to Connemara. The case for a land mass lying to the west and northwest of the Dalradian outcrop has recently been debated by Watson (1963), Knill (1963), and Roberts (1966*a*), but a satisfactory solution to this and other problems in Dalradian paleogeography must await further studies.

As we have seen, this type of sedimentation changes dramatically at about the general level of the Middle Dalradian, becoming dominantly turbiditic in character upward. At first, this change seems to have been localized in a narrow trough which differed in location with time. Thus, in southern Donegal, the Lough Eske psammites, which we interpret as being of this type, formed in a trough which lay a considerable distance to the south of that in which the succeeding Crana Grits were deposited (Fig. 1-7). Finally, however, this type of sedimentation became widespread with the establishment of a much broader area of turbiditic sedimentation in the Upper Dalradian.

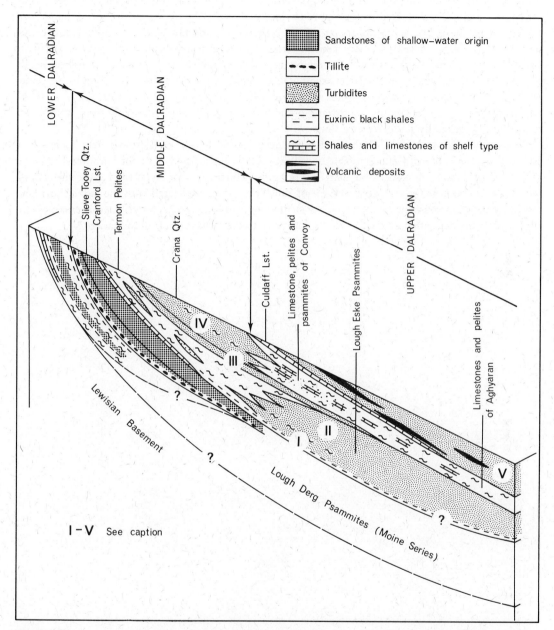

FIGURE 1-7. Schematic section through the Dalradian in Donegal prior to deformation. Line of section approximately NNE–SSW and extending 60 kilometers: I, proximal turbidites of the Croaghubbrid Grits; II, distal turbidites of the Lough Eske Psammites; III, distal turbidites of the lower Crana Quartzite; IV, proximal turbidites of the upper Crana Quartzite; V, turbidites (types undifferentiated) of the Upper Dalradian.

There is some evidence of the migration of the shelves of these early established troughs. Thus, in detail, it seems possible that the lower, fine-grained beds of the Crana Formation may represent distal turbidites—and the higher, coarser, and more obviously graded beds, proximal turbidites, the upward change being due to the lateral migration of an "apron" of turbiditic sediments in relation to the margins of a trough. Indeed, in more general terms, as we have seen, the Crana horizon gradually changes facies to the southwestward, becoming first increasingly more calcareous and then finally pelitic. It is as if we were seeing a classic example of the gradual passage from the margin to the center of a trough across such a marginal apron, a passage which changes its position with time—as if the apron sediments were onlapping southward onto the shelf, the material coming from a shoreline retreating in a generally *southerly* direction.

Finally, there is the relation of the contemporaneous igneous rocks to these different sedimentary facies. It is in accord with expectation that pillow-lavas are associated with the turbidites and that extrusives are lacking in association with the miogeosynclinal types of sediments. Nevertheless, we must draw attention to the very great volume of tholeiitic magma associated with the latter and represented by the numerous thick sills which we shall show in the next chapter to have been emplaced shortly after the deposition of the sediments and prior to the bulk of their deformation.

CHAPTER 2

THE DOLERITE SILLS: THOLEIITE IN THE EARLY STAGES OF OROGENY

CONTENTS

1. Introduction 43
2. The Quartz Dolerites 44
3. The Dolerites of Rosbeg Type 47
4. Age of the Dolerites 48
5. The Metamorphism of the Basic Rocks 48
6. Chemical Composition 49
7. Discussion 50

1. Introduction

Previous to or even during the early stages of their deformation, the Dalradian sediments were invaded by a great quantity of tholeiitic magma. This is represented by thick and thin sheets and sills of quartz dolerite, now metadolerite, which have been affected to varying degrees by the regional metamorphism and deformation, and, of course, by the later events associated with the emplacement of the Donegal granites. We concentrate here on the primary characteristics of these rocks, but give a brief discussion of their structural and metamorphic relationships later in this chapter.

The regional importance of this suite of intrusions is illustrated by Fig. 2-1 which shows the distribution of the more prominent bodies. Though this considerable episode of tholeiitic activity has its counterparts in other Dalradian areas, nowhere else, except perhaps on the seaboard of nearby Argyll, is the volume of magma emplaced anything like as large as in Donegal. These dolerites are therefore worthy of a special comment, and, as we shall see later, they raise a topic of considerable interest in the history of fold belts in general.

There are two main types, for in addition to the more usual *quartz dolerites* with simple contact aureoles, there are, in Rosbeg, a suite of dolerite sills with a distinctive chemical and mineralogical composition. These *dolerites of Rosbeg type,* which are rarely found elsewhere in Donegal, are now

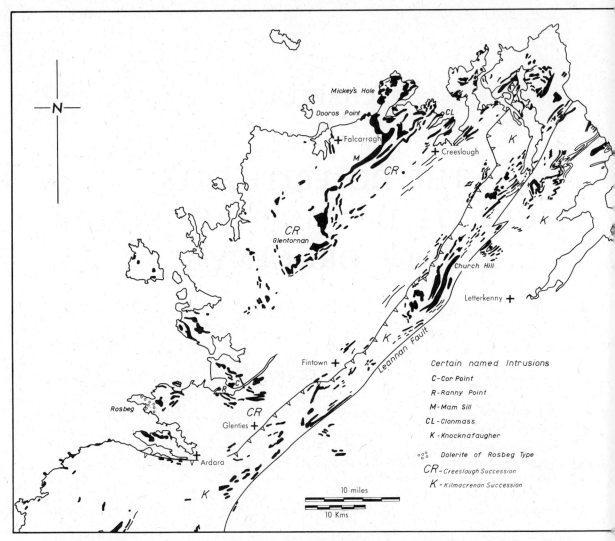

FIGURE 2-1. Sketch map showing the distribution of the metadolerites in northwest Donegal. The concentration of sills in particular parts of the Creeslough and Kilmacrenan Successions is well shown.

represented by garnet amphibolites which are so coarse-grained as to stand in metamorphic contrast to the phyllitic-schist forming the host rocks; further, each body is characteristically enveloped by a thin skin of white adinole.

It had always been thought that all of the Dalradian dolerites of Donegal were of one age and that their emplacement preceded the earliest phases of deformation. However, recent work by Obaid (1967T) has shown that while all the dolerites of Rosbeg type clearly predate the earliest deformation, there is evidence indicating that some sills of quartz dolerite might have been intruded between the first and second phases of deformation. We return to this important matter of chronology later, after giving some account of the rock types concerned.

2. The Quartz Dolerites

The quartz dolerites form by far the majority of the dolerite intrusions, and representatives are found throughout the whole of the Dalradian, though they are less common in the Central Donegal

Facies of the Kilmacrenan Succession (Fig. 2-1). As we shall see later, these rocks are now metadolerites composed mainly of hornblende, albite, clinozoisite, quartz, and ilmenite, but what we are immediately concerned with is the original form and composition.

Most of these intrusions are in the form of sills or slightly transgressive sheets. Some of the sills lie at the same stratigraphic horizon for many miles and thus make valuable markers; others can be shown to be gently transgressive over long distances, and still others cut sharply across the country rocks. Further, the branching of sills is not an uncommon feature. Sometimes possible feeders are observed, such as the pipelike bodies in the steep cliffs of Aranmore. Then, there is a crude relationship between the shapes of the intrusions and the lithology of the host rocks, for within the pelites, the metadolerites usually form regular sills, whereas in the quartzites they are more irregular, discontinuous, and even discordant.

The thickness of metadolerite sills and sheets varies from a few tens of centimeters up to 200 meters or more, the upper limit being the figure givn by Rickard (1962) for a particular measurement of the important Mam sill. Certainly, many of the sills are consistently from 60 to 150 meters thick and run so for many kilometers; the Mam, Ranny Point, Clonmass, and Knocknafaugher intrusions are all examples of such major sheets (Fig. 2-1). Rough estimates show that the maximum total thickness of dolerite in the Creeslough Succession is approximately 660 meters in the Errigal district, of which a total of 440 meters is found between the base of the Ards Quartzite and the top of the Lower Falcarragh Pelites. This estimate is of the same order as the 410 meters for the same stratigraphic range in the Maas area, and such a thickness represents about 20% of the total column of strata. Although the more complex outcrop of the Kilmacrenan Succession makes measurement of the total thickness of dolerite sheets there very difficult, it appears that the amount of magma emplaced into these rocks was considerably less than into the Creeslough succession.

Although all of the dolerite sheets have been deformed to some extent, it is chiefly the margins which are affected, and the primary structures and textures can often be found in central portions of the intrusives. The primary grain size of the dolerites varies irregularly with the size of the intrusion and the position within it; the smaller bodies are usually fine-grained throughout, while the larger ones often show fine-grained chilled margins that pass gradually into a medium-grained interior.

Primary ophitic textures are well preserved in the central parts of thicker sills, but the original minerals, on the other hand, are only occasionally seen, as in the dolerites of Rosguill. These primary constituents appear to have been, in descending order of abundance, plagioclase (chiefly labradorite-andesine), pigeonitic augite ($2V = 30-45°$), quartz, ilmenite, alkali feldspar, and apatite. Clumps of augite are often surrounded by laths of twinned feldspar, about 3 millimeters in length, between which is an often granophyric mesostasis of quartz and alkali feldspar. Rare pseudomorphs of bowlingite after olivine have been reported (Knill and Knill, 1961, p. 277), and in the chilled margins, andesine (An_{35}) phenocrysts, and small flakes of biotite are sometimes present.

A few dolerites show a mineral layering that has not yet been properly investigated. For example, in the large intrusion in the Horn Head peninsula south of Mickey's Hole in the Mam sill at Dooros Point, feldspathic bands alternate with less feldspathic (Fig. 2-2a), and there are even indications of grading.

Xenoliths are rare, but there are some pockets of coarse material similar in mineralogy to their more fine-grained hosts and which are clearly of cognate origin (Gindy, 1951b, p. 104; Pulvertaft, 1961, p. 265). Occasionally, fragments of mobilized hornfels have been caught up in the magma and sometimes plastically deformed by its continued movement (Fig. 2-2b; McCall, 1954).

Many dolerite sills in pelitic and semipelitic horizons are surrounded by narrow static aureoles in which their hosts have been indurated to a varying degree. These hornfelsed zones have largely restricted the subsequent deformation and metamorphism, and in them, as we have already mentioned (Chapter 1), primary sedimentation characters are often excellently preserved. Especially interesting is metamorphically reversed graded bedding, in which the new minerals are distributed in

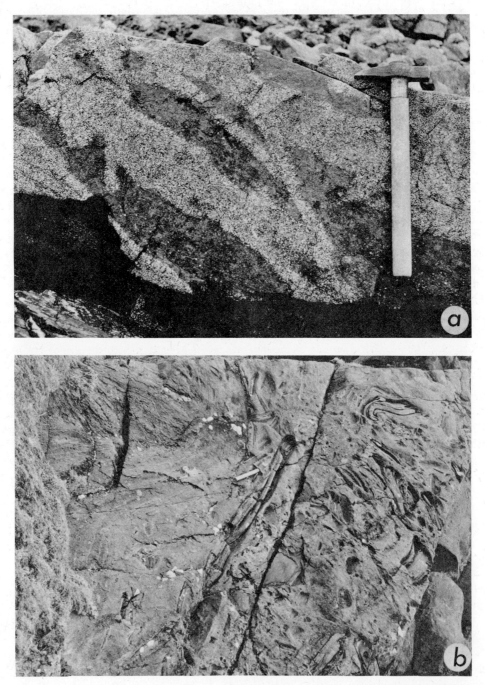

FIGURE 2-2. (a) Irregular banding in the Clonmass Sill, Clonmass Head. (b) Mobilized sedimentary inclusions within the margin of the Mam Sill, southwest of Dooros Point.

accordance with compositional changes in the original sediment (McCall, *loc. cit.*). As we have also pointed out earlier, the striped flags of the Termon Formation in the Church Hill area (the Gartan Lough Group of Pulvertaft, *loc. cit.*) owe the preservation of their original lithology to widespread induration by the swarms of dolerite intrusions which characterize this horizon. In a few places, there are minor folds of varying style preserved in the hornfelses close to the margins of the sills. These we think are largely the result of mobilization during intrusion of the sills (*cf.* Fig. 2-2b).

There are many other examples of this contact effect. Several of the larger dolerite sheets in the Dunfanaghy area, particularly the Mam and Clonmass sills (Fig. 2-1), are bordered by up to 12 meters and more of spotted hornfelses which pass outward into indurated pelites and thence into regional schists. The spots in the innermost parts of these aureoles consist of aggregates of unorientated biotite and muscovite which, from their polygonal or rectangular form, appear to represent original cordierite porphyroblasts reconstituted by later regional metamorphism (McCall, *loc. cit.* pp. 161–2).

3. The Dolerites of Rosbeg Type

This special type of dolerite occurs as sills almost completely confined to those outcrops of the Upper Falcarragh Pelites occurring along the shore of the Rosbeg Peninsula.* Within these sills, which range in thickness up to about 45 meters, all the primary structures have been completely obliterated by subsequent deformation and metamorphism, and the rocks now exhibit a strong schistosity—except where this itself has been obliterated by the main recrystallization.

These Rosbeg amphibolites are now characterized by intersecting laths of green-brown hornblende, with interstitial oligoclase-andesine, sphene, ilmenite, magnetite, calcite, and a little quartz. Porphyroblasts of garnet are common and form at the expense of hornblende, and the presence in them of helicitic trails indicates that they have overprinted an earlier crenulated fabric (*cf.* Howarth *et al.*, 1966, p. 142).

The characteristic marginal adinoles, in which the host pelites have a bleached and whitened appearance, pass by transition into the normal pelites. The width of this contact zone depends largely, of course, upon the width of the sill, but none were seen to exceed 4.5 meters, and many are but a meter or so across. The most important mineralogical change is an appreciable increase in the proportion of albite in the groundmass toward the sill. This is naturally reflected by an increase in Na^{1+} with a corresponding decrease in K^{1+} (Fig. 2-3, in terms of oxides); there is, however, no other consistent chemical difference between adinole and unaltered pelite.

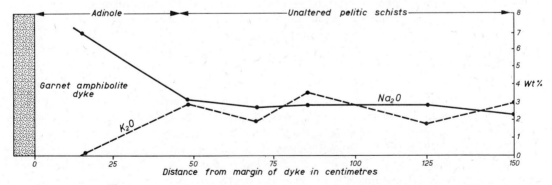

FIGURE 2-3. Chemical variation in adinolized contact zone adjacent to a metadolerite sill. Portabrabane, Rosbeg. After T. M. S. Obaid, based on analyses by M. Brotherton.

* Much of the information which follows is due to T. M. S. Obaid (1967T).

Similar variations have long been recognized as characteristic of adinoles (*e.g.,* Dewey, 1915; Agrell, 1939), though here, this metasomatism is associated with intrusions which are not especially rich in alkalis. The fate of the displaced potassium remains unknown; there is certainly no evidence of its outward displacement, such as recorded by Davies (1956) in an example from North Wales, and further, the potassium content of the igneous rock is even lower than normal for tholeiitic basalts, so that absorption into the magma seems unlikely. Of course, the dyke material may well have moved on, but in the final analysis, the fate of the displaced potassium remains unknown.

4. Age of the Dolerites

It had always been considered that the Donegal dolerites were emplaced prior to the onset of the folding and metamorphism of the Dalradian sediments, as shown, for example, by the preservation in minute detail of original lithological features in some of the marginal hornfelses. However, two distinct episodes of dolerite intrusion, separated in time by fold movements, have been recognized in the Dalradian of Scotland (Rast, 1958; Sturt and Harris, 1961) and western Ireland (Edmunds and Thomas, 1966; Cruse and Leake, 1968; Taylor, 1968; Phillips *et al.,* 1969).

There are indeed two lines of evidence which suggest that this may well be the case in Donegal. At Rosbeg, Obaid (*loc. cit.*) has shown that the dolerites of the Rosbeg type are affected by all the deformation phases, whereas the quartz dolerites there were emplaced between the first and the second structural events. Presenting considerable support for this idea is the hitherto puzzling situation at Dunlewy, where the Mam sill can be shown to cut across the Horn Head Slide (Rickard, 1962, p. 221), which we attribute to the first deformation phase; the fact that this same intrusion, like other dolerites, is elsewhere cut by this and other early dislocations (McCall, *loc. cit.,* Knill and Knill, *loc. cit.,* Rickard, *loc. cit.*) may only indicate a renewal of movement along these planes.

5. The Metamorphism of the Basic Rocks

It is instructive at this point to consider briefly the metamorphism of the quartz dolerites which, having a wide distribution and uniformity of composition, should provide an ideal index of the facies of regional metamorphism (for details see Kirwan, 1965T).

Everywhere north of the Leannan Fault, and indeed over much of northern Donegal as well, there is a great uniformity in the metamorphic state of the quartz dolerites. What differences there are stem from differing responses of the basic sills themselves rather than variation in the regional metamorphic conditions. Thus except along margins or specific shear zones, the thicker sills resisted deformation, and without the catalytic intervention of the latter, the boundary conditions of the retrogressive reactions were not reached. All stages of alteration are seen even in the same sill and are often clearly related to the minute cracks and slip planes which have formed preferred paths for the penetration of hydrous solutions.

Although traces of the original ophitic texture are common, the boundaries between the light and dark minerals become progressively blurred by the intergrowth of new phases. The augite shows all stages of replacement by stumpy prisms of green to blue-green amphibole and by felted patches of acicular actinolite, and the magnetite released during the conversion forms trellis patterns in the amphibole. Quite often chlorite enters this assemblage. The plagioclase is converted to a saussuritic mass of albite-oligoclase, with minute flakes of a white mica and granules of zoisite and clinozoisite. Sphene replaces the original ilmenite, and there is a new quartz. A brown biotite appears late in the assemblage.

It is not easy to determine exactly which parts of this assemblage formed during each of the several regional tectonic and metamorphic events. It seems likely that the formation of amphibole, andesine, and some epidote took place during the main recrystallization (MP_2 as defined in Chapter 3) and was followed by the growth of actinolite, chlorite, sodic plagioclase, and zoisite during the subsequent retrogression (RMP_3, *ibid.*). Scattered biotite flakes overgrowing these assemblages probably result from the late (MP_4) porphyroblastesis.

In all these quartz dolerites, there has been a considerable lag in the response to the several regional metamorphic events; sometimes, indeed, the original igneous mineralogy has persisted unchanged throughout their long history. Our experience is, in fact, that in all but the higher grades of metamorphism, coarse basic rocks, with their differential response to reaction, are particularly unsuitable for the identification of metamorphic facies. We restate this in another way to conclude that the mere identification of facies in relation to basic rocks is not a very useful exercise, not at any rate without a full knowledge of the chronology of deformation and metamorphism.

6. Chemical Composition

If we compare the available analyses of all the Donegal dolerites (Table 2-1 and Fig. 2-4) with those of the various established types of basalts, we find that the average composition of both groups falls well within the tholeiitic field as defined by Kuno (1960).

Further examination of these analyses (Table 2-1, expressed as oxides) shows some significant differences between the two types of dolerite, which we think are important in explaining the present differences in the metamorphic mineralogy. On the average, the normal quartz dolerites have lower Ti^{4+}, Fe^{2+}, and higher Al^{3+}, Fe^{3+}, Mg^{2+}, and Na^{1+} than the Rosbeg dolerites; the only consistent differences are in the lower content of Ca^{2+} and the higher content of OH^{1-}, H_2O, and $(CO_3)^{2-}$ in the Rosbeg variety. The higher degree of hydration and carbonation of the latter group might be due to the cumulative effect of an early autometasomatic alteration associated with the formation of the adinoles, and of the low temperature metamorphism associated with the first and the last of the regional deformation phases (see p. 54). However, we hesitate to ascribe the consistently low calcium content of the Rosbeg dolerites solely to these later events because the low values are found equally in rocks showing very different degrees of alteration. We think, therefore, that the contrast in calcium existed prior to the main prograde metamorphism, and that it may even represent a difference in original magma type, just as might be expected from the different times of emplacement suggested above.

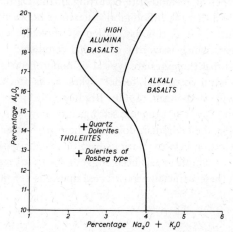

FIGURE 2-4. Average compositions of the two compositional types of metadolerite plotted in relation to the basalt field-boundaries of Kuno (1960).

TABLE 2-1. Analyses of Donegal Metadolerites (Wt %)

	Garnet Amphibolites Rosbeg[1]			Normal Metadolerite Rosbeg[2]		Normal Metadolerite Average of 18[3]
SiO_2	49.18	48.57	49.68	47.74	48.00	49.07
TiO_2	3.35	3.52	3.18	2.24	2.58	2.38
Al_2O_3	12.29	13.24	12.82	14.28	14.95	14.15
Fe_2O_3	2.28	2.32	2.40	3.33	4.25	3.43
FeO	13.17	13.42	13.53	11.45	11.30	11.16
MnO	0.26	0.19	0.14	0.23	0.23	0.24
MgO	4.48	4.90	5.23	5.56	4.23	6.13
CaO	6.93	5.65	6.60	8.24	8.23	9.73
Na_2O	1.49	1.66	2.06	2.59	2.98	2.01
K_2O	0.50	0.36	0.51	0.48	0.50	0.33
P_2O_5	0.23	0.11	0.31	0.19	0.22	0.23
H_2O	3.03	3.60	2.41	2.08	1.91	1.17
CO_2	2.52	1.63	0.82	0.85	—	—

[1,2] After Obaid, 1967T. Analyses by M. Brotherton.
[3] After Kirwan, 1965T.

It seems unlikely however that such variation in Ca^{2+} could by itself account for the present marked contrast in metamorphic mineralogy and texture between the two suites. A more probable explanation is based on the earlier age of the Rosbeg dolerites: that they were sufficiently altered and perhaps even rendered schistose as the result of the early retrogressions, so that they reached more vigorously to subsequent prograde metamorphism than the coarser assemblage of pyroxene and labradorite of the quartz dolerites.

7. Discussion

We have shown that both types of metadolerite occurring in the Dalradian rocks of Donegal fall into the category of tholeiites, and that the lithological differences between them are related to differences in their metamorphic history. It is this association of tholeiitic magma and a thick pile of miogeosynclinal marine sediments which is worthy of special comment.

It has always been considered that tholeiites are typical of nonorogenic situations, where they occur as great piles of flood-basalts with associated feeder dykes or as thick sills in some continental environments (*e.g.,* Turner and Verhoogen, 1960; Bradley, 1965; Joplin, 1964). Examples of tholeiites (both extrusive and intrusive) in orogenic settings are known (*e.g.,* Waters, 1955; Glikson, 1970; Dimroth, 1970), but unlike the Dalradian dolerites, these are located in eugeosynclinal environments. We believe that the distinction between various basaltic magma types and their relationship to constrasted environments is not so simple as has been suggested; careful research will almost certainly obscure the boundaries between these artificially constructed divisions (*cf.* Read, 1961; Turner, 1970).

CHAPTER 3

CALEDONIAN REGIONAL STRUCTURE AND METAMORPHISM: A POLYPHASE HISTORY

CONTENTS

1. Introduction 52
   (A) Preamble 52
   (B) Outline of the Chronology of Deformation and Recrystallization 53
2. The Structures of the Creeslough Succession 56
   (A) Early Signs of Deformation: the Problem of the Slides 56
   (B) The Major Folds 58
   (C) Minor Structures and Chronology of Deformation 61
3. Metamorphic Development of the Creeslough Succession 66
   (A) Preamble: the Early Recrystallization 66
   (B) MP_2 Events: the Evidence of the Porphyroblasts 66
   (C) Later Events: Biotite Porphyroblastesis and a Two-stage Retrograde Metamorphism... 68
4. Structure and Metamorphism in the Kilmacrenan Succession Northwest of the Leannan Fault 69
   (A) The Major Structures 69
   (B) The Minor Structures and Chronology of Deformation 71
   (C) The Metamorphic History of the Kilmacrenan Belt 72
5. The Structural Relationship between the Creeslough and Kilmacrenan Successions 72
6. An Outline of the Structural and Metamorphic Development of the Kilmacrenan Succession Southeast of the Leannan Fault 74
   (A) The Fold Pattern 74
   (B) Some Details of the Folding and Its Chronology 74
   (C) The Metamorphic History of South-central and Northern Donegal 76
7. Donegal in Relation to the Caledonian Mobile Belt as a Whole 76
   (A) In Terms of Major Structure 76
   (B) In Terms of Chronology of Deformation 79
   (C) In Terms of Degree of Deformation 81
8. Final Comment 82
A Geological Appendix: the Late Tectonic Lamprophyres 82

1. Introduction

(A) Preamble

The metasediments and metabasites we have just described suffered a complex history of deformation and metamorphism during the polyphase Caledonian orogeny, just as did their counterparts in Scotland. It is our intention in this chapter to outline this history, both as a contribution to the general knowledge of the Caledonian belt and as a necessary background to the studies of granite emplacement.

A first complication that arises in doing so is that the region is dissected and separated into a number of blocks by a system of pronounced NE–SW trending strike-slip faults (Fig. 3-1). Of these, the Leannan is by far the most important and provides a natural division for a description of the regional structures.

The basic structure of the Dalradian metasediments to the northwest of the fault is fairly straightforward (see Map 2 in folder and Fig. 3-1). There is a general NE–SW trend, with the rocks of the Creeslough Succession mostly younging to the northwest, and, in an opposing sense, to those of the Kilmacrenan sequence to the southeast of them. The intensity of deformation is fairly uniform throughout, but increases markedly near the Main Donegal Pluton, as does the metamorphic

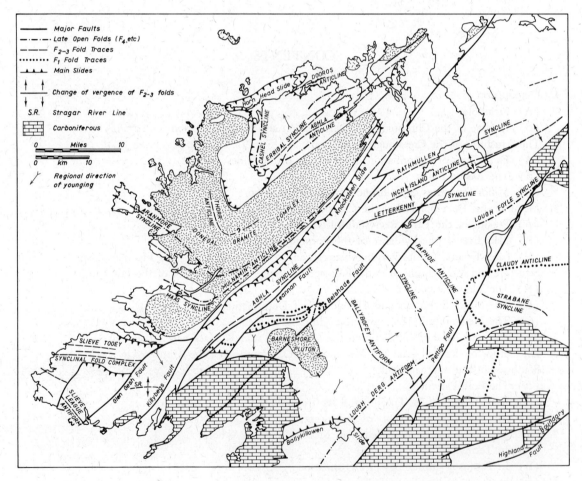

FIGURE 3-1. Outline structural map of Donegal showing the traces of the more important folds.

grade, which rises from regional phyllites and schists of the greenschist facies to schists and gneisses of the almandine-amphibolite facies; this situation we discuss at length in Chapter 12.

On the other side of the Leannan Fault, there is a progressive swing from NE–SW strike-trends in Inishowen and the Sperrins to NW–SE trends southwest of a line from Letterkenny to Strabane (Fig. 3-1). The greater part of this deviation is associated with a large-scale fold complex centered near the Barnesmore Granite, and which, despite its involved geometry and chronology, results simply in major repetitions of the Dalradian succession. In this region, the metamorphic grade rises southwestward from greenschist to almandine-amphibolite facies, though the picture is greatly complicated by the superposition of widespread retrograde changes.

In the following sections, we shall deal separately with the structural and metamorphic development in these several regions. Most attention will be directed to the Creeslough Succession, but some account will also be given of the Kilmacrenan in both its situations, to the northwest and to the southeast of the Leannan Fault. We then deal with the interrelationships of structure and metamorphism within these areas, and finally with the connection between them and other parts of the Caledonian mobile belt of the British Isles. Although the wide variety of lithologies presents a diversity of metamorphic assemblages, we shall restrict ourselves largely to pelitic and semipelitic rocks, since these, as in other polymetamorphic areas, provide the most complete evidence of the metamorphic history.

(B) Outline of the Chronology of Deformation and Recrystallization

As in many other metamorphic areas, the fact that there have been several phases of movement is easily demonstrated by the superposition of two or more structures, even within a single outcrop. Different metamorphic events can also be recognized by careful study of textural relationships in thin sections. A chronology of deformation and metamorphism can thus be constructed (Tables 3-1 and 3-2) to form a useful basis of description and correlation, though we must emphasize that the deformation events so defined are not equal in intensity nor are they necessarily reflected by important fold phases, which are, after all, only incidental products of deformation. Further, it must be realized that such a systematic chronology tends to simplify what is undoubtedly a continuum of overlapping growth and movement events.

We must, at this point, state our misgiving that the whole basis upon which such chronological schemes are constructed may not, in fact, be valid. In the establishment of a sequence of structures over areas of polyphase deformation, correlations of minor structures from outcrop to outcrop and area to area are based, to one degree or another, on similarities in geometric and metamorphic style, in attitude, and in local chronological sequence, though none of these by itself is considered an infallible guide. Thus, a tight, NE–SW trending fold in pelites which deforms a penetrative schistosity and has its own crenulation cleavage is correlated with a similar tight fold in schists on another outcrop, and this correlation given *time equivalence*; similar arguments apply to the correlation of metamorphic textures. In most cases, there are no independent means for establishing a time "plane" and correlations usually produce a coherent picture (*e.g.,* Rast, 1963; Zwart, 1964; Spry, 1963a; Offler and Fleming, 1968; Harte and Johnson, 1969). However, when the areas involved are large (more than several hundred square miles), problems often arise and the specter of diachroneity appears (*cf.* Den Tex, 1963); nowhere is this better illustrated than in the British Isles (*cf.* Dewey, 1967, 1969b; Simpson, 1968; Helm and Roberts, 1968; Anderson and Owen, 1968; Dearman, 1969). It may well be, as Park has argued (1969), that there are no valid criteria for establishing chronological correlations in structural and metamorphic studies, apart from obvious breaks in stratigraphic sequences or the use of acceptable time-rock units, such as suites of dykes (*e.g.,* Sutton and Watson, 1950, 1969; Wegmann, 1963; Watterson, 1965). Indeed, as we shall see in the following pages and later in the discussion of the structures in the envelope of the Main Donegal Pluton (Chapter 12),

TABLE 3-1. Chronology of Regional Deformation and Metamorphism[1,2]

	Deformation Bedding = S_s		Metamorphism
	D_1: Ubiquitous "bedding-schistosity" (S_1) axial plane to minor folds (F_1). Certain bedding-plane thrusts or slides: major F_1 folds at Ballybofey only.	MS_1:	Growth of minute phyllosilicates, quartz, and feldspar along S_1.
Predominant folds and cleavages are composite D_2–D_3 structures.	D_2: Major, generally recumbent, folds (F_2) and associated crenulation cleavage (S_2).		
		MP_2:	Peak of regional metamorphism, upper greenschist to amphibolite facies. General recrystallization overlaps D_3.
	D_3: General intensification and coplanar flattening of D_2 structures; separate D_3 folds and cleavages occasionally seen.		
		RMP_3:	Retrogression of MP_2 assemblages, recognized mainly in Slieve League and central Donegal.
	D_4: (a) One or more crenulation cleavages (S_{4a}), mostly without associated folds. (b) Open, upright major folds (F_{4b}), mostly without associated minor structures. *Relationship of S_{4a} to F_{4b} not known*		
		MP_4:	Biotite and chlorite porphyroblasts. Age relative to D_4 based mainly on evidence from Glencolumbkille.
		RMP_4:	Widespread retrogression. Growth of carbonate and [?] pyrite porphyroblasts.

Emplacement of Donegal Granites
(accompanied by various local movements, recrystallizations, and retrogressions)

[1] We are using here, for the most part, what has become standardized nonmenclature (*e.g.,* Mallick, 1967; Harte and Johnson, 1969). MS refers to a syntectonic metamorphic episode, MP_n to a static metamorphic phase following deformation D_n.

[2] In this book, we use the term *schistosity* to refer to a penetrative planar alignment of minerals, and *cleavage* to describe a nonpenetrative structure, whether marked by a mineral alignment or not. We follow Knill (1960) and Rickard (1961) in using the term *crenulation cleavage* to describe what others refer to as strain-slip or phyllitic cleavage.

TABLE 3-2. Chronology of Deformation Adopted in This Book Compared to That of Previous Writers

This book	Pitcher and Read (1959, 1960b)	Knill and Knill (1961)	Rickard (1962)	Howarth et al. (1966)	Cambray (1969a)
D_1		S_2* = Bedding-schistosity and slides "Major fold of Rosguill" [?]	Bedding-schistosity	F_1	F_1
D_2		Errigal Syncline with S_3 (and S_3' [?]) cleavage	Major folds (L_1), slides and crenulation cleavage.	F_2	
MP_2	Major Folding and Regional Metamorphism		Major regional metamorphism	Metamorphic Peak	F_2
D_3		Flat, S_4, cleavage.	Dunlewy cross folds [?]	F_3	
D_4 S_{4a} F_{4b}		Oblique cleavage S_5 [?]	Cashel Belt of Superimposed folds (L_2) due to Thorr Pluton	F_4 [?]	F_3 (Lough Ea fold) F_4
Granite Emplacement DMG_2	Folds due to forceful emplacement of Main Donegal Pluton				
DMG_3	Late shear zones		Devlin Belt of Superimposed folds (L_2') due to Main Granite		
Post-DMG_3				Later fold phases F_5 to F_8 as yet uncorrelated with late events in other areas.	F_5 [?] (late angular folds)

see Chap. 12 for full discussion

* S_1 = Bedding.

many such problems arise even within our own relatively small area. Nevertheless, we shall follow the standard procedures and utilize the chronological scheme of Table 3-1, if only because it presents a device for lucid discussion.

2. The Structures of the Creeslough Succession

(A) Early Signs of Deformation: the Problem of the Slides

The rocks of the Creeslough Succession are internally broken by a number of major dislocations which are located chiefly along bedding planes, especially those between competent and incompetent horizons. The most important of these are the bedding-plane dislocations or *slides* which were formed very early in the history of deformation.

The Horn Head Slide. The most striking example of these early dislocations is the Horn Head Slide, which, as we pointed out earlier (Chapter 1, Section 2), progressively cuts out various members of the Creeslough sequence above the Ards Quartzite in the Dunfanaghy–Dunlewy–Gortahork area (Figs. 1-3b and 3-2). In the immediate vicinity of the actual slide plane, the rocks have locally been converted to finely schistose mylonitic flags in which pinch-and-swell structures are sometimes seen (Rickard, 1962, Fig. 4d), but for the most part, it is only the cutting out of stratigraphy on a regional scale that identifies this junction as tectonic.

The timing of these movements on the Horn Head Slide presents a number of problems. In the Dunlewy area, where the effects of sliding are most pronounced, the dislocation lies along the lower limb of the recumbent Errigal Syncline (Rickard, *loc. cit.* and our Fig. 3-2). The core of the fold seems here to have broken away from an enclosing sheath of competent Ards Quartzite, and both Rickard (pp. 218–21) and McCall (1954, p. 166) concluded that the slide formed during the main (D_{2-3}) folding. According to this interpretation, the dislocation is a slide in the original sense of Bailey (1910, p. 593) and as recently redefined by Fleuty (1967), *i.e.*, a fault formed in close connection with folding and which is concordant to the major structures. However, on Rosguill, the Knills (1961, p. 279) found that the Horn Head Slide is *folded right around* the Errigal Syncline, and they concluded that it was therefore an earlier structure. A similar situation also exists at Horn Head itself, where the slide is folded by the Polnaguill Syncline (Fig. 3-2).

Here we have a problem of more than local significance (*cf.* Whitten, 1966a, pp. 213–4; 565–7; Badgley, 1965, p. 187). Can a dislocation which is folded around major recumbent structures be a product of the same movements which caused these folds? For example, can the absence of over more than a thousand meters of strata in the hinge of the Polnaguill Syncline, a structure which we consider coeval with the Errigal Syncline, be explained by sliding during the formation of this fold, as McCall originally suggested (1954, p. 165)? Shearing parallel to bedding planes in a buckling deformation can certainly cause considerable dislocations on the limbs of folds, but usually the degree of thinning decreases and dies out toward the hinges of the folds (*cf.* Donath and Parker, 1964, Fig. 7; Ramsay, 1967, p. 394). Likewise, large-scale thrusts can arise outside a buckling competent layer as the adjacent rocks accommodate themselves to the new shape, but the movements are again relatively restricted near the hinges, where, if anything, the rocks are thickened, not thinned (Ramsay, *ibid.*, p. 415–21). Folded bedding-plane thrusts are well known in the supracrustal rocks of the Canadian Rockies and in the southern Appalachians, for example (Badgley, *loc. cit.*; F. G. Fox, 1969; Price and Mountjoy, 1970; Jones, 1971), but there the effect is to thrust older over younger rocks and to thicken the stratigraphic pile, not to thin it as in Donegal. We think therefore that the Horn Head Slide existed *prior* to the main fold movements, though subsequent deformation may well have caused a reactivation of this structure. Nevertheless, we shall continue to use the term *slide,* partly out of deference to established precedents and partly because we cannot rule out the possibility that there were in north-

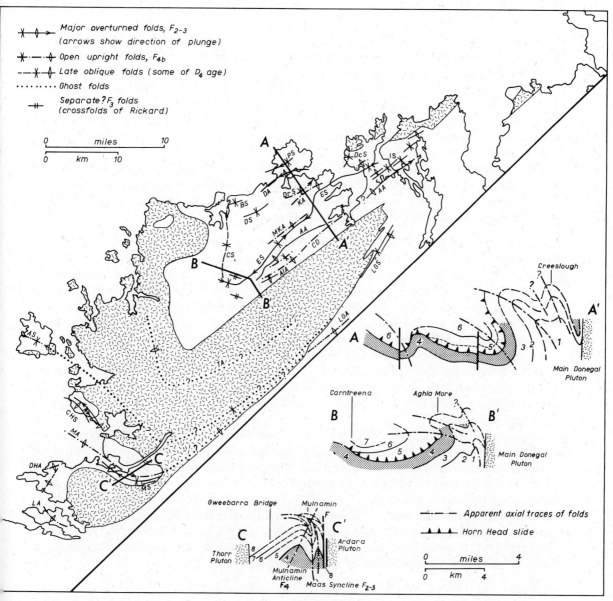

FIGURE 3-2. Folding within the rocks of the Creeslough Succession. Includes possible traces of major folds in the area now occupied by the Donegal Granite Complex. In cross-sections (not to scale) the Ards Quartzite is shown by stipple; other stratigraphic divisions are keyed to Map 2 in folder. *Major overturned folds*: AA, Aghla Anticline; AlA, Altan Anticline; BS, Ballymore Syncline; DA, Dooros Anticline; DS, Dunmore Syncline; ES, Errigal Syncline; IS, Island Roy Syncline; LGS, Lough Greenan Syncline; MS, Maas Syncline; PS, Polnaguill Syncline. *Open, upright folds*: AS, Aranmore Syncline; CD, Creeslough Downwarp; DHA, Dunmore Head Anticline; DrS, Derryreel Syncline; KA, Knockfadda Anticline; LA, Loughros Anticline; LGA, Lough Gartan Anticline; MA, Mulnamin Anticline; MKA, Muckish Anticline; TA, Thorr Anticline. *Late oblique folds*: CS, Cashel Syncline; DcS, Derryhassan Syncline. *Undesignated*: CHS, Crohy Hills Structure.

west Donegal early (D_1) major folds associated with these dislocations, as indeed there seem to be in central Donegal (see p. 74).

Other Slides. There are several other important bedding-plane dislocations in the Creeslough rocks. Some, like the medial dislocation of the Crohy Hills, are likely to be connected with the major folding, while others are difficult to date. For example, the breaks in the Fintown zone, along which various members of the Creeslough Succession are cut out, and the major dislocation which cuts out the whole of the Ards Pelites between the Creeslough Formation and the Ards Quartzite along the northwestern margin of the Main Granite, might well be comparable in age to the Horn Head Slide, or could be the result of the later intense deformation in this central region. At the moment, it is difficult to choose between these alternatives.

At this juncture, we may note a local detail: it seems that in Rosguill, the geological situation which Knill and Knill (1961) attributed to their Derryhassan Slide can now be explained more simply and without supposing a dislocation at all—by equating the quartzites of Derryhassan with the Ards Quartzite instead of with the Sessiagh–Clonmass Formation.

Minor Structures of Early Date: the Bedding-Schistosity. There are other structures which may have formed at the same time as these early dislocations. The most important of these is a penetrative schistosity, S_1, marked by graphite and ores and by minute flakes of muscovite and chlorite which generally lie parallel to bedding. Such a first-stage recrystallization, the MS_1 event, certainly involved other minerals like quartz and plagioclase, but apparently did not produce any entirely new metamorphic phases. This S_1 schistosity is generally obliterated by later structures and is only occasionally preserved in the hinges of later folds or as helicitic trails in later porphyroblasts (see p. 66).

Although earlier workers (Pitcher and Sinha, 1958; Rickard, 1962; Pitcher *et al.*, 1964) did not view this fabric as the result of a distinct deformation episode, the occurrence of infrequent small-scale folds to which S_1 is axial plane suggests its tectonic origin (Knill and Knill, *loc. cit.*, and our Fig. 3-4). Further, there is in many places a marked lineation (L_1) of single minerals and mineral aggregates which lies on S_1.

Although we have no direct way of correlating this early folding, schistosity, and lineation with the slides discussed earlier, the simplest model is that they together form the earliest event (D_1) in the structural development of northern Donegal (*cf.* Knill and Knill, *loc. cit.*, p. 280). As we have already hinted, the great slides seem to be comparable in many ways to the low angle bedding-plane thrusts which characterize gravity sliding (*cf.* F. G. Fox, 1969) and possibly the S_1 fabric developed at the same time, as a result, perhaps, of mimetic recrystallization along bedding planes, accentuating an earlier sedimentary fissility and aided by bedding-plane slip (*cf.* Rickard, *loc. cit.*, p. 173; Whitten, 1966 *a*, p. 258). If so, we might also speculate on the possibility that the associated lineation, L_1, is due to a "stretching" in the direction of tectonic transport (see p. 81).

(B) The Major Folds

All the major fold structures (Fig. 3-2) of northwestern Donegal, that is, those which materially affect the outcrop pattern, clearly postdate the D_1 structures just described. Although previous workers viewed these folds as essentially coeval, it is now clear that some were formed later than others. In the first place, however, we shall describe them together in purely geometrical and spatial terms so as to present a unified picture of the structural framework. The chronological position, as determined later from the interrelationship of the minor structures, is only noted at this juncture.

Folds Northwest of the Donegal Granite Complex. The area northwest of the Main Granite exhibits the most spectacular folds in the Creeslough Succession, of which the dominant structure is the northwest-facing* recumbent *Errigal Syncline* investigated by McCall (1954), Knill and Knill (1961), and Rickard (1962). This fold is well displayed in the quartzite mountains of the Errigal–Muckish range where visible overturns combine with good stratigraphical control to outline the structure. (Fig. 3-2 and 3-3).

FIGURE 3-3. Muckish Mountain viewed from two directions—(a) from the east-northeast, (b) from the north—to show Errigal Syncline and Aghla Anticline, respectively.

In detail, the plunge of the Errigal Syncline varies from gently southwest on Errigal and Crockatee, through horizontal in Muckish Gap and Rosguill, to gently northeast around Back Strand (Fig. 3-2). There is a variance, too, in the attitude of the axial plane from a general horizontal disposition on Errigal to moderately southeast dipping on Muckish and in Rosguill.

The complementary *Aghla Anticline* can be clearly seen on the Aghlas (Rickard, *loc. cit.*), at Dunlewy (Rickard, 1971), and by sharp eyes on the northern flanks of Muckish (Fig. 3-3); in all likelihood it (or an associated structure) is also responsible for the repetition of stratigraphy on either side of the Creeslough Formation (Fig. 3-2, AA') as is the case in western Fanad. In addition, there are several other northwest-facing folds in this northern district which appear to be associates of the Errigal Syncline (see Fig. 3-2). These include the *Polnaguill Syncline* and *Dooros Anticline* on Horn Head, the anticlines and synclines of Island Roy and northern Fanad, and the folds indicated by the local overturns at Dunmore and Ballyness.

Contrasting with these gently inclined or recumbent structures which result from the main D_{2-3} movements (see below), there are also several open, upright folds such as the *Derryreel Syncline*, and the *Knockfadda* and *Muckish Anticlines*; these, together with the important *Creeslough Downwarp*, will be shown to be of later development and are referable to a D_4 phase of folding.

The Folds of Western Donegal. Although the metasedimentary outcrop along the western seaboard of Donegal from Aranmore to Loughros is broken both by granite and by the sea, it is obvious from the stratigraphic relationships that the major structure here is an anticline whose axial trace lies under the Gweebarra estuary (Fig. 3-2) and indeed, the actual closure is exposed in the townland of Mulnamin where its axial plane is demonstrably upright and its plunge gently eastward. In this Maas area, Iyengar, Pitcher, and Read (1954) also mapped several other important folds, which they considered coeval with this *Mulnamin Anticline*; of these, the most important is the westward-plunging *Maas Syncline*, which is overturned to the north. However, as we shall see, examination

---

* For definition and use of the term "facing," see Cummins and Shackleton, 1955; Shackleton, 1958.

of minor structures indicates that the latter, a D_{2-3} structure, predates the upright D_4 Mulnamin fold (Fig. 3-2, CC') recalling the situation we have just described in northwest Donegal.

Farther west in the peninsula of Rosbeg, there is no major repetition of beds and the succession dips gently southward right up to the base of the Slieve Tooey Quartzite, though there are local, and relatively minor, upright anticlines at Dunmore Head and on northern Loughros.

In the Crohy Hills on the north side of the Gweebarra Bay, two major anticlines appear with the intervening syncline largely cut out by a major dislocation (see Map 1 in folder); we refer to these collectively as the *Crohy Hills Structure*. In the southeastern part of this area, all of these three folds are upright and open, but as they are traced northwestward they become tightened and their axial planes acquire a pronounced southeasterly tilt. This is particularly marked in the case of the northeastern anticline whose lower, now nearly recumbent limb is represented around Maghery by inverted Ards Quartzite. The closed form of the outcrop clearly indicates a culmination in the plunge of these anticlines. This complicated structure looks as if it may be due to interference of E–W trending recumbent folds with NW–SE trending open, upright structures. Since we do not fully understand the chronology of the minor structures here, we are not sure which is the earlier, but we suspect that the overturned folds are in fact older (D_{2-3}), since this is the situation in northwestern Donegal as a whole.

The Folds of Lough Gartan and Lough Greenan. To complete our review of the major structures of the Creeslough Succession, we turn briefly to the narrow strip of metasediments forming the southeastern envelope to the Main Donegal Pluton. Superficially, the structure of this zone appears to be very simple, with strata dipping steeply to the southeast around Fintown and Gartan Lough and more gently in the same direction farther to the northeast. Indeed, earlier workers (Pitcher and Read, 1959; Pulvertaft, 1961) viewed this as a consistently right way up sequence, but the recent stratigraphical correlations of Pitcher and Shackleton (1966) have established the presence, northeast of Gartan Lough, of major folds, overturned to the northwest (see Fig. 3-2), the *Lough Gartan Anticline* and *Lough Greenan Syncline*. We are not sure of the age of either of these folds relative to the other major structures, though we suggest that the first is of D_4 age, in comparison with the Mulnamin Anticline, and the second of D_{2-3} age.

However, the situation is in reality quite complicated, as is shown by the recognition of a chronological sequence of folding on the southeast flank of the Lough Greenan Syncline (in the townland of Meenlaragh; see Map 1 in folder). There, gentle open folds (F_{4b}) on NE–SW axes deform a recumbent anticline (F_{2-3}), which is probably the complementary fold to that of Lough Greenan. Further, the outcrop pattern of the metadolerites in the Falcarragh Limestones of this locality can be interpreted as outlining an even earlier recumbent structure (presumably F_1) and, if so, then this represents one of the few examples of a sizeable fold of D_1 age in the Creeslough area.

The Spatial Connection between the Folds of the Envelope and the Granite Complex. Before turning to other problems, it is instructive to consider the spatial connection between the major structures in these three main outcrops of the Creeslough Succession. Although they are now separated by considerable bodies of granite, the presence of ghost stratigraphies and structures in two of these intrusives, the Thorr and the Main Donegal Plutons, enables a crude reconstruction to be made of the situation prior to granite emplacement (Pitcher and Shackleton, *loc. cit.*; and Fig. 3-2). Despite the many difficulties inherent in such a restoration, especially when the chronology of folding is as complex as it is here, we are encouraged in the endeavor by the realization that few such attempts have ever been made to reconstruct the detailed structural pattern of an area subsequently invaded by a granitic complex.

In the Thorr district, the disposition of enclaves of rocks younger than the Ards Quartzite reflects a northward-plunging ghost anticline, the *Thorr Anticline* (Pitcher, 1953a; and see Chapter 5, Section 2). Again, on the northwest side of the Crohy Hills, the sedimentary relics in the Thorr Pluton dip northward and thus presumably lie on the southern limb of a major ghost syncline, now mostly obliterated by the non-xenolithic Trawenagh Bay Pluton. It is possible that the major upright syncline on Aranmore Island is an expression of this structure, the axial trace of which may formerly have continued southeastward through Dunglow (Fig. 3-2).

It will be recalled that the ghost stratigraphy in the Main Granite indicates, as Pitcher and Shackleton implied (*loc. cit.*), some kind of linkage between the anticlinal structure in the Maas area and the inclined anticline of Lough Gartan. It might be suggested, therefore, that the Lough Greenan Syncline is the continuation of the Aranmore-Dunglow fold, the complementary Thorr Anticline being completely eradicated by the Main Granite. There is, however, the complication that the Aranmore-Dunglow and Mulnamin folds both deform the main schistosity (S_{2-3}), whereas the Lough Greenan structure, at least, appears to be an F_{2-3} fold with this fabric as its axial plane structure (see p. 60 and Fig. 3-2). Nevertheless, in view of the close spatial relationship elsewhere between the F_{2-3} and F_{4b} folds (see p. 63 and Fig. 3-2), it is possible that the true connection may be between the Aranmore–Dunglow Syncline and the late open F_4 folds in the Lough Greenan area (see Fig. 3-5). We think, therefore, that the trend of the ghost structures through the Donegal Granite Complex is broadly as shown on Figure 3-2.

The Strike Swing. A salient feature of the structural pattern of the Creeslough Succession is the marked change in western Donegal from the normal Caledonide trend to a westward and northwestward direction (Fig. 3-2 and Map 2 in folder). This strike swing, which affects all the structures, is well brought out by the changes in axial traces of the major folds just outlined and predates the emplacement of the granites, as the presence of the ghost structures clearly demonstrates. It has, however, been accentuated by movements connected with granite emplacement, as is well seen by the deflection of stratigraphy around the northern region of the Ardara Pluton. Likewise, at Crockator, east of Thorr, where the swing is most acute, Rickard (1963) has shown that the granite emplacement was responsible for accentuating what was an original bend of modest degree (see p. 99).

We are dealing here with a phenomenon of considerable regional extent, for a similar swing can be traced in Dalradian rocks, as well as in later strata, all along the western coast of Ireland (Fig. I-1). Such large-scale strike swings have been called *oroclines* by Carey (1955, 1958), who ascribed them to the late bending of fold belts. In Donegal, on the other hand, the swing lacks associated minor structures and appears to have controlled the folding; the L_1 stretching lineation in the Creeslough Succession, for example, maintains a more or less consistent orientation throughout. Thus the swing was probably initiated very early in the orogenic process, and we subscribe to the growing body of opinion that many "oroclines" represent fundamental curvatures in the original depositional basins (*cf.* Poole, 1968; Knowles and Opdyke, 1968), which control the directions of subsequent major stresses along the fold belt (*cf.* Rickard, *loc. cit.*; Cambray, 1969*b*). Nevertheless, as we shall see later (p. 79), restoration of movement along the Leannan Fault suggests that this swing could also be related in part to what appears to be a later swing in eastern and central Donegal.

(C) Minor Structures and Chronology of Deformation

The Major Cleavage—a Composite D_{2-3} Fabric. We have already seen that the oldest small-scale structure throughout the Creeslough Succession is a schistosity, S_1, which is essentially parallel to the bedding. This is folded about a generally shallow-dipping crenulation cleavage (Fig. 3-4a, b, d) which is often intensified to a virtually *penetrative schistosity*, and which is described in detail by Rickard (1961, 1962). This structure is the dominant cleavage throughout the Creeslough Succession and is axial plane to the major north- and northwest-facing folds just described and to their accompanying minor folds (Figs. 3-4,-5,-6).

This surface often appears to be a composite structure resulting from two closely related phases of

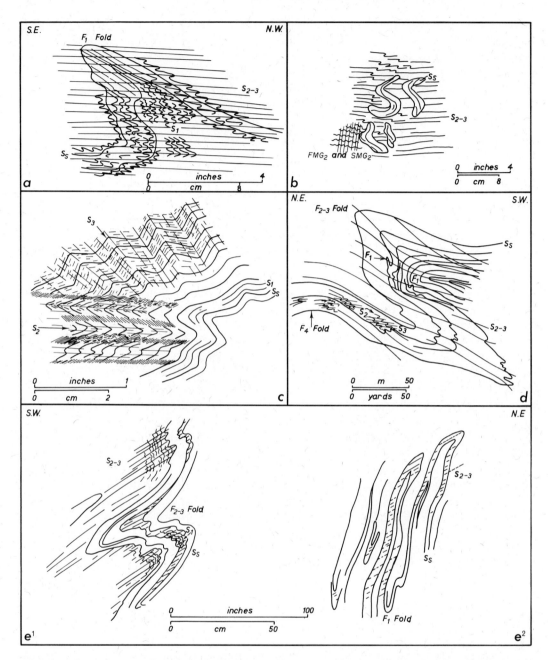

FIGURE 3-4. Some relationships between minor structures of the Creeslough Succession. (a) Profile of early (F_1) fold at Dunmore showing a penetrative axial plane cleavage (S_1) folded by a crenulation cleavage (S_{2-3}). (b) Folded and disrupted beds of psammite within the Ards Pelite at the summit of Crockatee. Shows that cleavage S_{2-3}, which is axial plane to the folds, is deformed by crenulation FMG_2 (see Chap. 12). (c) Detail of relationship between S_2 and S_3 crenulations, Ards Pelites on Crockatee. (d) Generalized profile of composite F_2-F_3 fold on Clonmass Head. Axial plane cleavage is mainly a composite S_{2-3} structure, yet in places S_2 and S_3 can be seen to be separate structures. Note the early F_1 isoclines and the late F_4 upright fold. (E_1, E_2) Portions of the same outcrop to show the relationship between D_1 and D_{2-3} structures. S_1 is presumably axial plane to F_1 but is obliterated in outcrop E_2. Near the southern end of Lough Agher, north of Muckish Mountain.

movement, D_2 and D_3. Its composite nature, which was not recognized by previous workers, is best demonstrated by its relationships with certain porphyroblasts, a matter we later discuss in some detail (p. 66). For the moment, it is sufficient to state that, following the formation of an S_2 cleavage now shallow-dipping, during D_2 movements, there was a widespread and dominantly static growth of porphyroblasts, MP_2, which have preserved within them D_2-crenulated S_1 trails. A subsequent deformation, which can be defined as D_3, resulted in a flattening of the S_2 fabric around the porphyroblasts and thus a discordance between the internal (Si) and external (Se) fabrics.

More direct evidence of the polyphase nature of this compound S_{2-3} structure is provided by the occurrence in some places, as for example, on Crockatee, of two distinct cleavages which intersect at low angles (Fig. 3-4c, d). Where this is the case, the crenulations produced by the intersection of these cleavages with the S_1 schistosity are generally coaxial and rarely produce more than faint interference patterns.

However, there are situations in Donegal where the two surfaces are geometrically more distinct. Thus, according to Knill and Knill, (1961, p. 280), the Errigal syncline on Rosguill where it is a more upright structure, shows an axial plane "slaty" cleavage (their S_3, see Table 3-2) folded about a flat crenulation cleavage (their S_4). If these are equated with the S_2 and S_3 structures of the area to the southwest, as seems likely, the greater angular difference might suggest an appreciable change in the generative stress pattern, thus providing a further reason for separate designation. It is also possible that the Dunlewy cross-folds, which Rickard (1962, p. 221) described as the result of the further twisting of the Errigal Syncline, could be correlatives of the D_3 structures, though the detailed work remains to be done.

At any rate, it appears that the major northwest-facing folds of the Creeslough Succession, and their accompanying minor structures are due to early D_2 movements accentuated during a D_3 event. The generally close geometrical relationship between the two sets of structures suggests that they are genetically related and that there was little difference in the directions of the causative forces.

It is of interest to note in passing the frequent variations in plunge of both major and minor F_{2-3} folds within a single unfolded axial plane surface (see Rickard, *loc. cit.*; and our Fig. 3-2). This is best shown by the rapid change in the Maas Syncline from down-dip plunges near Maas Upper to subhorizontal plunges within a half mile both to the west and to the east. The recumbent, N–S trending minor folds of the Portnablagh area (McCall, 1954, p. 166) and some of those in southwestern Rosguill appear to be D_{2-3} structures whose anomalous orientation results from sharp swings in axial trend within a consistently shallow S_{2-3} surface to the main F_{2-3} directions. The resulting "lobate" shapes are now widely recognized to be commonplace (*e.g.,* Voll, 1960; Dewey, 1967, 1969*b*; Chadwick, 1968; Ramsay, 1967), and are often considered to result from varying rates of transport during folding. The prevalence of markedly lobate folds in northwest Donegal is not surprising in view of the fact that subsequent flattening, as for example, during the D_3 event, would have accentuated any preexisting curvature of fold hinges.

The Deformation of S_{2-3} by D_4 Structures. The S_2 and S_3 surfaces, either separately or compounded, are commonly puckered by a minute crenulation (known to local workers as "Fred Across"), which is characteristically oblique, *i.e.,* with a high angle of pitch relative to F_{2-3}. This may be intensified to form a crenulation cleavage (S_{4a}) axial plane to steeply inclined, small-scale folds, as is best documented by Knill and Knill (*loc. cit.*) in Rosguill.

The S_{2-3} cleavage is also deformed over the whole area by the large scale open and upright folds (F_{4b}) referred to earlier, such as, for example, the Knockfadda and Muckish Anticlines and Derryreel Syncline, all of which fold the lower limb of the Errigal Syncline (see Fig. 3-2). Then, in the belt of country between the Errigal Syncline and the Main Donegal Pluton, the S_{2-3} cleavage is bent down sharply to the southeast, becoming locally vertical (Rickard, *loc. cit.,* 1971). It is this Creeslough Downwarp which is mainly responsible for the downward tilting of the Aghla Anticline between

Muckish and Fanad (Fig. 3-2). Likewise, in the Fintown belt, the S_{2-3} cleavage is folded by gentle upright structures, as is the (F_{2-3}) anticline complementary to the Lough Greenan Syncline.

There is here the complication that the rocks on either side of the Main Donegal Pluton were deformed after its emplacement by the DMG_2 and DMG_3 episodes (see Chapter 12). However, west of the Pluton, where no such structures are present, what we identify as S_{2-3} is folded around the open Dawros Head Syncline (immediately to the south of the Dunmore Head Anticline) and also steepens in several strike belts. Final confirmation that this folding is distinct from the granite events is found in the Maas area, where an intense S_{2-3} cleavage is clearly folded by the upright, pre-Main Granite, Mulnamin Anticline (Section CC'; Fig. 3-2); the relative chronology is thus obvious.

We assign all these and other large-scale upright folds which deform S_{2-3}, and which are strictly coaxial with the overturned F_{2-3} folds, to the one D_4 movement phase, although, because they are rarely accompanied by associated minor structures, we may be too emphatic in grouping them all together. Similarly, we really do not know the true relationship between these major folds and the oblique crenulations described above: since the latter are not concordant with any major structures, we are reluctant to elevate them to the status of a separate deformation event. For the sake of convenience, we shall therefore place these crenulations in the same category in our chronology as the large-scale upright folds, distinguishing the former when necessary as D_{4a} and the latter as D_{4b} structures.

Within the Thorr Pluton, the variation in dip of S_{2-3} in the enclaves suggests that the major folds outlined earlier are likely to be of D_4 age; indeed, they may have continued along the axis of the Main Donegal Pluton as shown on Fig. 3-2. If these speculations are correct, they suggest that the area now occupied by the Donegal Granite Complex was originally the site of a belt of major, generally upright, D_4 folds, as shown in the accompanying sketch (Fig. 3-5), which is a greatly simplified profile of the major structures in the Creeslough Succession.

Late Structures. There are other major and minor structures which are later than those due to the

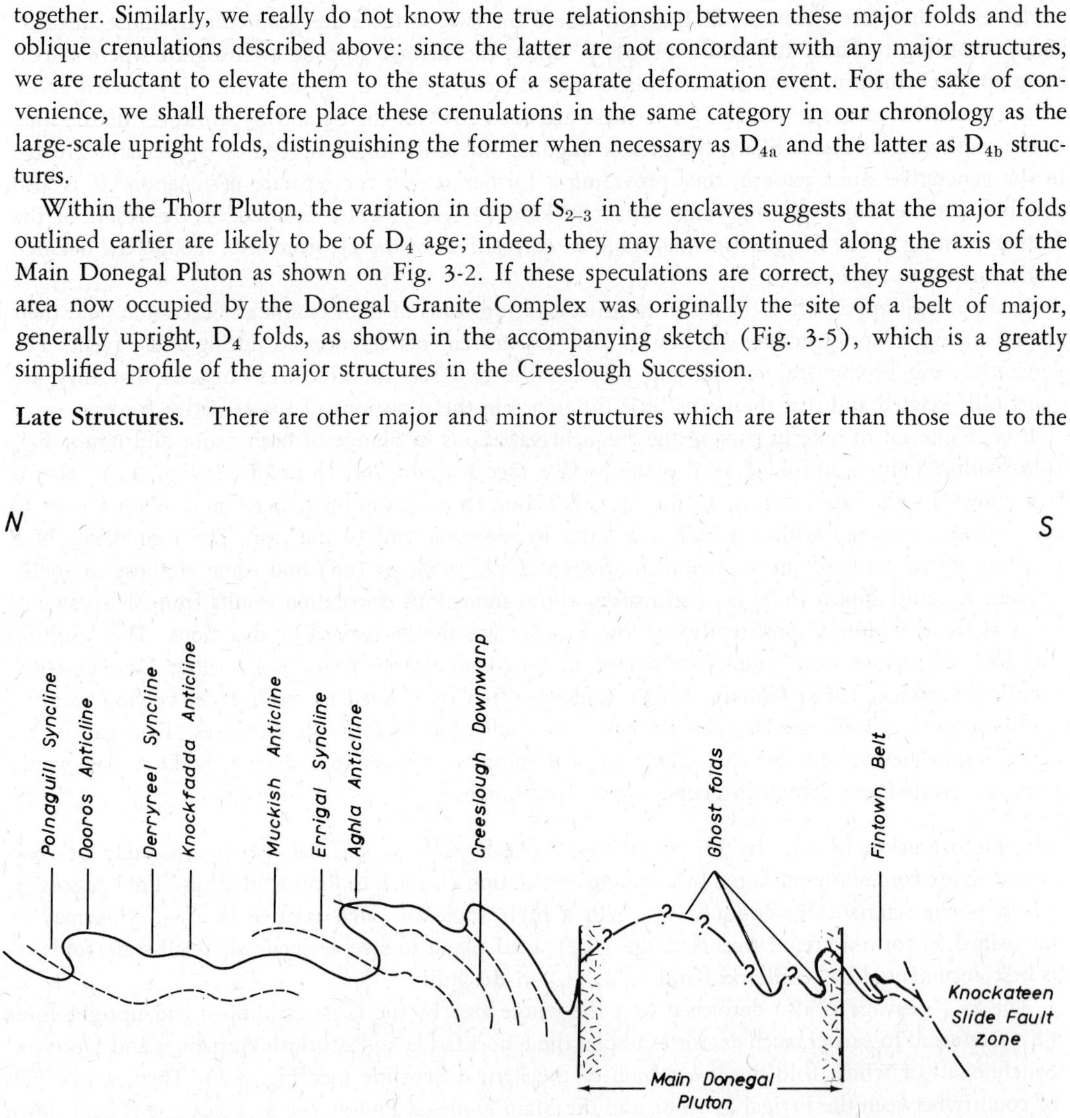

FIGURE 3-5. Profile of major folds in Creeslough Succession along line from Horn Head to Church Hill. Broken lines represent S_{2-3} surfaces, solid line represents bedding. Not to scale.

D_4 movements. In the Falcarragh area, for example, the Cashel Syncline and its associated cleavages and minor folds appear to be related to the doming of its cover by the Thorr Pluton (Rickard, *loc. cit.*; and Chapter 5). We have already alluded to the deformation of the Main Donegal Pluton by what we define as DMG_2 and DMG_3 events for reasons we shall discuss in Chapter 12.

In addition, there are a series of enigmatic weak crenulations on the main schistosity, and also systems of kink-bands, shear zones, and fracture cleavages, all of which appear to postdate the D_4 movements; some of these will also be dealt with in Chapter 12. We shall not label these as belonging to separate movement phases, partly because there is much work yet to be done in establishing their chronological sequence, and partly because we suspect that they are the result of innumerable localized stresses rather than widespread tectonic events.

A Summary of the Structural History. To summarize, the complex structural history of the Creeslough Succession, we present a composite diagram showing the mutual relationships of the important minor structures (Fig. 3-6). This shows to some extent, the general decrease with time of the intensity

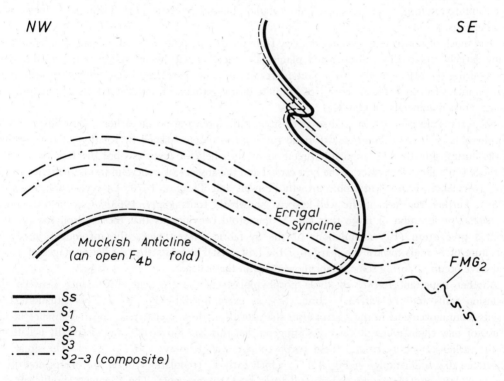

FIGURE 3-6. Schematic composite diagram showing the relationship between the minor structures as seen in the Ards Quartzite and Pelite on Crockatee. For explanation of FMG_2 see Chapter 12.

of deformation, but it brings out neither the remarkable coaxial nature of the deformation phases nor the even more remarkable coincidence of the swing in axial directions of both the major F_{2-3} folds and also of the later structures. Indeed, it appears to be a general feature of the higher levels of fold belts that the earlier and generally more penetrative structures control the attitude of later more "brittle" folds, producing a coaxial pattern that may even follow around major swings in the earlier axes—whatever the orientation of the later stress system (*cf.* Ramsay, 1967, p. 533; Tobisch, 1967).

3. The Metamorphic Development of the Creeslough Succession

(A) Preamble: the Early Recrystallization

Having determined the chronology of deformation, we turn to a discussion of the interrelation between this and the metamorphic history of the Creeslough pelitic sediments. (For discussion of the metamorphism of the dolerites see Chapter 2). We have already seen that the first act of recrystallization, which we defined as MS_1, resulted in the ubiquitous "bedding-schistosity," S_1.

(B) MP_2 Events: the Evidence of the Porphyroblasts

Where the S_1 fabric has survived, it is deformed by the S_{2-3} crenulation cleavage, whose metamorphic style is a result of the MP_2 event. This cleavage is generally picked out by the *new* growth of phyllosilicates along limbs of the microfolds and by the marked segregation of quartz and feldspar into their cores (Rickard, 1961). Indeed, the cleavage is recognizable over much of the area by the fine banding resulting from this small-scale metamorphic differentiation (*cf.* Talbot and Hobbs, 1968; and our Figs. 3-7, 8-7a, and 12-4b).

Over much of northwestern Donegal, away from the granite plutons, small porphyroblasts of garnet (now largely replaced by chlorite) and plagioclase, together with occasional ones of ilmenite developed during the MP_2 episode. An important feature of these porphyroblasts is that they occasionally contain inclusion trails (the S*i* fabric) of graphite dust together with minute crystals of phyllosilicates, quartz, and ore minerals (Fig. 3-7a).

These trails are generally curved so as to suggest the preservation of earlier microfolds within the porphyroblasts. It could be argued that the curving trails were formed by rotation of the porphyroblasts during growth—the usual interpretation of such helicitic textures—but the lack of symmetry of the S*i* trails about the center of the host crystal and the absence of noncyclindroidal patterns, provide strong evidence against syntectonic growth (Powell and Treagus, 1970; Dewey, 1967; Rosenfeld, 1968). Further, the lack of the well-known "snowball" trails (Spry, 1963*b;* Cox, 1969) and the occasional preservation of more than one S*i* microfold convince us that these porphyroblasts overprinted preexisting (F_2) crenulations of the S_1 fabric. Nevertheless, the infrequent increase in curvature of S*i* trails toward the margin of the host crystal suggests that the latter stages of growth may have occurred during some minor movements in the matrix.

Another important feature of these porphyroblasts is the frequent discordance between their inclusion trails and the external fabric (S*i*), as exemplified by Fig. 3-7a. This situation clearly results from movements in the matrix after the growth of those megacrysts, and the absence in most places of new crenulations or cleavages suggests that this late movement was essentially a flattening of the earlier schistosity around these porphyroblasts (*cf.* Ramsay, 1962, p. 323). The situation is illustrated diagrammatically in Fig. 3-8, in which early F_2 crenulations of S_1 are overprinted by the porphyroblasts and then tightened around them by D_3 movements. The frequent parallelism of S*i* from one porphyroblast to another is fully consonant with this model, as is the observation that the amount of "rotation" is involved in the infrequent syntectonic crystals is less than 90°. The general absence of more than one S*i* sigmoid may well be the result of the majority of the porphyroblasts having grown within the space of one microlithon, as is common elsewhere (*e.g.,* Spry, *loc. cit.*; Harte and Johnson, 1969; Powell and Treagus, *loc. cit.*).

If this interpretation is correct, then these relationships support our previous conclusion that the crenulation cleavage of the area is in fact a composite S_{2-3} structure. The evidence of dominantly static growth of the porphyroblasts indicates that the D_2 and D_3 movements, though producing coaxial and coplanar structures, were separate in time, as indeed is suggested by the occurrence in places of distinct S_2 and S_3 crenulation cleavages.

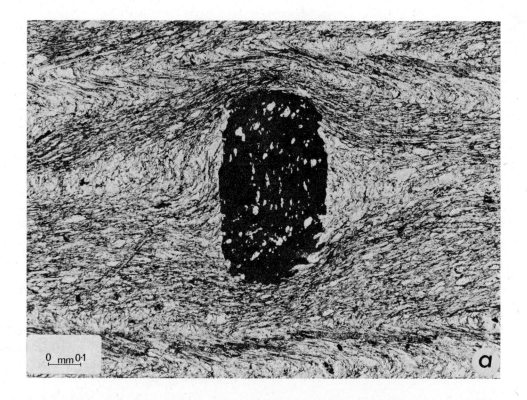

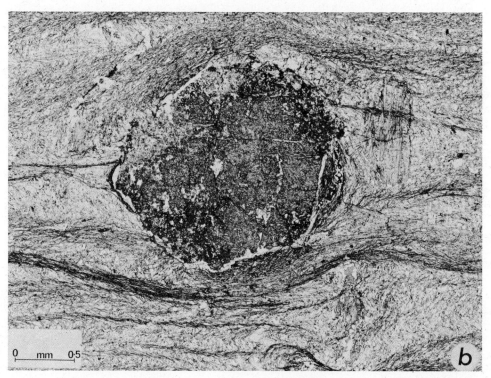

FIGURE 3-7. Photomicrographs illustrating the time-relationships between the growth of porphyroblasts and deformation in the Lower Falcarragh Pelites: (a) an MP_2 ilmenite porphyroblast enclosing an internal relict of S_2 which is also externally flattened around the crystal in the form of a composite S_{2-3} surface; (b) an S_2 crenulation cleavage is flattened about a MP_2 garnet and a late MP_4 chlorite porphyroblast (c) overprints this redeformed cleavage (S_{2-3}). Plane light.

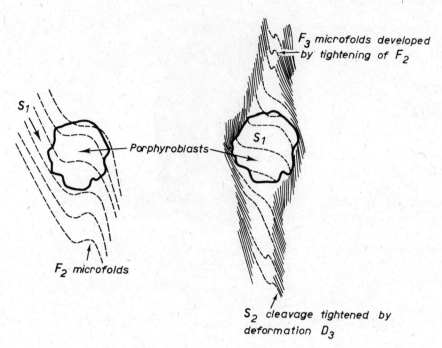

FIGURE 3-8. Schematic drawing of relationships between S_1, S_2 and S_3 and the MP_2 porphyroblasts.

There is little evidence of any appreciable recrystallization during the D_2 movement episode, but there is often an appreciable difference in grain size between the Si inclusions and their Se counterparts. This suggests that much if not all of the general recrystallization and differential segregation of minerals referred to earlier took place after the growth of the porphyroblasts. However, one can occasionally detect a slight coarsening of the inclusions as they approach the edges of the porphyroblasts, thus suggesting an overlap between the later stages of porphyroblast growth and the general recrystallization.

In summary then, the fairly widespread porphyroblasts of garnet and plagioclase appear to have grown during a static interval (MP_2) which followed the D_2 folding of S_1. Subsequent to the bulk of their growth, the relatively open S_2 crenulation cleavage was flattened around these porphyroblasts by D_3 movements, and broadly synchronous was the general recrystallization leading to a coarsening of the Se fabric. Finally, we may note here that there are relatively early quartz segregations in many horizons which are probably associated with this composite MP_2 metamorphic event.

(C) Later Events: Biotite Porphyroblastesis and a Two-Stage Retrograde Metamorphism

The timing of metamorphic events that took place between the MP_2 episode and the emplacement of the granites is not easy to establish. Although the S_{2-3} fabric is flexed by the minor F_{4a} crenulations and the major F_{4b} folds, it seems that there was little mineralogical response to these deformations, apart from some minor strain-healing of the constituent minerals.

The latest metamorphic events are, however, fairly distinctive. There is first a regional porphyroblastesis represented by the widespread growth of random biotite porphyroblasts (and sometimes chlorite), the phyllosilicate flakes overprinting the S_{2-3} cleavage (Fig. 3-7b). Grain growth of quartz is also involved, and according to Edmunds (1969T), the regional garnet was locally replaced by biotite at the same time. Although we are not sure of the time relationship of this static phase to the D_4 event in the Creeslough Succession, the evidence that it postdates D_4 structures in Slieve League seems fairly clear (see p. 72), and we therefore designate it as the MP_4 event.

Throughout the region, the biotite flakes, both those replacing the groundmass and the garnet—and indeed, garnet itself—are replaced to varying degrees by chlorite; and the fact that this regionally chloritized biotite and garnet are progressively overgrown by a *new* thermal biotite within certain of the aureoles (see p. 134) demonstrates that this retrogression, RMP_4, preceded the intrusion of the granites. This also seems to be the case with certain pyrite and carbonate porphyroblasts which may have formed during this retrogression, or which may represent an even later metamorphic event. However, the presence of pseudomorphs of new thermal amphibole after the regional carbonates in the aureoles of several appinitic complexes clearly indicates that even this event predated the emplacement of the Donegal Granites.

The production of chlorite in these rocks deserves further comment. It has long been thought that, apart from the phyllosilicate porphyroblasts just described, the widespread assemblage garnet-chlorite-muscovite-quartz was the result of *prograde* metamorphism; this is the usual finding in the classic Barrovian situation where pelites from the biotite and lower garnet zones may not carry biotite, simply because the bulk composition of the rocks was more appropriate to the production of magnesium-rich chlorites (Atherton, 1964; 1965, 1968). If this is indeed the case, then the groundmass chlorites are earlier than those produced during the RMP_4 retrogression and there is evidence in central Donegal and in Glencolumbkille (see p. 72) of the retrogression of MP_2 biotite and garnet prior to the MP_4 porphyroblastesis. If this is so, we may well have the situation where an early RMP_3 regional retrogression affected established chlorite-bearing assemblages instead of the biotite-bearing rocks as in Glencolumbkille; the difficulty is of course in recognizing a retrogression involving changes in chlorite alone.

4. Structure and Metamorphism in the Kilmacrenan Succession Northwest of the Leannan Fault

(A) The Major Structures

We now turn to that part of the Kilmacrenan Succession lying structurally above the Creeslough Succession and to the northwest of the Leannan Fault. Having already established the pattern of the structural history of the Creeslough rocks in some detail, we can be brief.

Within our map area, the strata of this portion of the Kilmacrenan sequence both dip and young to the southeast (see Fig. 3-1 and Map 1 in folder), though an important exception to this generality is presented by the long fault strip of Upper Crana Quartzite extending from Milford southwest to Sir Albert's Bridge and bounded by the Leannan and Lough Ea–Doon faults. The rocks *within* this fault zone are largely inverted, and the dominant dip direction is to the northwest. This contrast between a northwestern belt of southeasterly-dipping right-way-up strata and a southeastern zone of northwest-dipping inverted rocks suggests the presence of a major recumbent fold, the *Aghla Syncline* (Pitcher et al., 1964, p. 265; Cambray, 1969a).

Support for the existence of this major fold comes from associated smaller folds, such as the southeasterly-verging *Graffy Hill Fold* of Cambray, and from bedding-cleavage relationships, the sum of the evidence being that the rocks of this part of the Kilmacrenan Succession lie on the lower limb of a recumbent syncline which *faces downward to the southeast* as shown in Fig. 3-9. On this interpretation, the inverted strata referred to above are considered to represent portions of the upper inverted limb of this fold brought down on various fractures of the Leannan Fault complex (see Chapter 17).

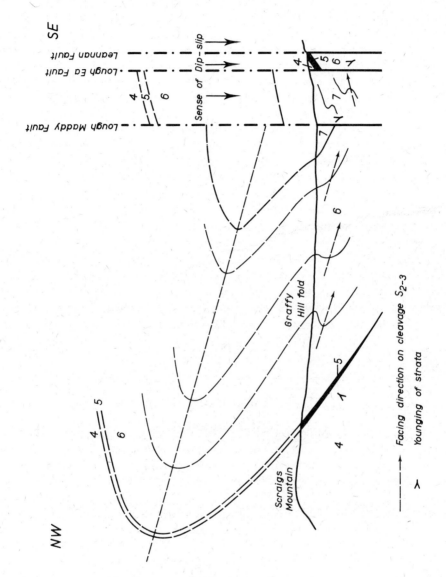

FIGURE 3-9. Cross section of the Aghla Syncline to show downward-facing character and disruption by faulting (mainly after Cambray, 1969a): 4, Slieve Tooey Quartzite; 5, Cranford Limestone; 6, Termon Formation; 7, Crana Quartzite. See also Fig. 17-3.

It is likely that this situation continues northeastward into Fanad, even into Malin. In the opposite direction, in Slieve League, the structural relationships are not yet properly understood, and we do not know exactly how the recumbent southeast-facing *Slieve League Anticline* and nearly upright *Slieve Tooey Syncline* (Anderson, 1954; Howarth et al., 1966) are connected to the Aghla Syncline just described, though the simplest hypothesis would make them all coeval, the Slieve Tooey Syncline being a down-warped extension of this Aghla Syncline.

Superimposed on these important structures are a number of open structures with a general Caledonide trend. These late folds are particularly common in the wedge of Crana Quartzite mentioned above, and other examples are the broad anticlinal *Fanad Dome,* the open *Stragar River Syncline* (Fig. 3-1) of the Slieve League Peninsula, and the *Lough Ea Fold,* a monoclinal downwarp which flanks the Lough Ea Fault (Cambray, *loc. cit.*).

(B) The Minor Structures and Chronology of Deformation

The structural chronology of the Kilmacrenan Succession is less well known than that of the Creeslough sequence. However, the detailed studies of Cambray (*loc. cit.*) in the Fintown area and of Howarth and others (*loc. cit.*) in Glencolumbkille suggest marked similarities between the two situations. There is again a bedding-schistosity (S_1) which, though axial plane to occasional tight folds, does not appear to be the product of a major fold event. Whether or not there were sliding movements at this early stage is not clear, but it is possible that part of the Knockateen thrust may have been initiated at this time, for where it is exposed it lies parallel or at low angles to bedding (see, however, p. 73).

Like its counterpart in the rocks of the Creeslough Succession to the northwest, the major cleavage in the Kilmacrenan sequence crenulates the earliest bedding-schistosity and is usually intensified to form a new schistosity. This surface dips gently to the southeast, except where it is folded, and is the axial plane structure of the major Aghla Syncline, as is shown by its relations to minor folds and particularly to the Graffy Hill Fold (Cambray, *loc. cit.*). Thus, in marked contrast to the situation in the Creeslough Succession, these folds *face downward to the southeast* on this cleavage. Unfortunately, the general absence of porphyroblasts of appropriate age prevents us from determining whether or not this cleavage is a composite of S_2 and S_3 surfaces as in the Creeslough Succession. However, in the Glencolumbkille district, Howarth and others (*loc. cit.*) have described a puckering, sometimes intensifying to form a distinct cleavage (their S_3), which intersects the major schistosity (their S_2) at low angles, these two structures being separated in time by a metamorphic episode (see following section). There is often, in fact, a strike-trending puckering, and this leads us to suspect that the major cleavage of the Kilmacrenan Succession may, yet again, turn out to be compounded of two structures; we shall thus refer to it and the folds it accompanies as D_{2-3} structures.

Although this S_{2-3} surface was very clearly refolded on all scales, it is again not easy to set up a precise chronological history of these secondary structures. As elsewhere, it is difficult to correlate these later and weaker phases of regional deformation which, by their very nature, may well be of only local significance. For example, quite widespread in the area southeast of Glenties is a second crinkling of S_{2-3} which lies at a high angle to all previous structural elements: this is commonly seen in outcrop to plunge down the dip of the S_{2-3} schistosity and is sometimes distinct enough to define a crenulation cleavage oblique to the main trend (Cambray, 1969*a*); we can correlate this tentatively with the F_{4a} crinkling to the northwest.

It is clear, however, that the S_{2-3} schistosity, and possibly even the F_{4a} crenulations, are folded around the large scale open folds we have already mentioned as occurring in Fanad, the Stragar River, and flanking the Lough Ea fault (Cambray, *loc. cit.*). Again we are going to take all these together to represent F_{4b} folds, another result of the D_4 deformation.

Following D_4, renewed movements gave rise to a series of sharp angular minor folds and kink bands (Cambray, *loc. cit.*; Howarth *et al.,* 1966) whose orientation varies widely from place to place, depending, of course, on the attitude of the structures upon which they are founded. Howarth and his coworkers have separated out even later minor folds (their F_6) which they connect with faulting, and two further sets of minor crenulation cleavages (their S_7 and S_8). While it is perfectly possible that these late structures may have correlatives elsewhere in the Kilmacrenan Succession, the detailed work which would allow such widespread correlations has not yet been done.

In summary then, we can see that there is a moderately good correlation between the early part of the structural chronology in the Kilmacrenan and Creeslough Successions (Table 3-2). In both situations, closely associated deformations, D_2 and D_3, produced the main structures which are coaxial, though the evidence for the later event in the Kilmacrenan rocks is admittedly weak. As regards the later and more "brittle" structures, oblique crinkles (F_{4a}) without associated minor or major folds are common in both sequences, as are the open upright major folds (F_{4b}) which are normally coaxial with the recumbent F_{2-3} folds. From then on in the structural history, we have insufficient evidence to permit correlation of the various minor structures, especially as regards the complex chronology of Howarth and others (*loc. cit.*).

(C) The Metamorphic History of the Kilmacrenan Belt

The overall metamorphic style of this part of the sequence also appears to be broadly similar to that of Creeslough. The pelitic lithologies are represented by chloritic phyllites and schists which show various stages of regression from garnet and biotite-bearing pelites, and the metadolerites are hornblende-andesine-quartz rocks also showing retrogressive changes. There is a slight rise in the metamorphic grade, or alternatively stated, a lessening of the retrograde effect, into the Slieve League–Glencolumbkille area, and this, as we shall see later, matches a similar change (though greatly offset) in the block to the southeast of the Leannan Fault.

As in the Creeslough metasediments, the S_1 schistosity, where visible in fold cores, is marked by minute phyllosilicates, and the major cleavage resulting from the crenulation of this S_1 fabric is followed by coarser flakes of biotite, muscovite, and chlorite in varying proportions depending on the locality and lithology involved; further, the segregation of quartz and feldspar along this S_{2-3} cleavage is common. Howarth and others (*loc. cit.*) have shown that the peak of the regional metamorphism in the Glencolumbkille area occurred during the interval between what we have here labeled the D_2 and D_3 movements, with a possible overlap with the latter. Garnet porphyroblasts there show the same textural relations as those in the Creeslough Succession, and indeed the whole situation is so similar to that existing to the northeast that we have no hesitation in making a correlation with the major metamorphic event, MP_2, in the Creeslough metasediments. Unfortunately, neither Cambray nor Pulvertaft (1961), working in the Fintown–Church Hill area, were able to link the infrequent porphyroblasts there to microstructures, but we suspect that the peak of the MP_2 regional metamorphism everywhere occurred during a static interval between two movement phases, D_2 and D_3.

Later metamorphic events in these Kilmacrenan metasediments have not been clearly defined, but there is evidence in Glencolumbkille of two separate metamorphic events following the D_{2-3} episode. Thus, Howarth and others (*loc. cit.*) noted chlorite growing from MP_2 biotite before, during, and after the D_4 movements (described in this book), again defining an RMP_3 retrogression. Then, just as in the Creeslough rocks, randomly oriented biotite and chlorite porphyroblasts are present on a regional scale, overprinting the "chloritic" S_{2-3} schistosity and even the oblique F_{4a} crenulation. It seems therefore that this MP_4 porphyroblastesis did everywhere follow a retrogression, RMP_3.

The porphyroblasts themselves show a retrogression to chlorite and epidote, so defining an RMP_4 episode. Howarth and his coworkers date both MP_4 and RMP_4 as being very late in the structural history, following their F_6 faults and preceding their F_7 and F_8 folds: a finding which we cannot accept for the rest of Donegal.

5. The Structural Relationship between the Creeslough and Kilmacrenan Successions

Having described the structural and metamorphic development and chronology of the two Successions northwest of the Leannan Fault, we now return to the structural relationship between the two sequences which we discussed earlier in Chapter 1. We showed there that the junction between the two is always marked by sliding or faulting. Thus, the major *Knockateen Slide,* albeit complicated by later dislocations such as the Mossfield–Lough Salt and Carbane Faults, is responsible for paring away whole stratigraphic units from both sequences, so that either the Slieve Tooey Quartzite or one of the beds below it is in contact with various horizons above the lower member of the Sessiagh–Clonmass Formation.

However, as we indicated earlier (p. 39), the stratigraphic succession in the Fintown-Loughros area suggests a natural upward passage, sometimes locally broken, from the Creeslough to the Kilmacrenan Succession. There are two major objections to this idea. The first, as we pointed out earlier, is that to the northeast of Lough Salt Mountain the Glencolumbkille Formation, the Port Askaig Tillite and the Slieve Tooey Quartzite together appear to form a flat to gently undulating sheet which truncates major folds in the Creeslough rocks below (see Maps 1 and 2 in folder). This suggests that considerable thrusting took place at a comparatively late stage in the structural history, though clearly predating the emplacement of the granites (see Fig. 11-2). On the other hand, it is equally possible that the structural contrast is simply due to the fact that a massive quartzite would respond to deformation very differently from an underlying pelitic series.

The other major factor to be considered in this matter is the facing directions of the major folds in the two successions (Fig. 3-10). We have seen above that everywhere the major folds in the Creeslough rocks *face northwestward* on their S_{2-3} axial plane surface, whereas the major structures in the Kilmacrenan Succession face *southeastward* on the equivalent cleavage plane. Cambray (1969*b*)

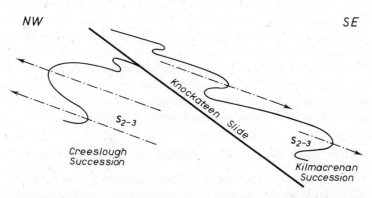

FIGURE 3-10. Opposing facing directions across the Knockateen Slide.

has suggested that this reversal of facing directions is the result of large-scale gravity slumping (*i.e.,* fir-tree folds) off a dome created by the rise of the Donegal Granite Complex. However, this hypothesis is hard to reconcile with the presence of northeast-facing folds in the Creeslough Succession in the Loughros–Fintown–Lough Greenan belt, and with the intervention in time of the open upright D_4 folds between the D_{2-3} structures and the emplacement of the granites (see Table 3-1).

Another possibility is that the regional structure of northwest Donegal prior to the coming together of the two successions was that of a great upright (F_1) anticline, whose limbs were later bent into major folds about a gently southwest-dipping cleavage (S_{2-3}) by D_2' and D_3 movements. This is comparable to the situation in the Scottish Dalradian where major folds face away from a central root zone (Rast, 1963; Johnson, 1966), the only difference being that these are labeled F_1 folds and not second phase as in Donegal.

We are not entirely satisfied with this latter explanation for the opposing facing directions in the

74   THE GEOSYNCLINAL PILE

two successions, and until further work is carried out, particularly in Inishowen, southeastern Donegal, and Tyrone, no firm conclusions can be reached. Nevertheless, it seems that the two successions have been moved together from their places of original deposition, but not necessarily very far. Although their original age relationships have thus been obscured, we still believe that the Kilmacrenan is simply the younger part of a formerly continuous upward succession (see p. 39).

6. An Outline of the Structural and Metamorphic Development of the Kilmacrenan Succession Southeast of the Leannan Fault

(A) The Fold Pattern

In this section, we follow in outline the structural pattern of Donegal up to and beyond the limits of our map area so as to broaden the scene for a wider discussion of the structural geology of the Dalradian as a whole. In doing so, the most notable feature is the change in the axial trace of folds crossing the Leannan fault; this provides the strong contrast which we later attribute to great strike-slip displacement along this line.

The markedly oblique trend to the Leannan fault on its southeastern side is related to the axial traces of three great overturned folds, the *Ballybofey Antiform* (Fig. 3-11), its complementary syncline, and the *Raphoe Anticline*, all three of which plunge to the northeast and face southeastward on the major east to northeast-dipping schistosity (Fig. 3-1). Then to the northeast of the Raphoe anticline are yet other folds which have an increasingly more easterly trend and which begin to steepen and verge northward: we know very little about these structures except the *Edenacarnan (or Rathmullan) Syncline* which brings the Crana Quartzite into the map area just to the east of Kilmacrenan, and which is but one of a number of related structures within the Crana outcrop of the Rathmullan District (McCallien, 1937). Later in this account, we shall refer briefly to certain major structures even farther eastward of our map area in Derry and Tyrone (Fig. 3-1): these are the *Lough Foyle Synclinal Complex* and the overturned folds on the northern flank of the Sperrin Mountains—the *Claudy Anticline* and the *Strabane Syncline*.

(B) Some Details of the Folding and Its Chronology

The Ballybofey Fold. We begin a brief survey of these structures by describing the Ballybofey fold complex, which, as we have already seen, controls much of the outcrop pattern of south-central Donegal (Pitcher *et al.*, 1971). This fold is a most impressive example of a large scale interference pattern produced by refolding; in this case, three major phases are involved (Fig. 3-11). An early deformation produced major isoclines which were then folded about a recumbent structure (actually a complex of several folds) whose axial plane dips ENE: this *Ballybofey Antiform* proves to be the dominant fold in the area. More open folding on almost N–S axes produced the *Ballard Antiform*, which, as the figure shows, warps the southern limb of this fold complex.

The main schistosity of this Ballybofey area is axial plane to the second phase major fold and to the northwest-verging minor folds which are associated with it: this is a situation well seen on the northern flanks of Gorey Hill near the town of Ballybofey itself. The schistosity is very likely to be the correlative of S_{2-3} in the northwest, so that we are able to identify the two primary fold phases as F_1 and F_2 in the terms we have previously defined. For the first time then, we find S_1 in axial plane relationship to major F_1 folds—though admittedly there are not many localities where this can be demonstrated.

In addition, and possibly associated with the main D_{2-3} movements, there is a strong SSE-plunging lineation marked by quartz rods. These are the result of the detachment and flattening of the cores of early (F_1?) minor folds produced by the deformation of early quartz segregations and veins.

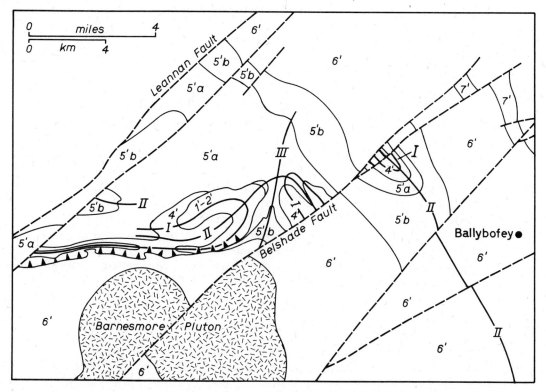

FIGURE 3-11. Structural map of the Ballybofey area showing axial traces of the several elements of the composite Ballybofey fold. I, II, III-successive fold phases (I-Croveenananta Anticline; II, Ballybofey Antiform; III, Ballard Antiform). Stratigraphy keyed to Fig. 1-5. After Pitcher, Shackleton and Wood (1971).

The open Ballard structure might equate in general terms with the F_{4b} folds to the northwest of the Leannan Fault, and as we have come to expect, there are one or two sets of post-S_{2-3} cleavages whose chronological relations have not yet been determined.

The Folds of the Raphoe-Strabane Area. Our preliminary map of the disposition of Dalradian strata in Donegal (Map 2 in folder) shows a number of fold structures analogous, and indeed, complementary to the Ballybofey fold. As yet we know little about them except that, if they are F_2 structures, then we might deduce from the associated minor folding and the low easterly dip of the regional S_2 schistosity that they are overturned, plunge at low angles to the northeast, with axial planes dipping gently ENE. Stratal dips are indeed very often low, and the association with late warping and extensive faulting leads to a confusing outcrop pattern which is difficult to assess in the very poorly exposed ground of eastern Donegal. However, we consider it useful to attempt a preliminary interpretation of the basic structure as shown in Fig. 3-1 and Map 2 in folder; we return later to a discussion of these folds (p. 77).

The Edenacarnan Syncline and Related Structures in the Rathmullan–Inishowen Area. In the northeastern part of Donegal, in the areas of Rathmullan and Inishowen, the geology has not been reexamined in any detail since the work of McCallien (1935, 1937). In both areas, there are a number of close to tight folds in the Crana Quartzite which are either *upright or overturned slightly to the NW or SE*. McCallien has stated that some of the folds have an associated axial-plane "slaty" cleavage which was the earliest minor structure observed by him (1937, p. 320). We are of the opinion, however, that these are second phase (D_{2-3}) structures because we have had the opportunity of examining the fold chronology on the Rathmullan coast section and also in the case of the Edenacarnan (or Rathmullan) syncline of the Kilmacrenan area. It is this latter northward-facing fold which has the important effect of locally terminating the outcrop of the Crana Quartzite and bringing the underlying Termon Formation into the northern districts of Donegal (p. 35).

The early part of the structural history, as determined from the associated minor folds, is one which is by now familiar, and indeed it was largely established in nearby Inishowen by McCallien (*loc. cit.*) long before our survey. An early schistosity (S_1) is folded around isoclines and deformed by the strong, penetrative axial-plane schistosity (presumably S_2), the intersection of the two planes being marked by a horizontal crenulation. Where these F_2 folds are upright, they can be clearly seen to be refolded and cut across by a regional, low-dipping cleavage which is probably our S_3, for in areas where the F_2 folds are gently inclined, S_2 and S_3 again become nearly coplanar; as in other areas, no major folds are associated with the S_3 cleavage. Finally, there are the usual later and more open folds sometimes associated with less conspicuous cleavages, but of these we know very little yet.

Most important for further research is the realization that there is a change in the sense of cleavage dip and in the facing direction of folds across a belt which runs southwestward across southern Inishowen into the Rathmullan area just described: a point we shall return to later.

(C) The Metamorphic History of South-central and Northern Donegal

In a general way, we can also outline the metamorphic history of the considerable area extending from Inishowen to the Blue Stack Mountains. Northeast of Letterkenny, the pelitic assemblages are composed dominantly of muscovite, chlorite, and quartz; garnet, though of sporadic occurrence, is widespread. The phyllosilicates follow S_2, and this has then (as S_3) been flattened and regrown around small garnets. Evidently, the main phase of metamorphism is again to be placed chronologically between the D_2 and D_3 movements, *i.e.*, it is MP_2 in our terms.

Farther southwest, beyond the Swilly, there is a progressive increase in metamorphic grade, leading to the production of coarse garnet-biotite-schists over a wide region betwen Boultypatrick and Silver Hill; even farther southwest, staurolite and kyanite appear sporadically in appropriate lithologies. Retrograde changes are again recognizable by the sporadic replacement of biotite and garnet by chlorite, and kyanite by micaceous pseudomorphs. Here, however, the textural evidence is quite clear that the partial conversion of biotite to chlorite predates the growth of the late porphyroblasts: the reality of the RMP_3 event thus appears to be confirmed.

Over the whole area, the final event in the mineralogical reconstruction—other than the growth of late pyrite and calcite—is recorded by porphyroblasts of randomly oriented chlorite and biotite which overprint the earlier fabrics (*cf.* McCallien, 1935 p. 415, 421). This episode is clearly the equivalent of the MP_4 of the northwest, so that it seems possible to correlate the metamorphic history over the whole of Donegal (Table 3-1). The final conclusion is that the two great blocks separated by the Leannan Fault have the same tectonic history, so that the fault is only important on account of its late displacement.

7. Donegal in Relation to the Caledonian Mobile Belt as a Whole

(A) In Terms of Major Structure

We now attempt to fit the geology of Donegal into the broader context of the Caledonides as a whole, examining first the main structural features. There is plenty of information about the Scottish Dalradian on which we may draw for comparative purposes, especially as the fundamental early work of Bailey (1922, 1934), and Bailey and McCallien (1937), has been much revised and discussed in recent years (see reviews by Watson, 1963; Rast, 1963; Johnson, 1965, 1969; Roberts, 1966b; Johnstone, 1966; Bennison and Wright, 1968; Anderson and Owen, 1968).

Broadly speaking, the Dalradian outcrop in Scotland is controlled by a major *southeast-facing* recumbent anticline, the *Tay Nappe,* whose downward-facing crest (the Aberfoyle Anticline) lies in a "vertical belt" close to the Highland Border and whose inverted limb—marking the "flat belt" or Loch Tay Inversion—extends from Deeside to the Mull of Kintyre (Fig. 3-12). This apparently first phase nappe was twice refolded, first on fairly tight structures with various directions of axial trace, and secondly by open warps with the familiar Caledonoid trend some of which are themselves polyphase, as in the case of the Cowal Antiform of the southeast of the Dalradian outcrop. Northward of this folded nappe, on the northwestern side of the outcrop, the basic structure is also broadly antiformal—the so-called Islay Anticline—consisting of both first and second phase structures which now face uniformly *northwestward*. Between the two antiformal structures of Cowal and Islay lies the Loch Awe Synform, which is in reality a complex synclinorium in which the sense of overfolding is outward from a medial steep belt (Fig. 3-12), the line of which passes through Loch Awe itself (Rast, 1963; Roberts, 1966b).

It is comparatively easy to follow these major structural elements into the northern part of Ireland (Charlesworth, 1963, p. 55; Anderson and Owen, 1968). The correspondence of the Lough Awe and Lough Foyle synclinal structures is particularly evident. It is also possible to trace into Ireland the steep belt which separates northwesterly from southeasterly vergence, but in Inishowen, unlike southwest Scotland, this belt does not lie in a medial position relative to the synformal structure (Fig. 3-12; *cf.* Thomas and Treagus, 1968).

On the northern flank of this steep belt, in Inishowen as in Argyll, the F_2 folds can be seen to verge northwestward, and successively older rocks gradually appear. Farther northwestward, in Donegal, the continuity of the Dalradian succession is then apparently broken by the important dislocation, the Knockateen slide. This is likely to be the correlative of the Iltay Boundary Slide of the mainland of Scotland (see Fig. 3-12) which lies in about the same general position (Bailey, 1934; see also McCallien, 1935), possibly even the Loch Skerrols Thrust of Islay (Wilkinson *et al.,* 1907; Bailey, 1916), though we prefer to regard this latter as a continuation of the Moine Thrust. We have already discussed this boundary between the Kilmacrenan (Islay) and Creeslough (Ballachulish) successions in some detail (see p. 73), concluding that, though it represents an important line of movement, there is not much evidence for a really fundamental structural break in the Dalradian Succession as a whole. We cannot, however, follow Voll (1964) in his conclusion that it is possible to correlate directly, albeit in different facies, the two successions separated by this slide, for in our view, it is likely that the Kilmacrenan Succession follows stratigraphically above the Creeslough, as is now again claimed for the Scottish situation by Rast and Litherland (1970).

On the southern flank of the Lough Foyle structure, the continuation of the Cowal Antiform has been recognized both in Antrim (Bailey and McCallien, 1937; McCallien, 1935; Knill in Wilson and Robbie, 1966) and in the Sperrin Mountains (Hartley, 1938; Charlesworth, 1963, Fig. 17A). Here the major component of the structure is again likely to be a great recumbent southwest-facing anticline, whose closure in Antrim is buried beneath the Tertiary Lavas, but which emerges north of the Sperrins as the Claudy Anticline, (Fig. 3-1 and 3-12, and see Charlesworth, *ibid.*). It would be attractive to regard this fold as the southwestward continuation of the Tay Nappe, and the rocks of the Sperrins as part of the broad inverted belt—the Tay Inversion.

It is of particular interest thus to follow these well-established structures westward into central Donegal, though our knowledge of this region is admittedly incomplete. In eastern and central Donegal, the regional plunge is to the northeast, and the successively lower tectonic levels so revealed (to the west) show a complex of large polyphase folds which we believe to be of D_{2-3} age. The problem is to discover what happens west of Strabane to the Claudy Anticline, a problem of especial importance if this is to be equated with the Tay Nappe, a structure generally accepted by Scottish

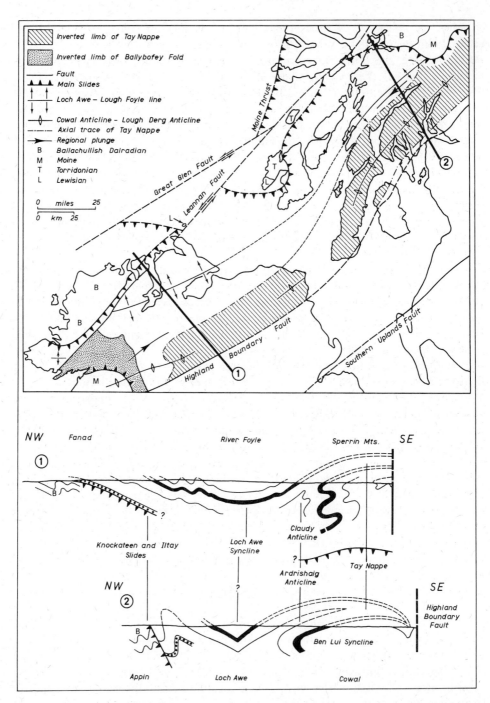

FIGURE 3-12. Extension of certain major structural features from southwest Scotland into Ireland. Comparative sketch sections across the Dalradian tract in these two regions; latter not to scale, lengths of sections approximately 50 miles. Tillite (dots) and Culdaff-Tayvallich-Loch Tay Limestone (black) shown on sections only.

workers as a first phase fold. The possibilities are, (a) that it dies out very quickly just as is shown by original sections produced by Bailey for Argyll (Bailey, 1922), (b) that its continuation is simply represented by the Raphoe anticline, in which case the Claudy fold would have to be of second phase age, (c) that this fold is deformed about the Raphoe anticline, its axial trace swinging either northwestward toward Lough Swilly or more likely southward into the country east of Castlederg.* Unfortunately, our knowledge of this critical area is insufficient to resolve this issue which is, we think, a problem central to the understanding of Caledonian geology: the kind of solution we envisage is, however, represented in Fig. 3-13.

Whatever its solution, the combined effect of the several fold phases in the Ballybofey structure is to invert the Middle and Lower Dalradian in south-central Donegal in such a way as to bring these strata over the Moinian of Lough Derg along a major dislocation, the Lough Derg (Ballykillowen) Slide (Pitcher and others, 1971).** This inverted relationship with the Moine provides the final term in our synthesis of the structure of the Dalradian of Donegal (Fig. 3-13). The hugh nappe so revealed seems to be rootless: can it be that the Iltay Slide is the continuation of the Lough Derg Slide and that the Dalradian as a whole is allochthonous in relation to the Moine?

In the Letterkenny area, it seems that the Loch Awe-Inishowen steep belt swings against the Leannan Fault, along which it is probably translated by a sinistral strike-slip of some 48 kilometers to the Slieve League peninsula, where it is represented by a change in the direction of cleavage dip along the Stragar River (Pitcher *et al.,* 1964, Fig. 1); the offset in the garnet-chlorite isograd supports this conclusion (Chapter 17). If this is so, then the Slieve League Anticline is probably the correlative of at least part of the Ballybofey fold, and the major strike-swing in western Donegal (see p. 000) would be intimately related to the swings in central Donegal. It is clear that further studies in central and southwest Donegal are badly needed, and we can only hope that these speculations will act as an impetus to future research.

(B) In Terms of Chronology of Deformation

We have just seen that most of the major structures of Donegal can be matched by similar ones in southwestern Scotland, but we must now examine more closely the correlation of fold chronologies. Of course, as we pointed out early in this chapter, the correlation of structural events from area to area is not an easy matter, for they may be unequally developed over any great distance (*e.g., cf.* Knill, 1959*b,* with Roberts, 1967), and the possibility of diachronism is often hard to dispute; further, there is the likelihood that erosion exposes different tectonic levels in different areas. Despite all these complications, we venture to compare our Donegal chronology with the sequence of folding in Scotland *as presently published* (Table 3-3).

In both areas, there are four widespread structural events, D_1 (F_1) to D_4 (F_4), though the directions of major folds do not always coincide. Further, in much of Scotland as in Donegal, the peak of regional metamorphism comes between the second and the third movement phases (see, however, Harte and Johnson, 1969). In contrast to this apparent similarity, there is the problem that the major

---

* Such an interpretation reminds us of the recent study of the Ben Lui fold of Scotland by Roberts and Treagus (1964), according to which this latter structure is a major second phase fold structurally beneath (and deforming) the Tay Nappe.
** We may note the presence in this area of the late, northeast-trending open Lough Derg antiform (Anderson, 1947; Church, 1969; our Fig. 3-1).

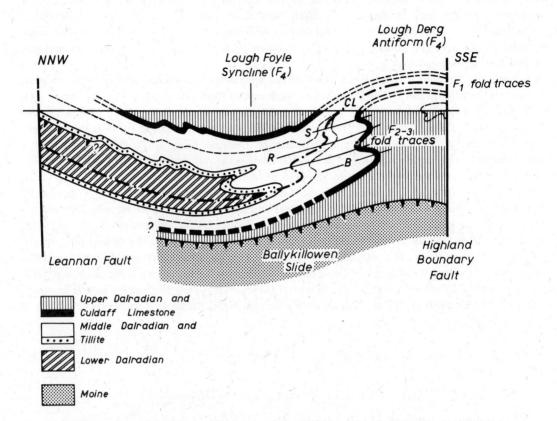

FIGURE 3-13. Highly diagrammatic section across the Dalradian outcrop between the Highland Boundary and Leannan Faults. Length of section about 40 miles (65 kilometers): CL, Claudy Anticline (F_1); S, Strabane Syncline (F_{2-3}); C, Raphoe Anticline (F_{2-3}); B, Ballybofey Antiform (F_1 and F_{2-3}).

recumbent folds in Scotland are F_1 structures while the vast majority of those in Donegal are D_2 ($-D_3$), according to us. Even if one reduces the status (as a separate deformation event) of the bedding-schistosity and early slides in northwest Donegal, there is the fact that the cleavage to the Scottish F_1 folds (and even, according to Roberts, *loc. cit.*, of the F_2 folds in Argyll) is a penetrative fabric, whereas in Donegal it is for the most part clearly a crenulation cleavage, however intense this may be. However, recent work in Scotland has emphasized the importance of major recumbent F_2 folds with strong axial plane fabrics (*e.g.*, Roberts and Treagus, 1964), and it may be that in our area large-scale D_1 folds are simply not shown, except for the major D_1 isoclines in central and eastern Donegal. The second and perhaps more important contrast is the apparent absence in Scotland of an event equivalent to our D_3 phase, that is, regional intensification and local refolding of the major

CALEDONIAN STRUCTURE AND METAMORPHISM

TABLE 3-3. Deformation Chronologies for the Dalradian of Donegal and Scotland

(No direct correlation implied: for discussion see text)

Donegal (see Table 3-1)		Scotland (after Rast, 1963; Johnson, 1965)	
D_1	Slides, and ubiquitous "bedding-schistosity." Major folds not easy to identify except in central Donegal.	F_1	Major recumbent folds, with NE–SW trends. Major regional penetrative fabric.
D_2 (MP_2) D_3	Composite movements separated by peak period of static metamorphism. Major regional folds and cleavages. NE–SW trends, except in W and SE Donegal.	F_2	Major "cross-folds." Variable trend. Peak of metamorphism.
D_4	S_{4a}—Minor cleavages and folds often N–S to NW–SE. F_{4b}—Relatively open, upright, generally NE–SW folds.	F_3	Relatively open folds with NE–SW trends.
	Various later minor folds and cleavages	F_4	Late minor structures.

S_2 cleavage. Because a correlation of our D_4 event with F_3 of Scotland seems reasonable,* either the Donegal D_3 is only a local event or it has not been recognized in Scotland as a separate deformation episode (*cf.*, however, Harte and Johnson, 1969).

In conclusion, then, we think that this conflict in the published chronologies is more apparent than real. The problem as to whether there is a real equivalence in time between widespread deformation structures, or whether there is a relative diachroneity, remains one of the most pressing in structural geology.

(C) In Terms of Degree of Deformation

Throughout this chapter, we have been concerned mainly with the geometrical and chronological aspects of deformation and have described the fabric largely in terms of folds and cleavages. There are many elements present, such as primary sedimentary structures, pebbles, groups of grains, and mineral alignments, which could be used to study the kinematic aspects of deformation (*e.g.,* Flinn, 1956, 1965, 1967; Watterson, 1968b; Stauffer, 1967). However, apart from local studies by Howarth and others (1966) and Kemp (1966T) such studies have not yet been made, though it is obvious that the amount of strain varies considerably from area to area, depending on the lithology involved and probably on variations in the generative stresses.

In order to carry out such analyses, particular attention would have to be paid to the chronology of deformation, for many of the fabric elements have been affected by more than one movement phase. For example, most psammitic lithologies in northwest Donegal possess strong linear alignments of pebbles and mineral grains which are an integral part of what is generally the S_1 surface (*cf.* Flinn, 1965; this we have earlier referred to as the "stretching lineation (p. 58). The Errigal Syncline appears to fold this lineation, but because the latter lies at high angles to F_{2-3} it is remarkably constant in direction, trending NNW–SSE. However, in many places, this lineation is parallel to later crenula-

---

* However, if Johnson's recent claim (1969, p. 156) is correct, that the F_3 folds in the Scottish Dalradian are of Silurian or younger age, this correlation breaks down, for the Donegal D_4 structures predate the granites which give a 470 million year age (*i.e.,* Ordovician).

tions (chiefly F_{4a}) and seems to have been intensified by post-D_1 movements: it is thus not entirely of one age (see Voll, 1960, for discussion of a similar situation in Scotland). Despite these difficulties, the overall strain picture is as follows:

D_1 strong linear and planar (L-S) fabric
D_{2-3} strong planar (S) fabric—the regional cleavage and schistosity
D_4, etc. weak and localized planar (S) fabrics

During the earlier and more important movements in Donegal (D_1 to D_3), there was a progressive though episodic recrystallization, but as is commonly the case, the earliest deformation (D_1), during which the rocks were ductile, was accomplished under very low metamorphic conditions. We speculate that the presence of intergranular fluids, however produced, was the main factor causing this low ductility. The flow of the rocks during the main D_{2-3} movements was a consequence of higher temperatures attained during the peak of metamorphism (MP_2), the most profound effects occurring in central Donegal where the rocks were buried most deeply beneath a pile of recumbent folds. The post-D_3 phases involved neither prominent recrystallization nor such fluids, since the latter were undoubtedly dispelled during MP_2 and subsequent metamorphic events; thus their relatively weak structures may have formed under "dry" stress regimes virtually as intense as those operating earlier in the structural history.

8. Final Comment

In Part I, we have dealt at length with the geological environment into which the granites of Donegal were emplaced, though in discussing the orogenic history, we have nothing to add to the *convenient* tectonic models of the Caledonides such as provided, for example, by Johnson (1965) and Dewey (1969a). In fact, we have merely scratched the surface as far as the structural studies are concerned. It is of paramount importance that future research be directed toward establishing (1) the true significance of the Loch Awe-Lough Foyle belt and its relation to the change in facing directions of the major folds, (2) the complete geometry of the Ballybofey fold and its connection with the Tay Nappe, (3) the nature of the major strike swing, and (4) perhaps most important of all, the proper correlation of deformation and metamorphic events within Donegal, and between it and the rest of the Caledonides.

Our primary aim in this book is to present and debate the character and mode of emplacement of the major intrusions into this metasedimentary pile. However, before we proceed to discuss these, we must refer briefly to a series of widespread minor lamprophyric intrusions which were emplaced at the end of this long tectonic history.

A Geological Appendix: The Late Tectonic Lamprophyres

All over Donegal, thin (1–2 meter) sills and dykes of lamprophyre and felsite were intruded near the end of the regional deformation history (Rao, 1951; Murthy, 1951; McCall, 1954). Doubtless their emplacement spans a considerable period. We have one record of a felsitic lamprophyre (at Breaghy Head, near Portnablagh) being deformed on an S_{2-3} surface; quite commonly, such intrusions have taken the impress of the very latest of the pregranite structural events; others are unaffected, yet predate the earliest phases of granitic activity, being included as xenoliths in the granites; still others postdate the earliest of these phases (see Chapter 7), and finally, a very few even fill late fault planes (*cf.* Reynolds, 1931). Nowhere can one better examine all these chronological relationships than along the shores of Rosbeg in the Dawros peninsula (Obaid, 1967T). It seems to us, therefore, that these lamprophyric rocks form a coherent petrological suite and help to link the orogenic with the granitic events.

There is a wide range of composition in respect to the proportion of biotite, hornblende, and plagioclase, and some rocks would be better classified as porphyrites than lamprophyres: the most diversified types can be labeled plagiophyres and malchites. A late-stage alteration commonly obliterates the original mineralogy and texture, but it can be discerned that the primary assemblage in the biotite lamprophyres consists of corroded flakes of biotite set in an intergrown mosaic of ill-defined crystals of oligoclase and some quartz; in the amphibole-bearing varieties, prisms of zoned, green-brown hornblende are set in a groundmass of plagioclase laths; many of the felsites (plagiophyres) consist of sodic plagioclase, quartz, and subordinate potash feldspar in an almost granular aggregate set with varying amounts of biotite and ore; a few others are characterized by a micrographic texture together with the development of spherules on a macroscopic scale.

Secondary effects are ubiquitous, and acicular amphibole, chlorite, epidote, and calcite form the usual assemblages. Such changes are usually attributed to an autometasomatism due to the concentration of hydrous phases in the late stages of crystallization. Wet magmas are implied, and indeed many intrusions are crowded with feldspar-filled amygdales representing the filling of gas bubbles by residual solutions (see Phillips, 1956 and Smith, 1946, for general discussion). Still others are crowded with xenoliths and xenocrysts, suggesting explosive action, and a few show metasomatic contact effects akin to adinolization.

We must underline the clear fact that in the field there are infinite variations in both the primary and secondary mineral assemblages, sometimes even along the length of a single dyke. The lamprophyres obviously represent a magma which was rapidy differentiating at the time of intrusion, and they probably came from a widely available "wet" magma of intermediate composition generated at a late stage in the orogenic rhythm.

PART II

THE GRANITES

CHAPTER 4

THE CALEDONIAN GRANITES AND THEIR ASSOCIATED ROCKS: INTRODUCTORY COMMENTS

CONTENTS

1. Introduction 87
2. Age Relationships 90
3. Notes on Nomenclature 91
   (A) Rock Classification 91
   (B) Rock Texture 93
   (C) Rock Structure 93

1. Introduction

Igneous activity in Donegal is recorded by four very unequal episodes, the first and second of which, the emplacement of rocks which are now metadolerites and altered lamprophyres, have already been dealt with in Chapters 2 and 3. It is with the third and major episode that we are most concerned in this section of the book—the emplacement of highly varied Donegal granites and their accompanying suites of minor intrusions—an event belonging to the late stages of the Caledonian drama. The fourth igneous episode, which is completely independent, is represented by the swarm of basic dykes of Tertiary age. In the present chapter, we shall introduce the main outlines of the granitic activity and discuss certain preliminary points, in preparation for a more detailed discussion of each of the separate units.

The dominant representatives of this magmatism are the eight granitic plutons already listed in the introduction and distinguished in the sketch map of Figure 4-1. These can be conveniently grouped according to whether they were emplaced by stoping, either active or passive, or by permissive or forceful means. The difference in the intrusion mechanism of these plutons is reflected in the variety of their contact effects, and the structural and mineralogical changes in the aureole rocks are appropriate to the mechanism involved; such modifications range from static hornfelses to thorough-going crystalline schists with varying degrees of metasomatic transformation.

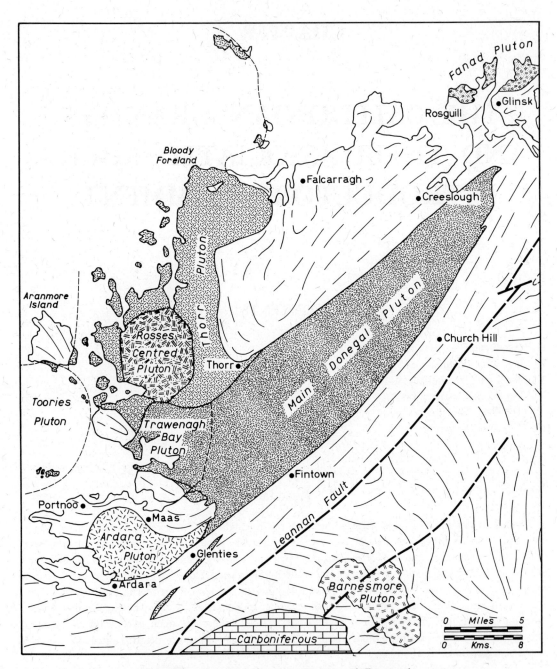

FIGURE 4-1. The granite plutons of Donegal.

Stoping is illustrated by the Thorr, Fanad, and Trawenagh Plutons. The Thorr Pluton consists of a transgressive granite with a remarkable roof zone of migmatitic quartz-diorite in which can be traced a pronounced ghost stratigraphy (Pitcher, 1953a). There is so much evidence of contemporaneous movement in the envelope and of reaction with the country rocks that we refer to the whole process as one of *active* stoping. The Fanad Pluton (Knill and Knill, 1961) shows somewhat similar relationships, although contemporaneous movement in the envelope was much weaker and reaction with the country rocks much less severe than at Thorr: the stoping is here then a largely *passive* process. Both of these stoped units are surrounded by broad static aureoles which are clearly superimposed on the regional metamorphic assemblages, but only in the Thorr aureole were metasomatic processes operative. Though the manner of emplacement of the Trawenagh Pluton is in doubt, it was emplaced at least partly by piecemeal stoping in an entirely static environment; here the contact effects are very slight.

Permissive emplacement is well exemplified by the Rosses and Barnesmore Plutons. Both are centered complexes in which a crudely concentric pattern of closely related granites has resulted from repeated cauldron subsidence of large dome-shaped or cylindrical blocks of the roof. Not unexpectedly with permitted intrusions, the contact effects are slight and in the case of the Rosses almost undetectable.

Forceful emplacement is best illustrated by the Ardara Pluton which was intruded diapirically by successive pulses of magma, which caused a radial distension of the wall rocks. Here, an early foliated monzodiorite mantle surrounds a structureless granodiorite core. The accompanying aureole, in its complexity, also provides evidence of the recurrent interaction of crystallization and deformation; metasomatism is restricted to a narrow inner zone. The least known member of the Donegal Granite Complex, the Toories Pluton, which is only exposed on off-shore islands, appears to be similar to the Ardara body in having a circular outline, a deformed envelope, and a strong marginal foliation.

The largest of our granitic masses, the enigmatic Main Donegal Granite, is difficult to fit into this scheme of emplacement mechanisms. Pitcher and Read (1959, 1960b) viewed it as the result of forceful magmatic wedging accompanied by near horizontal flow of partly consolidated magma, and the lateral distension of its country rocks. However, it is now clear that the salient features of the Main Granite are the result of its being strongly deformed at the present crustal level during the later stages of consolidation. Many of the original features produced during emplacement have thus been obliterated, yet it still appears that the Main Granite got into place by forcibly wedging into its country rocks. Perhaps the most remarkable feature of the Main Granite is its envelope of high grade schists of regional metamorphic character.

We may now mention the appendages to the Donegal granitic intrusions, the appinites and the dyke suites. Members of the appinite suite are well represented in Donegal, where the vast majority are associated with the early history of the Ardara Pluton. They appear as innumerable small bodies of all kinds, ranging in composition from ultrabasic to acid, and emplaced mostly by a combination of diapirism and an explosion mechanism; the larger bodies may exhibit narrow aureoles which are generally static. The petrographic character of these appinites, and their attachment to granodiorite diapirs and plutons throughout the British Caledonides, raise petrogenetic questions of more than local importance. In addition to these intrusions, most of the Donegal granites have associated with them swarms of dykes of varying petrographic character and mostly acid to intermediate in composition. These include the appinitic lamprophyres and felsites around the Ardara Pluton, the porphyritic microgranites radiating from the Rosses Complex, the microgranites of the Barnesmore Complex, and a substantial swarm of NE–SW trending acid and intermediate dykes perhaps associated with the Fanad Pluton. Lastly, the Main Donegal Granite has its own suite of pegmatites, microgranites, and felsites.

We consider that these granitic plutons and their associates are not separated by great time intervals, and all form parts of one magmatic episode. Further, their emplacement was later than the peak of the regional deformation and metamorphism. Certain of them were unlikely to have been emplaced under any great thickness of cover as, for example, the cauldron subsidences of Barnesmore and the Rosses, and it is our opinion that they are all relatively high level intrusions despite appearances to the contrary. The fact that certain units, particularly the Main Granite, have features characteristic of intrusions of deeper zones in the crust illustrates the difficulty of classifying plutons according to the *apparent* level of their emplacement as Buddington (1959) has done. Neither is there any neat order of events fitting some simple structural model, and it seems that Read's dire warning that "anyone who does not want to see his neat solutions of geological phenomena upset should keep clear of Donegal" (1961, p. 677) was timely indeed.

2. Age Relationships

The age relationships of the Donegal granites are not completely known, mainly because the various members are not all in convenient juxtaposition. The Thorr Pluton is certainly older than the Rosses Centered Complex and is likely to be older than the Ardara Diapir, since, in the Maas area, the Thorr aureole is cut by an appinite complex which is considered to predate Ardara. As regards the relationship between the latter and Rosses, we have only one observation to guide us—that dykes of the Rosses swarm cut the northern margin of the Ardara aureole. The Main Donegal Granite intrudes both the Thorr and Ardara plutons, but its relationship to the equally late Trawenagh Pluton is somewhat enigmatic; we shall later give grounds for believing that the latter may have arrived at its present level slightly after the Main Granite. We can only guess at the time of the Toories Pluton, and so present a probable age sequence for the units of the *Donegal Granite Complex* as: Thorr, Toories, Ardara, Rosses, Main Granite and Trawenagh.

There is no direct way of ascertaining the relative age of the isolated Fanad and Barnesmore Plutons either to one another or to the Complex, though we shall make certain suggestions in the appropriate place. Anyway, in what follows we shall not always adhere to strict chronology, but rather we shall be guided by the convenience of dealing with the plutons in groupings which are based on their mode of emplacement.

As regards isotopic age, the K-Ar dates so far available range between 404 ± 8 million years and 372 ± 6 million years for various members of the Donegal Granite Complex (Table 4-1), though as might be expected, the individual dates do not always agree with the relative ages of the plutons as determined in the field. Leggo and his coworkers (1969) have established a single Rb/Sr isochron, connecting points for the Rosses ring granites and the Trawenagh Bay Pluton at a date of 470 ± 1 million years,* which apparently reflects a time nearer emplacement. Surprisingly, aplites from the Rosses define a separate isochron at about 430 million years; this difference we comment upon later. The much lower K-Ar single mineral dates probably record the final cooling of the complex and indeed of the region as a whole. If so, it is intriguing to speculate what might happen in a system cooling over an interval of 70 to 90 million years!

It is not surprising that, considered in the broader context of the Caledonides as a whole, the granites of Donegal show many similarities to the post-tectonic plutons of the late Silurian–early Devonian "Newer Granites" of Scotland, affinities which we shall refer to throughout the main body of this work. In fact, apart from the Main Donegal Pluton itself, the similarities are so great as to throw some considerable doubt on the Rb/Sr dates recorded above. In one respect, however, there is a difference in that nowhere else is the variety of emplacement so remarkably diverse as in Donegal, giving good reason for this present work.

* These Rb/Sr dates were calculated using an ^{87}Rb decay constant of 1.47×10^{-11} yr^{-1}, but Leggo and Pidgeon (1970) using a constant of 1.39×10^{-11} yr^{-1} to bring their Irish dates into line with U/Pb dates on zircons from the Galway Granite, recalculated the isochron for the Rosses and Trawenagh Bay Plutons at 498 ± 5 million years.

INTRODUCTORY COMMENTS

TABLE 4-1. K-Ar Dates from Donegal

Pluton		Location	Mineral	Date (million years)	Source
Thorr	1.	650 yards E of Gweebara Bridge	Muscovite	391 ± 8	1
	2.	Owenator River near Lough Agher, Meencorwick	Biotite	390 ± 7	2
			Hornblende	392 ± 8	2
Ardara	3.	Ardara-Clooney crossroads, Hornfels at contact	Biotite	414 ± 8	1
	4.	Outer monzodiorite, Clooney, G/17373990	Biotite	394 ± 8	2
			Hornblende	392 ± 7	2
Rosses	5.	Greisen, Sheskinarone	Muscovite	404 ± 8	1
	6.	Greisen, ¼ mile W of Lough Nabrack	Muscovite	384 ± 8	2
	7.	Garnetiferous microgranite, Shore 1 mile SW of Dunglow B/17604106	Muscovite	382 ± 6	2
Main Granite	8.	Pelitic raft, zone 3b of Fig. 11-2, this book; Barnesbeg Gap.	Biotite	357 ± 8	1
			Muscovite	382 ± 8	1
	9.	"Mica hornfels," sheeted zone, N of Lossett.	Muscovite	384 ± 7	1
	10.	Granite, 1 mile SSE of New Bridge; C/20854263	Muscovite	378 ± 7	2
	11.	Pegmatitic granite, NE of Sand Lough	Muscovite	372 ± 6	2

$$\lambda_\beta = 4.72 \times 10^{-10} \text{ yr}^{-1}$$
$$\lambda_\epsilon = 0.584 \times 10^{-10} \text{ yr}^{-1}$$

[1] R. S. J. Lambert, personal communication, 1966.
[2] Brown, Miller and Grasty, 1968.

3. Notes on Nomenclature

Before proceeding to a detailed discussion of the plutonic events summarized above, we consider briefly here certain problems of nomenclature.

(A) Rock Classification

In what follows, we shall need to use an acceptable terminology for the considerable range of granitic rock types. It is not easy to be exact in this because such rocks are notoriously variable and gradational in outcrop and have an origin which is often the result of more than one process. It is natural, therefore, that there have been a variety of proposals for their classification.

Suggestions have been made for genetic classifications in which the granitic types are broadly related in time and space to the plutonic history (Read, 1957; Marmo, 1955, 1966; Buddington, 1959; Harpum, 1961; Raguin, 1965). There have also been attempts to relate the scheme of classification directly to the physico-chemical processes operating during their formation (Tuttle and Bowen, 1958; Szadeczky-Kardoss, 1960). Nevertheless, these methods are not yet sufficiently precise for day to day use, and the problem is essentially one of selecting the most useful of the many classifications, based on varying volume percentages of certain of the essential minerals.

In the classification of granitic rocks (*i.e.,* "acid" plutonic rocks in general), it is quite usual to follow Johannsen (1936) in the use of a system based on the potash feldspar-plagioclase-quartz content, using either modes or norms for the parameters (*e.g.,* Hutchinson, 1956; Chayes, 1957; Ginzburg *et al.,* 1962; Bateman *et al.,* 1963; Ronner, 1963; Streckeisen, 1967). Other systems utilize the three feldspar components (*e.g.,* Heitanen, 1963; Lipman, 1963), and yet others involve four or more components in various ways (*e.g.,* Roddick, 1965). Even more widely used are schemes which separate rocks into families or clans on the basis of the quartz content, further subdivision being made on the basis of the alkali feldspar/plagioclase ratio, sometimes modified by color index (*e.g.,* Shand, 1943; Nockolds, 1954; Jung & Brousse, 1959; Hatch and Wells, 1961; Joplin, 1964).

We could debate this matter at length, but fortunately, a very full discussion has recently been provided by Streckeisen (1967), and anyway many of the classifications mentioned are not radically different from one another in their definition of well-established rock types. We intend to follow a modified version of the modal classification of Streckeisen (*loc. cit.*), of which the pertinent portion is reproduced in Fig. 4-2. Our modifications are based on the fact that there are many examples of

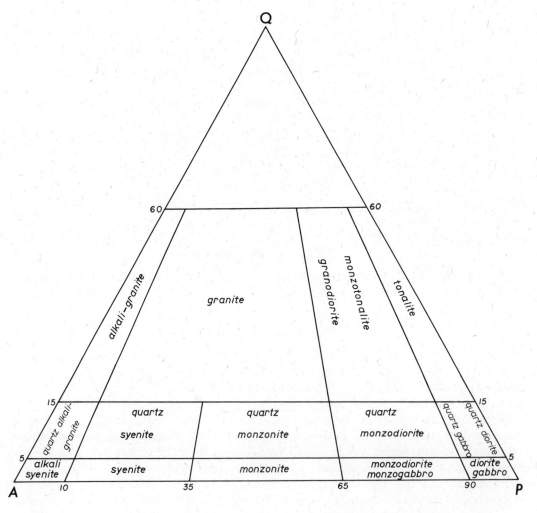

FIGURE 4-2. A Classification of granitic rocks: modified from Streckeisen (1967).

rocks which would normally be classified as quartz diorites on the basis of their high color index and high proportion of andesine, yet which have substantially more than 15% quartz and often up to 30% of potash feldspar. We contend that such dark rocks are not granodiorites in the presently accepted sense, and we prefer to record their special character by the use of the well-known rock name of "tonalite" (*cf*. Karl, 1966). We certainly do not regard this as a worthless term (see Chayes, 1957) or as a mere equivalent to quartz diorite (see Johannsen, 1936); in fact, we extend its application to all granitic rocks in which both quartz and mafics separately exceed 15%. We follow most workers in restricting the application of tonalite itself to rocks with less than 10% potash feldspar, using monzotonalite for the more potassic varieties.

When we come to discuss certain special rock types, particularly the variable dioritic rocks which characterize our "appinitic suite," we shall need certain additional terms which we can define in the section concerned.

(B) Rock Texture

Even more complex problems have to be faced in attempting to describe and explain the textural features. Detailed examination of such textures nearly always reveals a complex history of crystallization. What are possibly magmatic textures are invariably complicated by others resulting from subsequent recrystallization and deuteric (endoblastic, endometasomatic, autometamorphic, autopneumatolitic) reactions such as have been particularly well documented by Sederholm (1916), Drescher-Kaden (1948) and Erdmannsdorfer (1949, 1950).

In the case of a magmatic granite congealing over a long period, it is not hard to understand why this should be so, for a melt which starts to crystallize at, say 750°C is probably wholly crystalline at 650° to 700°C, leaving a further drop in temperature of some hundreds of degrees during which further adjustments of the mineral fabric must surely take place. Indeed, the only pertinent point of discussion is the extent to which magmatic textures can be identified at all. But because the use of textures as guides to petrogenesis is so prevalent in the literature we shall devote part of our work on granitic fabrics (Chapter 15) to a brief consideration of the petrogenetic significance of such features as a granitic texture (*sensu stricto*), zircon morphology, oscillatory zoning in plagioclase, growth twinning in feldspars, and concentric patterns of inclusions in potash feldspar crystals. For the moment, however, we simply state our belief that none of these characteristics can properly be used, as they so frequently are, to argue the case for the origin, magmatic or metasomatic, of the rocks in which they occur (*cf*. Mehnert, 1968).

The literature is also full of detailed studies of the sequence of crystallization in plutonic rocks, and the confusion caused by conflicting interpretations of individual textures, which have added so much fuel to the granite controversy, surely reflects the subjective nature of such explanations (*cf*. Cheng, 1944; King, 1947; Schermerhorn, 1956a; Hutchinson, 1956; Taubeneck, 1957; Crowder, 1959; Koschmann, 1960; Roddick, 1965; Gates and Scheerer, 1963; Wright, 1964). Modern studies have shown that many of the familiar criteria of idiomorphism and replacement, which led to the classic paragenetic sequences of Rosenbusch and Becke (see Savolahti, 1962, for a review), can now be more convincingly explained by recrystallization controlled by differences in interfacial energy and by other solid-state mechanisms (see Voll, 1960; Stanton, 1964; Brett, 1964; Rast, 1965; Spry, 1969). Thus, in our discussion of the petrography of the Donegal granites, we shall resist the temptation to erect definitive sequences of crystallization.

(C) Rock Structure

It is rare that granitic rocks fail to show some kind of preferred orientation, and the measurement of such structures has been standard practice since the pioneering work of Balk and Cloos: work later collated in some detail by Balk in his well-known book, *Structural Behavior of Igneous Rocks* (1937). We shall have much to say about granitic structures in the descriptions which follow, and indeed, we devote a separate chapter (Chapter 15) to a general discussion of their rheological and kinematic significance. Joints for the most part we have ignored, for, as we state elsewhere (Berger and Pitcher, 1970), we do not think that their study adds much to the understanding of the mechanics of emplacement, at least not in Donegal. Reference to the published works (*e.g.*, Akaad, 1956a; Pitcher and Read, 1959) will show that in many cases, as one would expect, the prominent joints form at right angles and parallel to the dominant planar fabric in the Donegal granites.

The distinction between primary and secondary structures in igneous rocks is not so clear as seems to be often assumed, especially if all structures which occur at the time of intrusion or as a consequence of intrusive energy are categorized as primary in origin. There must obviously be innumerable gradational stages in the evolution of a granitic fabric which is likely to be the product of flow extending from the fluid to the solid state, though the relative importance of the changes taking place at each point in this transition is not easy to recognize. In our own studies, we can rarely identify with confidence structures which formed when the rocks were essentially liquid, and what we see instead are structures which appear to have originated for the most part when the host rocks were largely crystalline. We prefer to discuss the structures in granitic rocks in terms of *deformation*, meaning simply change in shape, and we emphasize that the term *flow* should not be restricted to movement in fluids, since it describes any deformation without loss of cohesion on the scale observed. Indeed, even the fracture involved in cataclasis can be considered a mechanism of flow if the scale of the observation is large enough. In view of these and other uncertainties (see Chapter 15, and Berger & Pitcher, *loc. cit.*) we shall avoid the general use of terms such as primary and secondary and instead discuss the evolution of each structure as a separate case.

Perhaps the simplest way of discussing deformation, and one which we shall follow, is in terms of the finite deformation (strain) ellipsoid (Flinn, 1962, Ramsay, 1967). This is the "ellipsoid resulting from the deformation of an originally spherical portion of the rock being deformed" (Flinn, *ibid*, p. 386). Following Ramsay, the axes of this ellipsoid are designated as $X \geqslant Y \geqslant Z$, and the ratios $a = X/Y$ and $b = Y/Z$ provide convenient parameters for constructing graphical deformation plots (Fig. 4-3; see Hossack, 1968 and Burns and Spry, 1969, for more mathematically sophisticated methods of plotting strain). All ellipsoids can be characterized by the parameter k, such that $a = b^k$, and it is useful to distinguish three particular ellipsoids defined by the lines $k = \infty$ which connect all prolate ellipsoids, $k = 0$ connecting all oblate forms, and $k = 1$ which relate all ellipsoids whose intermediate axis remains equal to the diameter of the original sphere, assuming no change in volume during deformation. All other ellipsoids can be grouped as $\infty > k > 1$, representing the results of a constriction or stretching deformation, or as $0 < k < 1$ comprising ellipsoids due to flattening.

In the descriptions of preferred orientations of minerals, we shall use the noncommittal term *mineral alignment* rather than the more geometrically restricted terms foliation and lineation, for Flinn (1965) has drawn attention to the fact that many common mineral alignments, particularly those involving micas, define orientations which are neither linear nor planar, but some combination of both. This system of *L–S* fabrics Flinn has subdivided into L, $L > S$, $L = S$, $L < S$, and S alignments, depending on the relative strength of the linear and planar components (*ibid.*). We realize, of course, that it is possible that different minerals in the same rock (*e.g.*, mica and hornblende) could combine to produce separate fabrics of different character, or that the same mineral could contribute in distinctly different ways to form at the same time an L and an S fabric. However, since this classification of mineral alignments is so simple and easy to apply in the field, we shall follow it wherever possible. Further, as we shall see later, this classification provides a preliminary means of establishing the geometry of the deformation ellipsoid (Flinn, 1965a, 1967; for a more detailed discussion see Watterson, 1968b); thus, a perfectly linear fabric suggests the operation of a stretching

INTRODUCTORY COMMENTS

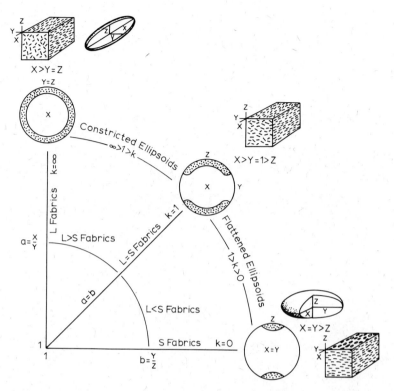

FIGURE 4-3. Deformation plot, showing the relation between the L–S fabric system and deformation ellipsoid. Stereoplots show poles to platy minerals (or other structural elements) shown in block diagrams. In part after Flinn (1965a, 1967).

deformation ($k = \infty$) while a planar (S) fabric suggests a flattening represented by an oblate ($k = 0$) ellipsoid (see Fig. 4-3).

In the following chapters we describe each of the Donegal granites and attempt to give simple models for their emplacement and subsequent history. Wider problems of chemistry, petrography, and structure will be left for summary chapters after we have presented the detailed evidence. We make a start with the largest and what seems to be the oldest of our plutons—that of Thorr.

CHAPTER 5

THE THORR PLUTON: AN EXAMPLE OF ACTIVE STOPING AND CONTAMINATION

CONTENTS

1. The Outcrop 97
2. The Envelope 97
   (A) The contact 97
   (B) External Structures Associated with Emplacement 99
   (C) Contact Metamorphism: a Static Recrystallization 100
3. The Magma 102
   (A) Internal Structures: the Mineral Alignment 102
   (B) Mineralogy and Texture: the Porphyroblastic Growth of Minerals 102
   (C) The Orbicular Diorite of Mullaghderg 106
   (D) Modal Variation: the Application of Trend Surface Analysis 108
4. The Reaction between Magma and Country Rocks: an Assimilation Model 111
   (A) Introduction 111
   (B) Reaction with Pelite 112
   (C) Reaction with Quartzite 115
   (D) Reaction with Calcareous Rock 116
   (E) Reaction with Metadolerites 117
   (F) The Mafic Clots and Schlieren: an Exogenic Origin 117
   (G) Summary and Discussion 119
5. Chemical Composition: the Need for a Contamination Model 120
6. The Distribution of the Inclusions: Ghost Stratigraphy and Structure 124
   (A) Overt Relict Structure 124
   (B) Cryptic Relict Structure 126
   (C) The Origin of Relict Structure 128
7. The Origin of the Thorr Rocks 129

1. The Outcrop

The study of what is apparently the oldest member of the Donegal Granite Complex, the Thorr Pluton, introduces some of the main problems of the origin and emplacement of granitic magma and throws special light on the processes of stoping, assimilation and late-stage recrystallization.

The Pluton consists of coarse-grained granitic rocks showing considerable local and regional variations in composition—from a homogeneous granite in the northeast which is almost free of xenolithic material, to a quartz diorite in the south which is often packed with inclusions, big and small, of the country rocks. There are, however, no internal contacts and the rocks of the Pluton form one continuous body (Fig. 5-1).

Despite the fact that this outcrop is limited by the sea, some idea of the former extent of the Thorr Pluton and of the form of the external contact can be deduced from the concentric nature of the trend of the foliation, especially when the traces of the outer contact on the off-shore islands and skerries are taken into account. The intrusion of later granites does not appear to have materially distorted the original form which, at the present level of erosion, seems to be a boomerang-shaped body consisting of a considerable western pluton with a peculiar prolongation to the northeast. It is the former and most granitic part with which we are immediately concerned; the prolongation of more basic rocks was so disrupted and deformed during the emplacement of the Main Donegal Pluton that it is difficult to decipher its early history.

There is, in addition to the above, just one completely separate mass of the same rock type—a narrow strip of tonalite which runs through the locality of Carbane Hill, near Glenties (Cole, 1902). There is little we can learn from this isolated outcrop, and the greater part of our knowledge of the field relations of the rocks of Thorr type is derived from the examination of the main body by Pitcher (1953*a*, 1953*b*) and Whitten (1957*b*).

2. The Envelope

(A) The Contact

The main body of the Thorr Pluton is emplaced in the metasediments of the Creeslough Succession, and it is the Ards Quartzite which makes up much of the envelope (see the geological map in the folder). Since the outcrop here is flanked by quartzite mountains and also since quartzite still forms a capping (possibly slightly rotated—see Whitten, 1957*a*, p. 29) to the isolated granite hill of Bloody Foreland, it might well be that quartzite originally formed a flattish roof to much of the northern part of the Thorr body. Elsewhere, the Thorr Pluton is markedly transgressive, especially along its southern margins where it is in contact with a variety of formations from the Ards Quartzite to the Upper Falcarragh Pelites.

The actual margin of the Thorr Pluton is easily located. In all northern and western areas, the contact is quite sharp, with occasional apophyses extending for short distances into the country rocks—these may be truncated by the main mass. In the southeast, however, in the Thorr district (Fig. 5-1), the margin is irregular on a small scale and is sometimes transitional within a few inches; here, local segregations of quartzofeldspathic material are found in the adjacent schists (p. 101). Farther south, in the Lettermacaward area, the contact is even more gradational, and earlier workers (Gindy, 1953*b*; Iyengar *et al.*, 1954) emphasized the transitional passage from veined and migmatized sediments to relatively homogeneous tonalite with numerous enclaves. Nevertheless, even in these areas, there is always a mappable boundary between highly contaminated granitic rocks on the one hand and veined and mobilized country rocks on the other.

Unfortunately, there is insufficient relief to determine directly the overall attitude of these main contacts. There are some outcrops which show shallow outward dipping contacts along the eastern margin of the Thorr Pluton, as around Meencorwick, suggesting local roof relationships (Pitcher,

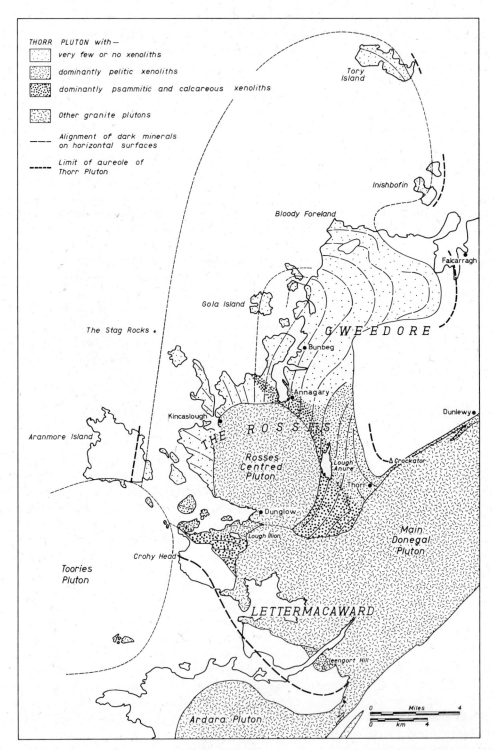

FIGURE 5-1. Outline map of the main part of the Thorr Pluton showing both the known and inferred contacts, the limits of the aureole and the regional trend of the foliation.

1953b, p. 428). This is not the general case, however, because everywhere else there are steep contacts, albeit variable and irregular, even in the south on Cleengort Hill where the relationship was formerly thought to be that of a gently dipping floor (Iyengar et al., loc. cit.).

Thus, a reconsideration of the contact relations of the Thorr Pluton seems to require the rejection of the earlier hypothesis of a great sheetlike mass emplaced between two great quartzites (Iyengar et al., loc. cit.; Pitcher and Read, 1963), and its replacement by a simpler model of a relatively steep-sided and cross-cutting body. It does seem possible, however, that the present level of erosion is near to an original roof, which fits well with the presence of innumerable pieces of the country rocks lying within the main mass of the Pluton.

(B) External Structures Associated with Emplacement

The Thorr rocks seem to be the oldest of the granites of Donegal yet the country rock xenoliths contain traces of all the primary deformations and even the open folding of D_4 (Fig. 3-2), though the evidence in respect to even later events is rather easily obliterated by a thoroughgoing recrystallization. Further, there are inclusions of foliated lamprophyres and of angular pieces (joint-blocks) of felsites. Clearly, the emplacement of the Thorr Pluton must have been very late in the structural history, if not strictly post-tectonic, a conclusion confirmed by the superimposed nature of the aureole. However, there do seem to be fold structures around the periphery of the Pluton which may well be connected with emplacement.

There is, for example, the evidence that along the northeastern margin, particularly at Keeldrum and Tievealehid, the Ards Quartzite is bent down sharply along the contact, suggesting a sagging towards the granitic "cauldron" (Whitten, 1957a, p. 28–30). Further, within this same part of the envelope, there is a gentle upright N–S trending fold—the Cashel Syncline (see Fig. 3-2)—a simple structure which deforms the earlier regional cleavages and is genetically associated with a steep crenulation cleavage (Rickard, 1962). This cleavage increases in intensity towards the Thorr margin, so that eventually (on Inishboffin) it parallels the contact and almost completely obliterates the earlier regional structures. Rickard concluded (ibid., p. 231) that the Cashel Syncline formed as the rising granite lifted the envelope, and the general radial disposition of dips around the great bulge in the granodiorite outcrop west of Gortahork certainly supports this contention.

Father south in the Thorr district, Rickard (1963) has also shown that part of a preexisting curving synclinal fold was bent from a WNW–ESE to a N–S trend at a relatively late stage in its structural history. These movements, during which the limbs of the syncline were steepened and thrust against an eastern block of gently dipping quartzite, were again attributed by Rickard, albeit tentatively, to the forceful intrusion of the Thorr Unit. In the west, on Aranmore, there are open N–S folds in the Ards Quartzite adjacent to the main contact, while in the Crohy Hills region, northwest of Lough Salt, the Ards Quartzite is again locally bent down against the tonalites of this region. Then, in the Maas and Lettermacaward districts, we consider that a localized, steep crenulation cleavage, lying roughly parallel to the boundary of the Thorr Pluton, is related to the emplacement history in the same way as in the more northerly exposures mentioned above.

In addition there is, in many places, evidence of a more local deformation recorded by contact zones of mobilized hornfelses (Fig. 5-2). Within these rocks, much of the cohesion between the bedding planes is lost; the more pelitic element is homogenized and forms a matrix enclosing the pieces of the dismembered competent layers. The implication here is that there has been partial melting of these contact rocks. There is a distinct subhorizontal alignment of these pieces, the quartz fibrolite segregations, and sometimes even the sillimanite prisms, together producing a streakiness on horizontal surfaces which parallels the contact. The importance of this mobilized contact-zone increases southward; local and subdued in the north, it becomes many hundreds of meters broad in Lettermacaward and Maas. This same kind of mobilization is often seen in the xenoliths, where there is

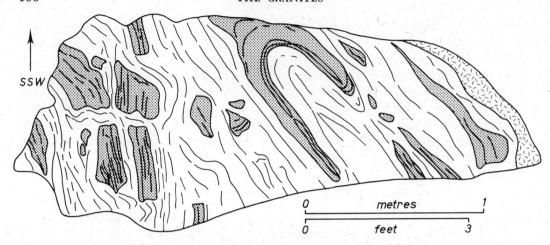

FIGURE 5-2. Mobilized hornfelses at the contact of Thorr Pluton (hatched) showing fragments of banded semipelites (stippled) in a base of flow-foliated, sillimanite-bearing pelitic hornfelses. Outcrop 210 meters west of north end of Lough Keel, near Thorr.

every gradation from newly imposed open folding on random axes to complete dismemberment as described above.

Thus altogether there is considerable reason to suppose the operation of directed stresses in the envelope during emplacement. There is also, as we shall see later, an essential unity between these external structures and the internal mineral alignment, which further strengthens our opinion that the intrusion of the Thorr rocks was not an entirely passive event.

(C) Contact Metamorphism: a Static Recrystallization

Although we propose to deal in more detail with the metamorphic effects of the various granitic bodies in Chapter 14, we need to summarize here the main features of the aureole attached to the Thorr Pluton.

As will be apparent from the geological map, the intervention of quartzite often prevents a full view of progressive changes in the pelitic rocks on any one traverse. Nevertheless, a fairly complete picture can still be assembled, in which there are increasingly reactive contact effects from north to south (Pitcher and Read, 1963).

In the north, near Gortahork, and within 1½ kilometers of the contact, what is a new biotite becomes abundant in a zone of true hornfelses. In a more advanced state of alteration, these rocks are differentially recrystallized in bands in which the segregation of quartz is strongly controlled by the existing planar structures, particularly by the main, S_{2-3} schistosity. Rosettes of andalusite grow on this schistosity and individual porphyroblasts overprint the late crenulation cleavage mentioned above (Fig. 14-4 and Rickard, 1962, 1964); furthermore, the regional chlorite-garnet knots are represented by aggregates of biotite, muscovite, and quartz. A more advanced stage of this static contact metamorphism is represented by the andalusite-bearing cordierite-hornfelses (particularly at Currans Point east of Bloody Foreland), in which sporadic fibrolite begins to appear in the immediate contact zone.

This weak development of fibrolite in the northern outcrops contrasts very strongly with the situation in the eastern (Thorr) and southern (Lettermacaward and Maas) areas, where there are wide zones of very thoroughly recrystallized rocks with an abundance of this mineral (Fig. 5-3). In these situations, it is clear that the general static recrystallization postdates all deformation structures (Pitcher and Read, 1963). An outer zone is characterized by a static growth of andalusite—often mimetically

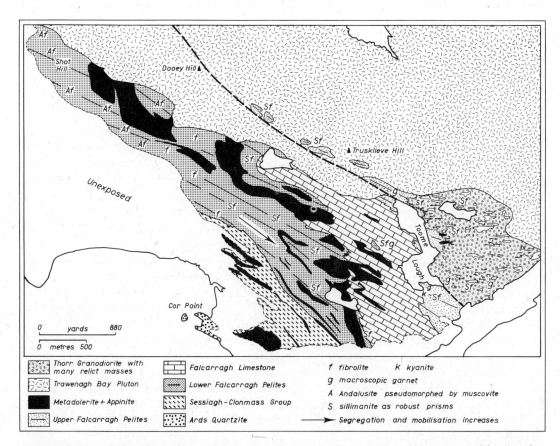

FIGURE 5-3. Geological map of the Lettermacaward area to show the metamorphic zonation of the aureole of the Thorr pluton. Note relicts of metasediments (white) and metabasites (black) in the Thorr granodiorite. Dashed lines in Thorr represent mineral alignment. After Pitcher and Read, 1963, Fig. 5.

oriented along the late crenulations—and a beginning of the segregation of quartz along both the new and old structures. In an inner zone, there was a particularly thoroughgoing recrystallization in which localized metasomatic processes were operative; here, there are segregations or "sweat-outs" of quartz together with concentrations of aluminosilicates, especially robust crystals of sillimanite. In this situation, fibrolite, sillimanite, and tourmaline appear abundantly in close genetic association with a new growth of porphyroblastic muscovite, which wholly replaces the earlier andalusite. From the distribution of these micaceous pseudomorphs, this metasomatic inner zone can be shown to have overlapped considerably upon that of the purely metamorphic andalusite-bearing zone.

Adjacent to the contact in the Lettermacaward area, mobilized hornfelses show an even more advanced stage of metamorphic segregation of minerals, and in these rocks a new oligoclase appears, suggesting that the partial melting was accompanied by the metasomatic addition of material from the intrusive (Gindy 1953*b*, but see p. 114).

Thus, the aureole of the Thorr Pluton records an early and rather subdued deformative episode followed by a static recrystallization in two stages, the first metamorphic and the second metasomatic, and culminating in wholesale mobilization. From north to south, the intensity of metasomatic reaction between the intrusive and its country rock increases markedly in sympathy with the greater degree of mobilization of the hornfelses, and as we shall see later, with a change in the composition of the igneous body as a whole.

3. The Magma

(A) Internal Structures: the Mineral Alignment

Despite the considerable modal variation exhibited by the Thorr rocks, there is an overall structural uniformity brought out by the common alignment of the minerals on horizontal surfaces (Fig. 5-1). For the most part, this is due to a crude parallelism of single crystals and small aggregates of biotite, and also of hornblende where present. In places, there is also an alignment of microcline metacrysts, though it is important to note that there are always some individuals which lie across the general trend.

Most horizontal surfaces show this alignment, though of course it is most apparent where dark minerals are abundant. On vertical surfaces, it is sometimes possible to detect a steep mineral parallelism, indicating that a planar structure is involved, but in most places the horizontal component of this mineral alignment is dominant. Indeed, in the northern parts of the Thorr Pluton, where the alignment is chiefly due to the preferred orientation of large microclines, Whitten (1957b, p. 274) noted that individual feldspars were oriented with their longest axes in a subhorizontal direction. The general finding is, in fact, that the mineral alignment is an $L > S$ fabric, in which a roughly horizontal lineation is linked to a weaker steeply dipping planar structure.

We shall refer later in some detail to the structural relationships of the enclaves within the Thorr Pluton, and we need only point out here that there is usually an obvious parallelism between the longer dimensions of these inclusions and the mineral alignment. This is especially true of the numerous dark schlieren and the small discoidal or ellipsoidal mafic masses which we consider to be of xenolithic origin (p. 117). They have a mica fabric in common with the host, and appear to have been flattened in many places, their longest axes generally lying parallel to the mineral lineation (Whitten, 1957b, pp. 276, 278).

Fig. 5-1 shows that the mineral alignment conforms in general to the margins of the Thorr Pluton, though there are local discordances, as in the Crohy Hills, Lettermacaward (Fig. 5-3) and Gweedore areas, where the trends in the granodiorite seem to abut against, or even cross, the contact: the same kind of discordance occurs in association with the Bloody Foreland quartzite enclave.

The origin of this mineral alignment presents several problems. The general parallelism with the margins of the pluton suggests at first sight that this fabric is the result of flow during intrusion, as in the classical Cloosian interpretation. However, the way in which this alignment crosses the margin of the Pluton in places, passes across pegmatite dykes in others (Fig. 5-4), and on a large scale cuts across the ghost stratigraphy (Fig. 5-1), is extremely difficult to reconcile with flow in the liquid state: it is more likely to be due to deformation in the near solid state (see Chapter 15).

The problem which must finally arise is whether the deformation is the direct result of the forceful emplacement of a granitic body into a relatively static environment, or whether emplacement occurred during the operation of regional stresses. But we have to assemble much more of the relevant data before returning to this fundamental problem.

(B) Mineralogy and Texture: the Porphyroblastic Growth of Minerals

The Thorr Pluton consists of a gradational sequence of granitic rocks, ranging in composition from a homogeneous granite in the north through heterogeneous, highly xenolithic, dark-colored quartz-diorites and tonalites in the south to even more dioritic rocks in the northeasterly prolongation. How-

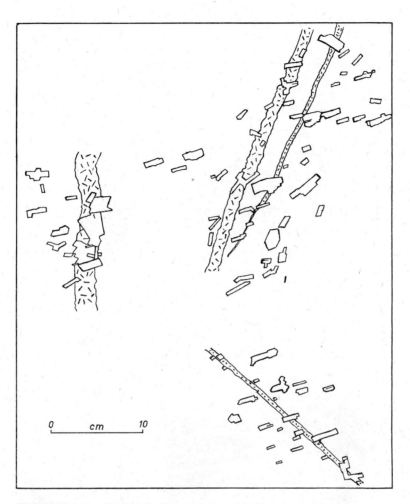

FIGURE 5-4. Monzodiorite (blank) of the Thorr Pluton cut by aplo-pegmatite veins (stippled). Potash feldspar megacrysts, showing a crude alignment, overgrow both vein and host. Outcrop 800 meters east of Meenatotan.

ever, all are coarse-grained rocks, and everywhere too the dark minerals occur both singly and as small aggregates. In addition, the large individual potash feldspars often give the rock a distinctly porphyritic appearance.

Detailed petrographic descriptions have been provided by Pitcher (1953b) and Whitten (1957b), and only the salient features need be repeated here. A green-brown *biotite* occurs together with a bluish-green *hornblende,* though the latter is chiefly restricted to rocks with high color indices. These minerals are frequently intergrown in aggregates which enclose little patches of quartz and small crystals of plagioclase. Generally associated with these mafic clots are the most frequent accessories, ilmenite, sphene, epidote, and apatite, the latter occurring as swarms of acicular crystals in biotite. The hornblende often contains a darker central zone laden with opaque grains and rodlike inclusions, possibly indicating a derivation from pyroxene. Biotite frequently encloses epidote-clinozoisite which may itself contain a core of allanite—a relationship we shall have more to say about in our discussions of the Ardara and Rosses granites.

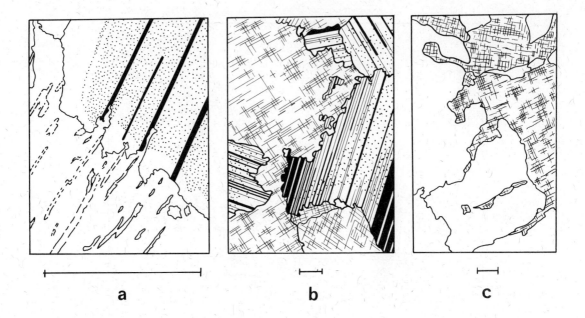
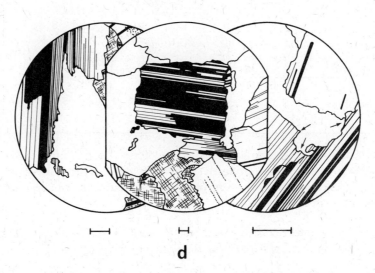

FIGURE 5-5. Microcline-plagioclase relationships in the Thorr Pluton: (a) interdigitation interpreted as microcline microperthite invading weakly zoned plagioclase; (b) microcline interpreted as replacing and enclosing relicts of plagioclase. Microcline-quartz relationships: (c) one interpretation is that quartz is invading microcline; (d) the fretting along contacts against twin lamellae, the cuspate boundaries of plagioclase and the apparent enclosure of islands of plagioclase in quartz are interpreted as due to replacement of plagioclase by quartz. The left and right wings of this drawing are magnifcations of parts of the center. All after Whitten (1957b). Bar scales equal 0.01 inch, 0.25 millimeters.

The *plagioclase*, which occurs both as single crystals and as large composite grains, is for the most part a sodic oligoclase, though there is a slight increase in the anorthite content as the color index increases. Faint regular zoning, both normal and oscillatory, is fairly common, typically with cores of sodic andesine and mantles of albite-oligoclase. Both simple and polysynthetic twinning are common, thin rims of albite and myrmekitic intergrowths fringe plagioclase crystals where they adjoin potash feldspar, and antiperthitic intergrowths are also present. Plagioclase is generally anhedral against the mafic minerals and, in particular, the biotite often appears to be corroded or replaced by plagioclase, leaving apatite crystals and tiny clusters of sphene granules in the feldspar to mark its former presence and outline.

It is worthy of special note that plagioclase, growing porphyroblastically in the xenoliths and the contact hornfelses, shows all of these features—even the zoning (*e.g.*, Gindy, 1953*b*, p. 392).

Potash feldspar, a microperthitic microline, often appears to be euhedral, but in thin section has highly irregular outlines (Fig. 5-5a) filling intersititial areas between relatively euhedral plagioclases and frequently enclosing islands of other minerals (Fig. 5-5b). Plagioclase inclusions, often mutually coaxial, are particularly common and are easily distinguished from the coarser patches of perthite by their lack of coherence with the host microcline, relict zoning, and twinning, and their occasional parallelism with adjacent external plagioclases.

These features of potash feldspar suggest a late and largely replacive origin, as Pitcher (1953*b*) and Whitten (1957*b*) argued in some detail, and this conclusion is reinforced by the many occurrences of individual microclines crossing both dyke walls and contacts with inclusions (Fig. 5-4). The replacive nature of potash feldspars, which is characteristic of a great many—perhaps the majority—of granitic rocks (see discussion, p. 333), also makes it easier to understand their role in contributing to the mineral orientation if, as we have already seen, this is a relatively late structure (see p. 102). The presence of zonal arrangements of inclusions in the microcline crystals, of perthitic intergrowths, and the usual association with myrmekite and albite rims, indicate the complex history of development and exsolution of these potash feldspars, a history which, as we shall see later, is also characteristic of the other Donegal granites. We shall also show that the source of the potassium-rich fluids lay in the intrusive itself, a conclusion in line with the recent work of Mehnert and Willgallis (1961), Boone (1962), and Mehnert (1968).

The *quartz* of the Thorr Pluton shows an even more complex development, and there is evidence of late activity with replacement textures formed particularly at the expense of the feldspars (Whitten, 1957*b*). For example, quartz often veins both plagioclase and microcline, and appears to replace both minerals along twin lamellae and intercrystalline boundaries (Fig. 5-5c, d). Here, however, we have an excellent example of the ambiguous nature of many textures, for the irregular peninsulas of plagioclase surrounded by quartz suggest to some workers replacement of plagioclase by quartz (*e.g.*, Whitten, 1957*b*; Schermerhorn, 1956*a*), and to others quite the reverse (*e.g.*, King, 1947; Koschmann, 1960). Nevertheless, despite these difficulties, we are confident that various features of quartz, such as its frequent unstrained appearance adjacent to bent feldspars, the presence of intimately sutured mutual boundaries, of triple-point junctions, and of large areas of quartz in which individual grains have similar optical orientations, all indicate that it was one of the last minerals to finish its crystallization or recrystallization.

Even from this brief summary of the textural relationships, it is obvious that the Thorr Pluton has a complex history of crystallization and recrystallization. Some of the textures could be interpreted as the result of crystallization from a magma; examples are the zoning in plagioclase, the simple (growth) twinning in both feldspars, the presence of concentric zones of inclusions in microcline, and, in the crudest sense, the order of crystallization itself. On the other hand, many of the features described above argue for a late stage replacement, and indeed these textural considerations are so central to the whole granite debate that we shall discuss them in detail after we have dealt with the other Donegal granites.

Despite our reluctance to draw up a detailed paragenetic sequence of crystallization and recrystallization (see p. 93), we can construct a simplified model of the development of the Thorr fabric. As we shall see later, the mafic clots represent almost the final stages in the digestion of country rock material—a process accompanied by the wholesale recrystallization of the dark minerals. Some biotite and hornblende, however, may well be of primary origin and much of the plagioclase likewise appears to have crystallized early from a liquid. There was continued growth of plagioclase at the expense of the mafics, and it is clear that the bulk of the quartz and potash feldspar grew or recrystallized when the rock was largely solid, during and perhaps even after the formation of the planar fabric. Such late-stage processes have, of course, gone a long way toward obliterating any textural evidence for precipitation from a magma, and it is, therefore, of interest to turn to a possible relict of such early stages as seen in the orbicular diorite of Mullaghderg.

(C) The Orbicular Diorite of Mullaghderg

Forming essentially part of the Thorr Pluton at Mullaghderg, this small body, 4·5 to 6 meters across, has been mentioned frequently in the literature because it was particularly well described very early by Hatch (1888) and later by Cole (1916).

The mass is formed of a cluster of orbicules (Fig. 5-6b) ranging individually up to 15 centimeters in diameter, the whole body being contained in a sacklike, thin sheath of the same nature as the mantle of the orbicules themselves. In addition, there are a few isolated orbicules outside this sheath and occurring in the normal coarse granite of the host. The feldspathic cores (7 centimeters across) of the orbicules are composed of a coarse association of oligoclase, potash feldspar, quartz, and accessory sphene, while the peripheral mantle is made up of single crystals of oligoclase radiating outward from the core to the boundary. Within this mantle, biotite and magnetite occur as concentric shells. Occasionally, the core consists of granitized pelitic material, and it seems likely, as suggested by Cole, that the alteration of xenolithic material plays an essential part in nucleating the growth of feldspar in the familiar radial form. If this is the case, it is most puzzling that the phenomenon is so unique in Donegal, where examples of the reaction between xenoliths and their granitic hosts are commonplace.

As with many other reported occurrences of orbicular rocks, there are here distorted orbs which suggest that there was some kind of mutual interference between them at some stage during their growth, and there are also examples of partial resorption of orbicules (Cole, *loc. cit.*).

The problem of the origin of orbicular rocks has received considerable attention (see Leveson, 1966, for a comprehensive review: also Raguin, 1965, pp. 67–69; Theime, 1965; Simonen, 1966; Kulish and Polin, 1966; Palmer *et al.*, 1967; Van Diver, 1968, 1970). The structure has been recorded from metamorphic and migmatitic, as well as igneous, environments so that, not unnaturally, both metasomatic and magmatic origins have been proposed.

The radial growth of crystals from a liquid is not difficult to explain, for it is a necessary result of the restriction of the area of precipitation of material by mutual interference of crystals growing outward from a nucleating surface (see Bathurst, 1958). It is comparable to the now familiar harrisitic texture of many layered ultrabasic rocks (see Wager and Brown, 1951, 1968) which originate by growth of crystals from a magma floor or wall. Clearly, a xenolith or even an early formed phenocryst may equally provide a surface on which growth may start.

The main problem is to explain the rhythmic banding. A common opinion is that a diffusion mechanism is set up during reaction between xenoliths or early formed crystals and their magmatic host. Within the confines of such a system, there may have been established a Liesegang type of rhythmic precipitation of ferromagnesian material diffusing outwards from the center. Presumably, this process must have preceded crystallization, because the ferromagnesian shells are distorted by contact with adjacent orbicules yet overgrown by the radially disposed crystals of plagioclase. An

FIGURE 5-6. (a) Xenolithic contact faces of the Thorr Pluton, west of Curran Hill, Tubberkeen. (b) Orbicular granite of Mullaghderg beach, east side of the Mullaghderg peninsula.

alternative explanation can be based on the possibility of the rhythmic and alternate precipitation of biotite and plagioclase. It seems that rhythmic precipitation can apparently occur in conditions of intense undercooling of a highly fluid magma; according to Taubeneck (Taubeneck and Poldervaart, 1960; Taubeneck, 1967a), this implies crystallization from a magma that was virtually free of crystals. In either case, the especially closed and protected environment which would be required in order to maintain such localized and delicately balanced systems might well be provided by a closely packed swarm of xenoliths (cf. Cole, loc. cit., p. 145).

We find such magmatic explanations of the origin of the orbicular structure at Mullaghderg convincing, if not complete in all respects. Indeed, the distortion of individual orbicules and the formation of an orbicule-like outer skin to the whole mass seem inexplicable on any other basis than growth in a liquid (or gel) environment. Indeed, a picture is built up of a kind of bag of immiscible globules of viscous magma, which early on, mutually interfered with one another, though finally consolidated by separately undergoing radial crystallization. The rarity of the occurrence is, however, puzzling, though it may be a consequence of the rather delicately balanced conditions required and of the possibility that subsequent destruction may be a common feature, as Leveson (1963, p. 1038) has indeed suggested.

The inference of this discussion is that at least part of the Thorr Pluton was originally in the magmatic state.

(D) Modal Variation: the Application of Trend Surface Analysis

The main mass of the Thorr body is very clearly a single unit, for there are no internal contacts between different types of the host rock. There is, however, as we have already stated, a wide range in modal composition which is well illustrated in Fig. 5-7. This variation follows a fairly systematic geographic pattern in which relatively dark rocks (quartz-poor tonalites) flank the contacts in the south and east, and light rocks (quartz-rich adamellites) occupy what may be a central position around Gola Island. An attempt is made later to provide an explanation for this kind of mineral

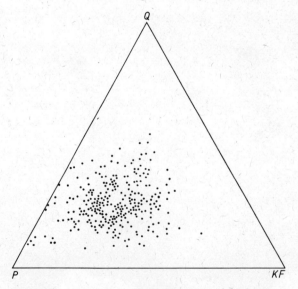

FIGURE 5-7. Q–P–KF diagram showing the wide variation in composition of the Thorr Pluton (after Whitten, 1966b).

distribution—a common feature in granite terrain—but we shall first comment on the nature of the variation itself, particularly as the Thorr Pluton has been the subject of a series of exhausting researches into the determination of regional variations of rock modes by E. H. T. Whitten (1953, 1957b, 1959b, 1960a, b, 1961a, b, c, 1962, 1963, and 1966b).

Aspects of the geographic variability of granitic masses have been studied sporadically since 1924 (see Lee, and review in Whitten, 1963); it is only in recent years that they have played an important role in granitic studies, thanks largely to Whitten's work on the Thorr rocks. In simplest terms, the method involves the statistical fitting of a smooth *trend surface* to the plots of a mapped variable quantity. In effect, surfaces with equations of linear, quadratic, cubic, or even higher degree are fitted in turn, until an equation is found which minimizes the sum of the squares of the departures of the observed points from those represented by the computed surface. A useful application of this method is that it permits the observation of marked deviations between the observed values and the computed trend surface, thus giving an indication of local concentrations over and above the regional trend. Fig. 5-8, taken from Dawson and Whitten (1962), gives a simple example of this method, which is a more rigid variation of the sort of calculations used for many years by geophysicists to separate local from regional anomalies.

In computing trend surfaces, it seems that one is making an assumption that there is indeed a single smooth surface which is reasonably approximated by the observed data (see Link and Koch, 1962, p. 413); that there is, in geological terms, a *single* petrogenetically significant regional parent variation, as distinct from local variations (residuals) produced by subsidiary processes. If this is indeed the case, then how does one decide which trend surface represents the parent trend, since one

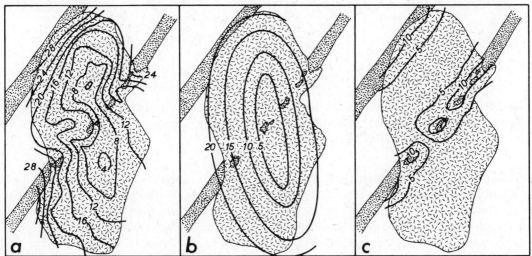

FIGURE 5-8. Synthetic model of a pluton to illustrate one of the ways in which deviations from trend surfaces can have geological significance. The granite is indicated by hatched ornament, and within the envelope rocks two ultramafic bands are stippled. Roof pendants of ultramafic rock are shown within the pluton, and it is to be assumed that, adjacent to these pendants and the ultramafic wall-rocks, the granite is more mafic as a result of contamination. (a) Isopleths for color index based on observed values within the granite. (b) Computed linear plus quadratic partial trend surface based on observed values of color index within the granite. (c) Deviations from the trend components. Isopleths drawn for positive deviations only; these deviations indicate that close to the contacts with the ultramafic rock the granite is more mafic than would be anticipated from the regional trend component. From Dawson and Whitten, 1962, Fig. 6.

can continue to draw trends of successively higher degree until a surface is reached which is near to that produced in the original contour map (*cf.* Chayes, 1970)? Further, the method of computation and the significance placed upon apparent deviations have been severely criticized by Link and Koch (1962), Chayes and Suzuki (1963) and Howarth (1967), mainly on the basis that the underlying assumption is untenable and that the fit is often too poor to have significance, especially when account has to be taken of the possibility of trend surfaces being generated by randomly distributed data.

Whitten has himself considered other possible drawbacks, (1966b, p. 184); the micrometric analyses may not be adequate estimates of the actual mineral content of the rock (see especially Solomon, 1963, p. 381); the computed surfaces refer to the sampled population, and any inferences concerning the actual target population must be based on geological interpretations; further, inhomogeneities in the rocks at local level—a few inches, perhaps—could mean that the computed surfaces merely reflect variations on this small scale rather than significant departures from the regional trend. As a result of all these possible errors, the question arises as to whether the trend surface map has any particular advantage over the standard contour map based on unweighed averages of the observations. In fact, in his earlier work, Whitten (1957b) announced his results in the standard form of isopleth maps of the modal variation, which are reproduced in Fig. 5-9.

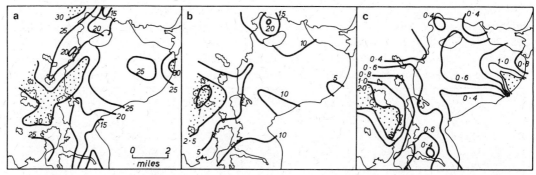

FIGURE 5-9. Modal variation of the Thorr Pluton: simple isopleth maps of the northern outcrop of the Thorr Pluton (after Whitten, 1957b). (a) % modal quartz, (b) modal color index, (c) ratio of microcline to plagioclase.

We note that the values derived for color indices, total quartz, and the ratio of potash feldspar to plagioclase are mutually consistent, although those for quartz and feldspar are less significant statistically than for the color index. These results naturally confirm the changes seen in the field—that there is, toward the Gola center, a steady reduction in the dark minerals, which is almost exactly compensated for by the increase in quartz. Further, the potash feldspar to plagioclase ratio increases in the same direction. In view of this very clear series of modal changes, it is perhaps not surprising that Whitten (1960a) was able to fit reasonably simple trend surfaces to the data, with the results shown in Figs. 5-10, 11. In further work on similar lines, Whitten (especially in 1966b) has provided much additional modal and chemical data and collated this by increasingly sophisticated techniques.

We think that these variation maps are essentially valid, but what is contentious, though of very great interest, is the significance that Whitten places upon the deviations derived from these trend surface maps—this we return to in the discussion of ghost stratigraphy. An equally significant fact of modal variation which is not brought out by Whitten's work is the importance of local changes in the mode, involving a *contact facies,* and this we now proceed to investigate.

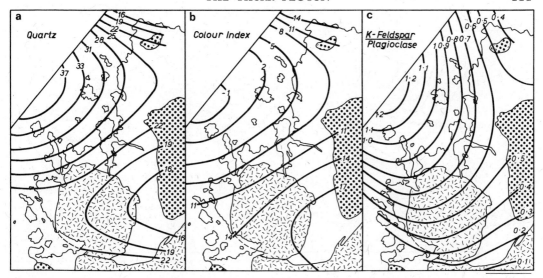

FIGURE 5-10. Modal variation of the Thorr Pluton: degree 3 polynomial trend surfaces for modal attributes (volume percentages). After Whitten (1966b). Quartzite and other country rocks—dotted, later granites hatched.

4. The Reaction between Magma and Country Rocks: an Assimilation Model

(A) Introduction

As in many granitic terrains, the relationships between the Thorr Pluton and its country rocks are infinitely variable in detail, but it is necessary to generalize. In the northern and western (more granitic) portions of the Pluton, there are few inclusions and contact relations are relatively simple. Those parts to the south and east provide a complete contrast, particularly in the Lough Illion and Lettermacaward areas, where there are thousands of xenoliths strewn throughout a very inhomogeneous granodiorite with "gradational" main contacts (Fig. 5-6a). These very mixed zones (marked "M" on Map 1 in folder) are what Read would call "contact-migmatites" (1961), and in them, as is commonly the case in migmatites, psammitic, calcareous, and meta-igneous rocks persist as relicts much more readily than pelitic lithologies.

In main contact situations involving large xenoliths, large groups of xenoliths or stretches along the margin of the pluton, particularly where pelites are involved, the host Thorr rocks pass by rapid but imperceptible gradation into a highly variable, though distinctive, *contact facies* (Pitcher, 1953a; Whitten, 1955); indeed, in such migmatic zones (marked "M" on Map 1 in folder), swarms of enclaves lie in a "sea" of this rock type. The variety of the contact facies depends on the lithology of the adjacent country rock, but as pelite and semipelite predominate in the envelope, it is with their particular contact rocks that we are largely concerned.

It is of great importance to emphasize the difference between the xenoliths in this contact facies and those in the external tonalite–quartz-diorite. In the former, they are usually recognizable, albeit highly altered, country rocks while in the latter, they are mafic aggregates of generally smaller size, which are the relicts, we think, of nearly completely transformed country rock material (p. 117).

In the following discussion, we shall deal first with this most common situation, referring then to other lithologies. Further details may be found in papers by Gindy (1953b), Pitcher (1953a) and Whitten (1955, 1957a, 1957b).

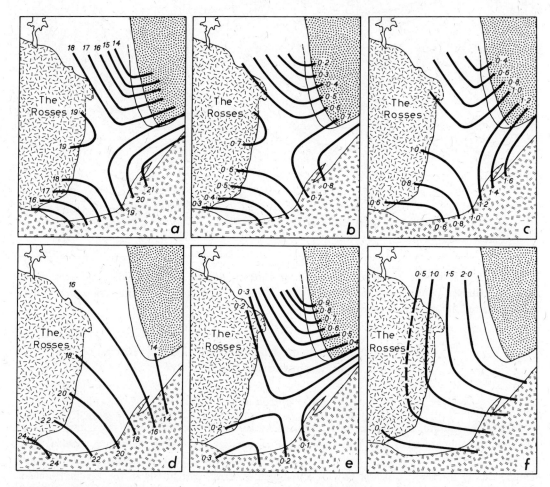

FIGURE 5-11. Modal variation of part of the Thorr Pluton: comparative linear plus quadratic partial-trend surfaces for the Thorr district (After Whitten, 1961b). (a) color index, (b) sphene, (c) epidote, (d) quartz, (e) K-feldspar/plagioclase, (f) hornblende.

(B) Reaction with Pelite

A glance at the map shows that considerable lengths of the contact around Thorr in the east and the Gweebarra estuary in the south are bordered by rocks of dominantly pelitic lithology. In the Thorr district, this contact, like that with the larger xenoliths, is in general quite regular and easily delineated, though in detail there are many irregularities, and granodioritic apophyses may penetrate the pelitic hornfelses in a variety of complex ways (Fig. 5-12). Transitions, even where present, are

FIGURE 5-12. Some contact relationships between the Thorr Pluton and its envelope showing the contrast between the reaction between pelite and more competent rocks: (a, d), Agmatites from metadolerite (stippled); (c) group of semipelitic xenoliths (lined) in the contact facies; (b, e, f) various relationships between pelite (lined) and contact facies, varying from highly irregular contacts to contact permeation, contacts in these cases rather more transitional than depicted. Various outcrops in Thorr District. After Pitcher (1953a).

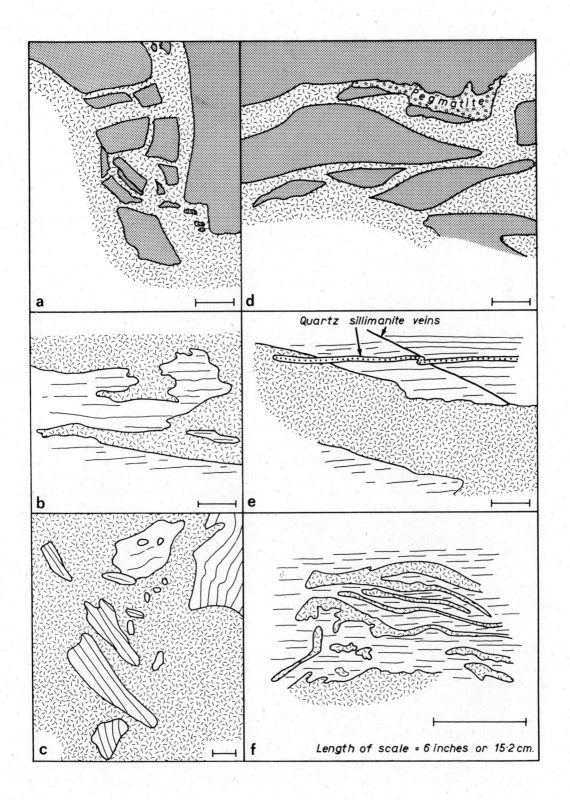

rarely more than a few inches across, and indeed the junction is often sharp enough to be traversed in a single rock slice, where it can be seen to be firmly welded and crossed by feldspar crystals. There are many places where local mobilization has occurred over a few tens of meters, the more competent beds in the metasediments being fractured and rotated in a base of once mobile hornfels.

At the main contact and in the larger xenoliths, the pelites are cordierite-bearing sillimanite-quartz-biotite-hornfelses. Within a few inches of the host, small granitic patches may appear, coalescing in places to yield a heterogeneous migmatitic selvedge. Here, oligoclase-andesine grains increase in the hornfels groundmass, as do irregular replacive microclines; biotite recrystallizes to larger plates, and sillimanite is usually replaced by muscovite. Despite these local reactions, changes in bulk composition are undetectable (Mercy, 1960a, p. 118).

Farther south, around the Gweebarra estuary, the impression is gained, as we have previously noted, of a rather greater degree of reaction at contacts (Gindy 1953b). Thus, the Thorr Pluton here is a particularly heterogeneous streaky contact facies abounding in enclaves (even isolated prisms of sillimanite) and, further, the zone of mobilized contact hornfelses, referred to earlier, is especially wide (up to $\frac{1}{2}$ kilometer). Indeed, anyone visiting these areas for the first time would immediately be struck by the highly xenolithic nature of the Thorr Pluton and by the mobilized and veined character of the hornfelses, and it would not be surprising if the visitor regarded this as an excellent example of granitization, as indeed Gindy (*ibid*) concluded. However, on closer examination, the situation is much less straightforward.

Gindy described the appearance of veins and segregations of quartzofeldspathic material, often containing fibrolite, in pelites not far away from the contact, as the "Zone of Veins." He viewed these segregations as the first liquid manifestation of the migmatization process, and considered that their frequent deformation indicated an overlap between granitization and regional orogenesis (*cf.* Gindy, 1958). In a sense, we agree with the first of these contentions, but otherwise we consider that the deformation referred to was part of the general mobilization of the contact hornfelses during the emplacement of the Thorr Unit.

The salient mineralogical feature of the more thoroughgoing contact effects here is the growth of oligoclase porphyroblasts, sometimes showing oscillatory zoning (Gindy, 1953a); again, these increase in size and coalesce to form segregations near the tonalitic host. Sillimanite, however, decreases near the contact, in sympathy with the appearance of interstitial microcline.

Gindy viewed these changes as part of a granitization involving the exchange of alkalis, but Mercy's chemical analyses (1960a) of pelites at the main contacts provide no evidence of metasomatism *on a significant scale*. It is true that, on Cleengort Hill, analyses of the Upper Falcarragh (Cleengort) Pelites show a progressive increase in Si^{4+} across their strike towards the Thorr Unit and a corresponding decrease in Ti^{4+}, Al^{3+}, Fe^{2+}, Mg^{2+}, and K^{1+}, with other major elements remaining constant. However, we prefer, with Mercy (*loc. cit.*, p. 119), to view these changes as original sedimentary variations, such as those demonstrated by Pitcher and Sinha (1958) in the same formation, where it occurs around the Ardara Pluton (see Chapter 8).

These chemical data and the fact that one can always determine a contact between the heterogeneous contact facies of the Thorr Pluton and the zone of mobilized metasediments within a distance of about a meter—and very often over much shorter distances—clearly indicate that we are not dealing with an example of wholesale granitization. Nevertheless, we must emphasize that although many pelitic inclusions have sharp contacts, others have been affected by metasomatism to a much greater degree than the pelites at the main contact (see Pitcher, 1953a, Table 1, analyses 2–4; Mercy, 1960a, Table 6, analyses E45, E56, E60). These reactions, of which a soda feldspathization is the most common, are important enough to change the composition of some xenoliths significantly nearer to that of the host, leading finally to the production of the indistinct dark patches and streaks (p. 118).

In the host rock immediately adjacent to contacts with pelite, there is often a narrow zone of quartz-enriched granitic rock characteristically lacking in large feldspars; this obviously hybrid rock grades rapidly outward into the contact facies proper, usually in the form of a halo of biotite granite, itself passing, by almost insensible gradation, outward into the normal quartz-diorite or tonalite. In contrast to the latter, this granite of the contact facies contains a reddish-brown biotite exactly similar to that in the adjacent hornfelses; further, hornblende and sphene virtually disappear, while muscovite appears and quartz increases, the latter in the form of coarse-grained patches speared by needles of fibrolite. Some of these mineralogical features strongly suggest an origin of at least some of the components by contamination and, appropriately, every outcrop shows an abundance of small patches of xenolithic origin in all stages of disintegration, recrystallization, and incorporation.

(C) Reaction with Quartzite

In all localities where quartzite forms the country rocks, there is an appreciable recrystallization for some hundreds of feet from the contact. Otherwise, the contact phenomena are very variable in nature and in scale, ranging from the situation where coarse-grained rocks of the Thorr Pluton are in sharp contact with little altered quartzite, to examples of the considerable addition of feldspar, to the quartzite across a contact with intermediate contaminated phases.

Such additions of material are not easy to prove objectively because the original quartzite contains potash feldspar in varying proportions up to about 15%, and increases at the contact may not be attributable to metasomatism. There are some examples (Gindy, 1953a; Whitten, 1953, 1957a), however, of rather intense concentrations of potash feldspar in the immediate contact zones; these sometimes extend outward from the margin of the Thorr Pluton for hundreds of feet, but are usually far less extensive. This probably represents true feldspathization because the concentrations are particularly marked where the quartzite is cut by networks of microgranite veins, and also because a new oligoclase appears, replacing both the potash feldspar and the quartz of the groundmass. Indeed, these few cases provide examples of the now rather familiar phenomenon of potash metasomatism preceding soda metasomatism.

Across what is usually a sharp contact, both at the main junction and with the xenoliths, the quartzite may be represented over several centimeters by a rock with the appearance of a microgranodiorite, this passing outward into a special type of contact facies.

Thus, Whitten (1957a) has described in some detail one example from the flanks of Bloody Foreland mountain, where the host rock, in sharp contact with the altered quartzite, is a porphyritic microgranite, having a micrographic quartzofeldspathic groundmass in which lie tiny areas of recrystallized quartzite. This contact facies is modally and texturally different from the granodiorite proper into which it grades and, in particular, has a lower content of potash feldspar. The impression is gained that this contact facies is largely restricted to protected reentrants into the country rocks, and, while it could represent the remnant of a marginal chill phase, we suspect that it may be the result of the local remelting of feldspathized quartzite.

Within this particular contact facies against quartzite, there are small mafic aggregates which we consider to be highly digested xenoliths of pelitic, calcareous, and meta-igneous lithologies (see p. 118), implying transport and mixing, so indicating that the rocks of the Thorr Pluton have here flowed past their walls. Together with the angularity of some contacts and the blocky nature of adjacent enclaves, this suggests that the environment in this northern area was one of mobility and stoping, with metasomatic processes playing only a minor part.

Whitten, although in general in accord with this conclusion, lays some stress on the local metasomatic features, pointing out that the compositions of the rock types involved conform to the common pattern as detailed by Reynolds (1946) in her standard work on the geochemical processes leading to granitization. While we agree that there is local evidence of metasomatic "fronts" with culminations in various elements across the contact zones (Fig. 5-13 and Whitten, 1957a, p. 36; cf. Saha 1958 for a similar analysis), we consider that these replacements are mere local, marginal phenomena

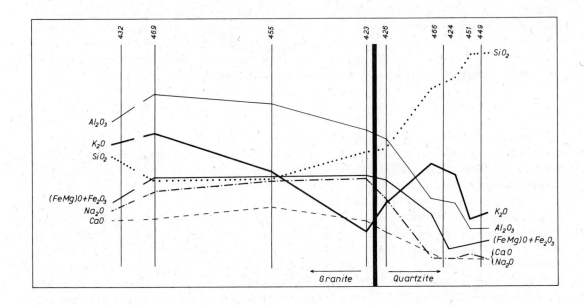

FIGURE 5-13. Relative changes in chemical composition across the granite-quartzite contact: represents a total distance of approximately 3 meters. Calculated from the modes. After Whitten (1953).

which cannot support the contention that extensive metasomatism was involved in the origin of the Thorr rocks.

The situation is not greatly different where smaller enclaves of quartzite are found within the body of the granodioritic rocks. Though these are highly recrystallized, extensive feldspathization is not the rule, and contacts are invariably sharp; we repeat that it is common experience to find that quartzite is far more resistant to processes of metasomatism and dissolution than the pelitic rocks.

(D) Reaction with Calcareous Rock

Other lithologies contribute to the production of the mixed rocks of the Thorr Pluton, and although calcareous metasediments are far less commonly represented at the main contacts than pelites and quartzites, they are abundant as enclaves of limestone and banded diopside hornfels in the Thorr district (Pitcher, 1950; Whitten, 1951). We can briefly mention that the limestone xenoliths are always armored by a zoned skarn in which an inner zone of grossular, idocrase, and wollastonite is followed outward by a zone of clinozoisite and diopside, commonly replaced by actinolite and plagioclase. There is then a passage into the normal host rock through a contact facies of coarse hornblende-plagioclase-sphene rock. In discussing such a situation, Gindy (1953a, p. 157) considered that the different zones were never static, but advanced into the limestone until arrested in their present positions; this provides a mechanism by which limestone inclusions dwindle in size.

The depletion of potash feldspar in the contact facies immediately adjacent to the calcareous enclaves is easily understood on the basis that calcium contamination will lead to a preferential precipitation of anorthitic plagioclase. As a consequence, the rest liquid is enriched in the components of potash feldspar and silica, which might well migrate outward from the contact to enrich the surrounding tonalite (for various discussions and examples of similar situations see Muir, 1953; Townend, 1966, p. 174; Barker, 1967, p. 608; Thompson, 1968). Further, the addition of CO_2 to aqueous magma will have the effect of increasing the vapor pressure, and in an isothermal system, we may expect a rising pressure gradient to be set up toward the contact; this would both favor the crystallization of plagioclase over potash feldspar near the contact and provide a reason for the transport of K^{1+} away from the contact area.

We have not ourselves noticed such zones of enrichment in potash feldspar, but it must be admitted that the necessary detailed modal work has not been done. However, as we shall see later, Whitten has made a most interesting correlation on a large scale between a potash-rich facies of the Thorr Pluton and the former presence of calcareous enclaves.

(E) Reaction with Metadolerites

Metadolerites are, in general, little modified at their contacts with the Thorr Pluton, which is, however, commonly enriched in hornblende and sphene. Hall's detailed study (1959b) of one contact shows that the metadolerite is converted at its edge to a narrow (5 centimeters) zone of biotite-rich rock, while the normal host granodiorite becomes finer grained and passes inward to an equally narrow zone in which microcline is very subdued. These changes, which are nicely matched by Hall's chemical analyses, are clearly in accordance with Bowen's reaction principle; in order that equilibrium be established, hornblende in the metadolerite has been converted to biotite—involving an influx of K^{1+} and expulsion of Ca^{2+} and Na^{1+} into the adjacent granodiorite. However, we must point out that this case of Hall's is not typical of the general relationship; biotite skarns are not common.

While there are small xenoliths of all the country rock types mentioned above, there are many more dark mafic clots and patches throughout the Thorr Pluton, the origin of which is not immediately discernible and which we describe below.

(F) The Mafic Clots and Schlieren: an Exogenic Origin

In the Thorr Pluton, the vast majority of these fine-grained mafic patches are discoidal or ellipsoidal, and generally range in size from a maximum length of 10–25 centimeters and width of 2–7 centimeters down to the tiny aggregates of biotite and hornblende which are so characteristic of the host rock. In addition, elongated streaks (schlieren) rich in biotite and hornblende are frequent. The number of such patches and schlieren is directly proportional to the color index of the host, greatly increasing where swarms of xenoliths occur, and the deduction is obvious. Such clots are essentially composed of fine-grained aggregates of biotite and hornblende with both feldspars and quartz, though the latter is often subordinate or even absent. The mode is very variable, but the ferromagnesian minerals are commonly in higher proportion than the host, and apatite is characteristically abundant. Some patches have so similar a composition and texture to pelitic hornfelses, even to the occurrence of sillimanite prisms or their muscovitic pseudomorphs, that their origin is hardly in doubt; that this is so is confirmed by the field observation that there is *every gradation between them and hornfels fragments* recognizable from the retention of original structures. Other patches are more dioritic and hornblende-rich, and could have been derived from metadolerite; indeed, the hornblendes often contain clusters of magnetite grains, just as they do in the metadolerites—a consequence, we think, of the replacement of pyroxene by amphibole. Other patches again resemble calc-silicate derivatives, and in all, it is reasonable to claim that these mafic inclusions are endmembers in a sequence of the progressive modification of sundry country rocks, with mineral assemblages reaching equilibrium with the host, in conformity with the ideas very well established by Bowen (1922), and Nockolds (1933). There is, of course, always the complication that the reciprocal exchange of material between the xenolith and the host may produce a relative basification of the former, as has been detailed by Reynolds (1946), and this might make it difficult to ascertain the parentage of any particular enclave. Nevertheless, we repeat our conclusion that all were derived from country rocks and that all lithologies contributed.

There has certainly been reaction between these mafic inclusions and the enclosing host rocks, for all stages in their "digestion" are displayed (Fig. 5-14). Many develop new feldspar, both in the groundmass and as porphyroblasts: in the latter case, the mineral is usually microcline. Other enclaves are so

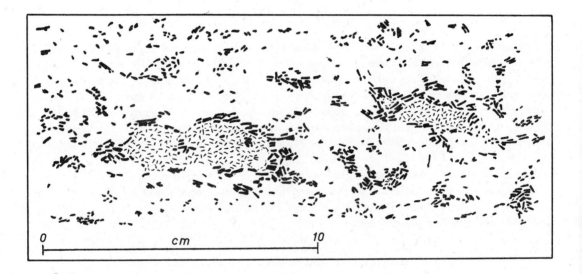

FIGURE 5-14. A surface of the quartz monzodiorite showing the distribution of the dark minerals, largely biotite and hornblende. There is a complete gradation between obvious xenoliths and mineral aggregates, and the coarsening of the biotite on the peripheries of the xenolithic patches is well shown. Shore west of Dunglow.

digested as to become merely greyish ghosts. Always, the contacts are transitional over small distances, and large feldspar plates may penetrate the mutual boundaries, leaving a coarsened biotitic mosaic fingering into the coarse host rock. These "fingers" appear to become separated from the main enclave and rapidly increase in grain size in their new environment. All that is left as evidence of their origin —particularly in the more dioritic rocks—is a patchy aggregation of the biotite (and hornblende), often enclosing small grains of quartz and plagioclase. Thus, these clots (at Thorr) are almost certainly residual fragments of disintegrated inclusions which are themselves of country rock origin, and this is a common finding in such situations (*cf.* Reesor, 1958, p. 44–47).

Dark inclusions of the kind described above are ubiquitous in granodiorites and diorites the world over, and their nomenclature and origin has often been debated (see for example, Grout, 1937; Didier, 1964; Piwinskii, 1968a). One must always take into account that there are "inclusions and inclusions," that there are several possible origins. But we believe that the Thorr examples are not cognate (endogeneous) xenoliths such as often envisaged, for example, by Pabst (1928), for basic clots with certain Californian diorites, but are true accidental (exogenous) xenoliths. If so, the point then arises that the dark clots and schlieren may not be of local origin, but represent a residuum from a partial melting process by which the magmas have been generated at *depth*. This is indeed possible, but this transitional relationship with xenoliths of recognizably *local* origin, coupled with the observation that dark material is mostly abundant where there is most evidence of stoping of blocks from the roof and walls, convinces us that some part of this material must be of high-level origin.

(G) Summary and Discussion

The nature of the reaction, and hence the form of the contacts, between the Thorr Pluton and its country rocks, whether at the margin of the pluton or in enclaves, depends on their lithology. Transitional contacts are most common with pelitic and semipelitic rocks, rare with psammitic and calcareous rocks, and almost never found with metadolerites. In parallel with this, the latter types are generally more persistent as enclaves than pelites and semipelites, presumably because pelitic rocks contain a relatively high proportion of low melting-temperature constituents, so that xenoliths of such rocks would more easily disintegrate. The monomineralic and basic rock types are left to form the "resisters" of Read (1957).

Although there is much evidence of recrystallization and segregation of minerals as the main contacts with the host Thorr Pluton are approached, chemical analyses show clearly that these features are not due to large-scale metasomatism, and the introduction of material must have been limited to a scale of a few feet. There is no evidence of the massive operation of geochemical fronts such as those postulated by Reynolds (1946; see also Raguin, 1965, pp. 93–98; Mehnert, 1968), nor is the frequent mobilization in contact zones dependent in any way on chemical introductions (other than, perhaps, water) as some writers believe.

Within the body of the Thorr Pluton, xenoliths less than a half a meter in maximum dimension are usually very greatly modified and exhibit every gradation to the mafic patches. It seems that there is a particular limiting volume—presumably different for each lithology—at which the reciprocal exchange of material would be just sufficient to produce homogeneity in the included fragment. Below this limiting volume, reactions between the xenolith and the granitic host would be at a maximum, producing mafic aggregates which eventually dwindle in size as they are progressively disrupted and incorporated into the host rock.

We think it is probable that these enclaves were being modified during the whole of the period from their incorporation in the granitic host to its latest stages of crystallization and even recrystallization. Indeed, if we follow Lazarenkov (1962), we may envisage these reactions continuing between enclaves and residual liquids after the host rock was essentially solid, even to the extent that xenoliths may be preferentially enriched in elements of late concentration, an attractive explanation of the geochemical culmination in phosphorus which is indicated by the frequent concentration of acicular crystals of apatite in the mafic clots. These late changes probably also account for such features as the growth of late microcline crystals across contacts.

Much work remains to be done on the distribution of the contact facies; indeed, it would be rewarding to prepare a detailed map of its occurrence in relation to xenolith density and xenolith lithology from what is already known. However, we consider that the hybrid origin is beyond question, and certain other observations suggest that this contaminated product was capable of differential flow;

the halo of contact rock is locally missing and, in places, the type of contact facies normally associated with pelites can be found in contact with metadolerite or calc-silicate rock.

We shall discuss later the full significance of this facies in the evolution of the Thorr rocks as a whole, but we need first to consider the chemical evidence as the next stage of our enquiry.

5. Chemical Composition: the Need for a Contamination Model

We have seen that the Thorr Pluton shows a considerable modal variation and have examined the contribution of contamination to this: it is of interest to see how the pattern of these changes is reflected by the chemical composition. The chemistry of the Thorr Pluton has been studied in detail by Mercy (1960a), who first utilized the rapid methods of silicate analysis introduced by Shapiro and Brannock (1952, 1956) and Pitcher (1955); not, it should be stressed, without much preliminary testing of their precision and accuracy (Mercy, 1956). In addition, we have the results of some 110 analyses in the Bunbeg and Thorr area published by Whitten (1966b).

From a collation of all this data, the broad features of the chemical variation can be clearly stated. Thus, from south and east to north and west, Si^{4+}, K^{1+}, and O^{2-} increase while all the other major elements decrease. Mercy's Harker-type variation plot (Fig. 5-15)* illustrates, in a general way, the varying rates of change in the amounts of these elements, from the darker tonalitic facies in the Thorr District to the light-colored adamellitic facies of Gola Island: yet another way of representing these variations is depicted in Fig. 5-16. Stated in relative terms, the Thorr Pluton in the south and east is enriched in Mg^{2+} and total iron relative to Na^{1+} plus K^{1+}, and in Ca relative to K^{1+}, while the reverse situation exists in the north and west. These changes obviously reflect the modal variations outlined earlier (Fig. 5-9); as an example, the increase of K^{1+} and the decrease of Ca^{2+} in the northwesterly direction is reflected by the increase in potash feldspar and the albite-anorthite ratio of the plagioclase. A point of particular note is that the ratio of Mg^{2+} to $Fe^{2+} + Fe^{3+}$ is remarkably constant (about 1:1) throughout, except for all groups of samples from the northern part of the pluton, which have a slightly higher ratio.

Even though there are these variations, there is an overall homogeneity in the norms which is worthy of special note. The majority of the Thorr rocks contain between 10% and 20% normative quartz, 70% and 80% normative feldspar and about 10% of mafic minerals (Fig. 5-17).

Considered in a general way, the chemistry of the Thorr Pluton, as illustrated above, is consonant with the evolution of a magma of intermediate composition by a process of fractional crystallization. The general trend of evolution, with Si^{4+}, K^{1+}, and O^{2-} increasing and other elements decreasing, would be considered by many petrologists as a strictly magmatic phenomenon and the result of normal *fractional crystallization,* even though there are sound arguments against such a conclusion, both from mathematical considerations (Chayes, 1962) and on the grounds that the composition of inclusions in the process of being transformed in the solid state also commonly converges with that of the host rock, as Lacroix pointed out long ago (1900; see also Raguin, 1965, pp. 125–7; Mehnert, 1968).

There are a number of other objections to the traditional model of magmatic differentiation which arise from the chemical data. It has been stated, for example (Turner and Verhoogen, 1960, p. 379; Mercy, 1960a, p. 123), that a characteristic of magmatic granites is a relatively low ratio of magnesium to total iron (*ca.* 0.5:1), yet in this present case, this ratio lies near to 1:1 (Fig. 5-16a). But this is so in the whole suite of "Newer" Caledonian granites—the magmatic origin of many of which can hardly be denied (see Mercy, 1963). This hypothesis is also unconvincing because, like many other petrogenetic arguments about granites, it implies that the origin of all plutons for which chemical data is available is known with certainty.

---

* We are, of course, well aware of the difficulties inherent in variation diagrams (Chayes, 1962; Tuominen, 1964), but there is an obvious need to represent the data graphically.

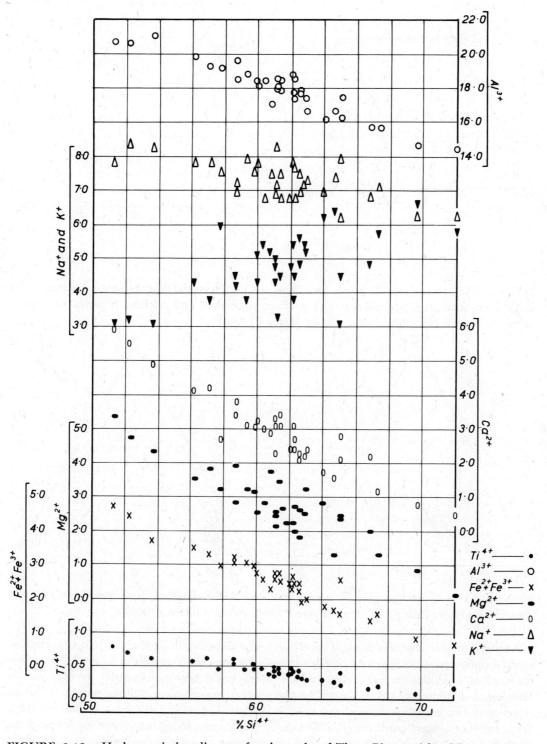

FIGURE 5-15. Harker variation diagram for the rocks of Thorr Pluton. After Mercy (1960a).

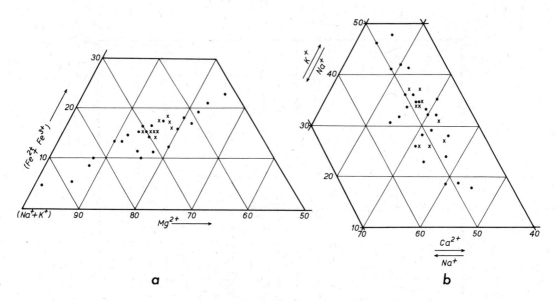

FIGURE 5-16. Variation diagrams for Thorr rocks. Plots showing relations between: (a) magnesium, total iron, and alkalis; (b) sodium, potassium, and calcium. Samples near or at pelitic contacts indicated by crosses. After Mercy (1960a).

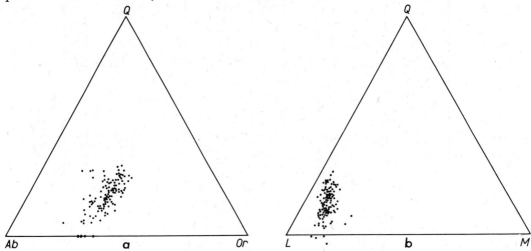

FIGURE 5-17. Q–Ab–Or plot of norms and von Wolff diagram, both based on 114 new chemical analyses of the rocks of the Thorr Pluton. After Whitten (1966b).

A more pertinent objection is based on the lack of enrichment in iron which seems to preclude crystal accumulation as an important process in the formation of the Thorr Rocks. Logically, all models for fractional crystallization require some measure of iron enrichment (see, however, the objections of Hamilton and Anderson, 1967), though homogenization by continuous diffusion might cancel the effects of this on the composition of the minerals. We think, however, that the operation of such a homogenization process, particularly for the cafemic constituents, acting throughout the *whole* of the Thorr Pluton and over its *whole* time span of crystallization, is unlikely (see, however, p. 131 for a different opinion!).

Other facts inconsistent with the straight differentiation hypothesis are the lack of a clearly defined trend in Na^{1+} and the very considerable scatter in the values of K^{1+}, even though these latter values increase statistically as Na^{1+} increases (see Fig. 5-15). It is, of course, well known that potash is frequently the most variable oxide in granitic rocks (*cf.* Mehnert, 1968), a variability which indicates that K^{1+} behaves to some degree independently of the other elements—an independence obviously reflected in the late replacive nature of the potash feldspars.

Reservations of this kind lead us to return to the possibility that *contamination* on a large scale was involved in the evolution of the Thorr rocks not only in the case of the contact facies, but in the more basic varieties of the main rock type. The mixed nature of the former is very clear from the field evidence, yet rather surprisingly, its chemical composition is not significantly different from the median type of the Thorr rocks as a whole (Fig. 5-16; also Pitcher, 1953*a*, Table 1; Mercy, 1960*a*, Table 6). If, however, comparisons are made (albeit with far too little statistical control) between individual examples of contact facies and the *adjacent* normal tonalite-quartz diorite, then there do appear to be significant differences—most important are a relative increase in the former in Si^{4+}, a decrease in Mg^{2+} and Ca^{2+} and changes in the iron oxidation ratio. On the basis that these differences are the result of mixing of two or more components, it is possible to envisage a number of chemically based models for the origin of the special contact facies associated with pelites.

At first sight, this contact facies might well appear to be the result of the addition of pelitic material to a tonalite host, exactly as described in detail by Nockolds (1934, p. 308) in the case of the contaminated tonalites of Loch Awe (see also MacGregor, 1937, p. 472). We find, however, that at Thorr, this simple addition model cannot fully account for the contact rocks because a consideration of the various possible mixtures fails to produce the required compositions.

This is a matter which we are now examining quantitatively, so that we can only make preliminary qualitative proposals here. It seems, in fact, that simple calculations based on mixing various proportions of possible parental components are not going to provide any real support for straightforward addition. Indeed, we would hardly expect it to do so in a complex process which probably involves both contributions to moving magmas and reciprocal exchanges.

Thus, the contact facies might be the result of a kind of dilution of the host tonalite by the continuous addition of a siliceous liquid derived by partial melting from the envelope rocks: similar suggestions have been made by Kennedy (1956, pp. 491–2) and Saha (1958). There is, indeed, a good deal of experimental evidence for the production of partial melts of such a composition from pelitic lithologies (Winkler, 1966; von Platen, 1965) and certainly the most mobile components in such systems would be silica (Tuttle and Bowen, 1958, pp. 89–91; Walton, 1960)—for which water derived during the dehydration of the hornfelses would provide an effective carrier. As we have seen, there is much segregation of quartz in the pelitic hornfelses, yet strangely, the chemical composition even of the envelope rocks very near to the contact shows nothing that we could properly interpret as a bulk loss of silica (Mercy, 1960*a*). However, the quantities involved (a few per cent by weight) are not great, and the abstraction of the required amounts of silica from a broad zone of semipelitic hornfelses might not be easily detected, except by most carefully controlled sampling. We conclude that a relative dilution by silica, coupled with a change in oxidation consequent on crystallization in a hydrous environment, might well go part of the way to explain the changes in the mode and the mineralogy.

There is, however, a third possible model for explaining these relationships. In chemical models of the kind we are discussing, it is nearly always assumed that there is a straightforward series of transitional rock types between host and magma, yet there is a distinct possibility that reactants have moved away from their place of origin, as, for example, by differential flow. Indeed, we have already suggested (p. 120) that this occurred in Thorr District, and it may well be that the haloes of contact facies are not directly related to the tonalitic and quartz dioritic rocks of the Thorr Pluton which now surround them. Relative movement between the tonalitic host, on the one hand, and the country rock material with its attached contact facies on the other could have brought contrasted Thorr rocks into juxtaposition. If so, we are led to enquire what was the nature of the parent magma which originally reacted with the pelites to produce the contact facies. Rather surprisingly, the preliminary calculations show that a simple genetic series can be established between pelitic hornfels, contact facies, and a granitic facies similar to the central (Gola) part of the Thorr Pluton—as if this granite had formerly reacted directly with the roof rocks.

Later, we discuss these three contamination models further in the light of the field evidence and conclude that there is no need to make a single choice—all were operative. Meanwhile, we turn to a consideration of the origin of the tonalitic facies itself, for which we also expressed certain reservations regarding the efficacy of differentiation processes to produce the observed variations.

It is possible that the tonalites are the result of an even more thoroughgoing contamination of the Gola type granite, and, appropriately, the xenoliths in the tonalites show a much greater degree of modification than those in the contact facies (p. 111). In fact, simple calculations on the basis of the admixture of Gola granite with country rock types in naturally occurring proportions can be adjusted so as to account for the differences between the most acid and the most basic of the Thorr rocks. The higher content of Ca^{2+}, Na^{1+}, and Ti^{4+} in the tonalite might well, therefore, be the result of the assimilation of metadolerite and calc-silicate, in addition to pelite. Although any summation of this kind can hardly represent the true situation, it does indicate that a relatively simple chemical model is available for the production of the tonalitic facies itself by contamination of an original *granite*—incidentally offering an explanation for the relatively fixed ratio Mg^{2+}: total iron, for, as Mercy (1960*a*) has shown, this particular ratio is common to both the Thorr rocks and their pelitic envelope.

Endomorphic modifications of granitic rocks enclosing country rock fragments are commonplace, though in virtually all examples, the effect of the contamination is to produce a more basic variety (*e.g.,* Baker, 1935; Exley and Stone, 1964, p. 169; many references in Didier, 1964). We are here, however, beginning to develop a rather complex model in which magmas are contaminated to different degrees according to the length of time of immersion in them of country rock material, and even involving the possibility that contaminants at one level are displaced by even more thoroughly contaminated magmas from a lower level. Indeed, if, as we shall later suggest, the magmas originated by partial remelting, it is likely that some of the basic material that they carry represents but a residuum entrained in the melt. It is, therefore, time to return to the field evidence in order to assess the validity of this kinematic contamination hypothesis and determine the nature of the mixing process involved.

6. The Distribution of the Inclusions: Ghost Stratigraphy and Structure

(A) Overt Relict Structure

We have already seen that in the south and east, the Thorr Pluton is crowded with fragments of the country rocks, the distribution and alignment of which have important bearings on the genesis of the Pluton as a whole. The inclusions range from several centimeters to 1600 meters in length, and are mostly composed of clearly recognizable lithologies which can be identified as belonging to that part of the Creeslough succession lying above the Ards Quartzite (Chapter 1). The shapes of the inclusions are often controlled by the bedding and early jointing, especially in the case of the more competent lithologies: the less competent pelites tend to be irregular because of the mobilization they have undergone.

In both the Thorr and Lough Illion (Dunglow) districts, there is a very clear pattern in the distribution of these inclusions, and individual horizons can often be traced along the strike from enclave to enclave (Pitcher, 1953*a*; Gindy, 1953*b*), as if the original continuity still held. Figs. 5-18, 19 give examples of this relict or ghost stratigraphy on different scales; Fig. 5-19 shows that the trend is generally parallel to that in the country rock envelope.

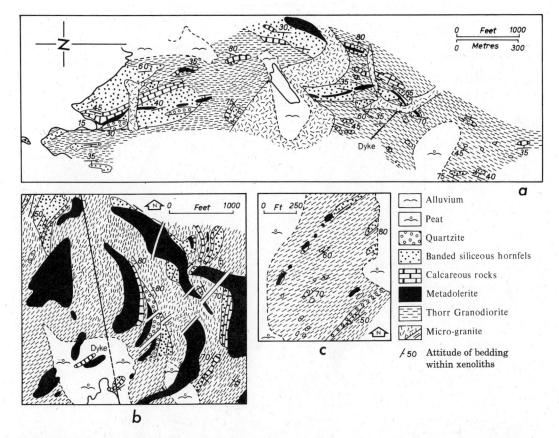

FIGURE 5-18. Some detailed examples of the structural and stratigraphic relationships between enclosures in the Thorr Pluton: (a) Meennamarragh; (b) Lough Anoon; (c) Meenderryherk. After Pitcher (1953a).

In the Thorr district, a broad zone of psammitic inclusions, associated metadolerites and calcareous rocks, is flanked by zones of largely pelitic and semipelitic enclaves (Fig. 5-19), the former representing fragments of the Sessiagh–Clonmass Formation and the latter the lower Falcarragh Pelites. The way in which the pelitic zones converge northward (Fig. 5-1) suggests the closure of a fold, and it is not unreasonable to assume that, in broad outline, the distribution reflects the former presence of a northward-plunging anticline, the axial trace of which paralleled that of folds in the adjacent envelope (see p. 60 and Fig. 3-2). We have here, in fact, an excellent example of ghost structure.

In detail, many inclusions also exhibit a remarkable uniformity of strike and dip, as if reflecting an original structural attitude. This does not necessarily follow, however, because the enclaves are often thin slabs which have resulted from the preferred splitting apart along the bedding planes. The orientation of such slabs would be likely to be controlled by flow and, indeed, it is certainly the case that they are generally aligned parallel to the mineral orientation of the host.

Despite the overall alignment of enclaves, there are many examples of local disorder (Pitcher, 1953a), even mixing, where inclusions have clearly been moved relative to each other, sometimes while preserving the overall ghost stratigraphy. Particularly good examples of disorder are the cluster of randomly oriented quartzite inclusions near to the main contact on Tory Island and at Bloody Foreland (Whitten, 1957a, p. 31; 1957b, p. 277).

Nevertheless, despite these local discordances the framework of the original stratigraphic relation-

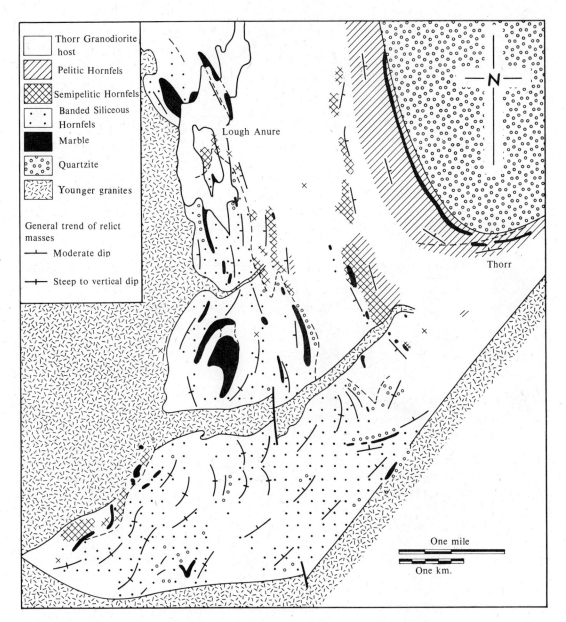

FIGURE 5-19. The distribution of the main rock types remaining as relicts in the Thorr district of the Thorr Pluton (After Pitcher, 1953a).

ships has clearly been preserved over a large part of the Thorr Pluton, even though it may have been modified locally by movements within the host rocks. The implication is that the xenoliths here have not moved far from their place of origin and have not been greatly displaced relative to one another.

(B) Cryptic Relict Structure

We can now return to the modal work of Whitten (p. 110), which has a direct bearing on the problem of ghost stratigraphy. It concerns the deviations, that is the differences, between the observed values and the corresponding points on the computed trend surfaces. Such deviations are important because they contain both the residual terms and also some trend terms of higher degree than the completed trend surface. They are, therefore, not random and may reflect underlying geological factors which account for the actual departure of the observed data from the regional trend surface (Whitten, 1960b, p. 189). On this assumption, Whitten produced a series of deviation maps (Fig. 5-20) for various modal and chemical parameters, and compared them with the ghost stratigraphy in the field (Whitten, 1959b, 1960a and b, 1961a and b, 1966b).

Of the deviation maps for the Thorr district, those for the color index data and for the microcline: plagioclase ratios are the most significant. They both produce patterns which parallel the northward closing fold in the ghost stratigraphy (Fig. 5-20). Thus, over and above the regional pattern of modal variability obvious in the field and in the trend surface maps, the local modal composition of the

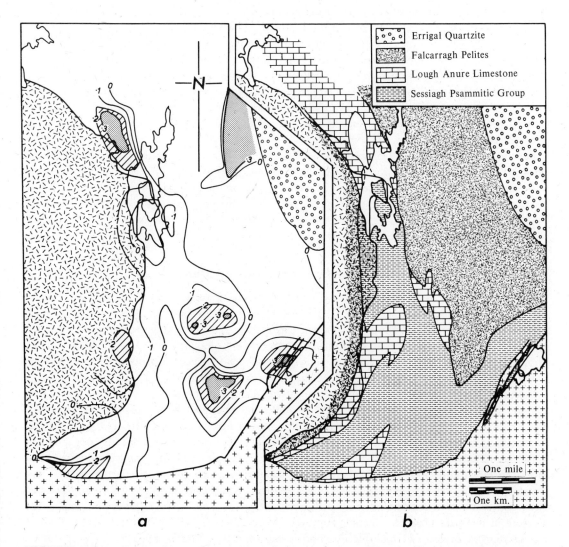

FIGURE 5-20. Comparison of a positive deviation map with the ghost stratigraphy of the Thorr Pluton in the district of Thorr. Positive deviation isopleths for the microcline/plagioclase ratio. Ghost stratigraphy based on Whitten's interpretation of maps by Pitcher (1953a).

granodioritic host seems to reflect the nature of the country rocks now represented by the isolated enclaves. It is not easy to understand the reasons for these distributions, for Whitten intentionally avoided sampling the granodiorite near "obvious xenolithic material" (1960a, p. 190), and would thus have avoided any complications involving the contact facies. The largest area of high deviations in color index seems to correspond broadly to areas of pelitic relicts, while a smaller central anomaly may reflect the presence here of large metadolerite enclaves, the implication being that these lithologies have contributed most of the biotite and hornblende content of the host rock—a reasonable conclusion in view of the previous discussions. The distribution pattern of the microcline:plagioclase ratio appears at first glance more difficult to understand, but there seems to be a general correlation between high ratios and the presence of calc-silicate enclaves, such as would exist if, as argued earlier (p. 117), a local enrichment in potassium can result from the reaction of calcareous rocks with the host. Whatever the reasons for these patterns, and there are many of them, Whitten (1960b, 1966b) seems convinced that the modal composition of the Thorr rocks is connected in some way to the ghost stratigraphy.

Whitten also applied these methods to the northern and western parts of the Thorr Pluton and found what he regards as well defined north-south trending patterns, constituting a "palimpsestic ghost-stratigraphy" fainter than that of the Thorr district (1959b, 1960) and providing the only relicts of ghost stratigraphy in the more homogeneous granodiorite that comprises this part of the Thorr Pluton. This fits in well with the dying out northward of recognizable xenoliths as the Thorr rocks become more granitic, but it is, nevertheless, remarkable that the heterogeneity due to reactions between the country rock material and the host should persist in what, in the field, seem to be homogeneous rocks.

In summary, then, these statistical techniques have demonstrated a direct relationship between the ghost stratigraphy and the composition of the host granite. The great importance of this relationship is two-fold: first, it implies that country rocks were involved in the petrogenetic process at the *present level*; secondly, it suggests a lack of mobility, even ionic migration of material, during and since emplacement. Circulation of magma cannot have occurred to any important extent (*cf.* the opposite views of Taubeneck, 1967a and Shaw, 1965).

(C) The Origin of Relict Structure

The preservation of relict stratigraphy and structure in granitic rocks is not uncommon and may be explained in several ways (Pitcher, 1970) which we shall discuss in some detail again in Chapter 15. First of all, we must discard the most common explanation, that these fragments are *in situ* relicts in a body produced by granitization (Read, 1957, pp. 346–7, Raguin, 1965, pp. 75–77), for the whole of the evidence so far presented has shown that progressive metasomatism and transformation was not an important mechanism in the formation of the Thorr Pluton.

It may be argued that the country rock fragments represent roof pendants, as has often been done in many previous studies. Some of the larger xenoliths, for example, the cap of Crovehy Mountain, might be so interpreted, but in vertical sections, the smaller masses can often be seen to be covered by tonalite, and total enclosure can be directly inferred in the case of a number of larger relicts. In fact, most masses of country rock are considered to be totally enclosed, for otherwise, the roof must have been fantastically irregular with pendants in the form of long slender "stalactites."

Successive intrusion of granitic sheets can also isolate fragments of country rock *in situ* between them, and as we shall see later, some mechanism of this sort has led to a ghost stratigraphy and structure in the main Donegal Granite. However, despite Mercy's claims (1963, p. 210), there is not the slightest evidence of multiple injection within the Thorr Pluton. Even if this were not so, it seems highly improbable that successive intrusions of material could have had the complex shapes necessary to preserve the irregular relict patterns exhibited in the Thorr district.

If, however, the abundance of large inclusions is taken as evidence of a roof situation, it leads us to consider the classic model of stoping. It was thought by early workers that stoped blocks would sometimes be fixed nearly *in situ* by congealing of the magma. We think that, in principle, this simple model will hold for the Thorr Pluton, and that it is possible that a fair proportion of the larger enclaves could maintain their spatial relationships with adjacent inclusions, and perhaps even their general orientation, during the stoping process, assuming that the final distance from the roof was not too great. There would, of course, be complications introduced by the varying degrees of flow during emplacement, and even varying velocities of sinking in the magma due to differences in the sizes of the blocks (Lovering, 1938); we now debate these possibilities.

7. The Origin of the Thorr Rocks

We have reviewed at some length the evidence for the nature and origin of the Thorr Pluton. The relatively late date of the emplacement structures, the local character of the contact metamorphism, and the generally clean-cut nature of the outer contact convince us that we are dealing with a fairly high-level type of emplacement—the mesozonal type of Buddington's classification (1959, p. 705). Nevertheless, the early stages of this emplacement seem to have been associated with a general doming and the formation of accommodation structures in the country rocks—a deformation phase which clearly ended before the general recrystallization of the contact hornfelses. Even so, sufficient stresses were still being applied to the intrusive body to continue its upward movement, cause mobilization of the marginal hornfelses, and produce a mineral fabric during the congealment—*the emplacement of the Thorr Pluton is certainly not to be regarded as a static event.*

However, these accommodation structures seem too trivial to account for the whole space requirement by updoming or forceful opening, and we turn to the other possibilities of granitization in place and piecemeal stoping. The lack of extensive *in situ* metasomatism and the abundance of sharp contacts seem to preclude a wholesale making over of the country rocks in place. On the other hand, it could be argued that transitional zones of metasomatic origin might well be swept away, only to be preserved in protective reentrants; if so, then we can only state that it is remarkable that they are so often removed. We are, of course, aware that sharp contacts are not, by themselves, completely valid criteria of intrusive relationships, but we feel on surer ground when such contacts cut straight across lithological boundaries and even early granitic apophyses. Despite the long history of crystallization, the latter part of it a recrystallization in the near solid state, reasonable evidence for early crystallization from a liquid is provided by the zoned cores of the plagioclase and the formation of orbicular structures.

The actual physical state of this magma during emplacement, the degree of its crystallization and mushiness at any stage, is now impossible to determine in view of the late-stage changes; we can only point to the complete lack of any grain-size changes in any of the mineral components which might be interpreted as evidence of chill phases.

That the Thorr rocks were once mobile (Pitcher, 1953a, p. 434) also seems clear from the disorientation of some great enclaves and the local intermixing of others, perhaps also from the way in which intermixing of others, perhaps also from the way in which biotitic streaks are sometimes deflected around xenoliths without any regard to the internal structure of the latter. The fact that the type of contact facies is occasionally out of sympathy with the adjacent country rock (see p. 120), and more often with the surrounding host rock, can be explained by invoking transference of reaction products by local flow movements. But even though mobility of the Thorr rocks seems reasonably established, this in itself is no evidence of the amount of flow; and indeed, the presence of unmoved or only slightly moved isolated country rock masses is only explicable if flow was limited, at least in the late stages of emplacement. This is the essential enigma.

This leads naturally to the discussion of the real significance of the ghost stratigraphy. Though probably somewhat less perfect than originally claimed by Pitcher, this ghost stratigraphy is certainly a feature of the southern sector of the Thorr Pluton, and, if Whitten's contentions are correct—on the possibility of tracing relict stratigraphy on a modal basis—then a considerable part of the outcrop of the Thorr Pluton shows relict features (see Fig. 5-19). Thus, the material between the enclaves must either have been converted to granitic rocks or bodily removed.

We have given reasons above for discarding granitization *in situ,* so that we are forced to appeal to the actual physical removal of material by stoping. Of course, a great deal of this material might be rapidly and locally incorporated into the host, especially if it was disrupted into small fragments, as is indeed shown by the clusters of small, highly altered xenoliths around the larger masses. We are perhaps making here a rather fine distinction between granitization in the strict sense and rapid incorporation of material, but the important point is that, in the latter case, an intrusive phase is implied. On a more general point, we consider that the fact that the Thorr Pluton is emplaced in a low energy environment is not in keeping with a hypothesis of wholesale granitization.

There is, as we have shown, very good evidence of contamination, and a model in which increasing basicity is directly related to the degree of assimilation of xenolithic material fits well with the observed regional modal changes. It is, therefore, likely that the central or Gola-type granite represents the composition of the parent magma, and that the wholesale assimilation of parts of the Creeslough succession by this magma led to the production of rocks of tonalitic and dioritic aspect near the contacts. Even so, we have some reservations: the rocks of the prolongation are relatively even more basic than at Thorr (Mercy, 1960a), and we could not claim to have established that this basicity is wholly due to contamination. If we are to develop the concept of granitic parent magmas, it must not be as an exclusive model.

Concerning the contact facies, there is probably no need to make a single choice between the three models we have discussed for its production—all were probably potent in producing this variant. Thus, the demonstrable break-up and digestion of pelitic material shows clearly that whole pelite fragments were added to the magma; the simultaneous addition of siliceous partial melt is also likely, both from simple chemical considerations and from the fact that a proportion of the xenoliths are basified; further, it would seem inevitable that different zones or levels of thoroughness of reaction would be set up because, near the roof, pelite would be the first to be assimilated by an invading magma, leaving the final digestion of the more resistant and generally more basic material to be completed in those parts where the magma had retained heat longest. Thus, we can imagine such resisters sinking to deeper levels, and there being assimilated to produce the dioritic facies. Relative movement of these differentially contaminated magmas would result in their being brought into juxtaposition at the same level. Of course, all these processes were simultaneous and interrelated, for this is an essentially kinematic model.

Whatever the details of such processes of contamination, they cannot explain the whole of the space problem so that any complete hypothesis must allow the sinking of blocks to depth, that is, the transference of mass downward. Nevertheless, assimilation must have led to the local quenching of the host magma, so that we might explain the xenolithic portion of the pluton as the relatively quickly cooled marginal zone in which the whole complex process of stoping and assimilation was frozen during its operation, so preserving the evidence. In the roof zone, ghost stratigraphy would simply result from the stoped blocks and associated hybrid rocks being held near to the site of origin, the increasing viscosity of the host effectively preventing removal and dissipation. A stage must have been reached in the late stages of congealment when there was a near balance between the slow sinking of blocks (together with their haloes of contact facies) and the upward movement of the magma—only thus could hybrid rocks of contrasted compositions be brought together and yet a measure of relict stratigraphy be retained.

It can easily be envisaged that for some considerable time there would be exchange of materials between the solidified periphery and the liquid interior—the latent heat of crystallization would provide energy enough. It seems likely that interstitial liquid would still exist in the peripheral rocks facilitating the movements of ions, and possibly leading to a degree of homogenization in the composition of the minerals (compare p. 122). During the long cooling history, the remaining magma would become the source of the late potash felspar, hence the greater lack of correlation between K^{1+} and the chemical variation in the other elements.

Many petrologists dispute the efficacy of contamination to provide the dark rocks which so often form the peripheral zones of plutons, mainly, it seems, on the basis that the energy requirements are thought to be quite inadequate. As a result, the concept of peripheral crystallization is generally preferred as an explanation of concentrically zoned plutons (*e.g.,* Larsen and Poldervaart, 1961; Taubeneck, 1967*a*). We can hardly deny the theoretical possibility that crystallization differentiation of this kind may have contributed to the production of the tonalites of Thorr—even despite the difficulties of long distance diffusion. Nevertheless, the occurrence of a more granitic contact facies, out of sympathy with the adjacent tonalite or quartz-diorite, seems quite opposed to such a hypothesis, for the acidification of the marginal rocks could hardly follow *after* the peripheral crystallization. Further, the intrusive fails to develop a basic zone along a substantial part of the outer contact—especially when quartzite forms the dominant rock of the envelope. Again, where basic rocks are present, xenoliths and reaction products are in abundance. Thus, we restate our opinion that country rock material has been added in substantial amounts to the magma and that contamination has led to the marginal basification at Thorr. Further, if we regard the present outcrop of the Thorr rocks as forming a roof to a sizeable pluton we are not then short of sufficient energy for local reactive assimilation in the roof zone.

The model we are proposing explains a considerable body of the facts available at the present level of erosion, though it must be admitted that stoping alone does not provide a wholly satisfying explanation of the space problem posed by the great area of seemingly homogeneous granite. Indeed, it is difficult to envisage blocks sinking and yet leaving no evidence of their dissolution at depth, especially when such good examples of high-level contamination are preserved in the roof of the pluton.

The most serious reservation lies in the proposition that a ghost stratigraphy, albeit cryptic, exists throughout a considerable part of the pluton. To explain this in terms of specific contamination by roof material sinking to intermediate depths may seem to many an unwarranted extension of the proposed model, requiring, as it must, little or no flowage in the magma. But for good reasons, we have rejected the hypothesis of granitization, whether by melting or metasomatism, for the formation of the Thorr rocks at the present level, and we are left with the alternative either to reject the evidence for a cryptic ghost stratigraphy or to accept the proposed model. We also repeat the reservation concerning the exclusive character of this model: some part of the basicity of the Thorr rocks may have been inherited from the original process of magma generation which, we suggest, involved differential and partial remelting (see Chapter 16).

Finally, we emphasize again the kinematic nature of the emplacement of the Thorr Pluton, which involved uplift and compression of the envelope, active stoping combined with reaction with the country rocks, and internal deformation of the congealing rocks of the pluton. The final problem, which we must return to later, is whether to regard these contemporaneous movements as the effect or the cause of emplacement, whether or not the Thorr Pluton is synkinematic in the strictly regional sense.

The above model is obviously incomplete, and we have only explained in limited measure the space problem, the energy problem, and probably not at all the wider question of the ultimate origin of the Thorr rocks at depth. We prefer to leave these and other matters until the final discussion of the origin of the Donegal granites in general.

CHAPTER 6

THE FANAD PLUTON: AN UNCOMPLICATED EXAMPLE OF STOPING

CONTENTS

1. Introduction 132
2. The Outcrop 134
3. The Contact with the Envelope: an Example of Varying Contact Effect 134
4. Composition and Texture: Contamination of a Magma 136
5. The Fabric 139
6. The Distribution of the Larger Xenoliths: Ghost Stratigraphy Again 139
7. The Mode of Emplacement: Piecemeal Stoping 139
8. An Appendix: a Related Dyke Swarm 141

1. Introduction

A fragment of what is probably a considerable body of granitic rocks outcrops on the three adjacent peninsulas of Rosguill, Fanad, and Inishowen in northern Donegal. It forms part of a distinctive unit which is spatially quite separate from the Donegal Complex proper.

The whole outcrop (see Map 1 in folder) has not been reexamined in detail since the original mapping by the officers of the Geological Survey of Ireland (1891). However, a sufficient part of it (Fig. 6-1) has been studied during the present research to enable this brief account to be written; to do so, we draw heavily on the work of Knill and Knill (1961) and the unpublished work of M. Fernandes-Davila, P. J. Hill, and W. M. Edmunds.

In a number of respects, the Fanad Pluton bears a resemblance to that of Thorr, and the petrogenetic processes involved are sufficiently similar for a relatively brief account of the former to suffice. Nevertheless, the Fanad Pluton is different from the Thorr Pluton, particularly in being more dioritic, locally even appinitic (Chapter 7), in composition, also in having a simpler crystallization history, and in providing a less complicated example of emplacement by a stoping mechanism.

The geographic isolation from other granitic masses naturally results in some difficulty in deter-

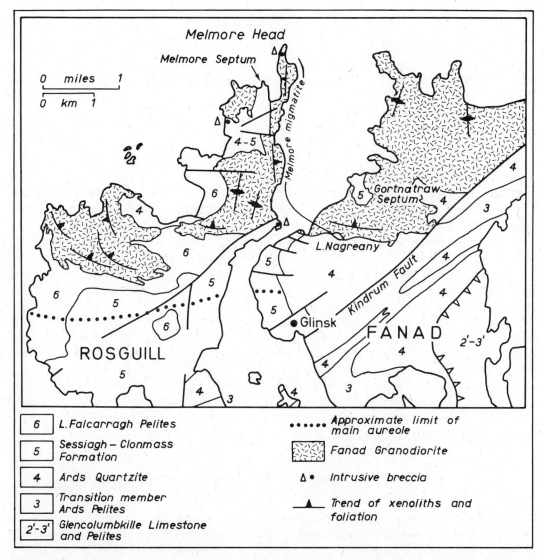

FIGURE 6-1. Outline map of Rosguill and Fanad to show position of the three main outcrops of the Fanad Pluton and certain other features. The limit of the aureole shown here marks the appearance of andalusite, fibrolite and cordierite. Modified from Knill and Knill (1961).

mining the time of emplacement relative to the latter. It is clear, however, that the Fanad Pluton reached its present level very late in the regional structural history, because xenoliths of country rock within it contain all the early deformation structures (D_1 to D_4). Further, it postdates certain of the late-tectonic lamprophyres and felsites and even such relatively late minor intrusions as members of the appinite suite, including an intrusive breccia. On the other hand, representatives of the swarm of microgranite dykes which cut the Fanad body are never seen to cut the Thorr Pluton, suggesting the earlier age of the Fanad intrusion, and, further, these dykes are deformed in the envelope of the Main Donegal Granite, indicating that the Fanad Pluton certainly predates the later stages of the structural evolution of the latter. Thus, it seems probable that, while the Fanad Pluton is of the same general age as the members of the main complex, it may represent an early phase, possibly closely associated in time with the appinites.

2. The Outcrop

The several inlets of the sea and the intervention of the Kindrum and Leannan faults break up the outcrop in such a way as to obscure the relationship of the Fanad rocks to their envelope. As a result, we can learn very little of the form of the Pluton outside the areas of Rosguill and the western part of Fanad.

Here, the Pluton is divided up by irregular septa of metasediments which seem likely to be true roof pendants, separating the outcrop into areas of slightly different lithology and contrasted foliation trends. Thus, from west to east, the Melmore septum separates an arcuately foliated quartz monzodiorite (in west Rosguill) from the almost structureless granodiorite making up much of the northwestern part of the peninsula of Fanad. This granodiorite outcrops as far as the Gortnatraw Septum (see Fig. 6-1), eastward of which, foliated tonalite appears again. In what follows, we restrict our remarks largely to that part of the pluton exposed on Rosguill because this has been the subject of considerable study.

In detail, there is some debate whether or not the migmatitic rocks forming the eastern side of the Melmore septum (Knill and Knill, *loc. cit.*, pp. 286, 193) are really part of the Fanad Pluton, but we interpret the evidence (at Melmore, and on Fanad at Lough Nagreany) as indicating a transition between this *Melmore migmatite* and the main mass; accordingly, the migmatite is simply the highly altered and injected remnant of a composite roof pendant.

The separate parts between the septa are similar enough to suggest that they are directly connected at no great depth and that the present level represents a roof zone. Thus, we believe that we are dealing with a single cross-cutting intrusive, lobing and fingering up to different levels into a roof consisting of rocks of the Creeslough Succession.

The only rocks distinct in both time of intrusion and composition are small masses of granophyric granite which clearly cut the monzodiorite and are often located in contact situations. However, even these bodies show welded contacts with their host, so that the time interval between them and their host cannot have been very great.

3. The Contact with the Envelope: an Example of Varying Contact Effect

Though there are traces of a weak crenulation cleavage at a distance from the contacts, and folding on steep axes in the border zones, deformation structures attributable to the emplacement are poorly developed. Much more obvious is the development of a broad static aureole (Fig. 14-3)—up to a mile wide in the pelites—the effects of which are clearly superimposed on the regional metamorphism (Pitcher and Read, 1963; Edmunds, 1969T; Naggar and Atherton, 1970).

The pelitic rocks are converted to biotite-muscovite-quartz hornfelses. In them, the regional garnet, which had earlier been partially chloritized during episode RMP_3, is replaced by biotite, with granules of a new garnet nucleated on the older. Further, the regional chlorite and biotite porphyroblasts of MP_4 age are converted to a new biotite, and the pyrite to pyrrhotite (Chapter 14). In the simplest case, where the changes can be followed in a single pelitic lithology, as at Glinsk (Fig. 14-3), a clear zonation exists. Significant recrystallization begins some 1750 meters from the contact, new garnet first appears at 1700 meters, andalusite at 1350 meters, fibrolite and cordierite at about 370 meters, and robust sillimanite at 200 meters; muscovite shows a steady diminution towards the contact, and potash-feldspar appears sporadically. In other (semipelitic) lithologies in Rosguill, andalusite is less abundant, and staurolite may accompany the cordierite. Here, too, the hornfelses can be spotted with cordierite–microcline-mica segregations up to 4 centimeters across, which have a remarkable microcline-rich peripheral zone (Knill, 1959). Perhaps this association is to be explained by a reaction of the type:

Chlorite + Muscovite + Quartz → Cordierite + K-feldspar + Water

coupled with a reciprocal process of centripetal diffusion. Yet another feature, first noted by M. P. Atherton, is the sporadic appearance of corundum, suggesting a rather higher temperature of formation than is usually the case in the Donegal aureoles. Further, within the more pelitic lithologies, in the inner zone of the aureole, there is locally a considerable degree of mobilization, accompanied by a marked segregation of quartz and feldspar. We later examine the many features of special interest in this aureole (Chapter 14), but we must record here that we find little evidence of Knill and Knill's proposition (1961) that two phases of contact metamorphism are recorded in the aureole, the earlier of which was considered by these authors to be connected with the formation of the Melmore migmatite.

As a whole, contacts are steeply dipping, surprisingly sharp, and uncomplicated. The most common relationship involving quartzite always provides knife-sharp junctions with only occasional evidence of transfer of material, shown by a very local increase in the potash feldspar content of the country rock. Nevertheless, there is some veining of the quartzite, and the adjacent intrusive may be loaded with xenolith material.

The response of dominantly pelitic lithologies is, not unexpectedly, more variable. Against the Falcarragh Pelites (Fig. 6-4), the contact of the Rosguill lobe is generally both sharp and uncomplicated by any new structure in the envelope. Locally, however, especially in the west, the country rocks become highly mobilized and veined by segregated material, but even here, there is a definite contact without any transition.

These differences are most apparent on the opposite sides of the Melmore septum. On the western side, there is a simple sharp contact, but on the east, as we have already mentioned, a most complex situation exists within the septum, for, as the eastern contact zone is approached, the members of a mixed assemblage of metasediments are locally thrown into irregular folds, and, in a more advanced stage of this mobilization, the more competent bands become separated in a streaky pelitic-granitic matrix.

Nowhere is this better seen than in the eastern coastal exposures of Melmore headland where competent horizons, such as metadolerite sills and psammitic layers, are remarkably brecciated and preferentially injected by granitic material (Fig. 6-2a). It seems that these broken zones have offered preferred passageways for invading fluids, and in them, it is quite certain that replacement has been more important than dilation. The final stage of the mobilization is marked by the streaking out of the sheets, veins, and lenses of granodiorite, together with the ribs of pelite, quartzite, calc-silicate, and metadolerite, producing a remarkable gneiss (Fig. 6-2, -3)—the Melmore migmatite mentioned above. Within the latter, all stages of reaction can be seen: pelites are granitized, calcareous rocks yield coarse hornblende–plagioclase associations, metadolerites form spectacular agmatites, and, at one locality north of Melmore Strand, a remarkable streaky diorite gneiss is produced by reaction with a hornblendite. The latter is worthy of special note in that dioritic knots forming in the hornblendite are finally left as orbicules in the dioritic gneiss.

Such mixed rocks were plastic and mobile, apparently even capable of differential flow because the xenolithic element is often chaotically mixed (Fig. 6-2b), yet at the same time, they suffered a deformation producing boudin-like structures and vertical plunging folds (Fig. 6-3). We have then, on Melmore, a very clear case of a migmatitic gneiss produced by a combination of intrusion, metasomatism, and marginal deformation in the very *local* situation of the contact of a high-level granite: in Read's terms it is an example of a contact-migmatite.

Apart from this Melmore situation, along all the other main contacts, the smaller xenoliths of different lithologies are mixed up and do not necessarily reflect the lithology of the adjacent country rock; the host must have been mobile.

These variations in the contact phenomena seem best explained on the basis that any parts not

FIGURE 6-2. The Melmore migmatite, Melmore Strand: (a) Preferential permeation along brecciated component bands within the metasediments; (b) highly xenolithic facies of the contact migmatite.

being continuously stoped and flushed free of reaction products would now record the greatest degree of reaction.

4. Composition and Texture: Contamination of a Magma

Away from the immediate contacts, the pluton is composed of a rather uniform, medium-grained quartz monzodiorite, the quartz content ranging up to 15%. The color index is as high as 22%, and the proportion of biotite to hornblende is very variable indeed.

This hornblende may have cores speckled with iron oxide granules and blebbed with quartz, both relationships suggesting an origin by replacement of augite. The plagioclase, which is more nearly euhedral than in any other of the Donegal "granites," is highly zoned with broad central cores near An_{43} and outer zones of An_{28-32}. Microperthite and quartz are interstitial to all the other minerals.

It is fairly easy to establish an apparent order of crystallization, albeit with considerable overlap, *viz.*, zircon and apatite, plagioclase, hornblende, biotite, potash feldspar, and quartz, with the opaques and sphene in some intermediate position outlasting the growth of plagioclase. The late endometamorphic effects are subdued and not so obviously displayed as in the other plutons. Indeed, we suspect that many petrologists would regard the texture characteristics of the monzodiorite as almost wholly igneous in origin.

This may be generally so, but we must record that the mafics are in the form of tiny aggregates intergrown with abundant sphene and allanite, speared by needles of apatite. Just as in the Thorr Pluton, there is every gradation from these aggregates to larger clots of essentially similar composition, and to the dark, rounded patches (usually less than 0.3 meters in greatest diameter) which are sparsely

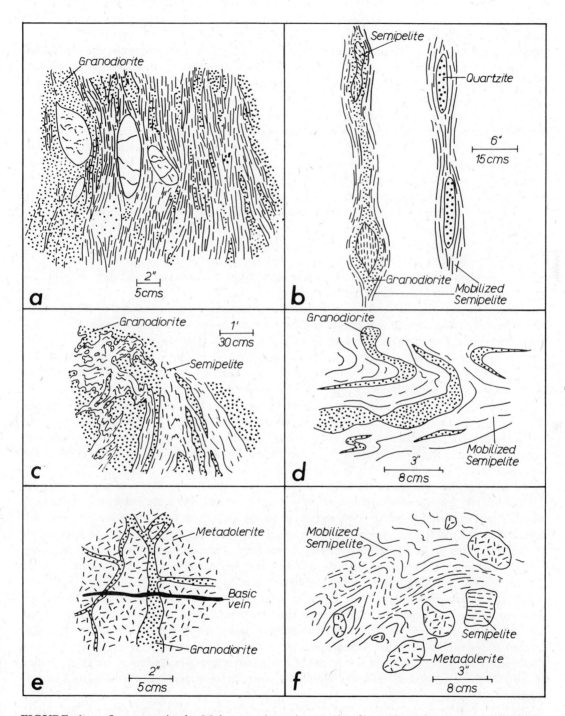

FIGURE 6-3. Structures in the Melmore migmatite; (a, b) disruption of competent lithologies [(a) patches of granodiorite (stipple) and lenses of quartzite (blank) in a semipelitic matrix]; (c, d, f) folding; (e) agmatised metadolerite. After Knill and Knill (1961).

distributed throughout the monzodiorite. There are the usual features of digestion of xenolithic material by the host, and we strongly suspect that some of the mafic element has been derived from external sources. It is not, however, easy to identify the source of the xenoliths; many are meladioritic with affinities with the appinite suite, while others may be the result of a convergence of mineral transformations acting in several different country rock lithologies; unfortunately, the transitional types necessary to establish the latter are not very common. However, in some areas of particularly high xenolith concentration (as near Magheradrumman School in north Fanad), it is possible to recognize metadolerite as contributing significantly to the transformed material.

An important variant within the Pluton is a distinctive granodioritic *contact facies*. Thus, near main boundaries with the envelope rocks, the monzodiorite becomes variable in texture, somewhat finer in grain, poorer in hornblende, and richer in quartz and potash feldspar. Along with this change, there is invariably an increase in the density of clearly recognizable country rock xenoliths of all shapes and sizes and often chaotically mixed up, as we mentioned earlier. There are also nebulitic patches produced by the granitization and partial digestion of this xenolithic material.

In Rosguill, this facies is clearly marginal, clinging to the main contacts as a narrow zone a few hundreds of meters wide, but elsewhere, in areas interpretable as being near the roof, the width increases, and in the area between the two septa of Melmore and Gortnatraw (see Fig. 6-1 and Map 1 in folder), the outcrop consists wholly of the biotite-grandiorite. This suggests that the present level is everywhere near the roof, which is in keeping with the localized margin of migmatite and the local abundance of xenoliths, large and small.

In detail, we note that the breakup of metadolerites sometimes gives agmatitic contact zones, in which the host is flooded with hornblende and dark clots of biotite derived by reaction therefrom—with a consequential diminution in potash feldspar (*cf.* Andrew, 1928*a*). At contacts with siliceous lithologies, the host shows an expected enrichment in quartz, and against pelites a foxy-brown biotite appears, just as in the case of the Thorr Pluton. These mineral changes, coupled with the presence of partially digested xenolithic material, provide clear evidence of contamination.

Thus, at the present level of erosion in Fanad and Rosguill, we have an example of the contamination of a dioritic magma by the assimilation of country rocks—of the *granodioritization* of dioritic rocks. The details have yet to be worked out in Fanad, but they must be very similar to those recorded in the Thorr district and to the other cases we have noted during the general discussion of the Thorr rocks.

There is, then, the problem of the origin of the dioritic magma itself. We have already suggested that the heterogeneities in the texture may have resulted, in part, from the imbibition of envelope material, even of early-formed appinitic rocks, just as we shall see is the case in the genesis of the Ardara Pluton. However, we suspect that the original magma of the Fanad Pluton was intermediate in composition before its emplacement at the present level. From the available compositional data, it seems that this magma was of appinitic affinities and derived from a different source than the Donegal Granite Complex, a matter which we debate again later (Chapter 16).

Further, we are dealing here with a simpler crystallization history than in other comparable situations in Donegal, especially in respect to the growth of potash feldspar. The texture is not crystalloblastic, and indeed, for the bulk of the dioritic rocks, the only reason to doubt a straightforward magmatic history is the presence of the peculiar mafic aggregates, whose origin we have already discussed. Otherwise, there is evidence of a relatively simple overlapping of the crystallization of the mafics and plagioclase, followed closely by interstitial precipitation of potash feldspar and quartz.

5. The Fabric

There is a rather weak alignment of the mafic minerals on horizontal surfaces of the outcrop (Figs. 6-1, -4). This is best seen in the several smaller lobes on Rosguill, but is only faintly shown in the main outcrop of the tonalite in eastern Fanad and is hardly discernible in the granodiorite of western Fanad. It seems to be a steeply dipping and dominantly planar structure which also involves trains of small xenoliths and mineral schlieren. Though sensibly parallel to the main contacts, this structure is seen in detail to be slightly oblique to the contacts and even to the sense of elongation of the larger enclaves. Further, the same structure is present, though weakly, in the late granites which cut the monzodiorite. These facts suggest that this alignment is not strictly a "primary fluidal" structure (see Chapter 15) but is again, as in Thorr, the result of forces applied to the rock in a near solid state. However, such a deformation must have been relatively weak over most of the Fanad Pluton because the smaller dark patches are very little flattened. Only in the *Melmore migmatite* is there evidence of any considerable deformation (Fig. 6-3), but even here, the mineral orientation is not strong. In fact, this marginal migmatite zone seems to us to represent plastic flow in a softened contact zone.

6. The Distribution of the Larger Xenoliths: Ghost Stratigraphy Again

In Rosguill and western Fanad, quite apart from the major roof pendants, there are numerous separate masses of metasediment and metadolerite within either the monzodiorite or the granodiorite (Fig. 6-4). It can easily be demonstrated that many of these represent true inclusions and are not attached to the roof and walls: exposure is excellent in three dimensions. We repeat that reaction between host and inclusion is very limited, for contacts are generally sharp, and only in the case of the pelitic rocks is there the slightest modification. The form of the larger inclusions is, indeed, controlled by the lithology, but only in the sense that the latter controls the nature of any fracturing. Thus, in inclusions of quartzite, semipelite, and limestone, outlines are often rectilinear, with the contacts clearly and directly controlled by the attitude of the bedding, schistosity, and early joints. There are even examples of dioritic apophyses interfingering along the bedding and so separating two great metasedimentary masses (for example, on the cliffs 1 kilometer southwest of Doagh), as if wedging them apart. Clearly, the inclusions have been broken away from the envelope, perhaps by thermal spalling (*cf.* Richey, in discussion of Pitcher and Read, 1959).

Most interesting is the observation that while some xenoliths lie in a host of granodioritic contact facies, others lie in monzodiorite, yet have a narrow selvage of granodiorite which we interpret as a remnant of the contact facies (see, however, Knill and Knill, *loc. cit.* p. 292). Some blocks have apparently sunk to deeper levels, carrying the early formed contact phase with them—a situation very much like that at Thorr.

Despite all this evidence of stoping and local movement of the blocks, it is, nevertheless, true that some degree of ghost stratigraphy can be demonstrated by mapping these masses, and a succession established which fits with the detail of the Falcarragh and Loughros Formations. It is important, however, not to overemphasize the perfection of this relict structure, as we believe has been done by both Knill and Knill (*loc. cit.*) and Reynolds (in discussion of Knill and Knill). It is, in reality, extremely crude, and while the main lithologies are grouped in areas, there is, nevertheless, obvious mixing and disorientation of the fragments (Fig. 6-4).

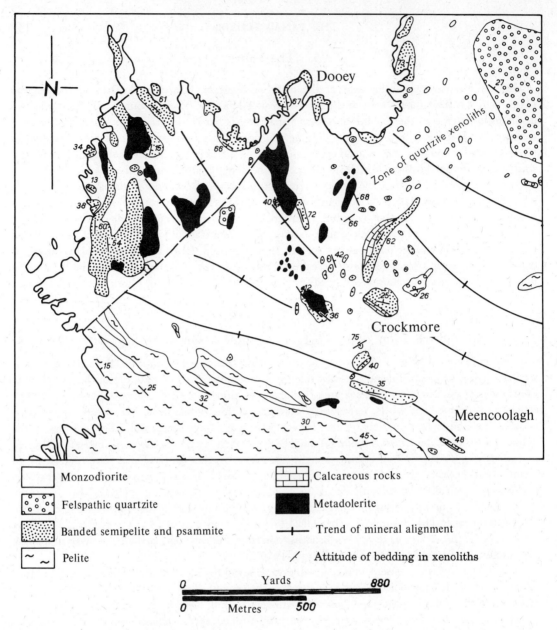

FIGURE 6-4. The Fanad Pluton in northwestern Rosguill. Shows the poor degree of structural and stratigraphic correlation existing between the xenolithic masses. After Knill and Knill (1961) with modifications by Fernandes-Davila (1969T).

7. The Mode of Emplacement: Piecemeal Stoping

It is especially important to understand the setting of the Fanad Pluton—that it was emplaced at a very late stage in the plutonic history into a relatively cold and structurally static envelope. Such an environment is in keeping with the presence of a simple static aureole, the abundance of sharp, cross-cutting contacts, the lack of metasomatic transfer across many contacts, and the relative homogeneity of the rock types of the pluton as a whole. This combined evidence surely indicates an intrusive origin, and, further, the simple textural history strongly suggests that the rocks crystallized from a magma.

There is some evidence, however, for the incorporation of country rock material on a relatively small scale, but there is only one localized example of a thoroughgoing making-over of metasediments into granitic rocks (*i.e.,* at Melmore), and this we are content to interpret as an exceptional case of extended reaction with the envelope. Contamination seems a likely explanation for the modal changes, especially for the contact facies, and provides yet another example where the addition of relatively small amounts of country rock radically changes the mode. We have already considered whether monzodiorite itself is of hybrid origin: while there is some evidence of addition of envelope material, the comparative homogeneity of the rock type suggests that, whatever its origin, the dioritic character was established elsewhere than at the present level—that, in fact, the rocks of the pluton were intruded as a monzodiorite magma.

Doubtless this magma invaded its own contact facies, sometimes all but sweeping it away from contacts and taking up xenoliths which had formerly been immersed in this contact facies. Such local mobility is confirmed by the disorientation and mixing of xenoliths; even in the migmatite, there are sheets and veins carrying xenoliths foreign to the adjacent rocks. Flow in this crystallizing magma could have produced the alignments of schlieren and dark patches, but we suspect that the final orientation of the minerals, especially in peripheral contact situations, was due to strain set up by the pressures involved in intrusion.

Having derived so simple a model, we are immediately faced, once again, with this problem of explaining the preservation of the weak ghost stratigraphy. Knill and Knill fully realized this problem, and considered that metasomatic replacement of the country rock (witness the Melmore migmatite) was followed by a limited degree of mobilization and intrusion—resulting in a partial disruption of the relict stratigraphy by flow. Our objections to this model are the general lack of evidence of replacement (including the displacement of certain constituents) in nearly all contact situations, and, in broader terms, the high-level structural context of the Fanad Pluton.

The metasomatic hypothesis was strongly upheld by Reynolds (in discussion of Knill and Knill, *loc. cit.*). She claimed that the ghost stratigraphy was even more perfect than the Knills stated; the metadolerite "enclaves," for example, were thought to form a series of arches separated from one another by steep thrusts; below these arches of metadolerite, the granodiorite outcrops as if it had taken the place of the former pelitic host of the metadolerite sills. The alignment of minerals was believed to have been inherited from structures in the pelite and even the former path of the thrusts was represented by local discordances in the orientation of these planar structures. Finally, some degree of partial mobilization and upward intrusion was suggested by the distortion of the relict structures.

The present authors find this advanced hypothesis difficult to accept, yet they are aware that it is based on very detailed field work. We particularly dispute the inheritance of the planar structure from the pelitic host (see p. 139), the relicts of which are nearly always sharply demarcated, with angular blocks lying in a weakly foliated host.

There is little evidence of major distortion of the envelope during emplacement, and neither, as far as one can tell, have the relict masses been forced apart to any great extent, so that forceful intrusion seems unlikely. If we then dismiss wholesale granitization as the explanation, we are left with a model involving piecemeal stoping, with blocks reflecting the lithology of the adjacent roof because they descended no great distance before being frozen into their present position—as at Thorr. One conclusion that is even more certain than in Thorr district—where, after much research, we embraced the same classic hypothesis—is that the present level of exposure is very near the roof. Thus, we consider that the only reasonable interpretation of the situation in both northwestern Fanad and northern Rosguill is of great, steep-sided apophyses fingering up into a roof and clutching lumps of country rock, which cannot have been removed far from their place of origin.

8. An Appendix: a Related Dyke Swarm

At this point, we must return briefly to a further consideration of the Caledonian dykes of Donegal, because, in addition to the numerous late-tectonic lamprophyres and felsites previously mentioned (Chapter 3), there are certain swarms of dykes having a more distinct and uniform trend, and including some thick individuals which extend for considerable distances.

One such swarm trending NE–SW through the Creeslough area (McCall, 1954) has been previously mentioned (p. 133) with reference to the relative age of the Fanad Pluton. Indeed, these particular dykes seem to be closely connected in time and space with this intrusive, just as the NE–SW trending dykes in Argyll are related to certain Caledonian plutons (Richey, 1938).

Petrologically, these dykes show many resemblances to those of the southwest Highlands (Knill and Knill, *loc. cit.* p. 295), and include rocks of lamprophyric affinity, best designated as porphyritic and aphyric microdiorites, and some porphyritic microgranodiorites. In composition, they are also more nearly related to the Fanad Pluton than to any other of the major intrusives of Donegal.

CHAPTER 7

THE APPINITE SUITE: BASIC ROCKS GENETICALLY ASSOCIATED WITH GRANITE

CONTENTS

1. Introduction 143
   (A) General 143
   (B) Nomenclature of the Appinite Suite 144
   (C) Some Textural Features 145
2. General Distribution in Space and Time 148
   (A) Space 148
   (B) Time 148
3. The General Form of the Intrusions 149
   (A) The Bosses and Their Highly Fluxed Magmas 149
   (B) Some Specific Examples 151
4. The Intrusive Breccias 156
5. The Associated Lamprophyres 161
6. The Origin of the Appinitic Suite 162
   (A) The Intrusive Breccias: Explosion, Fluidization, and Rock Bursting 162
   (B) The Magmas 164

1. Introduction

(A) General

Certain of the granites of Donegal, like their counterparts in Scotland, are closely connected in time and space with a highly characteristic assemblage of basic minor intrusions. These vary in form from small bosses to dykes, sheets, and sills, all of which are intimately associated with intrusion breccias occurring either marginal to the igneous intrusions themselves or as individual breccia pipes or diatremes.

One of the central types of this most variable suite of rocks is a dark, medium to coarse diorite which is largely composed of idiomorphic hornblende set in a groundmass of plagioclase and quartz, sometimes with a little potash feldspar. Such a rock was defined as an appinite (from Appin in Argyllshire) by Bailey (1916, pp. 167–8) who rightly associated with it other rocks such as monzonite, augite-diorite, cortlandtite, and kentallenite (Hill and Kynaston, 1900) to form an easily recognized, uniform geological and petrographic group—the *Appinite Suite* (Bailey and Maufe, 1960; see also Anderson, 1935; Read, 1961). This is, indeed, a most useful and comprehensive term for the rock suite we are about to describe, even though appinite, in the strict sense, is not the most dominant of these rock types in Donegal.

An understanding of the petrogenetic importance of this appinitic suite has grown in recent years, and new work on the Scottish examples, described early by Bailey (*loc. cit.*), Read (1925, 1926, 1931), and Anderson (1935, 1937), has recently been added to and reviewed by Bowes and Wright (1967) and by Platten and Money (personal communication). The Irish equivalents, which had been briefly dealt with by Hull and his fellow surveyors (1891) and by Grenville Cole (1902), have also been redescribed by French (1966) and Hall (1967c; see also Brindley, 1970). What follows below is in the nature of a summary; we first comment on the nomenclature of such rocks, then summarize the main features of the intrusions and breccias, and finally comment on the origin of both in more general terms.

(B)   Nomenclature of the Appinite Suite

The dominant mineral assemblage of the appinite suite in Donegal is a coarse-grained association of hornblende, plagioclase, sometimes quartz, and with biotite and pyroxene as alternative mafic minerals. The color index varies widely and patchily, and transitions from one modal type to another are the rule; as a result, it is not easy to classify such rocks, though it has been found useful to establish a twofold division into a pyroxene-bearing and a hornblende-bearing series. Further, and according to French (1966), it is possible to subdivide the hornblendic rocks on the basis of the habit of the amphibole: those rocks with long prismatic hornblendes are conveniently designated the "appinite series," while those with squat amphiboles, or even amphibole in aggregate form, belong to a "dioritic" series. Within each of these subdivisions, the color index can be used to define the rock species (Fig. 7-1).

We should perhaps note here that, although rocks as acid as granodiorite form an integral part of the suite, the great majority of the appinitic intrusions are thoroughly basic or ultrabasic in character; also, when quantities are taken into account, there is a distinct compositional hiatus between the appinites and granites.

Grain size is as variable as the mode but, for the most part, the rocks are coarse grained even in the smallest bosses, and only in the dykes are fine grained matrices at all common. In the latter, the mafic mineral is often so abundant as small phenocrysts that the rocks are lamprophyric in aspect; indeed, many could be specifically labelled as vogesites and spessartites. We hasten to add, however, that these particular hornblende-lamprophyres are, in Donegal, only part of a dyke suite which includes all gradations into light colored, biotite-plagioclase rocks, with or without quartz, for which we find some difficulty in providing an acceptable name, especially as the original mineralogy has almost always been greatly modified by late-stage changes. A convenient field term is "felsite"—which we shall often use in this book—though it may be that this should be restricted to rocks in the volcanic and cryptovolcanic environment: a more appropriate term is probably plagiophyre, though this is perhaps not a very widely accepted term.

Whatever the nomenclature, however, we are sure that study of the appinite suite as a whole can make a significant contribution to what Smith (1946) and Eskola (1954) have called the lamprophyre problem.

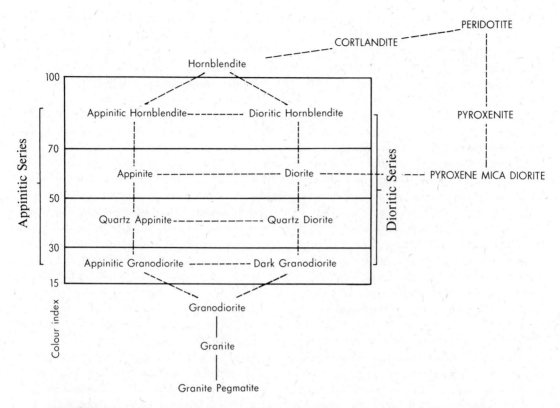

FIGURE 7-1. The nomenclature of the appinitic suite of Donegal according to French (1966).

(C) Some Textural Features

The crystallization histories of different minerals can only really be compared when they come from the same rock. Nevertheless, there are some textural features which are common to appinitic rocks as a whole, and we now comment on these in general terms.

Although, as we shall see, hornblendic rocks are much more important than those with pyroxene, there are still many rocks in which the latter mineral is poikilitically enclosed within the hornblende and many others in which pyroxene is in the process of conversion to amphibole. Thus, in members of the dioritic series, aggregates of hornblende, or even single crystals of that mineral, are often cored by felted masses of pale-green amphibole which may be seen to be replacing pyroxene. With reduction of the color index of the host rock, both the clots and the included pyroxene tend to disappear, their place being taken by biotite. The discontinuous reaction series is, in fact, beautifully acted out in these rocks, for the order of crystallization is very clearly olivine, pyroxene, hornblende, biotite (Fig. 7-2a, b, c). The relationship of the felsic constituents to this familiar pattern is that the crystallization of plagioclase feldspar clearly followed that of the brown hornblende, though the growth intervals overlapped in the case of the green variety, especially that which mantles the pyroxene. Biotite, it must be admitted, crystallized over a wider interval, even extending to the end of plagioclase growth. The quartz was generally latest to crystallize, either forming rather patternless intergrowths with the feldspar, or providing a granophyric texture, or filling embayments in the plagioclase laths in a manner suggestive of corrosion.

This order of crystallization has, of course, been determined by the conventional method involving relative euhedralism and enclosure of one mineral by another, relationships which might well be ambiguous if replacement is involved. However, we take comfort in the fact that this very matter has been very fully discussed in an elegant work by Wells and Bishop (1955), who, in dealing with the appinitic facies associated with the granites of the Channel Islands, come to similar conclusions as to the order of crystallization—a finding confirming the earlier determinations of Deer (1950) and Nockolds (1941, 1946) in Scotland.

Textures are, in fact, typically igneous, though they are not exactly those of basaltic rocks. Possibly, the wider spacing of nuclei—due to the fluxed nature of the magmas—reduced the possibilities for ophitic intergrowths, as well as producing a general coarseness of grain of the mafic minerals.

Another characteristic feature of this appinitic environment is a general change in the color of the hornblende from brown to clear green, either with decrease in color index in a rock series or in the zonal features of a single crystal. Unfortunately, the causes of this color change are not yet known in detail (*cf.* however, Nockolds, 1941; Deer, 1938), but we suspect that in general terms, the brown varieties represent the high temperature form. Then there are the rhythmic overgrowths on the larger hornblendes, where alternating zones of light and dark green varieties of actinolite-tremolite grow in crystallographic continuation with the core; these are particularly well described by Bowes and his co-workers (1964) in connection with identical rocks in Argyll. Yet another characteristic is the hollow hornblende with cores of felsic material (Fig. 7-2d), just as figured by Wells and Bishop (*loc. cit.*, p. 146 and p. 152)—a type of crystallization which is thought to be diagnostic of precipitation from a rapidly cooled, liquid phase (Wyllie, *et al.*, 1963) enriched in volatiles (Bishop, 1964, p. 311).

There are other more complex textures, some of which can be variously interpreted. Thus, the hornblende crystals of some intrusions appear to be fractured and even split along the cleavages (Fig. 7-2e, f). In other rocks, large ragged-margined amphiboles form the center of a cluster of smaller "fragments" intergrown with feldspar; this relationship is interpreted by French (*loc. cit.*, p. 305) as being due to the disintegration of the central crystal, but even if, as is more likely, it is merely a special kind of amphibole-feldspar overgrowth in which amphibole and plagioclase crystallized together, it certainly represents a second stage of growth.

In all the rock types of the suite, the minor constituents, apatite, sphene, rutile, calcite, chlorite, and pyrite are abundant, especially in pegmatitic varieties and the intermediate members of the appinitic series (see also Hall, 1965*a*); relatively, the dioritic series contains a lower proportion of them. Calcite is frequently associated with quartz; it is clearly a primary mineral, although it must belong to a late stage in the crystallization history, for it often forms thin veinlets lying along and dilating the cleavages of the amphiboles and the twin planes of the feldspars.

These various textural features indicate unusual conditions of crystallization. The very coarse grain of the amphibole, even in intrusions of small size, the abundance of hydrous minerals, the relative abundance of apatite, pyrite, and even calcite, suggest that the appinitic magma had a high volatile content (*cf.* Bowes *et al.*, 1964). The overgrowths probably represent growth under alternating physical conditions arising from the intermittent release of a high gas pressure (for the details see Bowes *et al., loc. cit.*). The fracturing, even partial disintegration, of the amphibole framework may well have been connected with rapid expansion on release of such volatile materials. Of course, the very presence of abundant amphibole in the appinites, where it takes the place of pyroxene and crystallizes over a greatly extended range of temperature, is almost certainly due to the crystallization of a water-rich magma at high vapor pressure: this is a matter we return to later, after examining much more of the evidence.

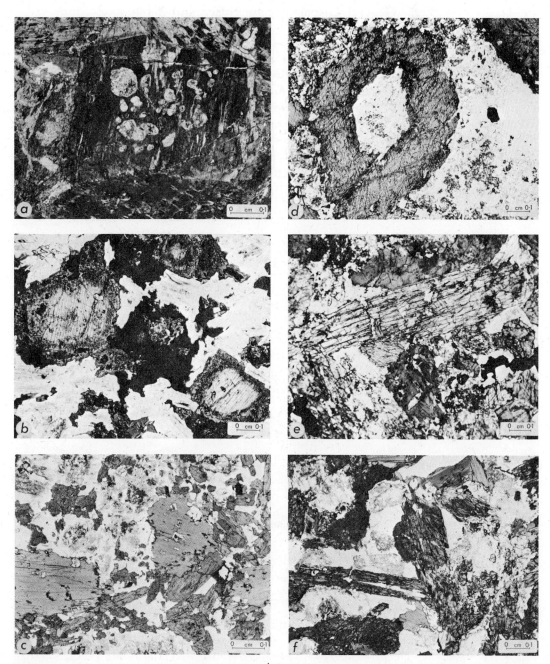

FIGURE 7-2. Some textural features of appinitic rocks. Reaction relationships: (a) Hornblende prism enclosing small pyroxenes and rounded pseudomorphs after olivine. Hornblendite transitional to cortlandtite, Kilrean. (b) Pyroxenes rimmed with amphibole on which are moulded large flakes of biotite set in a base of andesine. Pyroxene-mica diorite, Portnoo. (c) Biotite intergrown with (? replacing) hornblende in base of andesine. Biotite-hornblende diorite, Narin. Special features of the amphiboles in appinites: (d) Cored hornblende, Liskeeraghan. (e, f) Contemporaneous splitting across and along hornblende crystals, Portnoo. Plane Light.

148   THE GRANITES

2. General Distribution in Space and Time

(A) Space

Although the small basic bodies of the appinite suite are very widely distributed in Donegal, there is a tendency, as has long been evident in Scotland, for these intrusions, both igneous rocks and breccias, to occur in *clusters*. Thus, around the Ardara Pluton is clustered an assemblage of minor intrusions which includes about forty appinitic bosses, a swarm of lamprophyre-plagiophyre dykes, and several intrusion breccia pipes (Fig. 7-3). This close association in space can hardly be coincidental, and underlines the problem of the association of basic and ultrabasic rocks with granites.

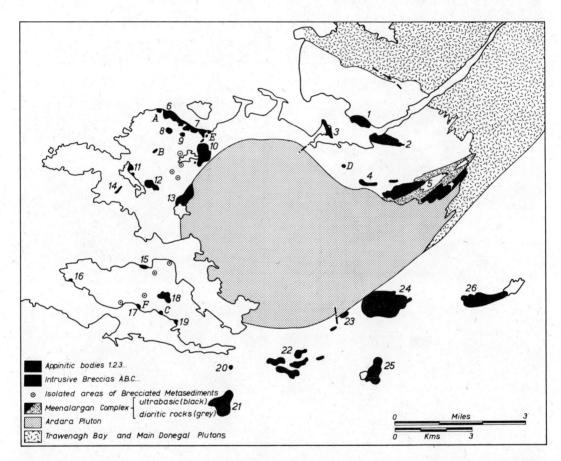

FIGURE 7-3. The Ardara cluster of appinitic intrusions: 1, Cor; 2, Mulnamin More; 3, Roskin; 4, Crockard Hill; 5, Meenalargan; 6–9, Portnoo group; 10, Naran Hill; 11, 12, Kiltooris Lough; 13, Summy Lough; 14, Rosbeg; 15, Liskeeraghan; 16, Loughros Point; 17–19, Southern Loughros group; 20, Maghera Road; 21, Glen Gesh; 22, Monargan; 23, 24, Kilrean; 25, Owenea Lough; 26, Lough Anna. Intrusive Breccias: A, Dunmore; B, Birroge; C, Crannogeboy; D, Kilkenny; E, Portnoo; F, Cloghboy.

(B) Time

All the intrusions which we include in the appinite suite clearly postdate the D_4 regional structures and within their aureoles the regional carbonate porphyroblasts (RMP_4) are converted to calc-silicate;

their late tectonic date is thus unequivocal. Although many of them cannot be dated more precisely than this, there are sufficient clear-cut individual time relationships available to determine the relative age span of the suite as a whole—we do not, of course, pretend that they are all of the same age. We first note that very few members cut any of the granites, a possible exception being certain tuffisite veins (Whitten, 1959a) cutting the Thorr Pluton on Tory Island (we have noted similar veins at Currans Point on the mainland), and a clear exception, the several breccia pipes and associated hornblende-lamprophyres cutting the Barnesmore Pluton (Chapter 10). Appinitic rocks of all kinds occur as xenoliths in the Ardara, Fanad, and Main Donegal Plutons: indeed, at Ardara, there is a close similarity between the outer member of this pluton and one of the rock types of the appinitic intrusions (Hall, 1966d). Even more critical evidence of the closeness of the time linkage is, in fact, shown by the way in which Naran Hill appinite body (No. 10 of Fig. 7-3) cuts an arcuate dislocation associated with the early stages of the emplacement of the Ardara Pluton, yet still acts as a rigid, preexisting body in regard to the deformation in the aureole (p. 179). This linkage is more generally confirmed by the presence, within this same aureole, of both deformed and undeformed members of the hornblende-lamprophyre suite which, in other situations, can be seen to cut and be cut by appinitic bosses and even intrusion breccia pipes.

Thus, the time of emplacement of the appinites overlapped that of Ardara, and though there are some which cut the Barnesmore, there is no evidence of any postdating the Rosses, Main Donegal, and Trawenagh Bay Plutons. This situation, where appinites are broadly synchronous with the earliest members of granite complexes, is a finding general to Scotland and Ireland (*cf.* Read, 1961).

3. The General Form of the Intrusions

(A) The Bosses and Their Highly Fluxed Magmas

The forty or so bosses which make up the most substantial part of the Ardara cluster vary in outcrop size from 100 meters across to masses measuring about 1500 meters by 500 meters. The outcrops have a generally rounded outline with a tendency to be lenticular, due to the control exercised by the structural trend of the country rocks. In three dimensions, nearly all have the general form of steeply inclined pipes which either cut cleanly across the regional structures or force them aside; some are more irregular, and a few of these have sheeted contacts.

These pipe-like bodies are fairly evenly distributed within the area of the Ardara cluster. They are found in all the metasedimentary groups (see Map 1 in folder), and their character varies somewhat according to the particular lithology in which they occur—as a generalization, the pelites seem to have been relatively easily plastically deformed and opened, while intrusion through more competent rocks was more easily accomplished by sheeting and veining associated with brecciation. In fact, many of the bosses are accompanied, along their margins, by masses of disrupted metasediments, here termed *intrusive breccias* (see p. 162 for definition); these breccias, like the igneous material, have remarkably sharp, regular, and vertical contacts with the surrounding country rocks. The importance of forceful opening is shown by the common deflection of the strike around the intrusions, and the actual opening up of the strata can be clearly demonstrated around the Mulnamin Intrusion (No. 2 of Fig. 7-3 and Fig. 7-6). On the other hand, a gas drilling mechanism seems indicated in cases where the bedding is sharply truncated and where there are marginal breccias—we have much more to say about this below.

External to these intrusions, there is always a narrow contact aureole up to 20 meters in width. In this situation, pelitic schists are converted to biotite-quartz-hornfelses, sometimes bearing andalusite in an outer zone and fibrolite in an inner zone—a zone which is sometimes extensively mobilized. Calcareous rocks are metamorphosed to assemblages of garnet, members of the epidote group, and amphiboles. Near contacts and in the xenoliths, porphyroblastic amphibole appears whenever the composition allows, that is, in calc-silicate assemblages and in metadolerites; in short, amphibolitization is the rule in this particular contact environment, and results in some contact rocks gaining an appinitic appearance.

Internally, the structure of these intrusions often appears to be confusedly variable, yet the rocks are still capable of subdivision on the basis of textural and modal variations; indeed, on the mapping scale, fairly ordered patterns can be made out (Fig. 7-5, -6, -7) which are more or less controlled by the outer form of the intrusion. Internal contacts between the different types are usually gradational though occasionally sharp; obviously, differentiation processes were operative *in situ,* or magmatic pulses followed rapidly one on the other. What are probably true fluidal structures are shown in some intrusions by mineral foliations which invariably lie parallel to the walls.

The rock types themselves vary from ultrabasic to acid, from cortlandtite and hornblendite, through various diorites, to granodiorite, and even occasionally to true granite. Within this assemblage, the melanocratic hornblende-bearing rocks are by far the most abundant and important, and pyroxene-bearing rocks, such as pyroxene-mica-diorite, form only a small proportion of the various intrusions; hornblende both crystallized preferentially and grew as a replacement of preexisting augite. Interestingly enough, the reverse is the case in Argyll.

We are shortly to give specific examples of the broad distribution of such rocks within the intrusions in which, as we have said, it is possible to discern meaningful patterns. The variation within the single outcrop is, nevertheless, very great. Thus, a general patchiness and streakiness is usual (Fig. 7-4a), and it is not uncommon to find examples of more sharply bounded blocks, from a few centimeters to several meters across, of the darker rocks within the lighter (Fig. 7-4b), blocks which are often veined by their host. There are also vein structures of a peculiar type in which appinites and hornblendites are cut by a pegmatitic facies of the same general composition as the host and with amphibole commonly extending across the vein walls; in fact, the veins can only be distinguished from the host by the better orientation (normal to the walls) and larger size of the amphiboles, coupled with a greater abundance of the normal accessory minerals (Fig. 7-4c). Besides these veins, there are irregular lenticles of similar pegmatitic material, and from such pockets, stringers of felsic minerals arise and mingle with the groundmass of the host rock (Fig. 7-4d). We consider that both veins and pockets represent the sites of local high concentrations of the volatile constituents, with the veins marking the preferred pathways of volatiles moving through a crystal mush.

A related phenomenon is the net veining where light-colored quartz-plagioclase veins form ramifying networks in all types of basic rocks, being particularly spectacular when in a black amphibolitic host. Such veins often seem to fill a breccia type of fracture system, just as do net veins in many cryptovolcanic complexes (*e.g.,* Wells, 1954; Windley, 1965); they are clearly later than the fabric of their host, yet as far as the felsic minerals are concerned, the veins are often continuous with the matrix of the host and may pass into small pools of pegmatitic appinite. In terms of the cooling history of the host, there are even later, more discrete, and sharply margined veins, and these are closely associated with cross-cutting lenses and dykes of granodiorite, granite, and granite-pegmatite. Indeed, in detail, there appears to be a sequence of vein systems, beginning with mere leucocratic patches formed when the rock was only partly crystalline and ending with sharply defined and more regular nets of veins, fed by discrete intrusions and cutting a solidified host.

We think that all these various phenomena are connected with an original high water content of the magma: in the early stages of consolidation, this is responsible for the widening of the stability field of the amphibole and the fluxing of the crystallizing magma. In an intermediate stage, this is also responsible for the disruption of the crystal frameworks, and in the late stages, when consolidation was well advanced, for the buildup of pressures leading to the brecciation and forceful injection of the residual, highly fluid acid magma.

FIGURE 7-4. Structures in appinites: (a) phases of net veining in a diorite-granodiorite mélange, (b) diorite blocks in a granodioritic host; (c) net-vein system in appinitic hornblendite; note disruption of the host in a dykelike conduit. Photographs a–c by French from the Liskeeraghan intrusion, No. 15. (d) Stringers of pegmatite appinite in hornblendite, Mulnamin intrusion, No. 2.

(B) Some Specific Examples

We choose four main examples of these intrusions (Nos. 24, 5, 2, and 10 of Fig. 7-3) to illustrate their main features, drawing mainly upon the descriptions of French (1966) and Iyengar, and Pitcher and Read (1954).

The steep-sided boss of *Kilrean,* SW of Glenties (No. 24 of Fig. 7-3, also Fig. 7-5) was emplaced in the Upper Falcarragh Pelites mainly by a dilational mechanism. It is of particular interest in that it shows well the zoned relationship between cortlandtite, hornblendite, and appinite, the darker rocks forming a central core to the intrusion. Each of four separate masses of cortlandtite has a discontinuous rim of hornblendite averaging 30 meters wide, and these give way to appinitic hornblendite and finally to appinite, so that there is a general decrease in color index away from the cortlandtite. In detail, there is a gradual reduction and eventual elimination of olivine in the cortlandtite, and the pyroxenes become smaller and rounded toward the hornblendite. In the latter, stumpy dark-brown hornblendes enclose rounded, colorless amphibole pseudomorphs after pyroxene, together with a few relict pyroxenes. The hornblendes have hackly margins, are separated from one another by a very sparse background of sodic plagioclase, and are transected by very thin veinlets of this groundmass. A few meters further from the cortlandtite, the hornblendite lacks the pseudomorphs after pyroxene and is laced by irregular appinite veins, in which smaller, more ragged green prismatic amphiboles are set in more abundant feldspar; these veins are continuous with, and seem to be the pathways of, the felsic material which is spread sparsely throughout the cortlandtite. It would appear at first sight that

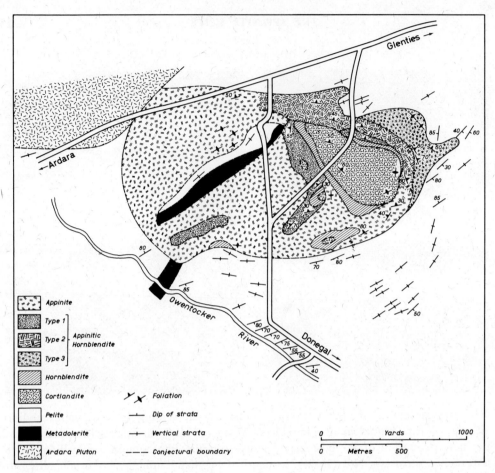

FIGURE 7-5. The appinitic intrusion (No. 24) of Kilrean, S.W. of Glenties. After French (1966).

the appinite was produced from cortlandtite, by way of hornblendite, through the agency of a permeating feldspathic fluid, just as Gindy (1951b) has suggested in the case of another similar intrusion within this Ardara cluster (No. 1 of Fig. 7-3). We shall argue below, however, that this process was secondary to an *in situ* crystal differentiation which, while still explaining the gradation of rock types, yet dispenses with the need to introduce considerable amounts of felsic material.

A similar example, where ultrabasic rocks, largely hornblendites, form a core to various kinds of diorite and granodiorite is the *Meenalargan Complex* (No. 5 of Fig. 7-3). It is actually made up of two adjacent ultrabasic masses, which, together with the associated dioritic rocks, form the largest appinite body in Donegal (1 × 3 kilometers). Briefly described by Iyengar, Pitcher, and Read (1954), it has been the subject of much research by Curtis in a yet unpublished work. We mention here only one special feature—that the calc-silicate schists which strike into the contact (at Lough Laragh) are progressively converted to appinitic amphibolite with the preservation of the bedding, showing that at least some appinitic rocks are of metasomatic origin.

A somewhat different form of intrusion is that of *Mulnamin More* (No. 2 of Fig. 7-3, and Fig. 7-6), which has very clearly dilated and opened up its country rocks—the calcareous facies of the Sessiagh–Clonmass Formation and the Lower Falcarragh Pelites. In fact, there is some evidence from the attitude of the contacts that it forms a northerly-dipping sheet.

Within a clear-cut and unchilled contact, there is a varied display of hornblendites and diorites again showing a crudely concentric arrangement, this time, however, with a core of diorites surrounded

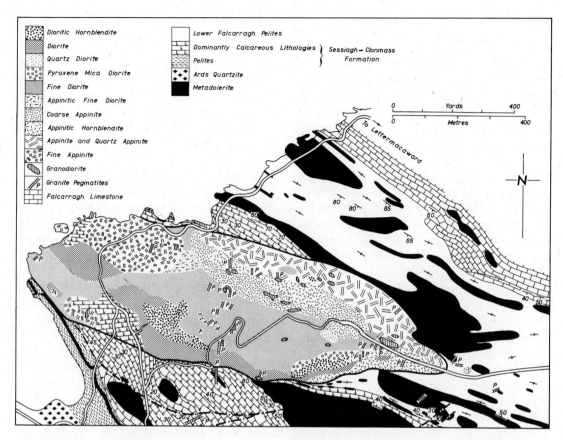

FIGURE 7-6. The appinitic intrusion (No. 2) of Mulnamin More. After French (1966). Outline of intrusion marked by heavy line.

by a partial ring of hornblendite-appinite rocks. Between the several rock-types making up the ring, there is complete and gradual transition, *viz.*, dioritic hornblendite–appinitic hornblendite–appinite, and a similar gradation exists between the diorites where pyroxene-mica-diorite passes, by way of a rock with aggregates of green amphibole after pyroxene, into a biotite-diorite. In each of the two parts of the intrusion, the hornblenditic and the dioritic, the members are obviously closely related, and were probably differentiated one from the other by processes operating *in situ* (French, *loc. cit.*). However, between the parts, *i.e.*, the diorites and the hornblendites, the gradations are usually so rapid and compositions so contrasted that they are more likely to represent different magmatic pulses—even though related at depth.

In the Mulnamin intrusion, only the granitic pegmatites and granodiolites occur as distinctly separate masses which are clearly not part of either gradational series at the present level; they are, however, highly contaminated in a manner suggesting that they were emplaced before the close of the cooling history of the host rocks.

There are two other features of this intrusion worthy of special mention. First, we note that the veining phenomena described earlier are particularly well displayed in association with the hornblendic rocks. Second, it is in this body, and its continuation at Cor, across the Gweebarra estuary (No. 1 of Fig. 7-3), that the reaction between the appinite and calcareous country rock can be especially well studied. Thus, Gindy (1951*b*) describes the production of contact skarns in which epidote-diopside-actinolite rocks develop large amphibole porphyroblasts near the contact. Within the intrusion, there are abundant small xenoliths, the commonest of which is essentially composed of actinolite or diopside. In these, a marked zoning is often apparent, and, outward from a greenish core of fibrous actinolite, the color of the amphibole deepens and the mineral takes on a stouter form; finally, plagioclase appears in the rim, along with a hornblende indistinguishable from the enclosing appinitic amphibolite. Where the core is composed of diopside, this mineral is converted to actinolite near the rim, whence the changes proceed just as described. Such evidence prompted Iyengar and his coworkers (1954) to suggest that the metasomatism of calc-silicate rocks played some part in the production of amphibolitic rocks (see also p. 149). We believe, however, that the relationship is restricted to limited reaction and assimilation, and that the bulk of the variation in these intrusives is the result of a special type of differentiation of basic magma.

Our last example is the *Naran Hill Intrusion,* lying to the south of the village of Naran (No. 10 of Fig. 7-3, and Fig. 7-7). In this body, there are no particularly basic rocks, and the essential form is a core of biotite-diorite passing out by gradation into hornblende-diorite, a relatively simple pattern broken by marginal areas of biotite-granodiorite which show much less of a gradation into the diorites. These granodiorites are closely associated with the highly contorted pelite, which represents softened and mobilized wall rocks—we think these might easily have provided siliceous partial melts, which, by mixture with diorite, could have formed the dark granodiorite.

Possibly the most interesting feature of this intrusion is the marginal breccia, which is quite typical of its kind. It is composed of rectangular blocks of country rocks, with a mean size of 50 centimeters by 15 centimeters, but ranging up to blocks several meters across; these and smaller fragments are remarkably tightly fitted together, though there is a sparse matrix of comminuted pelite (Fig. 7-8). Some blocks and rounded cobbles of metadolerite are mixed in with this breccia.

It is clear that these breccias have been moved from their original position because in one place a large metadolerite in the country rocks is transected by breccia lacking in amphibolitic fragments. On the other hand, there are no fragments representing the strata which must lie structurally below the present level, and, indeed, there is such a close comparison between the rocks found inside and outside the margin that the amount of movement cannot have been more than a few tens of meters. Because the contact between the breccias and the igneous rocks is flat, while against the country rocks it is steep, it seems likely that the breccias in this particular case may form a capping to the intrusion.

We may note in passing that, in general, there is little possibility of confusing breccias of this kind with their tectonic analogues. The former are characteristically chaotic, the disorientated, highly angular blocks often showing a remarkably neat fit, the necessary "voids" being filled with a structureless matrix which must have been intruded into its present position. All these features are well seen, not only at Naran, but in several other marginal breccias, such as that associated with the Portnoo intrusion (No. 7 of Fig. 7-3), also at Liskeeraghan (No. 15 of Fig. 7-3 and Fig. 7-11a), and intrusions numbered 11, 18, and 25 in Figure 7-3. As we shall see later, the best explanation of these rocks is that they result from the disruption and partial entrainment of material in a gas-fluidized system.

The immediate problem at Naran, however, is not so much the formation of the breccia, but to what extent its presence helps to explain the space problem. As it does not seem to have moved far, one can hardly hold that any material was ejected at the surface; on the other hand, there is not much evidence that the disrupted blocks have been stoped away by the intrusion below. As French shows (*loc. cit.* p. 319), the field evidence in Donegal indicates that some breccias form the roofs to large intrusions; he postulates that the intrusions, especially into pelite, are normally diapiric, but that toward the end of their uprise, gases escape into the fissures in the metasediments, disrupting and fragmenting the roof rocks.

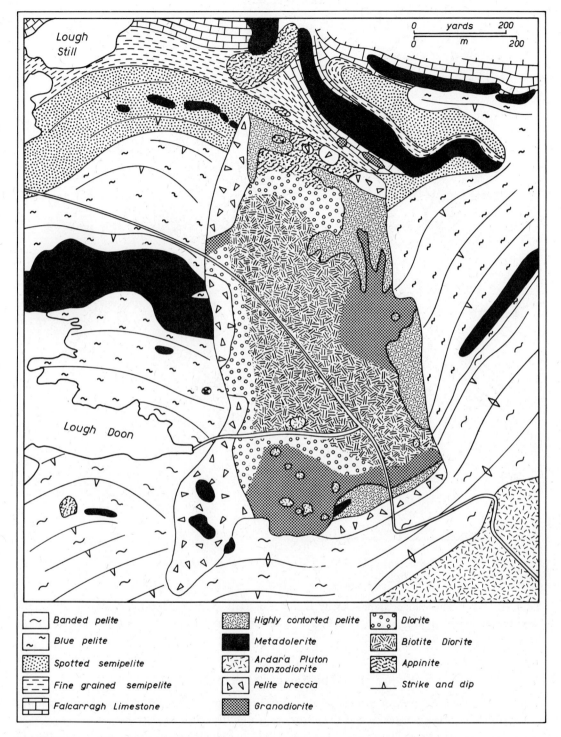

FIGURE 7-7. The Naran Hill appinitic intrusion (No. 10) and its marginal breccia. After French (1966).

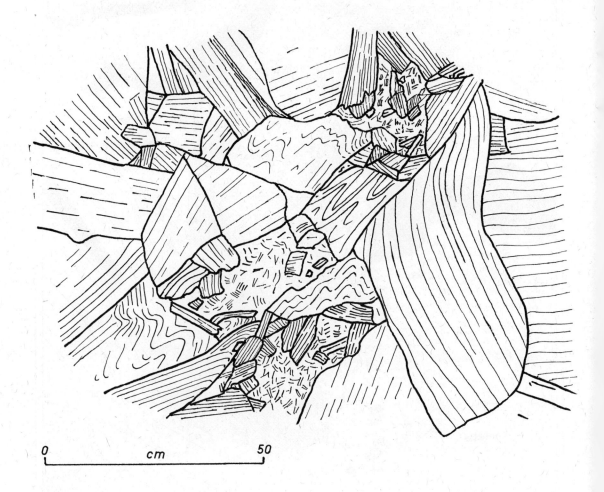

FIGURE 7-8. The marginal breccia of the Naran Hill intrusion located mid-way along the western contact of this intrusion (see Fig. 7-7). Note the remarkable "Inca-wall" nature of the fitting together of the blocks.

4. The Intrusive Breccias

The brief reference just made to a marginal breccia leads naturally to a discussion of examples where breccias form all or the greater part of the intrusion. Within the Ardara cluster, there are seven small areas (Fig. 7-3) of breccia and mobilized pelite, of which only two are sharply bounded outcrops of fragmented pelite, the remainder being composed of breccia and mobilized metasedimentary material in various proportions, forming a transition into the surrounding metasediments; within the mobilized rocks, the degree of contortion is progressively reduced outward. It is tempting to regard these occurrences as representing various denudation levels in the higher part of the roof zones of appinitic intrusions.

There are, then, a number of intrusive breccia pipes in which the material of the blocks has been transported upward for varying distances. As an example where transport was at a minimum, we refer to the breccia of *Dunmore* (A of Fig. 7-3), described by French and Pitcher (1959). This small, steep-walled pipe cuts sharply through the Falcarragh Limestone, and in it a core of breccia of blocks of this limestone is nearly contained within a ring of vesicular "felsite" (Fig. 7-9). This felsite is itself separated from the wall by a thin discontinuous skin of breccia, the fragments of

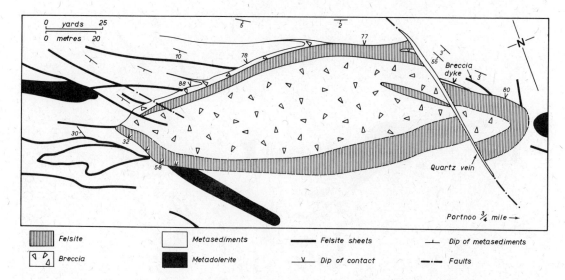

FIGURE 7-9. The Dunmore breccia-pipe (A of Fig. 7-3). After French and Pitcher (1959).

which are obviously in the process of being torn off the wall: from this peripheral breccia skin, breccia dykes penetrate out along the bedding of the country rocks, where they are accompanied by numerous felsite sheets closely associated in time with the pipe.

The breccia core consists of largely disorientated limestone fragments (*cf.* Fig. 7-11a), from small chips to blocks meters in extent, and either tightly packed together or separated by thin stringers of fine-grained material of felsitic aspect. In the latter, the crystals are fractured and veined; further, there is a very considerable range in composition, from felsite with a few grains of calc-silicate to a rock indistinguishable from the thin reaction zone of hydrous calc-silicate which margins the limestone blocks. These felsitic veins cannot represent the product of the simple intrusion of magma, and the evidence as a whole, especially the fact that the degree of metasomatism of the blocks seems inconsistent with the small amount of presumably magmatic material, suggests that the breccia was permeated by hot gases, probably rich in water and surely in carbon dioxide, in the van of a gas-charged magma.

In this pipe, then, country rocks have been shattered, fragmented, and mixed, without however moving very far and without suffering appreciable abrasion. As the outcrop of the pipe is very near that of the Portnoo appinitic intrusion (No. 6 of Fig. 7-3), it could well be directly connected with it at depth. Thus, we have a picture of a gas-charged, acid differentiate of an appinitic body coring into its roof—possibly representing the early stage of the formation of a vent-like intrusion which would eventually reach the surface.

There are, in fact, several pipes with far-traveled material which may well have been *open to the*

158   THE GRANITES

surface. Good examples are provided by the six breccia pipes which cut the Barnesmore Granite (Chapter 10, and Walker and Leedal, 1953): pipes which are very closely associated in time with the Barnesmore dyke swarm. The largest of these, the one at *Pollakeeran* (Fig. 7-10), actually consists of ten separate breccia pipes, all within an area 220 × 270 meters, over which the host granite is much

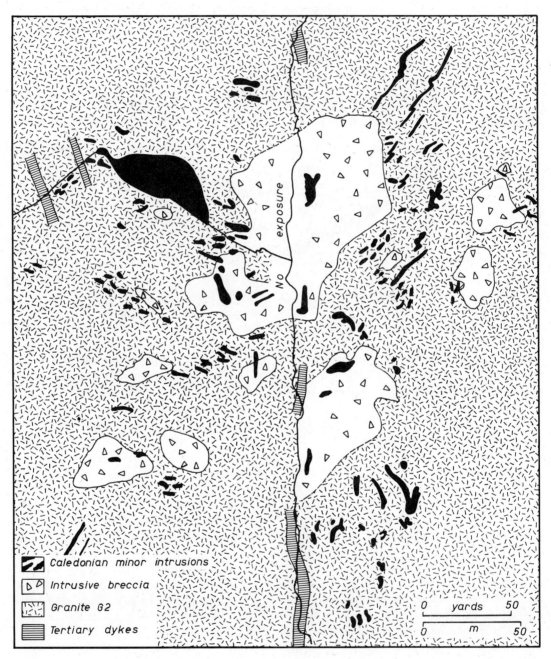

FIGURE 7-10. The Pollakeeran breccia-pipe complex cutting the Barnesmore Pluton. After Walker and Leedal (1953).

fractured. The pipes contain blocks of semipelitic schist and metadolerite, ranging in size up to 1.5 meters in length, the average being 15 centimeters, embedded in a matrix of comminuted granite and schist. Numerous magmatic intrusions of a wide variety of composition—though often lamprophyric in general character—cut, or are cut by, the breccias. These intrusions form dykes, bosses, and vein networks which show a roughly radial arrangement in relation to the pipes. The close association existing in time and place between the minor intrusions and the breccia pipes is thus very clear.

An important point concerning all the six areas of brecciation is that they lie along a line which is nearly coincident with the longer axis of the host granite: a strong structural control is thus indicated.

Clearly the pelitic material of these pipes has been entrained and transported a considerable distance, presumably from below, and has been abraded and comminuted in the process. Moreover, the very presence of such country rock material suggests a floor to the granite host, which may indeed represent the block which presumably subsided in order to permit the emplacement of this particular granite unit (see p. 204). If so, the source of the material is likely to have been thousands of meters below the present level—though we cannot rule out the possibility of the sinking of blocks in such tuffisite pipes.

There are quite a number of other examples of this type of breccia pipe in Donegal, and there is considerable variety in the lithology of the fragments, the degree of mixing, and distance of transport. We mention particularly two breccias where the blocks are dominantly of pelite because this is of some interest in consideration of the origin: one is a particularly fine line of pipes on Clogher Hill, southwest of Barnesmore (see Map 1 for localities); another single pipe is exposed on the cliffs south of Crannogeboys school in Loughros. In nearly all cases, the inclusions show some degree of alteration, and comminuted metasediments, variously recrystallized, dominate over igneous material in the matrices—an igneous fraction which is variously lamprophyric, felsic, or dioritic.

A very good example of long transport and consequent abrasion of the fragments is provided by the remarkable pebble rock which fills the small breccia pipe of *Kilkenny* (D of Fig. 7-3, and Fig. 7-11), described early in these investigations by Pitcher and Read (1952). The main component consists of small quartzite fragments lying in a scanty granophyric matrix; there are, also, some larger fragments of calc-silicate schists belonging to the Sessiagh–Clonmass Formation, which forms part of the walls of the pipe at the present level—these particular fragments tend to lie in an appinite matrix. Such a compositional relationship between clast and matrix suggested to Pitcher and Read that the matrices were produced by metasomatism of the comminuted material derived from the clasts. Another point of interest is the coating of the quartzite "pebbles" by a thin skin of amphibole and biotite. While we are at a loss to explain the physical chemistry of this surface nucleation of the mafic materials, we can record that it is not an uncommon feature (see, for example, Reynolds, 1936, 1954; Gindy, 1951a).

Perhaps the most interesting feature of this little intrusion is that the quartzite pebbles, the stratigraphic horizon of which is likely to be in the Ards Quartzite or low in the Sessiagh–Clonmass Formation, can be shown to have been transported upward at least 300 meters, suffering such considerable abrasion on the way that peculiar "dreikanter" shapes developed from the fragments. From these observations and the petrography of the matrix, the authors envisage quartzite fragments being entrained in a scanty gas-charged medium and rushing violently upward, suggesting perhaps that the pipe may have been open to the surface.

The most extreme case of comminution of the country rock by gas action of this kind is provided by certain pseudo-aplitic veins which cut the granite of *Tory Island* (Whitten, 1959a). These thin veins (5–6 centimeters in width) consist of minutely comminuted granite and quartzite, mixed as if intruded as a dust and partially recrystallized, presumably by the continued passage of hot gas through the conduit; the resultant rock is best referred to as a tuffisite. In fact, many of the matrices of the various intrusive breccias we have mentioned above clearly belong in this category.

At this point, we are naturally led on to a consideration of pipes and dykes of highly xenolithic

FIGURE 7-11. Intrusive Breccias: (a) the Liskeeraghan intrusion (No. 15 of Fig. 7-3) showing a chaotic breccia made up of fragments of banded pelite; (b) the breccia of Kilkenny (D of Fig. 7-3) showing quartzite fragments lying in a scanty granophyric matrix.

appinite, hornblende lamprophyre, or even felsite. An exciting example is provided by the small pipe at *Birroge* (B of Fig. 7-3), exposed along the main road between Naran and Portnoo (French, *loc. cit.* p. 317). This intrusion, which clearly dilates the pelitic schists into which it is emplaced, contains separate layers of densely packed pebbly xenoliths of both quartzite and calc-silicate, set in a hornblende-appinite. Such xenoliths were evidently derived from a part of the local succession which lies an estimated 600 meters below. Rounding of the fragments is so excellent that they resemble water-worn pebbles—in addition, the calc-silicate pieces show, to perfection, the production of amphibolitic rims.

Almost exactly similar relationships are shown by an appinitic sill at *Tallabrista* on the south coast of the Crohy peninsula (Gindy, 1951a). This time, the intrusion lies in quartzite, and only the calc-silicate fragments can be taken to have been transported any considerable distance. The latter are certainly in an advanced stage of alteration and have probably contributed material to the host, or so we interpret the darkening of the appinite in the vicinity of the amphibotilized calc-silicate material.

As we have seen, all these appinitic intrusions are very closely associated in time and space with dykes and sills belonging to a very varied suite of lamprophyric rocks of which we now give some brief account.

5. The Associated Lamprophyres

The composition of these minor intrusions ranges from very dark rocks composed almost entirely of hornblende, to acid types (plagiophyres) with only a small proportion of biotite compared with the dominant alkali-feldspar and quartz. As is usual in such rocks, there is a complex evolution of mineralogy and texture; French (*loc. cit.* p. 315–6) gives some of the details. Often, the brown hornblende is pseudomorphed by chlorite and biotite, the feldspar is corroded and altered to white mica and epidote, and there is much late chlorite and calcite in microveins, vesicles, and druses. Typical is the association of fresh and strongly decomposed minerals.

We may learn something of the history of crystallization from the chilled margins. Here, hornblende still retains its optimum size, despite the fact that the felsic minerals show a gradual decrease in grain size toward the contacts. Further, the amphibole becomes progressively more unstable into the center of the dyke, being replaced by biotite, or biotite and chlorite. Seemingly, the mafic constituents crystallized early as hornblende from a volatile-rich magma: these volatiles were then concentrated during the subsequent crystallization of the felsic constituents, as a result of which, during cooling, biotite, biotite-chlorite, and chlorite-calcite assemblages became successively the stable phases—with replacement and recrystallization the rule. Such a history of crystallization is typical of such rocks, and the significant role assigned to the volatile constituents has been emphasized by Smith (1946), Vincent (1953), and Bishop (1964), so that we need not follow all the details here.

In outcrop, such rocks are often seen to be patchy—almost pseudo-xenolithic. Some are, in fact, truly xenolithic and there is an instructive example, near Portnoo, of a lamprophyre dyke passing laterally into a breccia dyke in which limestone blocks are set in a sparse matrix of coarse calcite and pyrite and accompanied by a small amount of lamprophyre. This relationship points to the operation of a very high gas pressure during emplacement, and emphasizes the close affinity with the other intrusions of the suite.

Although we have not carried out a detailed study of these lamprophyres and associated rocks, we are convinced that their wide range of composition is merely a reflection of the equally wide range of the appinitic rocks proper, so that we regard the dykes as the fine-grained equivalents of the larger intrusions with which they are so closely associated in space and time. There is no doubt in Donegal of the consanguinity of lamprophyres and plutonic rocks.

6. The Origin of the Appinitic Suite

(A) The Intrusive Breccias: Explosion, Fluidization, and Rock Bursting

The breccias we have described above were clearly connected with plutonic igneous activity, and have only been revealed by deep erosion. However, some were obviously on their way to the surface, for the uprise of material for hundreds of meters can only be envisaged in a diatreme open to the atmosphere.

Even outside the obvious clusters, these breccias are much more common than is perhaps generally realized, so much so, that it is rarely possible to map a square kilometer of the metasediments within the Moine and Dalradian outcrop of Scotland and Ireland without discovering either a breccia pipe, breccia dyke or a xenolithic lamprophyre of one kind or another. Individual examples were recognized in the past, but it is only in recent years that their petrogenic importance has been properly understood: this is largely due to the application, in a stimulating essay by Doris Reynolds (1954), of the knowledge of the industrial process of fluidization (see also discussion in *Proc. Geol. Soc. Lond.*, No. 1655, 1969).

There is, as always, some difficulty in the nomenclature of such rocks, which has given rise to debate (*e.g.*, Bowes and Wright, 1961; Wright and Bowes, 1963). Because the use of the term agglomerate is inappropriate for those volcanic breccias formed by deep-seated fragmentation of *country rocks*, Bowes and Wright have carefully redefined three possible alternative terms as follows:

(a) *intrusion breccia* (Harker, 1908, p. 69), where the matrix is of igneous material and the fragments but little traveled.
(b) *explosion breccia* (Tyrrell, 1932, p. 168), where an igneous matrix is lacking, but still with little sign of transport of the fragments.
(c) *intrusive breccia* (Reynolds, 1954), with or without an igneous matrix, but where the entrained material can be shown to have traveled a considerable distance.

It is recognized that considerable overlap must exist between these types, and this is particularly well shown in the example quoted from Dunmore, where a central intrusion breccia (as defined above) is continuous with the explosion breccia filling the peripheral breccia dykes.

Whatever the nomenclature, it is clear from the world literature that such deep-seated brecciation is a common phenomenon, and its importance is recognized in an excellent review provided by Wright and Bowes (1968). It seems that breccia pipes, often referred to as pebble dykes in the western United States (see Bryner, 1961, for review), breccia dykes, or plugs of breccia situated on dykes, are associated with a great variety of igneous situations, and there are many examples of transport with abrasion of the fragments, the production of matrices by comminution of the entrained material, and the introduction of mineralizing solutions.

Before embarking on further discussions, we need to establish the relationship between the breccias marginal to the intrusions and the breccia pipes. There is, in Donegal, a close spatial association between vaguely defined areas of brecciated country rock, more sharply defined areas, breccia roof zones to intrusions, and distinct pipes full of exotic material. Indeed, in one case, Dunmore, it is probable that a breccia pipe is directly connected with a roof breccia. Thus, it is tempting to regard the various types as forming a genetic series, the material in the pipes having been the most thoroughly entrained and farthest transported. We might note that only a small part of the space required for emplacement of the associated magma can be attributed to this brecciation, unless, of course, material could be carried to the surface, or the fragments engulfed by stoping.

The only obvious mechanism for *moving* the material of intrusive breccias, especially those lacking an igneous matrix, is by entrainment in a high velocity gas stream, such as might exist in the van of intruding magma. One consequence would be the erosion of the fissure. Further, prolonged transport in such a moving gas stream would result in the rounding of fragments, particularly those that had traveled farthest, with the consequent production of the fine comminuted material of the matrices. Cementation would result from the introduction of "igneous" material along with the gas phase, by the redistribution of soluble materials by the gases, or by recrystallization brought about by intense heating. There is little problem here, and the main discussion must focus on the cause of the initial brecciation, on the formation of the primary fissures, and on the origin of the gas or gas-charged magma.

A variety of hypotheses have been advanced to account for the breccias. Early suggestions that the Scottish examples represented remnants of screes on an irregular land surface were soon dismissed as untenable and Cunningham-Craig, Wright, and Bailey (1911, p. 34) interpreted the breccias of Colonsay as explosion breccias, in the sense that volcanically disrupted material had fallen back down vents. However, Pitcher and Read (1952), appreciating the deep-seated nature of the process, appealed to a process of gas streaming for the emplacement of the Kilkenny pipe, and later Reynolds (1954), after describing the industrial process of fluidization, interpreted Kilkenny and a great many other breccias as the result of this process. Fluidization mechanisms were considered important by French (*loc. cit.*) and Platten and Money (personal communication), and have also been employed by others, such as Dawson (1962) in his discussion of the Basutoland Kimberlites (see also Harris et al., 1970).

It seems unlikely, however, that the initial brecciation could be the result of fluidization, which can only be responsible for entrainment and abrasion in an open system (Wright and Bowes, 1968). The brecciation might be attributable to explosion—the rapid expansion of compressed gas—at depth, a notion which was originally advanced by Rust (1937) for certain breccias in Missouri, and which has recently been strongly advocated by Bowes and his coworkers (1961, 1968), particularly in reference to the Back Settlement Breccia of Argyll.

In this *explosion hypothesis,* volatile constituents derived from the crystallization of a hydrous magma are presumed to have been trapped during their upward passage by a relatively impervious formation, so producing a sudden increase in gas pressure. A competent horizon, such as a quartzite, is likely to be well fractured, especially in fold cores, and cushioned explosions are thought to occur when this high pressure is dissipated into the joints and fractures. The shock waves fracture and brecciate the country rocks (*cf.* Hardie, 1963), conduits are opened up, and gas streaming leads to entrainment of the fragments.

There are, we think, some objections to part of this model. There is some doubt in our minds whether explosions of this type would occur at all. Further, there is little evidence, in many cases, of a trap situation, and there are abundant examples of breccias dominated by pelitic fragments and emplaced in wholly pelitic country rocks. We have referred to several in the Ardara cluster, and Platten and Money (personal communication) record many others. As regards the central problem, we may ask whether explosions can occur spontaneously at any great depth.

The *fluidization hypothesis* of Reynolds is based on the fact that when gas is passed rapidly through a bed of fine-grained solid particles, the latter are violently agitated and churned up, the bed expanding as a consequence. On increasing the rate of flow of the gas, bubbles are developed, and the particles eventually become entrained and transported. An important point to note is that the intimate mixing of particles and gas greatly facilitates chemical reaction; further, erosion of the containers constitutes a problem in the industrial application of this process.

The operation of a fluidized system, despite reservations which have appeared in the literature (*e.g.,* McBirney, 1959, 1963), could account for many of the features of breccia pipes. The degree of transport would depend on the gas velocity—large boulders could even sink *down* the pipe; prolonged churning would lead to the characteristic pebble-like forms; the development of more or less circular pipes would be expected, though differences in resistance to erosion would clearly be exploited. Further, because the mechanical effects are independent of the temperature of the gas, the variation in the alteration of fragments presents no problems. However, we must bear in mind that the fluidized system could hardly exist without free passage of gas, and, further, while not denying the efficacy of such systems to erode—especially in the final stages, where the expanded bed gives way to solids entrained in a flowing gas stream—it seems unlikely, as we have already indicated, that the shattering of the country rock can be attributed to the process.

At this juncture, it is of special interest to refer to the literature on *rock bursts* (*e.g.*, Gates, 1959, p. 809–812). Bursting results when stress in rocks is relieved by formation of a free face to which the stress becomes directed, the resultant strain being relieved by fracture. Natural stresses are present in rocks due to the weight of the superincumbent material; there are also stresses residual from regional deformation, and thermal stresses can be set up by differential heating. In mining operations, rock bursts vary from the spalling of small flakes from a free face to the complete brecciation of large volumes of rock, especially around tunnels.

Now, cooling of a gas or gas-charged magma must necessarily occur during its upward penetration into the country rocks, perhaps leading to so rapid a fall in pressure that stresses would be set up in the walls of the diatreme: if so, bursting may well occur.

There are other possibilities, including the one that fracturing might be caused by sonic waves (Gates, *loc. cit.*; Hardie, 1963), but it is probable that all these processes interact. In the van of bodies of highly gas-charged magmas, the prising action of hot gases permeating into established fractures might well be coupled with thermal spalling and rock bursting, the sum of these processes leading to the opening up of a conduit. Fluidization would be responsible for transport and abrasion, even perhaps some drilling by abrasion. As magma nears the surface, actual explosions are likely to occur, due to the increased possibility of rapid dissipation into open fractures. In Donegal, a possible example where all these processes were at work is provided by the group of pipes at Pollakeeran, Barnesmore (Fig. 7-10), where, perhaps due to rock bursting, the host granite is extensively brecciated around the pipes—the additional formation of dyke networks and the transport of material can be attributed to the fluidization process.

It is useful to consider the control of structure on the initiation of these intrusions, just as has been done by Bowes and Wright (1967) in Argyll. Such a control is demonstrated in Donegal by the alignment of the Barnesmore breccia pipes in the same direction as the local dyke swarm; we have also noted thin breccias along joints. Whether the necessary fissures were opened by intrusion of magma or by independent tectonic forces (as we believe), the result is the same—the provision of an escape route to the surface for concentrations of gas. Then there is the close relationship in space and time of the main cluster of appinitic intrusions and the Ardara pluton, which indicates to us that the diapiric action of the latter opened up fissures in the crust above; we envisage, for example, gas drilling being particularly effective along opening joints and faults, particularly at intersections of such structures. We therefore suggest that the development of the initial fissure was a result of tectonic forces—that the appinite-breccia suite of intrusions were emplaced in a kinetic environment, just as, indeed, we shall finally conclude for the whole Caledonian igneous event in Donegal.

The origin of the gases and the buildup of the high vapor pressures is to be sought in the magmas which gave rise to the associated hornblende-rich rocks, and it is this facet of Donegal geology to which we now turn.

(B) The Magmas

The appinitic suite of basic rocks is not unique to Scotland and Ireland, and indeed, if some care is taken to interpret the varied and loosely applied nomenclature of diorites, it appears that the association of appinites with granodiorites is a common one in orogenic belts (*cf.* Joplin, 1959). It is not surprising, therefore, that there are various opinions as to the origin of the rocks of this varied suite, ranging from the contention of Reynolds (1943) that certain basic rocks associated with the Newry granite are of metasomatic origin, through a more widely held belief that magmatic differentiation is the most important process (Nockolds, 1941), sometimes with hybridization a complicating factor (Deer, 1950), to suggestions that they were formed wholly by reaction between basic rocks and granites (Joplin, *loc. cit.*), or between calcareous rocks and granites (Pitcher and Read, 1952), or even by late stage alteration of early formed basic rocks (Wells and Bishop, 1955).

In any discussion of the origin of appinites, it is necessary to explain the three main features, *viz.*, the variations from ultrabasic to acid rock types, the alternative mineralogies (pyroxene or amphibole) of the basic types and their influence on the course of differentiation, and finally the relationship between the basic and granitic rocks.

As we have seen, the rocks vary from ultrabasic to acid in a continuous manner—with the reservation that, in some intrusions, the granitic members behave intrusively—and the highly gradational nature of the contacts may be taken as evidence of differentiation *in situ*. In general, the ultrabasic members are earlier than all the others, and the diorites earlier than leucodiorites and granitic rocks, a sequence which corresponds with the order of crystallization, *viz.*, the mafic minerals before the felsic minerals.

Hall (1967c) has shown that the chemical variation within the appinitic suite is quite distinctive, and cannot be matched in any other igneous association (Fig. 7-12a and b). However, the reported range of composition does include basaltic types, and, in fact, the appinitic diorites only differ essentially from basalts in their high OH^{1-} content. On the other hand, the hornblende-rich

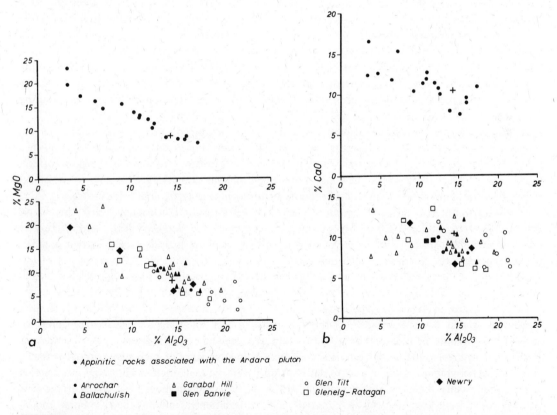

FIGURE 7-12. (a) Plot of MgO against Al_2O_3 for the Ardara appinitic rocks, and other Caledonian appinites. The average composition of basaltic rocks is marked by a cross. (b) Plot of CaO against Al_2O_3 on the same basis. After Hall (1967c).

ultrabasics are exceptionally rich in Ca^{2+} and low in Mg^{2+} in relation to the Al^{3+}, a feature most easily explained by assuming that, during crystal differentiation, the separation of hornblende took place, instead of the more usual olivine and pyroxene. That this is the true explanation is shown by the fact that analyses of the pyroxene-rich ultrabasics of the *same* suite show a greater enrichment in Mg^{2+} and lesser enrichment in Ca^{2+}.

The two types of ultrabasic rocks may represent, therefore, cumulates from amphibole and pyroxene-facies basaltic magmas, respectively, or possibly, as suggested by Hall, the early and later cumulates from an amphibole-facies magma—which would still initially crystallize olivine and pyroxene. It is important to emphasize that because such hornblendic and pyroxenic rocks represent end products with contrasted compositions, they cannot be regarded as the direct equivalents of one another (*cf.* Kennedy, 1935), and it follows that late-stage hornblendization cannot be the central process in the production of these particular amphibole-rich rocks. This is not to deny, of course, that the late-stage conversion of pyroxene to hornblende by the agency of residual solutions is an important secondary process; certainly, it is often invoked for just the association of rocks with which we are dealing (Wyllie and Scott, 1913; Nockolds, 1940), and was even proposed for the origin of the type cortlandtite itself (Shand, 1942).

In the appinitic suite, then, hornblende appears to be the stable phase throughout a large part of the history of consolidation. It is formed as a primary precipitate, and as a replacement in the solid of olivine and pyroxene; further, it can be wholly recrystallized, or occur merely as secondary overgrowths replacing earlier amphibole and even feldspar.

Though we may be able to accept a crystal accumulative model for these rocks in Donegal, there has been no serious attempt to explain how the differentiation so envisaged actually gave rise to the rocks seen in the field. Obviously, gravity can be of little influence in producing a zonation of rock types controlled by vertical walls. However, where basic rocks form the marginal phase, as at Mulnamin and Naran, a simple differentiation model of early-formed crystals, building inward from the cooling surface, will suffice: deviations from this simple picture may be due to renewed surges of the differentiating liquor producing the effect of successive intrusion, without however necessarily causing any marked break in the continuity of the rocks.

In the cases where the most basic rocks occur in a more central position, such as at Kilrean and Meenalargan, we are unable to suggest a convincing hypothesis, unless they represent accretion in a roof zone with the rest magma eventually surging upward and isolating portions of the early-formed, already consolidated, mafic phase. This may well have been the case at Meenalargan, where blocks of hornblendite show rapidly transitional contacts with rather leucocratic appinitic-diorites.

In contrast to the basic rocks, the leucocratic and acid types form a much more varied assemblage, and are, therefore, unlikely to have been formed as a result of any single process. Indeed, the very mixed appearance in many outcrops is good evidence that many are hybrids of one sort or another, as has been suggested for their equivalents in Scotland (Deer, *loc. cit.*). However, this facet of our study has not yet been followed up in Donegal in any detail, though it seems that Hall's crystal-fractionation model could account for the leuco-diorites, even some of those which show minor amounts of quartz in the last crystallizing residuum. There is, however, no marked trend of silica enrichment, so that alternative sources of the acid material must be sought. French has accounted for some of the variation by demonstrating that mixing processes take place between a wide variety of diorites and acid rocks, and no one can deny that there is an abundance of such acid material available in the associated granites. Thus, a picture develops of crystal differentiation modified by introduction of a second magma. There is also the possibility, as we have hinted above, that some of the new material could have been derived as a partial melt from softened and mobilized country rock— or so we interpret the close relationship of dark granodiorites and mobilized hornfelses in the Naran intrusion.

The relative importance of these several possibilities must be determined by future work, but we suspect that they all are involved to some extent, and that a complex picture will emerge—a combination of differentiation along two paths, with the direction depending on the volatile content of the magma, and hybridization by the addition of acid material from several sources; a history further complicated by late autometasomatic alteration and recrystallization.

The final problem concerns the explanation of the widened field of stability of hornblende. The fact that the composition range of the appinite suite spans that of the field of basalt, coupled with the evidence that volatiles under high pressure are present during the entire crystallization history, provides the clue. We know from the work of Yoder and Tilley (1962) something of the conditions of the crystallization of basaltic magmas under high vapor pressure—the essential point is that as the water vapor pressure rises, the field of stability of amphibole is progressively widened at the expense of olivine and pyroxene (Fig. 7-13), an effect which is probably amplified by the presence of a greater

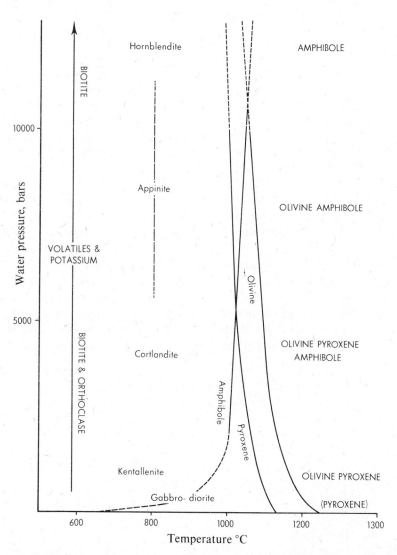

FIGURE 7-13. Representation of the P/T fields of rocks of the appinite suite based on Yoder and Tilley (1962). After Bowes and Wright (1967).

proportion of potassium in our rocks (*cf.* Bowes and Wright, 1967). Thus, alkali basaltic magmas crystallizing *either* amphiboles or pyroxenes, depending on the water vapor pressure, could be expected to give rise to rocks with a range of compositions similar to that of the ultramafic appinites. Apart from determining the course of crystallization differentiation, the high water content of such magmas, by lowering the viscosity of the melt, probably aided diffusion of ions to the cooling surfaces and widened the distances between crystal nuclei, also facilitated crystal accumulation, and increased the likelihood of metasomatic reaction on included material. Hall (*loc. cit.*, p. 168) even suggests that the release of pressure—presumably by opening of fissures to the surface—might provide a mechanism for the emplacement of some appinites in the form of crystal accumulates lubricated by a highly gas-charged magma (*cf.* Seifert and Schreyer, 1968), and this provides a possible explanation of the fractured and split crystals noted in some of our accounts.

Hall also finds that the composition of the Ardara Pluton, and other plutons with associated appinites, indicates that their magmas were generated under higher vapor pressure than plutons without such associations (p. 349).

From all these considerations, it seems clear that the differences between the appinitic and other basic rock series result from appinitic magmas having originated or differentiated under high water pressure (*cf.* Bailey, 1958). It seems unlikely that this water came from the mantle, which must have given rise to the great volumes of normal "dry" basaltic rocks (see, however, Vincent, 1953). Possibly, its source is to be found in the associated granites, for these rocks also originated under high water pressure (p. 347), and, as we have demonstrated, were in motion in the crust at more or less the same time as the appinites. While the evidence is against the appinites and granites forming a continuous comagmatic series, it seems possible that the latter might have supplied a low-melting acid fraction, together with much water. Water must also be driven off during contact metamorphism, so that a considerable cortege of emanations must of necessity precede a rising pluton.

We do not, however, exactly follow Joplin (1959) in envisaging these emanations and the granite magmas themselves actually remobilizing and hybridizing early basic rocks; rather, we think that basaltic magmas proceed upward faster than the granite magmas, the latter supplying the volatiles and opening the crust sufficiently for the gas-charged basic magmas to push to the surface. Perhaps the granites can sometimes overtake these basic magmas when the latter have lost their volatiles and therefore their mobility: clearly, the diapiric pluton of Ardara does just this, for, despite its providing the mechanical and chemical forces for the production of the appinites, it is itself, as we see in the next chapter, hybridized by mixing with the latter. The problem we must return to in our synthesis is why the two magmas can be in a depth order—acid underlying basic—which is opposite to expectation; perhaps we can begin to doubt that granite magmas are born in the crust.

CHAPTER 8

EXAMPLES OF DIAPIRIC INTRUSION

CONTENTS

SECTION A. THE ARDARA PLUTON	169
1. Introduction	169
2. The Components as Magmas	170
(A) The Outer Component	170
(B) The Central Component	171
(C) The Minor Intrusions	172
3. The Contamination Model	173
4. The Structure of the Pluton	174
(A) Foliation	174
(B) Fracture Systems	177
5. The Structure of the Envelope	178
6. Contact Metamorphism in Relation to the New Structures	180
7. The Ardara Pluton as a Diapiric Intrusion	182
(A) The Model	182
(B) Discussion	184
SECTION B. THE TOORIES PLUTON: A FRAGMENT OF ANOTHER DIAPIR	184

SECTION A. THE ARDARA PLUTON

1. Introduction

H. H. Read (1957, p. 332) says of diapiric granites that "the mechanism of emplacement of these circumscribed bodies is more often to be determined from the structures of their country rock walls than from their own," a statement which is particularly applicable to the study of the Ardara Pluton of western Donegal.

This member of the Donegal Granite Complex has a comparatively simple circular outcrop, some eight kilometers in diameter, with a northeasterly prolongation which we shall refer to as the "stalk" (see Map 1 in folder and Fig. 8-3). It is emplaced almost entirely in the Upper Falcarragh Pelites as a concordant body which has profound structural and metamorphic effects on its envelope. One of these effects, the bending of strata around the pluton, was clearly brought out by the original mapping of the Officers of the Geological Survey (Hull et al., 1891); nevertheless, it was left to Cole (1902) to suggest that forceful intrusion produced the remarkable structural conformity between the foliation in the country rock, the gneissose border of the pluton, and the strike of the "uplifted" schists. More recently, a very clear picture of the nature of this forceful emplacement has emerged from a detailed examination by Akaad (1956a and b), and it is this study on which we draw heavily for the following account.

Concerning the age of the Ardara Pluton relative to the other events in Donegal, its emplacement history overlaps but outlasts the period of intrusion of the appinitic rocks we have just described, and is certainly later than many of the undeformed felsites; it is therefore later than all the regional deformation phases. On the other hand, it is earlier than the Main Donegal Granite, which cuts across the stalk of the Unit; indeed, enclosures of the outer members of the Ardara Pluton are strewn along the southeastern margin of the Main Granite. Further evidence of the relative chronology is rather circumstantial. It is unfortunate that the aureoles of the Thorr and Ardara Plutons just fail to overlap (Fig. 14-1), but certain porphyry dykes comparable with those of the Rosses swarm cut the Ardara aureole, suggesting a time relationship *vis-a-vis* the Rosses (Akaad *loc. cit., a*, p. 275). Thus we are prepared to guess that the emplacement of the Ardara Pluton followed the Thorr and preceded both the Rosses and the Main Donegal Granites.

2. The Components as Magmas

The Ardara Pluton consists of just two main components, an outer ring of quartz monzodiorite, between 230 meters and 900 meters in width, which almost completely encloses a granodiorite core. The former is fairly uniform in composition, while the latter varies gradationally from a quartz monzodiorite through monzotonalite to a granodiorite, in the manner shown in Figure 8-1.

The outer contact is sharp, and there is no compositional change in the intrusive as it is approached. Between the two main components, there is also a sharp contact, though no chilling of either member is ever seen; indeed, this contact is locally interlobate as if the outer component had only partially consolidated before the emplacement of the central component (Fig. 8-5).

A dominating feature of the fabric of the rocks of the pluton is the gradual appearance and strengthening of a foliation as the outer contact is approached; it is this planar structure that we are particularly concerned with later in this account, but, first, we deal with the rocks as mineral aggregates.

(A) The Outer Component

The medium-grained, quartz monzodiorite or monzotonalite of the outer ring has a general modal composition 45–50% of plagioclase, 20% of potassium feldspar, 15% of quartz, 10–15% biotite, and up to 5% of hornblende.

The plagioclase, which characteristically lacks distinct zoning, forms grains of several types; the most distinct are the small, mutually interfering crystals (0.5–1.0 millimeters) of inclusion-free oligoclase, and the larger crystals (5–10 millimeters) of composite and inclusion-rich megacrysts of oligoclase. The microcline-microperthite, like the quartz, is usually interstitial and anhedral against all the other minerals. Microcline also occurs as a discontinuous mantle to oligoclase, or as small, antiperthitic patches scattered throughout its length. Biotite and hornblende occur intergrown in aggregates along with the accessories, and are commonly partially replaced by the plagioclase and quartz. In summary, it seems that none of these textural features are of straight magmatic origin, and indeed virtually every mineral relationship shows some evidence of recrystallization and replacement in the solid state (Akaad, 1956a; Hall, 1966d). It seems that the texture as a whole is crystalloblastic and "metamorphic" in origin.

All outcrops of the outer monzodiorite show some xenoliths aligned—and flattened—in the

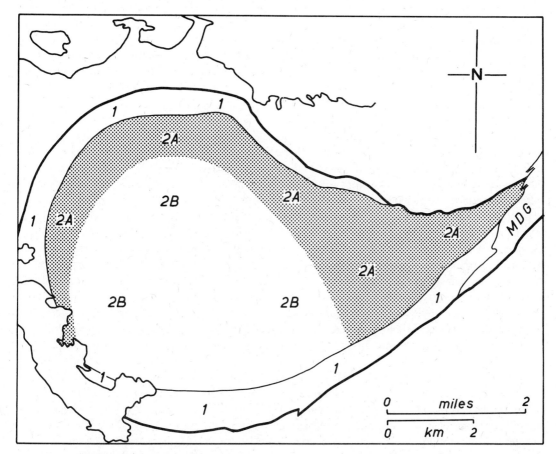

FIGURE 8-1. The components of the Ardara Pluton. (1) Outer Component, the quartz monzodiorite. (2) Central Component: (A) quartz monzodiorite or tonalite with many appinitic xenoliths; (B) granodiorite.

foliation plane (Fig. 8-4). These are mostly small, only a few being more than a meter in greatest dimension. The greater proportion were clearly derived from metasedimentary rocks with a contribution from the metadolerites, but there is a more restricted group representing the diorites and amphibolites of the appinitic suite.

(B) The Central Component

As we have already noted, the central component is much more variable in composition than the outer. The northern and eastern part (Fig. 8-1) is a quartz monzodiorite or monzotonalite with many xenoliths, and the western, central, and southern parts are granodiorites with but few xenoliths. There is a complete gradation between these facies, and although all rocks contain approximately 45–55% of plagioclase and 15% of potash feldspar, the quartz varies from 25% to 15%, with sympathetic changes in the biotite (5–15%), hornblende (0–5%), and epidote and other accessories (0–3.5%).

The more foliated types show an approach to the crystalloblastic textures just described, but for a large part, the mineral relationships are more normally igneous in character. Examination of the latter shows that there are also considerable textural differences between the extremes of the compositional varieties, though there is a continuous textural gradation between them.

The plagioclase crystals form clusters made up of a mutually interfering mosaic, in which the zoning within the individual crystal units is abruptly truncated at the mutual contacts. The range in zoning is from An_{25} to An_{19} in the most basic varieties, and from An_{17} to An_{14} in the most acid, where the zoning is also much more regular. Plagioclase is usually euhedral against the microcline and quartz, though there are the usual myrmekitic outgrowths from one feldspar into the other. In these rocks, again, the dark minerals tend to occur in clots, particularly in the more basic varieties. The biotite crystals show good faces against the plagioclases in the acid types, but are intergrown in such a manner that Hall concluded that biotite was here being replaced by plagioclase. The microcline-microperthite is usually interstitial, though in the granodiorites, it tends to form larger crystals enclosing euhedral crystals of biotite and plagioclase; nevertheless, there is no evidence here that microcline has replaced the adjacent minerals. Finally, quartz tends to occur in relatively large, distinct aggregates in the granodiorites, but is in somewhat more scattered isolated grains in the monzodiorites, where symplectitic intergrowths with biotite and hornblende are usual.

Hall (1966d) has attempted to explain these textural differences on the basis of a model involving the contamination of magma which then underwent a normal sequence of crystallization. The latter was explained in terms of the system An-Ab-Or-SiO_2-H_2O, in accord with the phase relationships predicted by Carmichael (1963), as shown in Figure 8-2. Thus, according to Hall, the composition of the most acid type indicated in Figure 8-2 requires that the magma must have started to crystallize in the field of plagioclase + quartz. Precipitation of oligoclase would move the composition toward the plagioclase–alkali feldspar–quartz boundary, fairly near to its low temperature end, where the three minerals would crystallize together until all the magma was used up. The clustering of the plagioclases and the quartzes, indicating early nucleation of both, the progressive and regular zoning of plagioclase with the production of more sodic margins, and the inclusion of euhedral plagioclase crystals in the microcline, never the reverse, seem (to Hall) to fit well with the experimental evidence, suggesting that the most acid parts of the central components were formed by magmatic crystallization. We are not, however, so convinced of the validity of this model because we can only see evidence for late coprecipitation of quartz and microcline, without plagioclase.

Hall further considers that the replacive features which appear in the more basic parts of the central component must be related to the contamination of the magma by basic material. He refers to the well-known hypothesis (Bowen, 1928) that reaction between sodic magma and a xenolith containing calcic plagioclase would result in the exchange of sodium in order to establish equilibrium; the resulting more sodic plagioclase might well be expected to increase in volume, not perhaps by causing the xenoliths to disintegrate, as suggested by Bowen, but by partially replacing the adjacent minerals. That equilibrium is so established is shown by the now identical pattern of plagioclase zoning in xenolith and host. Hall, (*loc. cit.*, p. 218) also argues that crystallization of a sodic plagioclase in the xenoliths must be accompanied by the new growth of quartz, explaining the intergrowths with biotite and hornblende, and if the amount of plagioclase crystallizing in the xenoliths were large enough to cause the composition to approach the ternary boundary, this might explain the apparent coprecipitation of microcline and plagioclase in the more basic rocks.

We are not entirely convinced of the applicability of these various explanations of the textures, and we shall have more to say on this matter later; our immediate concern is with the implication that a great deal of xenolithic material may have been incorporated into a magma of a composition near to that of the central granodiorite.

(C) The Minor Intrusions

It remains for us to note the presence, particularly in the outer part of the pluton, of a crudely radial system of minor dykes and veins of pegmatite-aplite, and also a few sheets and dykes of microgranodiorite. Even later, and perhaps having little genetic connection with the pluton, is the narrow zone of replacive tourmaline-quartz rock which sometimes occupies the outer contact plane.

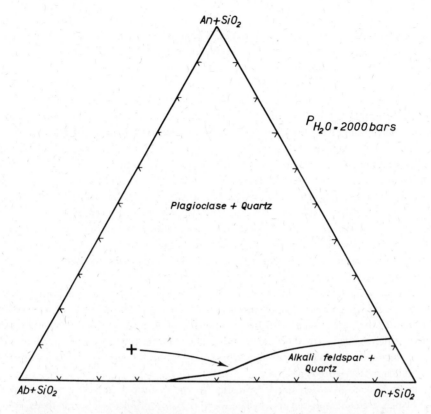

FIGURE 8-2. Suggested phase relations on the quartz-feldspar interface in the system An–Ab–Or–SiO$_2$–H$_2$O; (+) marks the composition of the least contaminated granite of the Central Component. Crystallization of the early anorthitic plagioclase would move the composition of the magma towards the phase-boundary, when alkali-feldspar would crystallize along with the plagioclase and quartz. After Hall (1966d).

3. The Contamination Model

The monzodioritic varieties of the central component always contain many xenoliths of dioritic and amphibolitic rocks. Although these sometimes occur in dense swarms, more normally, they form only a very few per cent by volume of the host; further, they are usually small, of the order of 5–10 centimeters. Even within the same outcrop, these xenoliths include a mixture of rock types which are *identical to the components of the appinitic amphibolite complexes* lying within the envelope (Fig. 7-1), and, in two cases, actually forming the wall rock. There are even some examples of adjacent angular xenoliths having contrasted lithologies which are so similar to those in the appinitic complexes themselves that it is abundantly clear that the variation was well established prior to their inclusion in the pluton.

It is particularly interesting to note that this might not be the first impression in the field, where the observed variation might be thought to represent a series of stages in reaction with the present host. The mineralogy of the xenoliths is, in fact, so similar to that of the host that it is probable that equilibrium would have been easily established early in the plutonic history, mainly by reconstitution of the plagioclase and conversion of hornblende to biotite, leaving the process of incorporation to be brought about mainly by mechanical breakup. That imbibition actually occurs in this way is shown by the disintegration of xenoliths to form the schlieren bands, rich in aggregates of hornblende and biotite (Akaad, *loc. cit.,* p. 273). We also consider that the mafic clots which are a characteristic feature of the darker rocks, are of xenolithic origin (*cf.* Chapter 5). Thus, contamination was certainly an important process in the formation of the monzodiorite, and, indeed, Hall has provided a semiquantitative model of the incorporation of country rock.

From what can be seen now, it is probable that the greater part of the xenolithic material in the *central* monzodiorite was derived from early basic rocks, though Hall suspects that the absence of pelitic types is simply due to the more thoroughgoing assimilation of the latter. Hall has considered the quantities which might have been involved, assuming that mixing occurred on a simple arithmetical basis. It seems that it would be necessary to add approximately 30% (by weight) of basic igneous rocks to 70% of granodioritic magma to account for the changes in Si^{4+}, Ti^{4+}, Fe^{3+}, Fe^{2+}, Mg^{2+}, and Ca^{2+}. However, in order to explain the K^{1+} concentration on this basis, it would be necessary to assume the addition of even greater amounts of pelite—though, in this event of doing so, it becomes impossible to get any calculations to agree with simple proportional additions.

The composition of the *outer* monzodiorite with its higher K^{1+} content, would require, it seems, a further increase in the ratio of pelitic to basic rock. Thus, making the same assumption as above that the granodiorite represents the parent magma, Hall computes a proportion of 15% basic rock, 75% pelite, and only 10% magma.

On simple energy considerations alone, it seems unreasonable to us that the added material should so greatly outweigh the contribution of the magma, especially as (at Ardara) the latter shows so little reaction with the walls, and even with a proportion of its xenoliths. This, together with the need to explain the time necessary for thorough reaction and mixing, prompts us to envisage incorporation taking place during the whole of the upward migration of the magma, such incorporation being presumably much more vigorous at depth.

Although we hesitate to accept exact estimates for the quantities involved, we are convinced that the monzodiorites are contaminated rocks. The more pressing problem, it seems to us, is the explanation of the relatively high potassium content of the outer zones of the pluton, which is a reflection of the high biotite content. This cannot easily, as in the case of Thorr, be attributed to a late, feldspar-producing, potassium metasomatism, but it still remains a possibility that there was a relatively early influx of this element, leading to the conversion of much hornblende to biotite.

4. The Structure of the Pluton

(A) Foliation

The steep outer contact is remarkably sharp, smooth, and free from irregularities. Inside it, the outer component shows a strong foliation, everywhere trending parallel to the contact and dying out progressively into the periphery of the central member (Fig. 8-3), a feature beautifully confirmed by the varying intensity of the magnetic fabric, as determined by King (1966). It is of interest to note that although there is no evidence in the field of a linear component to the mineral alignment— and recent studies have not confirmed Akaad's report (1956a) of a preferred horizontal alignment of long axes of xenoliths—the magnetic fabric defines a clear lineation with a generally steep plunge (King, *ibid.*).

There is some suggestion (Akaad, *loc. cit.,* p. 276) that the foliation in the central component is somewhat later in time of formation than the stronger structure in the outer component: in the latter, the foliation is certainly consistently steeper in attitude and *fails* to cross certain intrusive sheets, which are related to, and possibly even connected with, the central component. However, the completeness of the transition leads us to conclude that the foliation is really a single structure, the attitude of which was increasingly controlled by the contact as this was approached (see also Fig. 8-5).

EXAMPLES OF DIAPIRIC INTRUSION

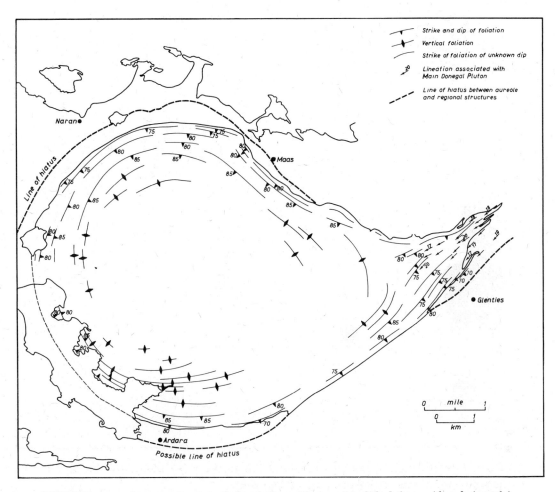

FIGURE 8-3. Structural map of the Ardara Pluton. Modified from Akaad (1956a).

The structure is marked by the biotite crystals, particularly by the lenticular nature of the little biotite-rich aggregates. In addition, in the outer parts, an augen texture is developed, the biotites forming a selvedge around aligned plagioclases; here, too, there are zones of granulated quartz which bend around the augen.

The many dark patches make up an essential element of this foliation. In shape, they are often oblate ellipsoids which become *progressively* more aligned and flattened as the main contact is approached until, in the gneissic border, they are represented by disc-like plates (Fig. 8-4 and Table 8-1). While many of these xenoliths have been made over from country rock material into mere concentrations of dark minerals similar to those of the host, there are also slabby pieces of much less altered metasediments which also mostly lie flat in the foliation. Some of these latter have clearly been boudinaged in this plane, and in such cases the foliation is invariably deformed around the Boudins, as indeed it is around any particularly lumpy enclave (Fig. 8-4a). The deformation involved is essentially a flattening ($k = 0$), and estimates of the maximum flattening in terms of axial ratios are shown in Table 8-1; measurement is, however, complicated by the fact that the xenoliths were probably originally very irregular in form (see Chapter 11 for discussion).

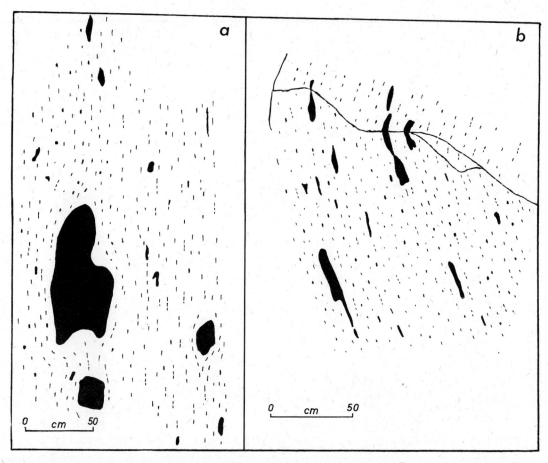

FIGURE 8-4. Foliation and flattening of xenoliths in an outer (a) position within the Ardara Pluton (b) and an intermediate (facies 2A of Fig. 8-1).

TABLE 8-1. Average Ratios Shortest Axis: Longest Axis of Amphibolite Xenoliths

Outer Component	Junction zone	Outer part of Central Component
0.29	0.41	0.63

Further evidence of deformation is shown by the way in which some of the pegmatite-aplite veins are disrupted (Akaad, 1956a, Fig. 4). The further fact that some veins are foliated while others are not confirms that the formation of the foliation was essentially synchronous with the intrusion history.

Another important observation concerning the nature of this foliation is that it is not concordant with minor irregularities in the contacts: for instance, it passes through and across the irregularities in the junction of the outer and central components (Fig. 8-5), and, along the main contact, can cross into the hornfelses. Obviously, these relationships could only occur if the structure were imposed while the host rock was essentially in the solid state (Chapter 15).

The crystalloblastic textures we have described, particularly those in the outer component, also indicate that deformation occurred at some time during the growth history of the minerals. It is even possible that a partial homogenization is represented by the paucity of zoning in plagioclases and this

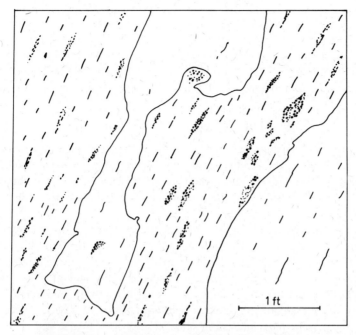

FIGURE 8-5. Contact of Outer and Central Components (the latter least xenolithic) showing the independence of the foliation where this contact is irregular. After Akaad (1956a).

may be the result of deformation activating the movement of the constituents (see p. 334 and Mueller, 1963, p. 768).

The conclusion seems inescapable that this planar structure was continuously imposed by deformation of the host throughout the history of the crystallization or recrystallization, and while the host was essentially solid. The disappearance of the structure inward and the restriction of cataclasis to the outermost parts indicate to us that the periphery of the pluton was consolidating and being deformed while the center was still mobile and capable of being added to by magma.

There is, of course, the possibility that this structure is a composite one and that an early "flow foliation" was followed, especially in the margins of the pluton, by a deformative "secondary foliation." Such a distinction has been made in some studies (*e.g.*, Phillips, 1956), yet it is usual to find that, even in association with flow foliation, the xenoliths are "drawn out"; also, that there was a thoroughgoing recrystallization during the imposition of the secondary foliation, producing the reorientation of the tabular minerals. It seems logical, therefore, to regard the two structures as successive stages of the same process, the differences being related to the changing rheological behavior of the host (see Chapter 15).

Foliated borders to plutons are relatively common, and we consider that such structures are, for the most part, formed in the solid state. Certainly this is the case at Ardara, where the complete identity in attitude and time of formation of the foliation, within and without the pluton, suggests that the pluton and its walls were deformed together—a thesis which we shortly develop below.

(B) Fracture Systems

Joint systems with a strongly marked preferred orientation are restricted to the foliated margins of the pluton. Here, there is commonly a vertical joint lying normal to the foliation, and sometimes another, still radial but with a dip between 45–50°. Both joints cross the contact into the envelope, and both are sometimes occupied by pegmatite-aplite veins. There is also a system of barren steep joints lying parallel to the foliation.

Such radial and tangential systems have been recognized in many plutons; they are, for example, respectively the B–C and A joints of Martin (1953) and they fit very well the later stages of the evolution of a distending pluton (see Chapter 15 for further discussion).

5. The Structure of the Envelope

The principal structural changes induced by the emplacement of the Ardara Unit were the deflection of the rock groups, the bending of the country rock structures into concordance with the contact, the tightening of preexisting folds, the tangential flattening, and the elimination of certain horizons by thrusting and faulting; all these taking place within about a mile from the contact (Akaad, 1956b).

Along the northern flank of the Pluton, the deflection, particularly of the outcrop of the Falcarragh (Portnoo) Limestone, is obvious from Map 1 in folder, and is particularly impressive in the field. The axial planes of the regional (F_{2-3}) folds are clearly bent into the upright and tightened. Around the southern and western flanks of the pluton, the contact takes advantage of the regional swing of structures, but local deflection of trend is again very evident westward of Ardara, where it is likely that the axial plane of the Loughros Anticline of northern Loughros has been deflected into a position parallel with the contact. The most interesting situation arises in the west and northwest (Fig. 8-6), where the regional strike is more nearly normal to the contact (Lunn, personal communication). Here, it is most obvious that the structural changes involved in swinging the strike into complete conformity with the contact do not show a gradual transition from the regional pattern; there is, in fact, an abrupt change along an arcuate line parallel to and some 800 meters from the contact (Fig. 8-6). Inside this line, the dip of S_{2-3} quickly steepens, in some cases by 40° in less than 50 meters, and the regional folds are truncated in a most remarkable way. Though there is no obvious fault along this line, nowhere are exposures continuous across it.

This structural *hiatus* may be of the same type as the dislocations recognized by Akaad as approximately limiting the structural effects of the pluton in the north. This worker shows that, where the Falcarragh Limestone enters the aureole (see Map 1 in folder), its upper members are progressively eliminated along what appears to be a bedding-plane thrust. He also considers that considerable strike-slip occurred at the same time on a Maas–Ardlougher fault, a movement which may have been connected with the rupture of the outer monzodiorite skin, which allowed a northerly protrusion of the central component to extend out to the main contact. This explanation again suggests a link between intrusion and deformation in the envelope.

The degree of exposure around the southern contact of the Pluton does not allow us to determine whether similar dislocations or lines marking especially rapid changes of structure are present all around the pluton, although we think that this is likely and have indicated such a situation in Figure 8-3.

From the hiatus to the granite margin, the regional structure is progressively modified (Fig. 8-8). The regional S_{2-3} schistosity is bent down into an upright, even slightly overturned, position to wrap neatly around the pluton. In an outer zone, the regional folds are recognizable; they then become distorted and tightened until, adjacent to the pluton, there is an impression of a swirling which may be due to the interference of refolded folds or even to partial mobilization.

In this inner position in the aureole, there is a single strong schistosity lying parallel to the contact, and this we consider to represent the regional structure (S_{2-3}), steepened and highly intensified. Locally, however, especially some little way from the contact, a new steep crenulation cleavage (Sg) is superimposed on the old fabric (Fig. 8-6), and the conclusion must be that the contact deformation was also capable of producing a new cleavage. It seems that where the old structures were in an

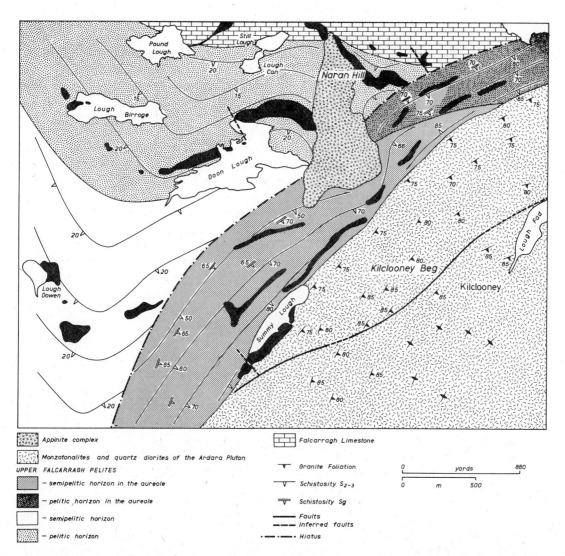

FIGURE 8-6. The northwestern segment of the Ardara Pluton showing the change across the hiatus line between the regional structures and those in the aureole; also the correspondence between these latter structures and the foliation in the Pluton. Metadolerites in black. Mainly after Lunn (unpublished work).

appropriate position to take up the newly induced strains—as in the inner aureole—then no new cleavage was produced, and the old schistosity was merely rejuvenated and reinforced.

Further evidence that flattening has occurred is provided by the occurrence of various boudinage structures, whose axes lie at random within the plane of the steep schistosity. Again, the metadolerite sills are far more strongly disrupted in the aureole than in the purely regional environment, and, in addition, the late-tectonic lamprophyre sills are also broken and separated into rhomb-shaped boudins, the shape of which is controlled by early cross fractures in the sills.

In addition to the rejuvenated S_{2-3} and the new cleavage (S_g), there are certain late, vertical shear zones which also trend parallel to the margins of the pluton. They have been attributed (Akaad, *loc. cit., b,* p. 284) to upward movement of the pluton after it had largely solidified, in which case they are probably closely related to the cataclastic phenomena within the periphery of the intrusive. There is the distinct possibility, however, that they represent the DMG_2 or DMG_3 phases of deformation connected with the late stage of emplacement of the Main Donegal Granite (see Chapter 12).

6. Contact Metamorphism in Relation to the New Structures

All the rocks of the aureole are completely recrystallized biotite-muscovite-andesine-quartz rocks which everywhere show a strong foliation: such rocks are best referred to, as a whole, as *contact-schists*. These metamorphic effects extend to a distance of just over $1\frac{1}{2}$ kilometer from the pluton, *i.e.,* over much the same distance as the structural changes are recorded, and in the latter connection, it is especially important to note that the hiatus forms an effective outer limit to the bulk of the thermal changes.

The mineralogical changes have been described by Akaad (1956b), and by Pitcher and Read (1963), particularly by the former, who established the following zonal sequence (see also Naggar and Atherton, 1970). In the outermost part of the aureole, biotite quickly replaces the chlorite of the regional muscovite-chlorite-phyllites, and, at the same time, the chloritic spots after regional garnet are converted into new biotite aggregates: the superimposed nature of the Ardara contact metamorphism is thus clearly established. Farther in toward the contact, comes an andalusite zone up to 500 meters in width, and this is followed inward again by a sillimanite zone up to 150 meters wide, with a narrow cordierite-rich zone at the immediate contact.

Concerning the porphyroblastic minerals, the andalusite is often accompanied by staurolite, and there is quite abundant kyanite in one particular pelite lithology (Naggar and Atherton, *loc. cit.*); further, sillimanite is accompanied by a new garnet. As we shall confirm later, we have no doubt that the staurolite and kyanite are aureole minerals, for they are most closely connected in time and space with the contact effects of the pluton. In the outer aureole, the andalusite porphyroblasts grew within the reinforced schistosity, overprinting the aligned micas and even the relict microlithons (Fig. 8-7a): there is, however, no sense of any preferred orientation in this plane (Fig. 8-7b). The kyanite and staurolite, sometimes in parallel growth, occur as small equidimensional crystals, rather evenly scattered throughout the rocks and overprinting the fabric. They are enclosed by the large andalusites, never the reverse, indicating a definite difference in time of nucleation (see p. 309 for furthr discussion). Nevertheless, there is not the *slightest evidence for any replacement or resorbtion of one aluminosilicate in favor of the other.*

The growth of porphyroblasts appears to have been static in character, but it was followed by a further flattening across the schistosity, as is shown by the way the micas of the matrix sweep around their porphyroblasts, particularly the large andalusites, producing an augen structure (Fig. 8-7a and 14-7a). This relationship is not seen in the rocks of the innermost aureole, where andalusite is simply overprinted on the single schistosity which we have shown to represent the completely reconstructed regional schistosity. This could be because the growth of the andalusite here outlasted this deformation or because these inner rocks were competent enough to resist the stresses. The presence of cored andalusites in this position (see Fig. 14-7b) possibly fits with this suggestion, the core being equivalent to the augen in time of growth. The sillimanite also overprints the schistosity here, and may even be sensibly later than the andalusite (see p. 324). Lastly, the micas and quartz appear to be able to recrystallize at each stage in this complex history.

Finally, a late-stage retrogression is represented by the growth of new muscovite and the replacement by sericite of the aluminosilicates—a common feature of all aureoles.

Whatever the complications, the Ardara aureole provides plenty of evidence that the growth of minerals and deformation have gone on together, albeit that the two processes form separate though overlapping phases. This is exactly the situation which appertained in the pluton itself, and simply

FIGURE 8-7. (a) Skeletal Porphyroblasts of andalusite clearly overprinting a crenulation cleavage (regional S_{2-3}) which has subsequently been slightly flattened around them. Upper Falcarragh Pelites on Clooney Hill. Photograph by W. M. Edmunds. Plane light. (b) Andalusite growing at random on a surface marking the rejuvenated schistosity S_{2-3}. Clooney Hill.

confirms what is obvious in the field, that the internal and external structures represent the same response to the forceful emplacement of the pluton.

7. The Ardara Pluton as a Diapiric Intrusion

(A) The Model

The model for the emplacement of the Ardara Pluton must involve two magmatic pulses fairly near one another in time. The almost total lack of veining and reaction at the outer contact of the pluton suggests that the outer component was emplaced as a semisolid crystal mush, the composition of which had been determined by contamination of a granodioritic magma at depth. Possibly, this material was emplaced before, and independently of, the central component, which then displaced the more mobile material of its interior. However, we prefer, like Akaad, a more kinematic model which envisages a skin of viscous monzodiorite being swept up and along with the second pulse, thus helping to seal in the magma of the central component from the confining country rock. The contamination of this central component was due to inclusion of material dragged away from the many basic bodies which must have existed at depth, just as they do at the present level—indeed, this process can still be seen, frozen in its operation, in the "stalk" area where the Meenalargan complex has been invaded and disrupted by the central component (Iyengar et al., 1954).

The emplacement of the pluton has obviously been accompanied by distension of its envelope. Possibly, as we have hinted, this occurred in two stages (*cf.* Akaad, *loc. cit.*, p. 286) commensurate with two upwellings of pluton material, but nevertheless, the total effect was to uniformly deform together the country rocks and the peripheral parts of the pluton. As there is no important repetition of strata, the major structure in the deformed zone *cannot be a simple rim fold*. Indeed, where exposures are particularly good around the northern contact, the dip, which is regionally toward the pluton, is seen to simply, but rapidly, *increase* in inclination; there is certainly no proof of pushing upward or downward, and the evidence is best interpreted as due to outward pushing with the progressive thinning of strata inward. Neither is there any field evidence for upward flow in the pluton, rather only of expansion, evenly "distributed" in the horizontal plane. There is, however, King's demonstration (1966) of a steeply plunging linear magnetic fabric (see p. 174). It seems that, although the pluton must have moved up and pierced the envelope at some time early during its emplacement, from then on the size was increased by radial extension: thus, a concept which can be likened to that of an expanding balloon (cf. Martin, 1953).

Distension was in all directions within the plane of the contact. Over a distance of 800 meters, the envelope was deformed together with the carapace of the pluton, perhaps suffering an overall thinning of the order of 50% in the process (see Table 8-1). When we attempt to discover to what extent the space problem can be solved by taking into account a combination of this thinning in the deformed zone and the regional deflection shown by the bending of the axes of major folds, we find that only perhaps a quarter of the present area of the pluton is accounted for by this distension. The remaining space cannot then be accounted for because the evidence is not present to guide us: xenoliths obtained from the present level do not make up any substantial volume, so that we hesitate to invoke an important degree of stoping.

The attitude of the contacts and foliations gives little clue to the shape of the pluton at depth: we can only gain the idea of a steeply inclined cylinder from these observations, and this is substantiated by the geophysical work (see p. 340). We prefer to think, however, that the Ardara pluton is not bottomless, but has a funnel shape, in the way that has been surmised for other examples of this kind (see Fig. 16-2).

The model we propose, therefore, is the rise of a turnip-shaped body into the crust. A capping of increasingly contaminated and therefore congealing magma is overtaken by more mobile material from below, which then sweeps this forward as a skin, deforming it and the envelope at the same time (Fig. 8-8). As this elongated bubble of magma rises, the country rocks are more and more distended, and after it has passed on upward, we might even conceive a partial closing in of the crust: perhaps somewhere there are aureoles without plutons (*cf.* Hamilton and Myers, 1967)!

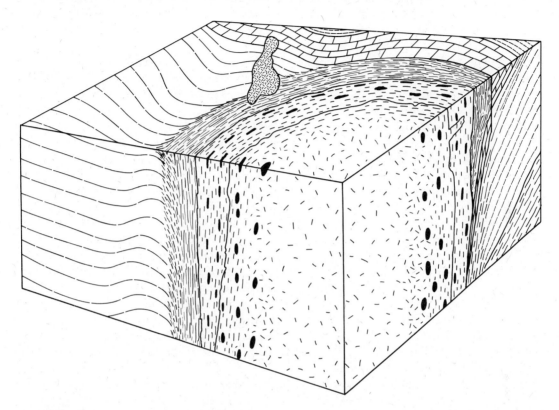

FIGURE 8-8. Generalized block-diagram of the northwestern part of the Ardara Pluton showing progressive deformation in the carapace and the peculiarly restricted deformation of the envelope. Diagram shows schematically the progressive flattening of the basic patches and the strengthening of the alignment of the minerals within the pluton.

The amount of distortion of the envelope must have been facilitated by the thermal contribution from the pluton; thus the close relationship in space between the zones of strong deformation and considerable recrystallization. The line of hiatus may represent some kind of plastic limit because, as we have seen, it also marks the limit of effective recrystallization. The fact that it is so sharply marked is possibly connected with the concept that the thermal zone adjacent to an intrusion is also sharply limited, the temperature falling off remarkably rapidly over a short distance (Jaeger, 1964). The kinematic picture is of an ovoid globule of magma, migrating upward in response to differences in density in a relatively mobile environment—a globule which, by loss of heat and mobility, eventually came to rest in the upper crust (*cf.* Grout, 1945). Magma from the root zone, possibly activated by the heat gained by partial crystallization, would continue to surge up into the now stationary roof zone, causing distension of the body and its congealed peripheral skin. Finally, the stresses associated with cooling gave rise to the radially oriented fractures, the earliest of which were filled with residual fluids.

(B) Discussion

Plutons with envelopes which have been deformed during emplacement are widely reported. While there may be difficulty (see Chapters 11 and 12) in distinguishing examples of contemporaneous contact deformation from those where plutons have been subjected to a later, superimposed metamorphism, it is even more difficult to determine whether such contemporaneous deformation is imposed by the intrusive forces themselves or by regional forces operating on the intrusion (*cf.* Berger and Pitcher, 1970).

We are not absolutely certain of the situation at Ardara. Clearly, the peripheral structures are later than the regionally imposed D_4, and they are so closely associated with the pluton in space and time that the intrusion itself seems wholly responsible for the forces involved. Nevertheless, it is probable that regional forces triggered the magma into motion at depth.

There are some very high-level plutons which have deformed their own carapaces, and, particularly in association with cryptovolcanic centers, uptilting and uplifting of strata is common (*e.g.*, Ardnamurchan, Richey and Thomas, 1930; Arran, Bailey, 1926; Mt. Alford, Stevens, 1959; Mt. Barney, Stephenson, 1959; and various examples in Victoria, Hills, 1959). In these cases, where there is little constraint on the uprise of crustal material, relatively little strain is involved, and new cleavages are only produced in rather restricted marginal zones.

In somewhat deeper-seated environments, where there is much more constraint and less loss of heat, radial pushing aside is the predominant effect and is combined with flattening and all its resultant structures, as at Ardara (other examples are Flamanville, already mentioned; also Adamant, P. E. Fox, 1969; Bidwell Bar, Compton, 1955; Chandos Lake, Saha, 1959; El Pinal, Duffield, 1968; Ellicott City, Cloos *et al.,* 1964; Killingworth, Mikami and Digman, 1957; Loon Lake, Cloos, 1934; Rogart, Soper, 1963; Strontian, Munro, 1965; Vradal, Sylvester, 1964). In fact, as Fourmarier (1959) points out, it is likely that the structural and metamorphic effects associated with diapiric plutons depend on the difference in temperature between the country rock and the pluton; the contact effects are a record of a *depth or denudation series,* which traces the rise of the pluton into a crust which is becoming more and more cold and rigid.

It was Wegmann (1930) who first pointed out that some granite plutons show indications of a mechanism of ascent similar to that of salt *diapirs,* a term first used by Mrazec (1915) for the salt domes which migrate upwards by piercing the overlying strata. This comparison has often been utilized to provide a model for the rise of forcefully emplaced granites (*e.g.*, Cloos, 1936; Cloos and Rittman, 1939; De Waard, 1949a, b; Ramberg, 1967, 1970), and, though properly criticized by O'Brien (1968), it has obviously influenced hypotheses for the emplacement of the Ardara pluton. However, in regard to the latter, there is more evidence of distension than piercement, and, further, there must be differences inherent in the fact that plutons are necessarily hot in relation to the enveloping rocks.

Despite such reservations, we prefer to retain the concept of a tailed globule rising diapirically into the crust for the Ardara Pluton, and in later pages, we shall try to fit this into a whole kinematic scheme for the *mis-en-place* of the Donegal Granites, and attempt to answer the important question as to what triggered off the ascent of magma—as to what extent, in fact, the emplacement is independent of regional deformation.

SECTION B. THE TOORIES PLUTON: A FRAGMENT OF ANOTHER DIAPIR

Off the west coast of Donegal, there are fragmentary outcrops of what we believe to be an intrusion very similar in type to the Ardara mass just described, and which we shall refer to as the Toories Pluton. Thus, the south coast of Aranmore (at Toories), the islands of Iniskeeragh, Illancrone, and

Roanish (see Map 1 in folder) are all constructed of a quartz monzodiorite or monzotonalite, the foliation of which conforms in strike to the inferred arcuate contact. In each of these isolated outcrops, this foliation is a well marked, steep, planar structure, marked by the alignment of the clots of hornblende and biotite, and in it, the xenolithic patches are flattened into discs (an S-fabric), just as at Ardara; indeed, it seems, from a comparison of the development in the several islands, that it also strengthens near to the margin, though never to the same degree. An observation of considerable interest is a small but significant angular difference between the dips of the mineral foliation and the plane of flattening of the xenoliths (see p. 247).

Only a small part of the contact is visible on Aranmore, where the monzotonalite interfingers with the mixed group of strata underlying the Ards Quartzite, a relationship best seen around Caladaghlahan Bay. The actual contacts here are sharp, highly regular, steeply dipping, and concordant with the strike of the country rocks, which are, however, sometimes intensely veined. There is an aureole—poorly seen—in which a general recrystallization is accompanied by the growth of andalusite and calc-silicate, but the interest lies not so much in the mineralogical changes as in the deformation structures. Thus, adjacent to the contact (on Aphort beach), marginal pegmatites are seen to be isoclinally folded and boudinaged with the mobilized metasediments, the axial planes of these folds lying parallel to the contact plane. Here again, as at Ardara, is evidence of synchronous deformation in the envelope; indeed, this is one of the best places to demonstrate this situation.

Despite this interesting observation, there is too little of the Toories outcrop left for it to provide much new data for our studies, except to add one more diapiric pluton to the catalog of Donegal Granites. As to its age relative to the associated intrusions, we have three useful observations, *viz.*, the pluton is cut by a porphyry dyke of Rosses type; it contains a xenolith which seems to be of Thorr type, and appears, by extrapolation from the map, to cross-cut the main contact of the Thorr Pluton beneath the sea between Aranmore and Iniskeeragh; these suggest a time of intrusion near to that of the Ardara analogue.

CHAPTER 9

THE ROSSES CENTERED COMPLEX: AN EXAMPLE OF CAULDRON SUBSIDENCE

CONTENTS

1. Introduction 186
2. The Components: Field Relationships 187
   (A) The Microgranite Sheets 187
   (B) The Centered Complex of Granites 189
   (C) The Porphyry Dykes 192
3. The Components: Composition and Texture 192
   (A) Petrography 192
   (B) Mineral Paragenesis 194
4. Petrogenesis 196
5. Relation of Form of the Contacts to Mode of Emplacement 198
6. The Rosses as a Centered Complex 199

1. Introduction

The Rosses Pluton forms one of the neatest and least enigmatic of the Donegal granites. We are fortunate in having a wealth of information about the mineralogy and chemistry of this relatively simple pluton, and we shall attempt in this chapter to make a synthesis of the detailed discussions of Mercy (1960*b*; 1962) and Hall (1966*b*, 1966*c*, 1967*a, b;* 1969*b*) in the light of the original account by Pitcher (1953*b*).

The complex occurs in the district of the Rosses in western Donegal where it outcrops on a low, well-exposed, coastal platform, partially bounded on the inland side by low hills. It is a roughly circular intrusion averaging 8.5 kilometers in diameter, and consists of a succession of four granite stocks, one within the other, three of which are separated from one another by sharp, steeply dipping contacts (Fig. 9-1). These main units are accompanied by several suites of minor intrusions.

The whole complex is emplaced within the Thorr Pluton and is cut by the Trawenagh Bay Pluton in the south. These are the only two clear indications of the relative age of the Rosses, though the fact that certain dykes, which may represent members of an associated dyke swarm, cut both the Toories Pluton and the aureole of the Ardara Pluton suggest that the Rosses complex lies chronologically between the Ardara (and Toories) and Trawenagh Bay Plutons.

The *order of emplacement* of the complex is as follows:

(1) A suite of variable microgranite sheets.*
(2) A medium-grained biotite granite, G1, which forms an outer ring. There is locally a marginal facies, G1A.
(3) A coarse-grained biotite granite, G2, forming an intermediate ring.
(4) A N–S trending swarm of porphyry dykes.*
(5) A medium-grained biotite-muscovite granite, G3, now forming a central mass and altered in places to a muscovite greisen.
(6) A small intrusion of muscovite granite, G4, near the center of the complex.

The essential characters of each of these components are briefly described in what follows.

2. The Components: Field Relationships

(A) The Microgranite Sheets

The sheets, which form the earliest members of the Rosses Complex,** are composed of fine- to medium-grained granites, chief among which are porphyritic and nonporphyritic microgranites and a pink, leucocratic microgranite. Though generally structureless, the more biotitic of these intrusions sometimes exhibit a planar fabric of micas parallel to the walls, and xenoliths, where present, usually conform to this pattern, though occasionally they may be assembled in crude swirling structures, as if by turbulent movements in the host (see Fig. 6 in Pitcher, 1953b).

The great majority of the intrusions exhibit ample evidence of dilation and are clearly intrusive. Walls frequently match, and the opening is seen in the smaller sheets to have been oblique to the walls, with a generally sinistral offset, thus indicating the existence of a rather puzzling shear couple during emplacement.

These sheets form a discontinuous fringe to the complex, trending NE–SW roughly tangential to its generally arcuate outer contact, and having a generally steep dip to the north or west. There is a range in thickness from one to a hundred meters, with the largest, the Crovehy Sheet, reaching 210 meters. Though these microgranites are generally free of inclusions, the Crovehy Sheet in an exception, for it encloses, near its base, blocky xenoliths of country rocks (Fig. 9-2a) whose angular shapes appear to have been determined by early fracture planes.

Representatives of all types of microgranites are cut sharply across by the main G1 granite, which may occasionally enclose xenoliths of the former near to its outer contacts. On the other hand, the general parallelism of the microgranite sheets themselves makes it difficult to decide on their relative ages. It is true that representatives of the porphyritic microgranites can be seen to cut across pink microgranites, but the age of both these types relative to the grey microgranites is unknown. Although all these microgranites are earlier than the earliest of the central granites, G1, they do not seem to be much earlier than the complex as a whole; indeed, the larger sheets of Crovehy and Annagary are

---

* Shown on Map 1 in folder only.
** There are certain small boss-like granites lying outside the periphery of the centered complex and these appear to be unrelated to the latter. All we know of the emplacement history of these bodies is that one of them, forming the hill of Croaghpatrick, occurs as angular, joint blocks in the Crovehy Sheet, thus establishing that some of them, at least, are earlier than the earliest phases of the complex.

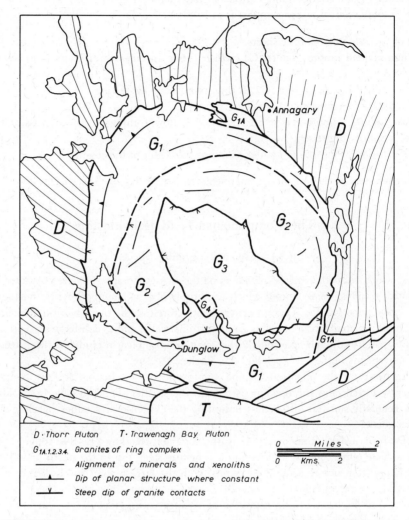

FIGURE 9-1. Outline structural map of the Rosses Centered Complex. Dykes omitted, for details see Map 1 in folder. After Pitcher (1953b).

identical to the marginal facies of G1 at Moorlagh, which we shall regard as the first phase of the emplacement of the centered complex.

A few rather distinct intrusions seem to be somewhat later than the rest, as for example, the thin grey dyke which runs south from Keadew Strand parallel to the granite margin and crossing all other microgranites in its path. There is also the composite porphyry-microgranite dyke running from Portacurry to Mullaghderg, which cuts a porphyritic microgranite sheet in one place. This steeply outward-dipping composite dyke is especially significant in that its outcrop curves through 70° over a distance of some 10 kilometers, the center of this arc coinciding neatly with the center of the main complex; it is, in fact, a partial ring dyke.

There are also some quite irregular marginal masses of biotite microgranite, as for example, the

FIGURE 9-2. (a) Angular xenolith of Thorr tonalite in microgranite of the Crovehy Sheet. Shape appears to have been determined by early joint-like fracture planes. Glacial pavement, Crovehy Townland. (b) Outer contact of Rosses Granite, G1, against the Thorr Tonalite to show how very sharp and planar is the contact. 300 meters east of St. Colomba's Church, Leckenagh Townland.

body at Moorlagh along the northern edge of G1, which we could regard as a marginal facies of the latter (as G1A). This microgranite locally contains large angular fragments of the Thorr country rock along its junction with G1, and the rotation of these xenoliths, as shown by the discordance between their internal foliation and that of the adjacent Thorr rocks, suggests that they are disrupted remnants of a central block of Thorr granodiorite which subsided to permit the intrusion of G1; we thus interpret the Moorlagh microgranite as an early ring fault intrusion.

Parallel to, and associated in time with, all these intrusions are simple and composite dykes of pegmatite and aplite (Fig. 9-3), many of which show a concentration of large crystals on the hanging wall, a phenomenon which is presumably due to the fluxing of the volatiles which naturally rise to the roof.

(B) The Centered Complex of Granites

The main body of the Rosses Complex consists of the four units, G1 to G4, each of which is internally quite homogeneous and easily separated from the others (Fig. 9-1). The only departure from this simple pattern is provided by a suite of N–S trending porphyry dykes which separates G2 and G3 in time—the significance of these we discuss later.

The outer contact of the medium-grained granite, G1, against the Thorr Pluton is very distinct and everywhere dips outward at 60–70° (Fig. 9-2b). It is sometimes emphasized by a marked decrease in grain size near the junction, and in other cases, by a pegmatitic growth of minerals on the contact

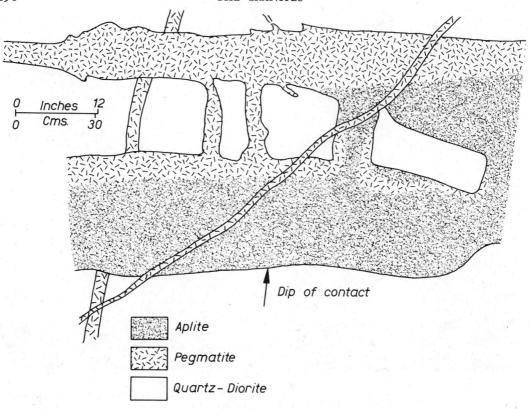

FIGURE 9-3. Pegmatite-aplite sheet of the type associated with the Rosses Complex. The concentration of pegmatite on the hanging wall is thought to be due to the collection and trapping of the volatile constituents in this position. Crovehy. After Pitcher (1953*b*).

surface. There is no veining, and the contact is remarkable in being a perfectly planar surface, at least on an outcrop scale (Fig. 9-2b) with an overall crudely polygonal outline. Neither the Thorr granodiorite nor the microgranite sheets are altered in any way near this sharp outer contact.

In contrast, the contact between the medium-grained G1 and the coarse G2 granite, though easily located, is often transitional over distances up to 100 meters, especially in the north and northeast. Sometimes, a zone of pegmatite again marks a sharper boundary, and in such cases, potash feldspars can be seen to project into G2; neither granite, however, veins the other. Fig. 9-1 shows clearly how the G2 outcrop breaches the ring of G1 in the east and makes direct contact with the Thorr unit, this outer sharp junction now dipping eastward at around 55°. Thus, despite the gradational nature of the junction between G1 and G2, it seems that the latter represents a clearly separate intrusion of new material; G2 is not simply a grain-size variant of G1, a fact which is confirmed by significant differences in chemical composition (see p. 346).

The boundary between the G2 and the finer grained G3 granite is absolutely sharp and unveined, and is another planar surface. It dips, for the most part, steeply outward at approximately 70°, and of particular interest is the way in which it is highly angular in detail (Fig. 9-4), and on a large scale, defines a polygonal shape with many reentrants. Here again, a pegmatitic fringe may be present with feldspars projecting into the younger granite but, in a few other places, a thin strip of microgranite (5–7 centimeters) separates the two units with a straight outer contact and an irregular inner one, as if a continually moving magma had resorbed an early chilled phase (Pitcher 1953b, p. 160). In the south, G3 breaks through the G2 ring to make sharp contact with G1; a boundary once again invariably fringed by a thin pegmatite zone. That this is so is fortunate, for to determine the junction between granites of similar grain size is not easy, even through the rocks can be distinguished by the presence of more obvious muscovite in G3.

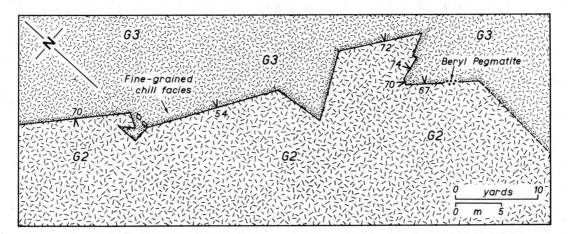

FIGURE 9-4. Details of the G3–G2 contact to show its angular character. Sheskinarone.

Certain proportions of G3 granite are characterized by large plates of muscovite, much of which is the result of a late replacement process clearly akin to greisenization. For the most part, these areas are gradational to normal G3 granite, but sometimes clear contacts are present (Burke et al., 1964), and the mass of muscovite-rich granite at Altcrin is certainly a distinct intrusion, here called G4. The presence of occasional garnetiferous muscovite-rich pegmatite and aplite dykes cutting G1 and G3 confirms the existence of such a late magma.

All these rocks are virtually structureless and weak alignments of minerals, and xenoliths are only seen in the G1 and G2 granites and not at all in G3 or G4. This planar mineral alignment (of biotite, chiefly) in the outer members of the complex is probably the closest we have in Donegal to a primary flow structure in the Cloosian sense, conforming in general to the margins of the complex. Even here, however, there are anomalous situations, such as on the western side of Lough Anure and farther south in the Lough Nageeragh–Lough Laragh region, where the alignment, weak though it is, appears to cross major contacts, though this is not demonstrable within the space of a single outcrop. We shall return to a discussion of such fabrics, which are much better shown by other Donegal granites, in Chapter 15.

Regular and irregular schlieren of relatively fine-grained biotitic material, usually less than a foot or so in length, are occasionally seen in the main granites. In G1, they are oriented parallel to the outer contacts, but in G2, a good alignment is rarely seen. These schlieren may take very irregular forms (Fig. 9-5 and see Pitcher, loc. cit.), and some are even saucer-shaped, appearing as rings in outcrop, representing, we think, thin plates suspended medusa-like in the magma. Occasionally, these biotite patches can be certainly identified as metasedimentary material derived from enclosures in the Thorr rocks, i.e., they have twice suffered immersion with its attendant recrystallization and metasomatism, so explaining their highly altered appearance. In fact, we interpret all the patches as having originated in this way—a conclusion of some importance, as we shall see later.

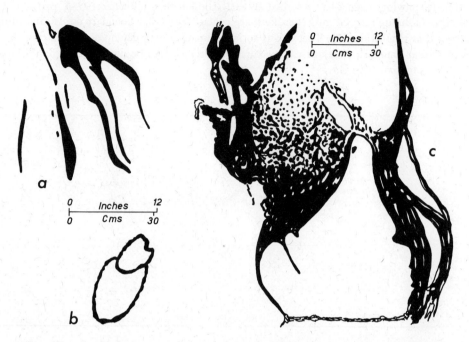

FIGURE 9-5. Various biotite schlieren in the G1 unit of the Rosses Complex. Medusa-like shapes are common and appear as rings on horizontal surfaces (b).

(C) The Porphyry Dykes

A swarm of N–S trending porphyry dykes with steep to vertical dips cut the Thorr granodiorite and the G1 and G2 granites, but never penetrate into G3 and G4. These dykes often show considerable variations in width, and may bifurcate or even die out, only to be followed *en echelon* by a new dyke with the same trend. The dykes have perfectly sharp contacts against unmodified granite or granodiorite, and the wider members frequently have fine-grained dark margins which we interpret as due to true chilling. Further, thin members may consist wholly of this dark, fine-grained material.

Alignments of mica are common in the porphyry dykes, and near the margins, invariably lie parallel to the contact, sweeping around projections in the wall; this structure is interpreted as due to laminar flow in the congealing margin. The centers of larger dykes, in particular, lack any orientation of minerals, either because a pre-existing alignment has been obscured by subsequent mineral growth or because this portion of the dyke never developed a foliation. A number of dykes in the southern part of the complex have oblique internal foliations in which the acute angle between the contact and the foliation points in the direction in which the dyke walls have moved (Fig. 11-6c). As we shall see later (Chapter 11), such internal fabrics are due to deformation during the period of crystallization of the dykes, and provide useful information about the local stress system operative at the time (Berger, 1971*a*).

3. The Components: Composition and Texture

(A) Petrography

The dominant rock types of the Rosses Complex are varieties of adamellite consisting of zoned oligoclase, microcline, quartz, biotite, and muscovite, together with minor amounts of zircon, epidote, apatite, sphene, allanite, magnetite, and hematite (Table 9-1). The details of their petrography have been given by Pitcher (1953b) and Hall (1966b, c), and only the broader features need be repeated here.

TABLE 9-1. Modal Composition of Granites

Averages*	Microgranite Sheets and Dykes (% Vol.)				Main Granites (% Vol.)			Muscovite Granites (% Vol.)		
					G1	G2	G3	G4		
	1	2	3	4	6	7	8	9	10	11
Quartz	26.4	27.5	35.2	31.3	32.1	34.3	31.7	28.3	35.8	44.2
Plagioclase	45.7	38.0	36.5	36.6	40.9	37.2	40.3	38.7	41.1	35.8
Potash feldspar	20.3	30.7	25.3	25.8	21.6	24.2	23.9	27.7	21.2	15.8
Biotite, iron oxides, and secondary minerals	7.5	2.8	1.1	6.3	5.1	4.1	2.7	1.3	0.5	0.3
Muscovite	0.1	1.0	1.9	Nil	0.3	0.2	1.4	4.0	1.4	3.9

* 1. Average of 5 dark grey microgranites. 2. Average of 5 light grey microgranite sheets. 3. Average of 5 pale microgranites. 4. Average of 2 grey porphyritic granites. 6-8. Average compositions of the main ring-granites. 9. Muscovite granite of Altcrin. 10. Muscovite granite from the Meenbannad area. 11. Partially greisenized granite from Sheskinarone.

We first refer briefly to the *microgranite sheets* which, as mentioned earlier, are mostly fine- to medium-grained rocks, though some members are distinctly porphyritic. These microgranites are more equigranular than the main granites, and the felsic minerals tend to show irregular boundaries with occasional crude micrographic textures; plagioclase is, however, much more regular in shape than microcline. Biotite is prominent in the darker types but may be subordinate to muscovite in the paler types, which are also richer in microcline and commonly bear garnet as an important accessory. In the porphyritic sheets, large plates of microcline show irregular margins and frequently vein or enclose the other minerals.

The *main biotite granites,* G1 to G3, are individually very uniform in mode (Table 9-1) and are very similar to one another, both in texture and composition; the chief differences are that G2 is appreciably coarser, and that the biotite content decreases with relative age of the granite from an average of 5.1% in G1 to 2.7% in G3. Plagioclase is generally zoned with sericitized cores (An_{18-19}) becoming progressively more sodic outward to unaltered rims of An_{11-12}, and the average bulk composition is $Or_{9.1}Ab_{82.3}An_{8.6}$ with most of the potassium being held in the sericite. A particularly interesting feature is the presence of aggregates of plagioclase sharing common outer zones (Hall, 1966b, Fig. 2)—an excellent example of synneusis texture. Microcline (average composition $Or_{87.9}Ab_{10.6}An_{1.5}$) is generally microperthitic, and though it is not as flagrantly replacive, as for example, in the Thorr Pluton, it does nevertheless appear to be of relatively late origin. Biotite (straw yellow to dark brownish-green) may be enclosed in any of the major minerals, and, as in the Ardara and Main Granites (Chapters 8 and 11), it frequently encloses epidote, which has euhedral junctions with its host, but is anhedral and corroded where it comes into contact with quartz and feldspar (*cf.* Fig. 11-9e)—a relationship of considerable importance, as we shall see shortly.

The *muscovite-rich granite* varieties, which include G4 and greisened portions of G3, are characterized by large irregular plates of replacive muscovite. These may enclose relicts of feldspar in the same crystallographic orientation as the surrounding feldspar, and often form a symplectic intergrowth with the latter. Similarly, microcline and quartz appear rather more replacive here than in the other granites. A culmination of this process of late stage modification is seen at Sheskinarone, where progressive alteration of the normal G3 biotite has led to the production of a beryl-quartz-muscovite rock with associated beryl-bearing pegmatites. In this situation, plagioclase becomes increasingly turbid and its calcic cores enriched with muscovite, biotite is altered to chlorite, and both minerals are fringed by flakes of white mica which are eventually replaced by robust muscovite. Replacive beryl appears sporadically in both pegmatitic patches and in greisens located along the G2-G3 contact, and is accompanied by minor amounts of garnet (almandine-spessartine, according to Hall, 1965b), ilmenite, and hematite, together with rare torbenite and uraninite. These beryl-bearing greisens have been described in some detail by Burke and others (1964), who estimated that 50 tons of beryl averaging 13% BeO are present in the Sheskinarone body.

The *porphyry dykes* are petrographically distinct from the rest of the complex, and show considerable variations in texture and in proportions of minerals present—from the chilled margins to the coarse-grained, porphyritic cores. Sometimes, a vague contact is present between the fine-grained margin and the center, as if magma had moved on and passed its congealed contact phase. In general, the dyke rock consists of phenocrysts of green to brown biotite, plagioclase, microcline-microperthite, and quartz set in a felsic groundmass of the same minerals; a crude graphic texture is often seen in the central zones. Plagioclase generally shows marked oscillatory zoning with cores of andesine and oligoclase rims. The margins of potash feldspar phenocrysts are commonly irregular and intergrown with the groundmass; but, in contrast to the uniformly low-temperature structural states of the alkali feldspar in the ring-granites, those in at least one of the porphyry dykes are of a higher temperature state, a fact which Hall (1966c, p. 980) attributed to a low water content in these dykes during cooling.

In passing, we may refer to a detailed study of one of the thicker dykes at Mullaghduff, where Mercy (1962) has shown that there are gradational changes in Si^{4+}, Mg^{2+}, and Ca^{2+} across the intrusion, Si^{4+} increasing inward from the walls, the other elements decreasing. Mercy dismisses the possibility that this variation is due to the introduction of new and different material into the center of the dyke—the composite dyke effect—and attributes it to ionic diffusion, supporting the hypothesis by a particularly interesting discussion on diffusion, temperature gradients, and thermal convection in magmas. We are not so sure that separate pulses were not involved, for there are some internal contacts in this Mullaghduff dyke.

(B) Mineral Paragenesis

It is clear from their textural relationships that most of the microcline and muscovite in the Rosses Complex is of relatively later origin. Both minerals replace plagioclase, broken crystals of which may occur immediately adjacent to completely undeformed microcline and muscovite. Furthermore, late veinlets of quartz and muscovite are common, and joints may localize the late greisenization, as described above. Additional evidence of the late activity of microcline has been provided by Hall (1967b), who showed that while the Ca^{2+} and Na^{1+} contents of plagioclase vary with the composition of the host rock, the plagioclase becoming more sodic with increasing silica, no such systematic variation could be found in the alkali feldspar. This situation, which is also reflected in the distribution of gallium, he attributed to "postmagmatic recrystallization" of the microcline, an event associated with exsolution and formation of myrmekite and late albite at feldspar-feldspar junctions.

The time of formation of biotite presents considerable problems. Hall (1966b, p. 206) concluded that its inclusion by any of the other essential minerals pointed to its early origin. Furthermore, the corrosion of epidote inclusions in biotite, where they come into contact with quartz and feldspar, suggested to him that epidote was not a stable mineral in contact with granitic magma, and that its presence was due to the fact that it had been carried by biotite from wherever the magma was generated (*ibid.*), the implication being that some of the biotite originated from a deeper seated source. This particular relationship between biotite and epidote has been noted in the Ardara and Main Donegal Plutons as well as in other epidote-bearing granites (*cf.* Keyes, 1895, Plates 38, 40; Exner, 1966, Fig. 3; Cloos *et al.*, 1964, Plate 34), and clearly reflects the instability of epidote along with quartz and feldspar. We think that this instability could just as well be due to late- or post-magmatic reactions (*cf.* Hickling *et al.,* 1970). The biotite is an essential mineral of the plumose pegmatitic fringes along some contacts, suggesting that it grew at the present level of exposure. Furthermore, as in the Ardara and Main Granites, there is the fact that the mineral alignment to which biotite contributes crosses major lithological boundaries (see, Fig. 9-1 and p. 330), implying that some biotite has either crystallized or recrystallized at its present crustal level and not at depth. Thus, although some of the biotite, as for example, that in the schlieren, may be of relatively early crystallization, some is certainly of later date.

However, on the basis that biotite crystallized out early, Hall (1966*b*) has discussed the sequence of crystallization in the Rosses Complex in terms of the relatively simple system An-Ab-Or-SiO$_2$-H$_2$O. The first point is that all the compositions of all the units fall near the minimum of the system (Fig. 9-6a). Secondly, Hall concluded that the crystallization history, as deduced from the textural evidence, is in accord with the experimental findings. Of the leucocratic minerals, the expectation is that plagioclase would crystallize first because the magmas started crystallization in the plagioclase field, as represented in Figure 9-6b. It would have been joined by the other two minerals when the composition of the magma reached the quartz-feldspar and plagioclase-alkali feldspar boundaries, respectively, and when finally the magma reached the quartz-alkali feldspar-plagioclase boundary, all three minerals would have crystallized together until the magma was used up (for further details see Hall, *loc. cit.*).

Petrographic evidence may be used to distinguish successive stages of plagioclase crystallization, and to follow quantitatively the change in composition of the magma. Taking as an example the G1 granite, Hall has concluded that the cores of plagioclase began to separate by equilibrium crystallization before quartz or microcline. According to Hall, reaction between this plagioclase and the magma would result in a local excess of Al^{3+} which would combine with Si^{4+} to assist in the formation of the sericite which occurs in most of the plagioclase core zones. Using various actual and normative compositions, Hall showed that after the crystallization of these sericitized cores (of composition A in

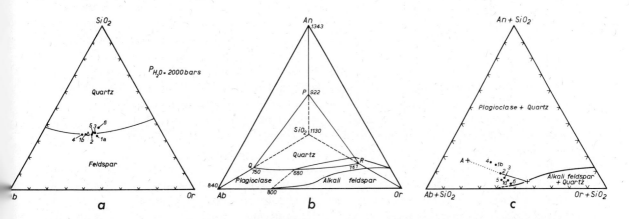

FIGURE 9-6. (a) The system Ab–Or–SiO$_2$–H$_2$O at a water pressure of 2000 bars; the position of the minimum is indicated by the small triangle. Shown are the normative compositions of the Rosses granites after removal of biotite and apatite. (b) Suggested phase relationships in the system An–Ab–Or–SiO$_2$–H$_2$O (after Carmichael, 1963) for water pressure of 2000 bars. (c) Suggested phase relationships on the quartz-feldspar interface of the system An–Ab–Or–SiO$_2$–H$_2$O projected onto the plane An–Ab–Or. A represents composition of the sericitized central parts of the plagioclase in G1. Points in a and c represent: (1) microgranite sheets (1a-muscovite-rich; 1b-biotite-rich); (2) G1; (3) G2; (4) porphyry dykes; (5) G3; (6) G4. All diagrams after Hall (1966*b*).

Figure 9-6c), the composition of the magma would approach the plagioclase-alkali feldspar boundary and that these two minerals would eventually crystallize together (Fig. 9-6c). Thus plagioclase may be enclosed by microcline but not *vice versa*. The minor growth of late albite at feldspar boundaries he interpreted as a result of alkali feldspar exsolution and recrystallization.

Hall also pointed out that various experimental researches indicate that muscovite can only crystallize from magmas under relatively high water pressures. During crystallization of the feldspars and quartz, which are all anhydrous, the pressure would rise as the proportion of liquid to crystals decreased, until it was sufficiently high for muscovite to form. This increase in pressure would also depress the liquidus and displace the quartz-feldspar boundary away from quartz, both changes leading to the resorption of feldspar and its replacement by quartz. Similarly, the replacement of feldspars by muscovite would release alkalis and silica, which would also result in some recrystallization of the feldspars. The greater replacive activity of microcline and of muscovite in the muscovite granites is consistent with theory.

While Hall's account of the sequence of crystallization of the Rosses Granites is a straightforward explanation, it is based largely on the assumption that the biotite ceased its crystallization prior to the onset of the other essential minerals and is therefore not involved in the chemical changes which follow. As we have previously indicated, we regard this assumption as unproven, though apart from this caveat, Hall's model is an appealing one.

4. Petrogenesis

In his original account, Pitcher concluded that the great uniformity of composition throughout the Rosses Complex confirmed its magmatic character. The slight decrease in biotite content from early to late members was the only mineralogical evidence of differentiation which might indicate crystal settling at depth. However, the wide range in composition of the early microgranite sheets proved difficult to fit into a simple pattern of differentiation; in the case of the Portacurry ring dyke, there is even some indication of the existence of two magmas simultaneously, as shown by the presence of two distinct components, porphyry and microgranite. Furthermore, the porphyry dykes, with their wide range of composition, suggested that some process of mixing of mobilized Thorr granodiorite with the Rosses magma may have operated at depth. The correlation between abundance of xenolithic material and the biotite content—the earlier, more biotitic members being richer in xenoliths—likewise suggested that contamination was a controlling factor in the evolution of the Rosses complex.

On the other hand, Mercy's (1960*b*) detailed chemical study, involving some 106 whole-rock analyses of the various components of the Rosses Complex, indicated to him that some degree of crystallization differentiation had occurred. In general terms, granites G1 to G4 show a progressive increase in Si^{4+} and a decrease in Fe^{2+}, Mg^{2+}, Ca^{2+}, Ti^{4+}, and Al^{3+}, the other major elements remaining constant (Fig. 9-7).

It is the chemical composition of the early sheets and the later porphyries which present the main problem as far as the differentiation model is concerned. While the grey and pink microgranites are chemically similar to G2, the biotite microgranites and porphyry dykes are generally outside the range of composition of the main intrusions, particularly in having a lower Si^{4+} and K^{1+} content and a higher one of Ca^{2+}, Fe^{2+}, and Mg^{2+}. In order to correlate the chemical variation and the order of intrusion, Mercy proposed a model involving two cycles of differentiation at depth from a magma with a composition taken as the average for that of the microgranites and porphyries. The first cycle of differentiation gave rise, in turn, to the biotite-microgranites, to G1, G2, and possibly, even the pale microgranites; the second cycle, to the porphyries, G3 and G4. In each cycle, the components evolved through an enrichment in Si^{4+} and K^{1+} and a decline in Al^{3+}, Fe^{2+}, Mg^{2+}, and Ca^{2+}, with Na^{1+} remaining invariable.

This may provide a convenient chemical model, but it is based on the dubious assumption that

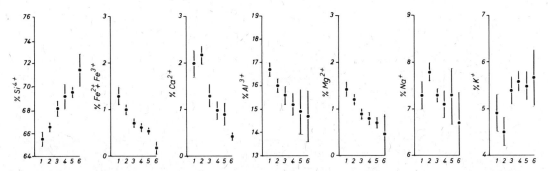

FIGURE 9-7. Variation in the composition of the Rosses granites with 95 percent confidence limits shown: (1) biotite microgranite sheets; (2) porphyry dykes; (3) G1, (4) G2; (5) G3; (6) G4. After Mercy (1960b).

members of a differentiation series are intruded in strict order of chemical composition. Even in the Rosses, this can be shown to be invalid, because on the field evidence, the pale microgranites were intruded before G1, so that a magma apparently more advanced in the differentiation series was clearly available at an early date in the intrusion sequence.

In contrast to Mercy's ideas, Hall (1966b) argued at length that differentiation is incapable of explaining the variation in the magma series, and proposed instead a model whereby the Rosses magmas evolved by partial melting and assimilation of the downward extensions of the Thorr Pluton. As melting began at depth, the magma would have been produced mainly from the low melting-temperature constituents, quartz, albite, and orthoclase (according to Winkler and von Platen, 1961), and would always have a lower proportion of high melting-temperature minerals than the surrounding unmelted rocks. As the degree of melting increased, higher melting-temperature constituents would be added to the magma. The microgranite sheets which include, as we have seen, the complete range of compositions found in the complex, represent the magmas produced when the degree of melting was gradually increasing as postulated above: the muscovite microgranites representing an early generation of magma, the biotite microgranite a later.

As the new biotite microgranite magma moved upward, it would incorporate increasing amounts of the stoped-off granodioritic wall rocks. During this encounter between the rising magma and the solid blocks, the magma would cool over part of its crystallization range, while the solid granodiorite would begin to melt. In fact, the sequence of crystallization in the Rosses granites does suggest that oligoclase, biotite, and possibly quartz crystallized from the magma, while the constituents of albite, orthoclase, and quartz—the minimum melting-temperature mixture—were added to the melt from the blocks of granodiorite. The net effect would be, remarkably enough, the acidification of the initial magma, and indeed, as Hall has pointed out (*loc. cit.*, p. 216), the *compositions of each of the main granites are equivalent to a mixture of the preceding granite plus material of a minimum melting-temperature composition.* Thus, dilution of the biotite microgranite parent magma by liquid derived from partial melting of Thorr Granodiorite roof would give rise to the G1 magma, and thence, by progressive enrichment in low melting-temperature constituents, to G2, G3, and G4. Such a development of the magma series with respect to time and composition is summarized, diagrammatically, in Fig. 9-8. In a further study of the distribution of trace elements in the various components of the Rosses Complex, Hall (1967a; 1969b) produces more evidence in support of this hypothesis; it is of interest to note that the pattern of distribution in the porphyry dykes makes it likely, as Mercy and Pitcher maintained, that they, too, sprang from the common source.

We see, therefore, that the detailed mineralogical and chemical research on the Rosses granite has generally supported Pitcher's original contention of a granitic magma contaminated by its

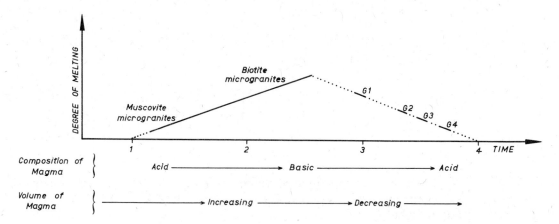

FIGURE 9-8. Diagrammatic representation of the relationship between the composition of the Rosses magmas and the processes of magma formation and emplacement: (1) start of magmatic episode; (2) roof of magma body cracks—intrusion of microgranite sheets; (3) roof of magma body founders—intrusion of main granite units G1–G4; (4) end of magmatic episode. After Hall (1966b).

granodioritic country rocks. Mercy concluded that an influx of Na^{1+} and K^{1+} ions in an aluminosilicate medium into the Thorr granodiorite could have produced the source magma of the Rosses Complex (represented by the biotite microgranites); subsequent crystallization differentiation produced the G1 and G2 units, the whole process being repeated again to produce the porphyry dykes and the G3 and G4 granites. Hall, on the other hand, derives the members of the Rosses Complex by a process of melting at depth of the Thorr Unit. The chemical variations noted by Mercy are explained in terms of this partial melting producing magmas which are further changed by partial assimilation of additional Thorr granodiorite during their uprise.

5. Relation of Form of the Contacts to Mode of Emplacement

The succession of stock-like intrusions having smaller diameters toward a common center is a familiar pattern, and along with the accompaniment of radial, tangential, and concentric dyke suites, clearly identifies the Rosses Complex with the ring-dyke type of intrusion, where overall arcuate fracture and cauldron-subsidence is thought to permit the emplacement of magma. We propose to group all the various types of intrusion which are clustered about a center, whether emplaced by ring-fracture stoping (Daly, 1933, p. 269) or by any other means of cauldron subsidence (see Hills, 1963, pp. 382–91), under the general term *centered complex,* extending the earlier term "central complex" (Kaitaro, 1956; Stephenson, 1959; Oen, 1960; Hills, *loc. cit.*) to cover both volcanic and plutonic situations. We must point out, however, that as in many other examples of this form of intrusion, there is no direct evidence in the Rosses of the subsidence of a central plug of older rocks, though the situation in the Moorlagh area cited earlier (p. 189) is suggestive of its presence.

What is particularly well demonstrated in the Rosses is that all the components were emplaced under a tight structural control. The polygonal shape of the margins of the individual granites, with long straight segments and highly angular reentrants, must surely be due to the control of the boundaries by an earlier structure, such as is the situation, for example, in the Castro Daire complex in Portugal (Oen *loc. cit.*) and in some of the centered intrusions of Nigeria (Jacobson *et al.,* 1958; see also Hills, 1959; Eggler, 1968; Berger and Pitcher, 1970). In the Rosses, Pitcher (1953b) showed that there is a coincidence between the directions of polygonal contacts and the present master joint system (Fig. 9-9), and, as we have already seen, angular junctions between different components are common, and xenoliths frequently have parallelogram shapes in cross-section.

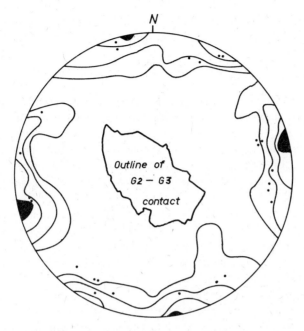

FIGURE 9-9. Stereoplot showing coincidence between the present joints and the outline of the G2–G3 contact. The poles to substantial parts of this contact are represented by individual points. After Pitcher (1953b).

There are also other indications of the early existence of fractures parallel to the present joint system, for many joints are coated with minerals or filled with quartz veins, while others act as loci for greisenization. Thus, while we may subscribe to the view that the present joints are relatively late in origin (Price, 1966), we consider that they follow an early established "grain," and one that could control the direction of fracture at any time.

An important point is that the joint directions are common to *all* units of the complex, and may even be related to a regional stress pattern operative over the whole of the time of emplacement of the complex. This is a problem central to the whole field of "granite tectonics" (*cf.* Balk, 1937), and one which we discuss in some detail elsewhere (Berger and Pitcher, 1970). For the moment, we merely emphasize our conclusion that in the Rosses there were early fracture systems, to which the present joints are related, which controlled the emplacement of various members of the complex.

6. The Rosses as a Centered Complex

We have seen how a contemporaneous fracture pattern controlled the detail of the contacts, and it is a simple matter to construct a model for the permissive emplacement of the Rosses granites by cauldron subsidence whereby the intermittent sinkings of a central block have permitted magma to flow into an upper cavity of decreasing volume, which is now revealed by erosion as a series of intersecting plutons.

The N–S trending porphyry dykes, with their slight outward divergence from the center of the complex, are not difficult to fit into this scheme, but the early microgranite sheets do present a problem, for their generally tangential relationship to the main body of the pluton prevents their correlation with cone sheets or shear-fracture fillings—at least, not according to the classic model of Anderson (1936). However, one must exercise caution in invoking explanations based on such a simple structural model, for recent, more rigid mathematical analyses of the stress fields associated with various shapes and densities of magma bodies suggest a wider variety of contact attitudes in centered complexes (*e.g.,* Robson and Barr, 1964; Durrance, 1967; Roberts, 1970); and all the possible situations have certainly not yet been investigated.

Another difficulty with Anderson's theory is the now familiar objection that many—perhaps the majority—of the outer contacts of centered intrusions are vertical or inward-dipping (*e.g.,* Billings, 1945; Reynolds, 1956; Oen, 1960; Smith *et al.,* 1961; Turner, 1963; Taubeneck, 1967*b*). Fortunately, however, the contacts of the Rosses, all well exposed, dip uniformly outward and we can, with Anderson, safely appeal to some sort of downward-operating forces related to the withdrawal of magma pressure, such as might result from the injection of the early microgranite sheets (*cf.* Williams, 1954).

Finally, it is of interest to point out that the Rosses Complex has the same order of diameter as the majority of centered complexes of this type (*cf.* Richey, 1932; Hills, 1959; Turner, 1963, p. 359). This uniformity of dimensions may find a general explanation when related to the thickness of the rigid crust at the time of magma generation, but for the moment it remains unresolved.

Concerning the magmas involved, we conclude that they were derived at depth by a process involving the progressive dilution of a primary "biotite microgranite" magma by liquids derived from the partial melting of the surrounding Thorr granodiorite. Because the granites are so homogeneous at their present level of outcrop, we think that each can be ascribed to a single upward surge of magma. Possibly, this is less true in the case of G3, which may have been built up as a multiple intrusion by pulses following rapidly one upon the other (*cf.* Harry and Richey, 1963).

Of the relationship between the units themselves, G2 probably intruded G1 when the latter was only just becoming capable of fracture and while it was still capable of limited flow; the crossing of the foliation from G1 to G2 suggests that G1 was still able to respond by mineral growth to stresses involved during the emplacement of G2. This concept that magmas can be fractured and faulted before final consolidation is not novel, and we return to a discussion of it in Chapter 15. On the other hand, the emplacement of G3 was late enough to chill against contacts and to postdate a dyke swarm in the host. These dykes injected in the interval G2–G3 represent, we suggest, the filling of N–S fractures opened as a result of an updoming which may have heralded the emplacement of G3.

Finally, the latest stages of the evolution of the complex were marked by the concentration of volatiles under sufficient pressure to permit escape into the walls and along contacts, giving rise to metasomatizing effects such as the growth of late muscovite and greisening.

Perhaps the most interesting finding of all is that the nature of the Rosses—a centered complex with fracture-controlled contacts—suggests emplacement into shallow crustal levels. Despite our awareness of the fact that whether fracture occurs or not is just as much likely to be a matter of rate of deformation as level in the crust (*cf.* Kaitaro, 1953), we consider that shallow crustal movement is consonant with the concept that centered complexes are but a lower level in a denudation series, of which calderas are the surface expression (Reynolds, 1956; Buddington, 1959). Yet, as we shall presently see, this Rosses Complex is associated in space and time with plutons of entirely different character and context.

CHAPTER 10

THE BARNESMORE PLUTON: AN EXAMPLE OF CAULDRON SUBSIDENCE AND THE FORMATION OF IGNEOUS ARCHES

CONTENTS

1. Introduction: Time of Emplacement 201
2. The Envelope: a Static Environment 202
3. The Major Components 203
4. Structural Relationships: Subsidence and Collapse 204
   (A) The Main Granite 204
   (B) The Sheet Complex: the Igneous Arch 204
5. Later Events: Desilication and Dykes 205
   (A) Desilication 205
   (B) The Dyke Swarm 207
6. Mode of Emplacement: Cauldron Subsidence and the Collapse of Roof Arches 208

1. Introduction: Time of Emplacement

With an outcrop making up the rugged Blue Stack Mountains of central Donegal, the Barnesmore Granite forms an isolated pluton some 52 square kilometers in area (Fig. 10-1). The Officers of the Irish Geological Survey (Egan *et al.*, 1888) who first mapped this intrusion did not discover its composite nature, but they did recognize that it was unlike most of the other granites of Donegal, particularly in its discordant relation to the envelope, in the total lack of any structural effect on the latter, and the absence of a directional fabric.

The earlier workers had also recognized that the Barnesmore Pluton formed the center for a swarm of basaltic dykes of Tertiary age (Haughton, 1857), a fact which, together with the apparent absence of granite boulders in the nearby Carboniferous basal conglomerate, led to the suggestion of a Tertiary age for emplacement. However, a thorough study by Walker and Leedal (1954) has conclusively demonstrated a Caledonian age; pebbles in the basal conglomerate can be matched with members of a swarm of acid dykes which also cut the Complex (see map in folder) and, furthermore, representatives of the widespread NE–SW Caledonian fault system—in particular, the Belshade Fault—displace the contacts.

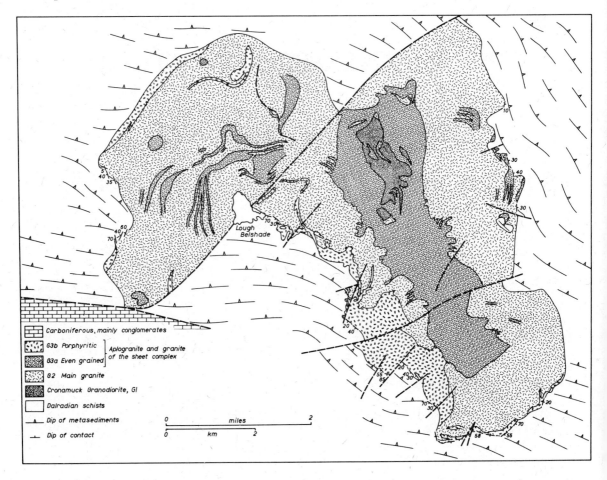

FIGURE 10-1. Map of the Barnesmore Pluton. Minor intrusions omitted. Simplified from Walker and Leedal (1954).

It is interesting that the hypothesis of a Tertiary age was given special credence by the fact that the Tertiary dykes become more abundant in and around the pluton. All that has happened, however, is that the Barnesmore granites have acted as a preferred path, in fact as a conduit, providing the easy ascent of basic magma along deep-seated joints (see p. 375).

As regards the age in relationship to the other granites of Donegal, we have no guide except that diatremes connected with the appinitic suite of rocks cut the Barnesmore Pluton, suggesting a relatively early time of emplacement.

2. The Envelope: a Static Environment

The country rocks of the Barnesmore Pluton consist of fine-grained psammites with pelitic interbands, the latter now represented by coarse muscovite-biotite schists with oligoclase augen. Together, these form a meta-greywacke lithology typical of the Central Donegal Facies of the Kilmacrenan Succession, as we have seen in Chapter 1. Within these rocks, there are absolutely no structural features which might be connected with the emplacement, and in fact, the only indication of the proximity of a considerable pluton are the effects of a subdued thermal metamorhpism (Smart, 1962).

The record of this metamorphism begins within 200 meters of the main contact where small, well formed but unoriented crystals of andalusite appear along with the patchy development of a hornfels texture (see Fig. 14-2). In the inner part, the recrystallization is much more complete, and andalusite forms larger spongy crystals, accompanied occasionally by cordierite. Rather strangely, and in contrast to all the other aureoles, the greenish-colored regional biotite shows little change in optical properties, even when regenerated. The regional oligoclase porphyroblasts are recrystallized and are sometimes partially replaced by the scanty microcline, which, together with fibrolite, enters the mineral assemblage near to the contact.

We would emphasize the simple nature of this aureole, in which, in contrast to the aureoles of many of the other granites of Donegal, there is not the slightest control by structure, either in the sense of mimetic or syntectonic growth. It is easy to agree with Smart (1962, p. 59) that "this simple aureole, narrow, completely static, lacking in metasomatic activity and characterized by andalusite and cordierite is in keeping with its connection to a relatively quickly cooled, low temperature magma, emplaced in the highest levels of the crust by cauldron subsidence."

3. The Major Components

As Walker and Leedal have shown, the Barnesmore Pluton consists of several distinct granite members, three of which were sufficiently separated in time of intrusion to be demarcated by sharp contacts. The rocks concerned are all rather similar in mineral composition and are distinguished by differences in grain size and modal proportions (Table 10-1).

TABLE 10-1. Typical Modal Compositions (Walker and Leedal, 1954)

	% by Volume		
	G1	G2	G3a
Quartz	25	31	30
K-feldspar	15	30	28
Plagioclase	55	37	40
Biotite, etc.	5	2	2

There is always a sodic plagioclase, often showing normal zoning and with cores as basic as An_{20}, a microcline which is usually a microperthite, quartz, biotite, and sometimes a little muscovite. In these leucocratic rocks, the micas rarely total more than 5% by volume and the accessories, sphene, allanite, and iron oxides, are unimportant.

There is certainly nothing remarkable about the texture—evidence of late-stage replacement is lacking, and none of the minerals show a preferred orientation. Perhaps the only point worthy of special note is the widespread, though subdued, cataclasis and reddening of the feldspars, both features connected, we think, with the NE–SW faulting.

The earliest member is a medium- to fine-grained biotite granodiorite with scattered megacrysts of quartz and plagioclase: this is the *Cronamuck Granodiorite,* G1, which is confined to a small area in the center of the complex. This is cut (Fig. 10-1) by a coarse-grained, nonporphyritic, leucocratic adamellite, the *Main Granite,* G2, which makes up well over half of the Barnesmore Pluton. A feature of this rock is that the quartz tends to be segregated into 0.5 centimeter sized patches, but otherwise, except for a very slight marginal basification, it is a remarkably uniform type both in grain size and composition. Both G1 and G2 are cut by the *Sheet Complex,* G3, in which two members can be distinguished, G3a in the interior and G3b along the contacts. The former is an even-grained, leucocratic granite with layers of pegmatite, but generally a little finer than G2, and the latter is a porphyritic aplogranite characterized by megacrysts of quartz and feldspar.

Closely associated in time with these major components are irregular and regular bodies of simple pegmatite in G3, and, in addition, there is a more widespread suite of discrete pegmatite and aplite sheets. There are also many quartz veins, some of which, however, are probably very late and related to the fracturing associated with the tear faulting.

The comparative uniformity of the main rock types, together with the presence of finer-grained marginal facies, aplitic in the case of G3, may indicate that the components consolidated from magmas. The lack of internal structures suggests that flow ceased before crystallization was at all complete: certainly they were never subject to deformative forces in the final stages of cooling.

4. Structural Relationships: Subsidence and Collapse

(A) The Main Granite

The Barnesmore Pluton flagrantly cross-cuts its envelope of Dalradian metasediments: the outer contacts of G2 are knife-sharp, and the country rocks maintain their regional attitude right up to the junction.

An outward dip of the contact of near 50° is common, and though in places the contact may be vertical, it never dips inward. On the other hand, this junction is sometimes so shallow that portions of the roof remain as true outliers, as, for example, the large mass on Pollakeeran Hill (see Fig. 10-1) and the smaller relict northwest of Lough Nacollum. It is interesting to note that as distinct from these fragments of the actual roof, true enclaves are rarely seen.

It is clear from these relationships and from the form of the outcrop that we are dealing with a dome-shaped body, elongated in the NW–SE direction (Fig. 10-2). The presence of the relicts of the envelope around the periphery of the pluton clearly suggests that there was originally a dome-shaped roof rising not far above the present level of erosion.

The complete lack of any evidence of shouldering aside, together with the rarity of enclaves and brecciated contacts, argue against both emplacement by diapirism or by piecemeal stoping, and we agree with Walker and Leedal that the "complex came into being as a consequence of the subsidence of a large block of schists, perhaps along a ring-dyke type of fracture" (*loc. cit.* p. 215). No part of this sunken block can now be seen, but there are a number of vertical breccia pipes (p. 158), cutting through all members, in which the presence of abundant, unaltered fragments of country rock brought up from depth provides strong evidence for a floor to the pluton. The occurrence of these natural "bore-holes" is perhaps the most intriguing feature of this pluton (see Chapter 7).

A possible depth to this block, and thus a "thickness" of the pluton, can be estimated from the gravity profile provided by Cook and Murphy (1952)—a figure of some 5,600 meters is given.

(B) The Sheet Complex: the Igneous Arch

The structural relationships of the sheet complex, G3, confirm this model of cauldron subsidence. As we have already seen, there are two more or less distinct structural components in this sheet complex—an inner mass of fine-grained aplogranite, G3a, and an outer discontinuous ring of porphyritic granite, G3b. Small fragments of G3a are locally enclosed in G3b, but although the porphyritic type thus seems to be the later, it is quite likely that the same type may not be everywhere of the same age. Indeed, the heterogeneity of the sheet complexes as a whole suggest multiple injection.

The inner component, G3a, forms in the south a thick sheet, inclined steeply to the northeast and giving off gently dipping, sheetlike apophyses. On Croaghconnellagh, this sheet links up with the southwest dipping sheet of G3b, referred to above, to form what Walker and Leedal term an *"igneous arch"* (Fig. 10-2, Section 2). A similar saddle-like structure is revealed in the outcrop of G3a itself, for, south of Cronamuck Hill (Fig. 10-2, Section 5), where the earliest member of the complex, G1, is sheeted by G3a, the outcrop form of G1 is evidently due to erosion cutting down through an arch within G3a. A most interesting feature of both these arches is that the apophyses always dip less steeply than the main contacts of the arch itself, giving a section like a cedar tree.

North of the Belshade Fault, G3a is represented by clearly separated sheets, having, however, the same general dip and a common arcuate outcrop pattern. There is some suggestion that these particular sheets may have formerly linked up with a larger mass at a higher level, a continuation of an upthrown and offset portion of the Cronamuck arch, which has now been eroded from the block to the north of the Belshade fault (Fig. 10-2, Section 6).

The outer partial ring of porphyritic granite, G3b, also varies in form from place to place. In the south, as we have already noted, it forms the southwest-dipping portion of the Croaghconnellagh Arch. Farther to the northwest, this same arch sends off long branching sheets, one of which follows the outer contact of the pluton. Similar marginal sheets are well-displayed in the southwestern and northwestern portions of the complex. Indeed, this marginal disposition might suggest, at first sight, that G3b simply represents an early chill phase to the main G2 granite. However, these marginal sheets are similar in all respects to, and often directly continuous with, undoubted sheets of G3 in G2. Then again, these sheets are themselves occasionally chilled at their contacts with the main G2 granite, and may even contain xenoliths of G2; furthermore, strips of G2 sometimes appear as a thin screen between G3b and the country rocks. There can be no doubt, therefore, that G3b is appreciably later than G2.

Walker and Leedal attributed the igneous arch of G3a to the limited subsidence, on crudely arcuate fractures, of a central part of the G2 pluton. This sagging provides an explanation of the preservation of the granodiorite, G1, which is thought to have originally occupied a high level in the pluton.

The outer sheet complex, G3b, may owe its form to simple collapse of the main G2 granite along its margins, that is, along the arcuate fractures that controlled its original emplacement. However, Walker and Leedal preferred the view that "the formation of the outer ring followed subsidence of the block below the Cronamuck arch; the consequent weakening of the supports of the surrounding granite led to a partial caving in of the latter." This subsidence was controlled by fractures now occupied by the sheets of G3 (the broken lines of Fig. 10-2) which were injected outward and downward from the central arches. In either case, the complex seems to have also failed along the line of weakness represented by the granite-schist contact. Fig. 10-3 shows this interpretation in a generalized manner and illustrates a cauldron subsidence controlled by successive collapse along sets of concentric fractures.

5. Later Events: Desilication and Dykes

Soon after the Barnesmore Granites had reached their present level, they were affected by two events directly connected with Caledonian magmatism. The first of these was a peculiar desilication, restricted chiefly to the G2 granite. This deuteric alteration affected the late aplites and pegmatites as well as the host granite, but predated the emplacement of the swarm of Caledonian minor intrusions.

(A) Desilication

This occurs chiefly in an area southwest of Lough Belshade, where it results in the formation of highly irregular patches of monzonitic rocks up to 30 meters across. These boundaries are marked

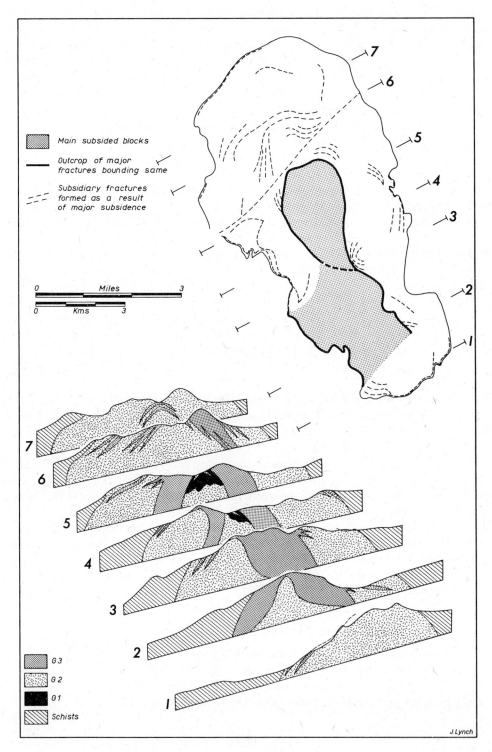

FIGURE 10-2. The structure of the Barnesmore Pluton: (above) structural map; (below) serial sections across the Pluton. After Walker and Leedal (1954).

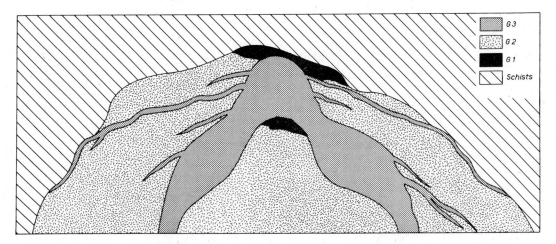

FIGURE 10-3. A generalized section of the Barnesmore Pluton to illustrate the essential features of the structural interpretation. The main mass of G3 was emplaced by subsidence of a large block of G1 and G2; the thin sheets of G3 were intruded along fractures resulting from the partial caving-in at the unsupported sides of the main subsided area.

by the very rapid disappearance of quartz from the host granite, which results in the contacts being fairly sharp. The most remarkable observation is that, even where such contacts cut across several types of granite, they always preserve the structure of the parent rock (Fig. 10-4) to the extent that aplite veins which had previously cut the host granite can still be traced without deflection as ghostly remnants through the quartz-free body. These masses have clearly been formed *in situ,* and have originated by late stage (post-emplacement of G3) alteration of the granite, as confirmed by the frequent enclosure of patches of unaltered rock. Their contacts are thus metasomatic and not intrusive.

Along with an obvious disappearance of quartz, there is a good deal of evidence of readjustment within the feldspars: indeed, there is some suggestion that feldspar crystals have enlarged themselves at the expense of the quartz. Other mineralogical changes include iron staining of the felspars and wholesale chloritization of biotite. The texture of the altered rocks is described as "crushed," and, further, in the outcrop, the outer zones of these desilicated patches have a cavernous appearance due to the leaching out of carbonate material.

A significant feature is the frequent elongation of these bodies in a NW–SE direction parallel to the major axis of the dome and to one of the directions taken by jointing and the later dyke swarm. This suggests a structural control, presumably in the form of a tensional opening of fractures, of the location of what seem to be pipelike conduits up which gaseous agents passed under pressure at a late stage in the evolution of the Pluton—dissolving the silica and partially disrupting the fabric of the host rock. This remarkable process might well be expected to result from the escape of water vapor from a granitic magma at depth: experimental work shows how easily silica can be removed from a charge of granite composition in an open system (MacKenzie, 1960). Indeed, the real problem is to understand why this desilication is not a more common late-stage feature of granitic plutons emplaced in a tensional strain environment.

(B) The Dyke Swarm

Clearly attached to the pluton is a particularly well-marked swarm of N to NE trending dykes, with a second lesser swarm trending at about S35°. Though widespread, these are comparatively

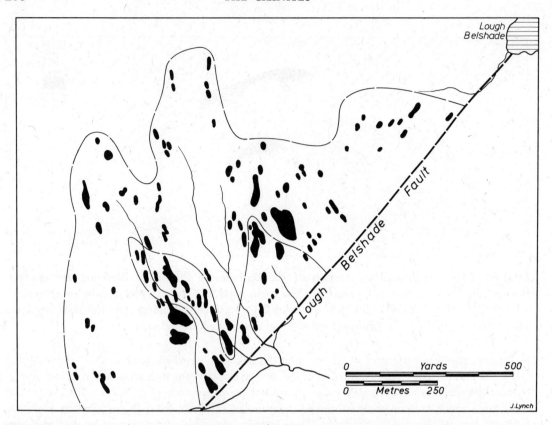

FIGURE 10-4. The G2 granite southwest of Lough Belshade, showing the desilicated patches. Those within the dashed line are non-cavernous, those outside cavernous. After Walker and Leedal (1954).

minor intrusions, averaging only one meter in width, the very largest being 15 meters. By far the most abundant are quartz microdiorites and microdiorites, of which the former are often markedly porphyritic.

These dykes were intruded into a relatively cool host, for they exhibit chilled margins; further, they trend parallel to a major joint system. Indeed, we consider that intrusion occurred along early joints opened during cooling, especially along the axis of the pluton.

Closely associated in time and space with these swarms are the groups of breccia pipes previously mentioned in Chapter 7.

6. Mode of Emplacement: Cauldron Subsidence and the Collapse of Roof Arches

We have followed the relatively simple hypothesis of Walker and Leedal, in which the subsidence of a large dome-topped block of country rock, separated on an arcuate, outward-dipping fracture, is envisaged as being the cause of emplacement of the two earlier members of the complex. Further subsidence, this time of a central block, allowed permissive emplacement of G3. The consequent weakening of the supports of the surrounding granite led to a partial caving in of the latter along fractures which are now represented by both the intrusions along the contact zone and the fingering sheets—a model which, as we have seen, is most easily appreciated by reference to Fig. 10-3.

The control of the main contact of stocks such as Barnesmore is generally assumed (*cf.* Anderson, 1936) to be similar to that of ring dykes, where the operation of ring fractures is well established. While this is a reasonable assumption, we have to remember that the evidence for such a fracture is rarely to be found in such plutons. Further, it would seem that any break where the Barnesmore contact is now located was unlikely to have reached the surface, but rather to have shallowed upward and closed to form a subterranean domical structure. This is a rather different type of fracture from that controlling many ring-dyke–caldera associations. Again, there is no suggestion at Barnesmore of the updoming usually found with the type of ring-fractures associated with caldera.

However, despite these reservations concerning the nature of the boundary fracture, it seems that a valid model for many plutons like Barnesmore involves the deep-seated subsidence of crustal blocks separated along domical fractures, yet a rational explanation of the formation and foundering of such domical blocks has yet to be provided.

CHAPTER 11

THE MAIN DONEGAL PLUTON: A STUDY IN SYNPLUTONIC DEFORMATION

CONTENTS

SECTION A. INTRODUCTION: THE PROBLEM	211
1. Preamble	211
2. Salient Features and the Pitcher-Read Model	211
SECTION B. THE INTERNAL FEATURES: A DEFORMATION HISTORY	215
1. The Banding: a Unique Structure	215
(A) General Description	215
(B) Discussion: Models for the Origin of Banding in Plutonic Rocks	218
2. The Mineral Alignment	219
3. The Dykes: a Chronology of Emplacement	220
(A) Introduction	220
(B) Microgranites: the Sederholm Effect	221
(C) The Felsites	223
(D) Pegmatites and Aplites	225
(E) The Late Veins	226
4. Discussion of Banding and Dykes	226
5. The Marginal Zones: Cataclastic Deformation	227
6. Petrography: the Importance of Late Mineralogical Changes	230
SECTION C. THE GRANITE DEFORMATION: ITS GEOMETRY AND PATTERNS OF STRESS AND STRAIN	234
1. Introduction	234
2. The Mineral Alignment: Kinematic Significance of the L-S Fabric System	234
3. The Inclusions: Important Parameters for the Analysis of Strain	234
(A) Preamble	234

(B)	Shape and Orientation	235
(C)	Internal Structures	235
(D)	Interpretation of Internal Structures: Ptygmatic Folds?	239
(E)	Structures in Inclusions as a Guide to Bulk Deformation of the Pluton	244
4.	Late Dykes and the Regional Stress Pattern in Central Areas of the Pluton	247
5.	The Deformation in Marginal Areas	249

SECTION D. A DISCUSSION SUMMING UP THE DEFORMATION MODEL   251

SECTION A. INTRODUCTION: THE PROBLEM

1. Preamble

We come now to a discussion of the largest, and in many ways, the most intriguing of all the Donegal granites, the Main Donegal Pluton. This extends in a northwesterly direction from the Gweebarra Estuary nearly all the way to Mulroy Bay and comprises, together with the closely associated Trawenagh Bay Pluton, an area of some 450 square kilometers. This Main Donegal Pluton (or simply the Main Granite) is characterized by a strong foliation and a largely concordant relationship with its envelope, the only exception to the latter being in the southwest, where the neck of the Ardara Pluton, the metasediments of the Maas area, and the southern part of the Thorr Pluton are cut across. Inclusions of both these latter plutons are common in the Main Granite, and its gradational boundary with the Trawenagh Bay Pluton, which itself cuts both G1 of the Rosses Complex and members of its N–S porphyry dyke swarm (see Chapter 9), indicates that the Main Granite is also younger than the Rosses, and by implication, the Fanad Pluton. Though their age relative to the Barnesmore is unknown, we are clear that the Main Donegal and Trawenagh Bay Granites are the latest members of the Donegal Granite Complex itself.

The major problem posed by the Main Granite is that, despite its occurrence as the last of a series of otherwise high-level, post-tectonic plutons, it possesses most of the features of an early, catazonal, syntectonic granite. It has a concordant outline, an intense internal fabric marked by a strong mineral alignment, a banding, a preferred orientation of inclusions, and a severely deformed envelope in which kyanite- and sillimanite-bearing schists of regional metamorphic aspect are found instead of normal contact hornfelses. All these features are hard to reconcile with the fact that the Main Granite postdates a dominantly static magmatic intrusion with a normal contact metamorphic aureole (Thorr), a high-level diapir (Ardara), and even a ring-complex (Rosses). Indeed, it was mainly the Main Granite which created so much confusion and debate regarding "intrusive" as against "metamorphic" granites among the 19th century workers in Donegal (see Chapter 1), and it was not until the intensive field mapping and research of the last two decades that the problem was at least partially solved.

2. Salient Features and the Pitcher-Read Model

In two papers read before the Geological Society of London, Pitcher and Read (1959, 1960b) summarized their own work and that of Gindy, Iyengar, McCall, Pande, Tozer, and Cheesman (see Fig. I-4). In the first of these papers, they gave an account of the internal features of the Main Granite, which we shall review as an introduction to our discussion.

Mineralogically, the authors established, the Main Granite is a highly variable body of medium to coarse-grained biotite-microcline granite varying to granodiorite; common accessories are muscovite and epidote. There is a tendency for a medium-grained biotite granodiorite to predominate in the southeastern part of the pluton, while a lighter colored, coarser-grained granite with a more variable biotite content and a higher proportion of microcline characterizes the northwestern portion.

The Main Granite exhibits a ubiquitous mineral orientation and in places a strong, steeply dipping banding marked by an alternation of darker finer-grained with lighter coarser-grained material, aligned everywhere in a NE–SW direction. Coarse bands may cut or enclose dark bands, and bodies of the southeastern facies occur as "rafts" in the northwestern variant, indicating that the coarse-grained varieties are younger than the fine-grained. The mineral orientation also trends NE–SW and passes from a crude subvertical planar alignment throughout central areas to a more intense foliation with a strong subhorizontal linear component in marginal portions of the pluton. Pitcher and Read (1959, p. 275) interpreted both the banding and the mineral alignment as primary flow structures in a "heterogeneous material of different consolidation points."

Contacts between the Main Donegal Pluton and its country rocks are sharp on a small scale. Nevertheless, there is everywhere a marginal zone—on average 800 meters wide—in which concordant or slightly transgressive sheets of granite and aplite-pegmatite, varying in size from small veins to the massive apophysis of Crockmore (see Map 1 in folder), increase in number inward to a point where they coalesce to form the main body of the pluton. These sheets provided the authors with a clue to the origin of perhaps the most striking feature of the Main Granite, the raft trains.

Throughout the pluton are numerous lenticular or slabby xenoliths, assimilated to varying degrees but on the whole easily recognizable as relicts of country rock material. These are everywhere aligned parallel to the "flow lines as beads on a rosary" (*loc. cit.*, p. 290), either as single individuals, in mixed groups, or in long trails of one or more lithological types—these are the "raft trains" (see Map 1 in folder and Fig. 11-1). The authors emphasized that these rafts were not roof pendants but were "free-swimming," and that the raft trains themselves could "be shown with certainty to be rooted in definite roof or wall horizons of country rocks" (*ibid.*, p. 284). Thus, the distribution of inclusions in the southwest corner of the pluton reflects with some accuracy the stratigraphy of the Maas and Fintown areas, as is the case in the Crockmore—Glen Lough area in the northeast. Pitcher and Read described each raft train in detail and concluded that the zone of inclusions along the northwest margin from Kingarrow to Brockagh was "obviously a smear-zone produced by viscous flow against a vertical wall" (p. 296), and the long trains from Glen to the Derryveagh and Glendowan Mountains were likewise attributed to laminar flow. Since all these raft trains would be rooted in or near the roof between Glen and Crockmore, and since they appeared to splay out to the southwest (see Fig. 11-1), they did not form a true ghost stratigraphy, but represented fragments plucked off both the walls and the tips of roof pendants and entrained in magma flowing laterally from the northeast and streaming them off to the southwest. The pendants themselves represented those parts of the envelope which had survived between the great sheets that tongue into and wedge apart the strata of the roof (Fig. 11-2). While the raft trains around Derkbeg at the southwest end of the Main Granite are rooted in the Maas area, Pitcher and Read considered that their relatively short lengths were consistent with a southwesterly direction of magmatic flow.

That this lateral flow of magma was forceful was also indicated by the strong mineral alignments, by the banding, and in particular by the structures in the marginal granite and the adjacent country rocks. In the marginal zones of the Main Granite, the normally weak mineral alignment of the pluton becomes stronger, and dykes and sheets are strongly sheared, in places developing "true ruptural slickensides" (*loc. cit.*, p. 279), and are frequently boudinaged on axes normal to the subhorizontal "intrusion lineation." The authors viewed this intense cataclasis as a true protoclastic deformation of a consolidating magma rasping against its carapace during a "late hydrothermal stage in the cooling history" (*ibid.*, p. 284).

In the adjacent envelope, this forceful wedging accounted for the generation of folds superimposed on the regional structures—folds which could be traced progressively from their earliest development

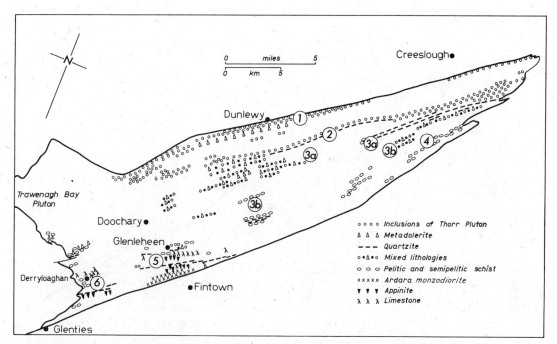

FIGURE 11-1. Ghost stratigraphy in the Main Donegal Granite. Modified from Pitcher and Read, 1959, Figures 9 and 15. The "raft-trains": (1) Border Zone consisting mainly of xenoliths of Thorr rocks; (2) Derryveagh Zone consisting mainly of xenoliths of quartzite and Thorr rocks; (3a, 3b) Glenveagh and Binaniller Zones consisting of xenoliths of mixed lithologies; (4) Crockmore Zone and Septum consisting of mainly pelitic lithologies; (5, 6) Glenleheen and Derryloaghan Zones consisting of xenoliths of nearly all the lithologies found in the southwest part of the envelope, including types from the Ardara Pluton and the Meenalargan Appinite Complex.

as minute puckers with an associated crenulation cleavage to the situation where sizeable folds develop a strong axial-plane schistosity close to the Main Granite (Pitcher and Read, 1960b; see also Rickard, 1961). Another related effect was the marked increase in metamorphic grade toward the granite contact, from regional chlorite-garnet schists to coarse marginal schists with andalusite, kyanite, staurolite, sillimanite, garnet, and plagioclase. The interpretation of the textural relationships of these porphyroblasts as syntectonic again fitted in neatly with the authors' model of forceful emplacement of magma with deformation and crystallization proceeding together. The occurrence of the three aluminosilicates without replacive relationships they attributed partly to temperature-pressure conditions near the triple-point and partly to compositional controls, a point commented upon at greater length in their later discussion of Donegal contact metamorphism (Pitcher and Read, 1963).

This model of forceful emplacement by "lateral wedging combined with horizontal stretching" (Pitcher and Read, 1959, p. 295) generated a considerable debate both in private and in public (see, for example, discussions to 1959 and 1960b papers). There was some doubt that a horizontally migrating magma could produce in the envelope folds whose axes were oriented parallel to the flow movement (cf. Rickard, 1958), and even though the authors emphasized that some of these structures represented older regional folds tightened by the granite emplacement, the geometry of the envelope structures remained a problem. Another major query concerned the southwestward flow of magma. What happened, then, in the southwest end of the pluton? Did this represent a "huge sink-hole with

214   THE GRANITES

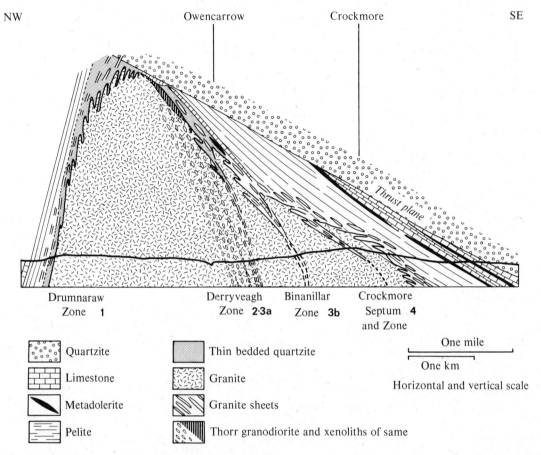

FIGURE 11-2. Diagrammatic section across the northeastern end of the outcrop of the Main Pluton to show upward connection of "raft-trains" to septa rooted in the room. References to numbered raft-trains as in Fig. 11-1. Folds in envelope and ghost folds in Pluton omitted. According to Pitcher and Read (1959).

a torrent of Main Granite magma pouring down one side and the Trawenagh Bay magma welling up the other" (Tozer, in discussion to 1959 paper)? Several suggestions were made that the Main Granite had intruded vertically and then been subjected to an "axial" extension due to confining pressure of the walls, and yet, the authors replied, the pluton was clearly later than the main fold structures and had apparently generated its own marginal deformation; there was no evidence of earlier vertical movement. Horizontal flow, indeed, seemed to be required in order to explain the southwestward splaying of raft trains.

These and other problems formed the basis for the doctoral studies of the junior author, and as a result of these new investigations, we now place emphasis less on forceful emplacement and more on the deformation *in situ* of a cooling pluton. Since most of the internal features of the Main Granite owe their character to this deformation, we shall amplify certain of the descriptions of Pitcher and Read in the following pages. For other details, we refer interested readers to the original papers.

So as to emphasize the chronological aspects of the granite deformation, we propose to deal, in the balance of this chapter, first with the internal features in the order in which they originated, and then with the geometric, kinematic, and dynamic aspects of this structural event. Apart from our conviction that a proper understanding of this unusual pluton can only come from an appreciation of the mechanics and effects of its deformation, we feel that the general lack of similar modern studies on granitic rocks also justifies a detailed treatment of these structural aspects. The structural and metamorphic development of the envelope to the Main Granite is treated separately in Chapter 12, because of the complexity of the chronology and the difficulty in making correlations between deformation events in the Pluton and those in the country rocks.

SECTION B. THE INTERNAL FEATURES: A DEFORMATION HISTORY

1. The Banding: a Unique Structure

(A) General Description

One of the unique features of the Main Donegal Granite is a banding of coarser-grained light and finer-grained dark granite, aligned nearly everywhere parallel to the long sides of the pluton. It was partly this widespread banding which led some of the early workers to argue a metamorphic origin for the Main Granite (Scott, 1862, 1864; Blake, 1862), while later researchers (Cole, 1902, 1906a, b; Andrew, 1928b) interpreted the banded granite as a "composite gneiss" resulting from the intimate mechanical intermingling of intrusive granite and the host schists, a kind of *lit-par-lit* injection modified by subsequent flow. Pitcher and Read, as we mentioned earlier, concluded that this banding was a flow structure due to the "drawing out" of relatively early fine-grained dark material in a host of the lighter material, though there was some disagreement among others of the Imperial College team (Pande, 1954T; McCall, 1954; Cheesman, 1956T).

Reexamination of the Main Granite banding by the junior author (Berger, 1971b) led to the conclusion that, as one might have expected, there are several distinct types of banding. In certain areas, particularly along the northwestern margin, the intrusion into the granite of a series of late parallel sills of dark microgranite gives a banded appearance. In this case, the compositional differences are usually distinct, and careful examination frequently discloses minor discordances between the sills and their host granite. In other places, a rather irregular banding results from the presence in the granite of a series of thin planar inclusions of country rock schist aligned parallel to the general NE–SW trend. The frequent partial assimilation of such pelitic strips by their host granite and the consequent contamination of the latter of course add to the banded appearance. These two easily recognizable types of banding are much more restricted in occurrence than a third type expressed by the regular alternation of bands of darkish, relatively fine-grained trondhjemite with bands of lighter, coarser granite. This *regular-banding,* which is for the most part that described by Pitcher and Read, forms the subject of the following discussion. The granite deformation caused an accentuation of this banding and in many places produced, by the deformation of preexisting textural or compositional inhomogeneities, an irregular, discontinuous "banding" (Fig. 11-3c; Pitcher and Read, 1959, Plate X, Fig. 2), particularly in the intensely deformed marginal areas of the pluton.

Regular-banding occurs in many locations throughout the northern facies of the pluton, and like other banding types, is steeply dipping and is generally oriented in a NE-SW direction parallel to the mineral alignment. The width of individual bands varies but is generally less than a meter. Although single bands can sometimes be traced along strike for thirty meters or so, the dark bands more frequently wedge out into a host of light granite (Fig. 11-3a, b), and whole zones of regular-bands fade out along strike into a host of light band composition (Berger, *loc. cit.*). On a small scale, fragments of dark bands may be found as inclusions in light granite, and though the junctions between bands are generally sharp in hand specimen, in certain cases transitional passages can be seen in which the dark granite passes into light granite by an increase in large feldspars (Figs. 11-3d, -4a). These

FIGURE 11-3. Banding in the Main Granite. (a) Interdigitation of dark and light regular-bands. Glenleheen. Note cross-band on right hand side of photo. (b) Regular-banding on the northwest limb of the Doochary arch. (c) Deformation of fine grained granitic inclusions to form an irregular banding. Glenleheen. (d) Fragment of a dark band enclosed in granite of light band composition, Doochary. Large crystals are microclines. Coin for scale.

relationships indicate, as Pitcher and Read suggested, that the dark bands are older than their lighter counterparts.

The strongest evidence for the relative ages of the band types, however, is provided by the *cross-bands*. These are cross-cutting bands or "dykes" similar in appearance and composition to the light bands, from which they can be distinguished only by slight differences in texture (Fig. 11-3a, 11-4b, c). Like their host bands, these discordant units have boundaries which may be sharp or gradational. Cross-bands are found chiefly in areas where the host banding is discordant to the main trend, in which case the cross-bands themselves lie close to this main direction; this is best seen in the Doochary area (Berger, *loc. cit.*). In some cases, the cross-bands do not offset their hosts, but more generally, the latter are so markedly displaced that it is impossible to match bands across the discordant unit. Cross-bands may occur by themselves or accompanied by others, and in some outcrops several generations of successively cross-cutting bands can be seen (*ibid.*). Although single cross-bands of dark band composition were not seen, some of the wider cross-bands contain discontinuous streaks of dark granite aligned paralled to their walls.

FIGURE 11-4. Regular-banding in the Main Granite. (a) Dark regular-band enclosed in host of light granite, Doochary. (b) Cross-band offsetting host bands, Doochary. Note transitional lower boundary. Large crystals are microclines. Pencil for scale. (c) Sharply-bounded cross-band with little or no offset of host bands, Doochary. Coin for scale. (d) Dark band offset along early healed shears. Note late pegmatite dyke in upper part of photo. Roadside between Doochary and Derryloaghan.

The similarity in texture and composition between the light host bands and the cross-bands thus supports the conclusion that the dark band material is older than the light bands; particularly convincing is the lack of single dark cross-bands. The cross-bands clearly occupy planar zones of differential movement, and their similarity to the light host bands suggests that the same conclusion may well apply to them. However, before proceeding any further in this analysis, we must consider the petrography of the bands.

Examination of regular-band pairs from several localities throughout the pluton clearly shows that the major difference between the dark and light bands is in the microcline content; the presence of of relatively large microclines gives the light bands both their color and coarse grain size, while the dark bands contain little or no potash feldspar. There are other mineralogical differences as well (see Berger, *loc. cit.*, Table 1), such as the frequent concentration of quartz in the light bands, but many of the modal variations seem to be a consequence of the addition of potash feldspar to the system. Certainly, there are no consistent differences between bands, as for example, in plagioclase composition (An_{10} to An_{40}) or zircon morphology, and the main textural contrasts relate to microcline. Thus, near microcline crystals in the light bands biotite is frequently chloritized, and myrmekite, antiperthitic intergrowths, and albite rims are all common in plagioclase crystals where they adjoin microcline. Microcline itself commonly veins plagioclase and sends out jagged tendrils along other grain contacts, and the larger crystals frequently enclose small grains of biotite, plagioclase, and quartz, which may be arranged in a zonal fashion within the host microcline (see Fig. 11-9 and p. 332). These relationships, again common to other Donegal granites and indeed to many microcline-bearing plutonics in general (see Chapter 15), clearly indicate that the microcline grew at a relatively late stage in the paragenesis of the banded granite.

The petrography of the regular-bands thus indicates that it was largely the addition of microcline to a parent composed mainly of quartz, biotite, and plagioclase (*i.e.*, a trondhjemite) that formed the granitic light bands. Scanty analyses support the conclusion that the major chemical distinction between bands is in the alkalis. The evidence from the cross-bands and from the mineralogy thus suggests a model by which potassium-bearing fluids were concentrated along planar zones of movement or differential shear, though other components such as silica were probably also mobile at this time. That the pluton was not then wholly crystalline is indicated by the complete lack of cataclasis in any of the bands or cross-bands (apart, of course, from those in the sheared margins of the granite), and we think that the banding developed at a late stage in the primary crystallization, when the proportion of crystals to liquid was fairly high.

There is abundant evidence that the regular-banding was considerably deformed after, as well as during, its period of formation (Berger, *loc. cit.*). On a small scale, there are scattered minor folds with varying styles and plunges, though their axial planes are consistently steep and NE–SW trending. In addition, there are occasional situations in which dark bands have been dragged out along *early healed-shears* which are sometimes occupied by granite of light band composition (Fig. 11-4d). These shears clearly demonstrate that under certain conditions, such as for example a lower strain rate, the dark bands could deform by highly ductile flow rather than by the fracture apparently indicated by most of the cross-bands.

The source of the potassium in the light bands is a problem of more than local significance. It is tempting to suggest that the bulk composition of the bands approaches that of the original parent magma, and that the banding formed by squeezing the potassium-rich melt-fraction from a granodiorite mush to form a granitic segregation and a trondhjemitic residue. However, there are no consistent lateral variations within individual dark bands which might support such a proposal, such as a progressive decrease of biotite toward their margins. Whatever the source of the potassium, the field evidence indicates that it became concentrated along the planar zones to form the light bands, and furthermore, that in many cases the potash feldspathization continued beyond the confines of these bands to convert whole sets of dark bands to light band composition (see Berger, *loc. cit.*).

(B) Discussion: Models for the Origin of Banding in Plutonic Rocks

Without going into great detail at this juncture, we wish to comment upon the *apparent* uniqueness of the regular-banding of the Main Granite among Caledonian intrusives and granitic rocks in general. We have searched through much of the literature for parallel situations but have found surprisingly few. Were this phenomenon similar to the layering formed in certain mainly epizonal plutons (*e.g.,* Harry and Emeleus, 1960; Emeleus, 1963; Ferguson and Pulvertaft, 1963; Taubeneck, 1964; Claxton, 1968; Sörensen, 1969; Aucott, 1970) by "high temperature magmatic sedimentation," then we could follow Wager and Brown (1968, Chapter 16 and p. 555) in arguing that its rarity results from an unusually low viscosity in the original magma. However, in contradistinction to layered granites, the Main Donegal unit is devoid of sedimentation features such as grading, washouts, and slumps; there is no evidence from the mineralogy that the melt was of low viscosity, the regular-banding is dominantly vertical and parallel to the steep sides of the pluton, and above all the banding results from a concentration of late and not early forming minerals. The presence here of the cross-bands rules out any parallels with banded gneisses of metasedimentary origin (*cf.* Dietrich, 1960), and although layered pegmatites are also frequently characterized by replacive potash feldspars, the Main Granite bands lack the delicacy of structure and texture customary in banded pegmatites (*e.g.,* Jahns and Tuttle, 1963; Redden, 1963; Stone, 1969).

The standard explanation of banding (gneissosity, schlieren) in granitic rocks appears to be derived from the general model of Scrope, who pointed out that flow of a solid-liquid system could produce a "more or less separation of the solid and coarser from the finer and liquid particles into different zones or layers" (1858, p. 84). This seems to have been the basic idea current in the heyday of granite tectonics (*cf.* Balk, 1937; see Harker, 1909, pp. 138–9, for an earlier review) whereby early crystallized portions of the magma become segregated and streaked out during "primary" flow to form schlieren, and it is still occasionally invoked, as for example by Pitcher and Read (1959) for the present situation (see also Sen, 1956; Hutchinson, 1956; Alper and Poldervaart, 1957; Wilshire, 1961; and recent studies by Bhattacharji, 1967, and Simkin, *in* Wyllie, 1967). Similar arguments have been used to explain banding in cataclastic rocks as a result of mechanical segregation of minerals with varying ductilities or response to recrystallization (*e.g.,* Schmidt *in* Turner and Verhoogen, 1960, pp. 585–6; Prinz and Poldervaart, 1964; Sclar, 1965; Ragan, 1963; Hibbard, 1964; Burch, 1968; Spry, 1969). However, the evidence from the discordant bands, the lack of cataclasis, and again the fact that it is the latest mineral, microcline, which delineates the banding, all argue against such an interpretation for the regular-banding.

We can, however, draw a closer analogy with a modified version of the filter-pressing model for differentiation first discussed by Harker (1909, pp. 323–7) and later elaborated upon by Bowen (1920, p. 161), who argued that "shearing of a crystalline mesh, particularly when it is still weakly knit together" could result in fractures that "instantly fill with liquid from the interstices of the mesh, and repetition of the action may give rise to banding, properly oriented with respect to the shearing stress and showing approximately that contrast between bands that obtained between liquid and crystals at the time of the action" (see also Mead, 1925, p. 697). Direct applications of this filter-pressing model to field examples are not uncommon (see, for example, Foslie, 1921; Emmons, 1940; Mikami and Digman, 1957; Gates and Scheerer, 1963; Watterson, 1968), and the modern tendency is to emphasize the role of diffusion, vapor transfer, or even partial melting in this segregation process (*e.g.,* Bennington, 1956; Gresens, 1966, 1967*a*; Carpenter, 1968; Shaw, 1969). Indeed, there is now available a whole spectrum of differentiation models based on the concept of migration of mobile material toward zones of relative low pressure, which can be applied to banding in both igneous and metamorphic rocks (*e.g.,* banding in pegmatites: San Miguel, 1955, 1969; Gresens, 1967*b*; planar zones of "synkinematic" granite in granodiorite massifs: Marmo, 1956, 1967; quartzofeldspathic veins and segregations in metamorphic terrains: Reitan, 1959; Ramberg, 1952, 1961*a*; Carstens, 1966; Watterson 1968*a*; banding in metamorphic rocks: Ward, 1959; Turner & Verhoogen, 1960; Cannon, 1964; Bowes and Park, 1967; banding in basic and ultrabasic rocks: Smith, 1958; Lipman, 1963, 1964; Wilshire, 1967; Burch, 1968). It is clear from these examples that structurally induced differentiation, whether purely mechanical, purely by diffusion, or by some combination of both, in a solid or a solid-liquid system, can lead at a wide range of PT conditions to the development of banded structures. The regular-banding of the Main Granite is a case in point, and its apparent uniqueness will, we feel certain, dissipate as more careful studies of plutonic areas are made.

2. The Mineral Alignment

We have mentioned that the Main Granite possesses a strong mineral alignment oriented everywhere in a NE–SW direction. The character and intensity of this alignment vary, depending partly on the lithology and partly on the location. Away from the marginal zones, in coarser lithologies, the alignment may be discernible only on weathered surfaces where prolonged exposure to constant wind and rain has resulted in the differential etching of minerals; in hand specimens and thin sections, it is often impossible to recognize. Geometrically, the mineral orientation in this central area is an S or an $L < S$ fabric (see Chapter 4) with vertical or steep dips, though in some finer-grained granitic types of the southern facies, the linear component may be stronger ($L = S$ to $L > S$) with gentle northeast plunges in the northeast and southwest in the southwest (see Map 1 in folder). In marginal zones, the mineral alignment, which is here much more intense and is clearly imprinted on all lithologies, is dominantly an $L > S$ fabric about which we shall say more presently. In this regard, it is interesting to note that the axes of maximum and minimum susceptibility of magnetic anisotropy in samples collected along the Fintown–Doochary road also define a composite L–S fabric with the planar component aligned parallel to the visible mineral foliation and the linear component with the mineral lineation (King, 1966, and see Berger and Pitcher, 1970).

The mineral alignment involves mainly biotite and quartz. In the finer-grained types, it is largely biotite which has the preferred orientation, but in coarser lithologies elongated single grains and aggregates of quartz make an important contribution, too. In some instances in central parts of the pluton, microcline tends to lie within the plane of foliation, but a plagioclase alignment has never been observed. Muscovite likewise is rarely aligned except in certain marginal zones.

Except where certain late dykes and shear zones are concerned, the mineral alignment is completely penetrative and crosses without deflection any lithological boundaries discordant to the general trend. Nowhere is this more clearly seen than in the Doochary area where the alignment cuts straight across the banding, defining a kind of axial plane "slaty cleavage" to a large synformal arch in the banding (see Pitcher and Read, 1959, Fig. 5; Berger, 1971*b*). It also cuts straight across certain pegmatite and microgranite dykes (see below), and at the northeast and southwest ends of the pluton, crosses directly into the adjacent country rocks without much deflection. These relationships clearly indicate that the mineral alignment is a superimposed structure and not the customary primary alignment due to magmatic flow in the classic sense of Cloos (see Chapter 15 for further comment).

Although the mineral alignment thus postdates the banding and certain dyke phases, other dykes and veins are unaffected by it and are thus presumably later. Indeed, the presence or absence of a mineral alignment parallel to the host fabric is one of the major parameters used to establish the chronology of dyke phases, as we shall now see.

3. The Dykes: a Chronology of Emplacement

(A) Introduction

A marked feature of the Main Granite is the presence of numerous dykes with varying structural relationships and of variable composition, though all are broadly granitic. These dykes all postdate the regular-banding and themselves span an appreciable range of time from relatively early pegmatites and partially mobilized microgranites to late, wholly post-tectonic quartz veins. Pegmatite dykes are especially prominent in marginal zones, a concentration interpreted by Pitcher and Read as "expressing roof phenomena" (1959, p. 271), but since the intense deformation here obscures their relative ages and original characters, we shall restrict ourselves in the following discussion to dykes away from these marginal areas.

There is usually no way of proving the mode of origin and emplacement of these dykes. The distinction between dilational and nondilational dykes is only rarely possible, and at any rate we share the doubts of Wells and Bishop (1954) and Chadwick (1958; see also Kretz, 1968) that nondilational relationships prove replacement origin, as has been argued by Goodspeed (1940), King (1948), and Cooray (1961). However, apart from certain felsite dykes which resemble members of the Rosses porphyry dyke swarm, there seems to be no difficulty in viewing these dykes as normal late-stage intrusives like those that characterize many granitic plutons.

In general terms, the Main Granite dykes may be grouped into four more or less distinct categories, partly on the basis of lithology and partly on structural grounds. These are (1) microgranites, (2) felsites, (3) pegmatites and aplites, and (4) late veins. Cross-cutting relationships in several selected localities indicate that the microgranites are older than the felsites, that the pegmatites and aplites range in age from pre-microgranites to post-felsites, and that the late veins are the youngest of these minor intrusives, but this chronology is only of the broadest kind and there are many overlaps.

(B) Microgranites: the Sederholm Effect

The microgranites comprise a variable group of dominantly dark-colored dykes with some leucocratic representatives. Although few of these are strictly microcrystalline, their grain size is sufficiently smaller than that of other fine-grained variants of the Main Granite to justify the term microgranite. It is sometimes possible to distinguish earlier from later microgranites on structural grounds, and we shall follow this twofold division in our descriptions.

The *earlier microgranites,* one of which has been described in detail by Pitcher and Read (1960a), are usually darker and more fine-grained than the much more common later microgranites. While these earlier members may in part be fairly regular dykes transecting the host granite and its banding, their salient characteristic is that they may also be cut by their host, remaining as a row of partially or wholly disrupted fragments (Fig. 11-5) preserving their original trend. They generally possess an internal mineral alignment which is parallel to and continuous with the host alignment. However in some cases, this internal fabric is inclined to both dyke walls and external structure (*e.g.,* Pitcher and Read, *loc. cit.,* p. 56), this obliquity resulting generally from movements of the dyke walls relative to each other, see p. 223).

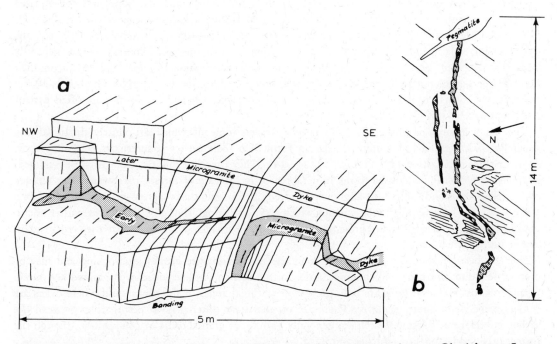

FIGURE 11-5. Microgranite dykes. (a) Early and late microgranites at Glenleheen: Large quarry on roadside. Dashes indicate mineral alignment. (b) Early microgranite dyke (stippled) at Sruhanavarnis in the Derryveagh Mountains. After Pitcher and Read (1960a, Figure 1). Irregular long dashes indicate pelitic xenoliths.

Dykes with similar structural relationships to these earlier microgranite dykes have been described at length from many areas as relict dykes, pseudo-dykes, synplutonic dykes, or reactivated dykes; indeed, Eskola (1961) termed this situation the "Sederholm Effect," and although Sederholm's original descriptions involved basic dykes, the term has been extended by Mehnert (1968) to include other lithologies (see Berger and Pitcher, 1970). Relict dykes (Goodspeed, 1955) are those preserved when the surrounding dykes are granitized under static conditions, that is, they are the "resisters" of Read (1957). Pseudodykes (Miller, 1945) and relict pseudodykes (Goodspeed, *loc. cit.*) are likewise resistant tabular enclaves of pre-granite country rock, though in this case not original dykes. Neither of these explanations can apply to the present situation, for the earlier microgranites clearly cut both inclusions of country rock and the granite banding. We follow the interpretation of Pitcher and Read (1960a) according to which these dykes represent synplutonic dykes emplaced into "embryonic fractures" in hot viscous granite still in the process of primary crystallization, a similar conclusion to that of Roddick and Armstrong (1959) for dykes in the Coast Range Batholith of British Columbia.

It could be argued that these dykes have more in common with certain intrusive dykes, which, together with their host rocks, have been subsequently reactivated from a completely consolidated condition during a period of high-grade metamorphism—a common feature of complex plutonic terrains (*e.g.*, Sutton and Watson, 1951; Wegmann and Schaer, 1962; Wegmann, 1963; Watterson, 1965, 1968a; Allaart, 1967; Watson, 1967). Such reactivated dykes likewise preserve their intrusive character in some places but are disrupted and "back-veined" by their hosts in others. Indeed, the only significant difference between the reactivated and synplutonic models is that in the former case the dykes were originally emplaced into a completely crystalline and relatively cooler host, whereas in the latter situation the dykes represent an intermediate stage in the primary consolidation of the host granite, though they were "remobilized" by movements in the surrounding granite after they were emplaced. In Donegal, there is no reason to suspect that the Main Granite had completely consolidated prior to the intrusion of the earlier microgranites; indeed, as Pitcher and Read put it (*loc. cit.*, p. 59), "the crystallization now shown by the banded granite, the (feldspathized) xenoliths in it, and the dykes cutting it is all one and the same crystallization." Furthermore, in contrast to the generally basaltic or lamprophyric reactivated dykes referred to above, which are petrologically distinct from their hosts, the earlier microgranites are similar in texture and composition to the Main Granite itself and are restricted entirely to the pluton and its immediate envelope, whereas the former are often part of a regional dyke swarm (*e.g.*, Watterson, 1965).

The much more widespread *later microgranite dykes,* some of which are clearly dilational, are generally coarser-grained and more regular in form than their earlier counterparts, and they show no signs of remobilization (Fig. 11-5a, c, -12c, -14). They usually exhibit a weak to strong mineral alignment which may be parallel to the host fabric, or in the case of certain narrower dykes, oblique both to their walls and to the host alignment. Indeed, it is largely the weakness of this internal fabric and the lack of "back-veining" relationships which serve to distinguish the later from the earlier microgranites. However, as we shall see presently, the later dykes are also deformed where they cross certain metasedimentary enclaves.

Contacts between the microgranite dykes and the host granite are always interlocking and may be more or less transitional on a microscopic scale. In thin section, the dykes are even-grained and rarely porphyritic; cataclastic textures are only present in marginal zones of the pluton and where dykes are folded in metasedimentary enclaves. There are no consistent differences in petrography between the earlier and later microgranites; both are generally granitic or granodioritic in composition, though the former may also be tonalitic (Pitcher and Read, *loc. cit.*).

(C) The Felsites

In its western and southwestern portions, the Main Granite is cut by a varied suite of sharply-bounded, leucocratic, fine to medium-grained, generally granodioritic dykes, which we shall distinguish as *felsites*. They may be even-textured or porphyritic, and on occasion decrease rapidly in grain size at their margins, perhaps due to chilling. In some places these felsites cut across members of the later microgranite suite, but in many cases they are difficult to separate from those leucocratic microgranites with an oblique internal fabric, and although their structural relationships generally indicate a post-microgranite age, there may well be an overlap in time between the emplacement of the two suites. We have already alluded to the close similarity in structure and mineralogy between these dykes and certain members of the Rosses porphyry swarm, a puzzling situation in view of the probable time relations between the Rosses and Main Donegal Plutons (see p. 187).

Felsite dykes may be quite regular in shape but may also have highly branching forms with non-matching walls (Fig. 11-6). Other felsites have planar walls and maintain a constant width for up to 230 meters along strike. Despite the lack of indications of external deformation, even these felsites have been subjected to structural modifications subsequent to their emplacement, as can be seen by the character of their internal mineral alignment.

Nearly all felsites possess a mineral foliation of varying intensity, which is best developed in their marginal zones and which may be lacking in the medium-grained core zones of some dykes (Fig. 11-6c). In some cases, particularly where the felsites are very thin, this foliation is parallel to the dyke walls, but for the most part it is oblique and follows an open sigmoidal path lying closer to the trend of the walls in marginal than in central portions. This fabric is generally deflected around irregularities in the walls and inclusions of the host granite (Fig. 11-6a). The dihedral angle between walls and internal foliation varies from 0° to 50°, with an average of roughly 30°.

In a few places, it can be seen that the felsite dykes occupy movement zones displacing elements in the host granite (Fig. 11-6b, c) so that the acute angle between the internal foliation and the dyke walls points in the direction toward which the latter has moved. In this connection, Fig. 11-6c shows a number of thin quartz veins cutting a member of the Rosses dyke swarm which is identical with the Main Granite felsites. The deflection of these veins indicates the same sense of displacement of the dyke walls relative to the internal foliation as stated above. We may note in passing that these dykes appear to have been more ductile in their marginal than in their central portions, though whether this is a cause or an effect of the fine-grained margin is difficult to establish.

It is clear from the field evidence that the felsite dykes were emplaced after at least the majority of the microgranites, and certainly after the formation of the host granite mineral alignment. There seems, from their mineralogy, no reason to view their fabric as anything but the product of normal primary magmatic crystallization, as exemplified by their highly zoned plagioclases (An_{26} to An_{55}) and by the presence in the central zones of some dykes of granophyric intergrowths. We regard the frequent fading out inward of the biotite foliation largely as a result of continued primary crystallization in these central portions of the dykes. However, the complete lack of cataclasis suggests that this internal fabric formed in response to stresses superimposed on the dykes after their emplacement and before their complete cooling.

Dykes with similar sigmoidal foliations have been recorded occasionally in the literature (*e.g.*, Kinahan *et al.*, 1871, p. 39–40; Miller, 1945; Kaitaro, 1953; Johnson and Dalziel, 1966), and similar fabrics have been recorded in deformed clays by Weymouth and Williamson (1953) and by Tchalenko (1968). However, the only detailed studies so far have come from the Precambrian basement of South Greenland (Weidmann, 1964; Harry and Oen, 1964; Allaart, 1967; and especially Watterson, 1965, 1968a). Apart from one study by Blyth (1949) in which the reversed pattern was noted in a dyke whose foliation was entirely cataclastic, there appears to be universal agreement that Z-shaped foliation patterns result from sinistral movement of the walls, while S-patterns indicate dextral offsets. This

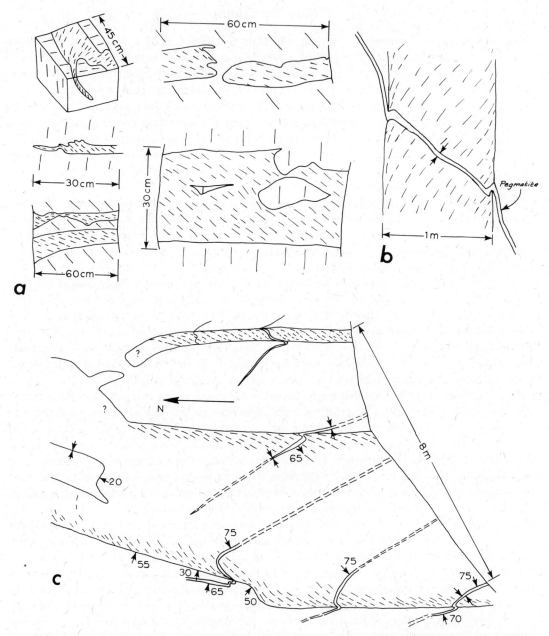

FIGURE 11-6. Felsite dykes showing relationships between internal foliation and dyke walls. (a) From the Doochary area. Note deflection of internal foliation around wall-rock irregularities and around inclusions of host granite. (b) Folding of a late pegmatite vein where it crosses a felsite dyke, Doochary. (c) Felsite dykes attached to the Rosses Centered Complex, Dunglow Baths. Note folding of late quartz veins which cut both host granite and felsite dykes. Central portions of large dykes are unfoliated.

simple situation, as we shall see later (p. 247), enables us to determine the direction of the regional stress system operative at this time (Berger, 1971a).

(D) Pegmatites and Aplites

Pegmatite dykes are by far the most common minor intrusives in the Main Granite, and are especially abundant in the marginal zones of the pluton. They appear to have been emplaced over a long period of time, some predating the earlier microgranites and others postdating the felsites. The much less abundant aplites, on the other hand, all appear to postdate the later microgranites. Although it is difficult to place all the pegmatites and aplites into a strict chronological order, it is possible to distinguish on a broad basis early pegmatites which predate the felsites and which are crossed by the host mineral alignment, and lated dykes which are synchronous with or later than the felsites, and which, apart from certain exceptional cases, show no evidence of having been deformed.

The *earliest pegmatitic bodies* are irregular, diffuse segregations and patches which seem to grade into patches of coarse pegmatitic granite (see Berger, 1971b, Fig. 2) and then to the coarse light regular bands. The presence in these patches of large replacive microcline crystals, which indeed are their main distinguishing characteristic, suggests that these early pegmatites may well represent end stages in the same potash feldspathization which was responsible for the formation of the light regular-bands. Whatever their origin, many of these irregular segregations possess a weak but detectable alignment of biotite and quartz (rarely of feldspar) which is parallel to and continuous with the host foliation.

The same applies to certain more regular and sharply bounded early pegmatite dykes which are often folded, with their mineral alignment and that of their host granite forming an axial plane fabric (see Fig. 11-22b). Some of these early pegmatites are cut by the later microgranites, or even by members of the earlier microgranite suite, while others cut the microgranites and are transected in turn by felsite dykes.

In contrast to their earlier cousins, the *later pegmatites* and *aplite dykes* invariably occur as regular dykes, which are not folded (Fig. 11-4d) and whose internal mineral alignment, where present, is generally parallel to the dyke walls, or on rare occasions, oblique to both walls and host foliation. These dykes may have matching or nonmatching walls, frequently branch off, lens out, or turn corners, and may be traced continuously along strike at fairly constant widths for distances up to 600 meters or more. In addition to the occasional crude zoning of composite dykes with aplitic margins and pegmatitic cores (or vice versa), many pegmatites exhibit the familiar concentric zoning with coarse-grained quartz-rich centers passing outward to finer-grained feldspar-rich margins. Some of these zoned dykes, together with some nonzoned pegmatites, contain large feldspar, mica, and quartz crystals aligned perpendicular to the walls, regardless of the attitude of these walls relative to the host granite foliation. We follow Jahns and Tuttle (1963, p. 84; see also Redden, 1963; Edelman, 1968) in viewing this alignment as the result of primary crystal growth inward from the walls; at any rate, it does not seem to be a superimposed fabric.

There is thus no evidence, away from marginal zones of the pluton, that these aplites and later pegmatites have been deformed. The only exception to this is provided by the shearing and folding of certain pegmatites which cut the felsite dykes (see Fig. 11-6b), but even here the deformation was highly localized and the pegmatites behave normally where they cut the host granite.

Mineralogically, the pegmatites and aplites are rather similar to the host granite, with pegmatites being quite rich in microcline and quartz. Microcline is more commonly perthitic, and there seems to be an almost continuous gradation from perthitic microcline to antiperthitic plagioclase. Plagioclase is rarely zoned. The only unusual mineralogical feature is the presence in some of the later pegmatites, particularly those in marginal zones, of small colorless garnets similar to those in the Rosses and Trawenagh Bay Plutons (see Hall, 1965c).

The field evidence thus indicates that pegmatites were forming more or less continuously from the time when the regular bands were produced to some point following the emplacement of the felsites. The earlier pegmatites seem generally to have been more replacive than later members (*cf.* Tozer, 1955T), and some patches of pegmatitic granite appear to represent end stages in the formation of the regular bands, a process involving mainly potash feldspathization. The later pegmatites and the aplites, on the other hand, appear to be normal intrusive fracture fillings. These dykes demonstrate nicely the decreasing effects of deformation with time, with earlier pegmatites often folded and all crossed by the host mineral alignment, while later dykes cut this oriented fabric and show signs of deformation only where they cut certain felsite dykes.

(E) The Late Veins

With the exception of the Tertiary basalt dykes (see Chapter 18), the last group of minor intrusives that cut the Main Granite are thin veins or lenses of quartz and occasionally of epidote. These veins, members of which cut all the other dyke suites, are variably oriented and may be parallel or inclined to the joint directions. Although they show no signs of deformation, quartz veins may sometimes be localized along small-scale shear zones in the host granite. This situation is particularly common in the margins of the pluton, and clearly indicates the operation of structural controls even at this late stage.

4. Discussion of Banding and Dykes

We have shown in the preceding pages some of the evidence for deformation of the Main Granite as it completed its crystallization. We can recognize no sign of any structural event earlier than the formation of the regular-banding, nor is there reason to view these features as formed during reactivation of a formerly completely consolidated and relatively cold body. On the contrary, the relatively late formation of potash feldspar and the emplacement of various granitic and pegmatitic dykes similar in texture and composition to the host granite are events encountered during the latter stages of primary crystallization of many granitic intrusives.

The regular-banding, which is largely aligned parallel to the long sides of the pluton, appears to be the result of the concentration of late alkalis along planar zones of differential movement, although throughout much of the Main Granite the crystallization of potash feldspar lacked any such structural control. The banding is penetrated by the alignment of minerals in the host granite, and the cross-cutting relationship, where the banding is discordant to the main NE–SW trend, shows clearly that this alignment formed after the banding.

Beginning some time during the development of the mineral alignment and continuing long afterward was the emplacement of a varied suite of dykes. Fig. 11-7 relates in a schematic way these minor intrusives to the formation of the banding and mineral alignment. We see here a progession from early remobilized or otherwise deformed pegmatites and microgranites, through later microgranites, felsites, and pegmatites in which the main evidence of deformation comes from the displacement of elements in the host granite and from the geometry of their internal alignments. The final intrusive phases, represented by unfoliated planar pegmatites, aplites, and veins of quartz or epidote, appear to have largely escaped the effects of deformation, although the emplacement of even some of these late veins may be localized along shear zones in the host granite. Paralleling this general progression from early deformed banding and dykes to late virtually undisturbed dykes and veins is a general decrease in "mineralogical" activity, with early phases involving the redistribution or alignment of minerals, especially microcline, and characterized by welded interlocking contacts, while later phases are more sharply discordant and clearly allochthonous.

Our discussion so far has dealt only with features of the Main Granite away from its highly

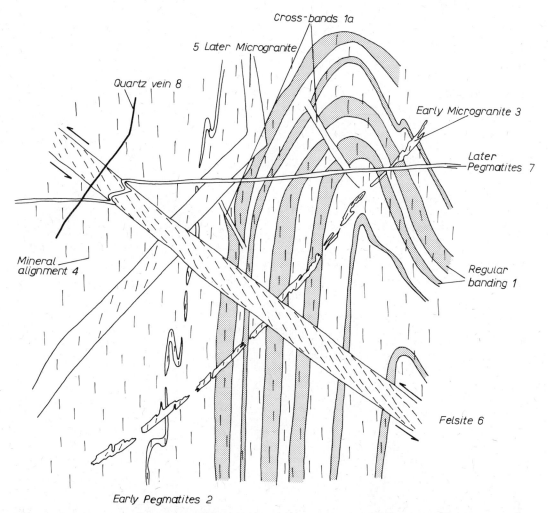

FIGURE 11-7. Schematic diagram showing chronological relationships of banding, mineral alignment, and the various dyke suites. In order of decreasing age: (1) regular-bands, (1a) cross-bands, (2) early pegmatites, (3) early microgranite, (4) mineral alignment, (5) later microgranites, (6) felsites, (7) late pegmatites, (8) late veins.

deformed marginal zones. Before proceeding to a consideration of the dynamic and kinematic aspects of the granite deformation, information concerning which is provided largely by the enclaves, we shall conclude the survey of the chronology of the deformation by examining these marginal areas.

5. The Marginal Zones: Cataclastic Deformation

As we mentioned earlier, the margins of the Main Donegal Pluton are marked by an abundance of granitic and pegmatitic sheets, which coalesce inward to form a boundary which is gradational on a large scale (Fig. 11-8a) though actual granite-country rock contacts, while interlocking, are sharp enough to be easily distinguished in thin section. The envelope structures (folds, lineations, schistosity) are parallel to the longer sides of the pluton, and the attitude of the granite sheets is controlled largely by the dominant schistosity in these metasediments. However, there are also numerous local discordances (see Fig. 11-8a, b) which, together with the presence in some sheets of rotated inclusions of country rock, provide indisputable evidence for the intrusive origin of these bodies.

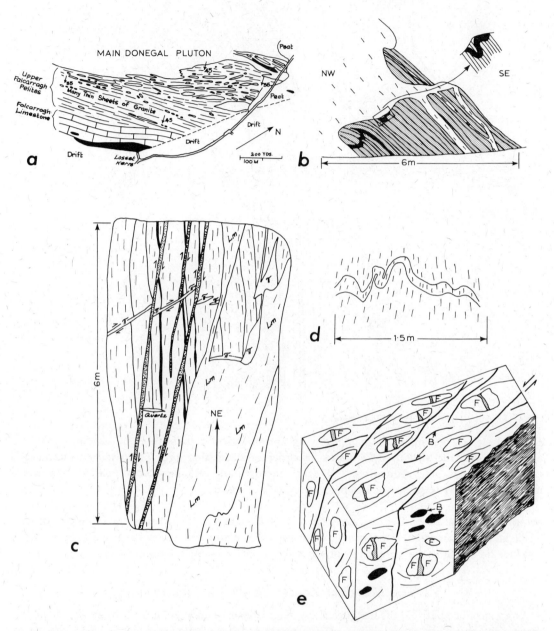

FIGURE 11-8. The margins of the Main Donegal Pluton. (a) Sheeted margin at Losset. After Pitcher and Read (1959, Fig. 6). Metadolerite sheets—black. (b) Edge of Crockmore sheet in railway cutting at Barnesbeg. Black—earlier granite veins (or regional quartzofeldspathic veins). Stipple—Falcarragh Pelites. (c) Dykes in marginal granite at Crocknawama: Arrows show sense of offset along dykes. Stipple—microgranite dykes; Lm—Leucocratic microgranite; black—dark bands; black—pegmatite. (d) Folded pegmatite dyke, Brockagh. (e) Schematic diagram showing relation between mineral alignment and cross-cleavage (heavy black lines). F—feldspar; B—biotite; stipple—quartz veinlets along tension fractures in feldspar.

Although the attitude of the actual margin of the pluton can only be determined by direct mapping in certain rugged areas, the dip of the sheets and their host schistosity indicates that the northwestern contact is dominantly vertical or has a steep dip to the southeast. The southeast side, however, ranges from steep southeasterly dips in the southwest to dips as low as 30° in the same direction in the northeast. The northeast and southwest ends of the Pluton, to judge by the outcrop pattern of granite sheets and certain contact effects, plunge at moderate angles away from the main outcrop.

As Pitcher and Read (1959) stated, the sheets and the marginal portions of the pluton itself have been subjected to an intense cataclastic deformation. Although the response of different rock types to this deformation varies considerably, in general the mineral alignment is greatly intensified, dykes and sheets are folded or boudinaged, narrow shear zones appear, and cataclastic textures are ubiquitous, though they may be hard to recognize in the adjacent country rocks. The width of this marginal zone, which passes gradationally inward to "normal" granite, varies (see Map 1 in folder) as does the intensity of deformation, which is particularly marked in the northwest, especially in the Galwolie Hill–Crockator area.

The marginal zones of the Main Granite proper are cut by innumerable sills and dykes, of which the vast majority are pegmatites and microgranites similar to those described in the foregoing. Their structural relationships, as we shall see in Chapter 12, suggest that they were coeval with their internal counterparts and represent, in these marginal zones, a second major influx of granitic material. Pegmatites in both regular and irregular bodies are especially numerous in these areas, and although some of the coarser pegmatites may appear in the field to be relatively undeformed, they, like all the other dyke phases, can be seen on close examination to be affected at least in part by the cataclasis. Only a few late quartz veins and certain felsite dykes at Meenalargan (see Iyengar et al., 1954, p. 227) appear to have escaped this deformation. These dykes, certain sheets, and the immediately adjacent envelope rocks, are variously folded, drawn out or boudinaged, depending on their attitude (Fig. 11-8) and as we shall see later, they provide valuable information about the nature of the deformation here.

Many dykes are disrupted by cross-cutting shear zones, while others are themselves the loci for displacements of their lost structures (Fig. 11-8c). Indeed, some of these shear zones in the Brockagh area are marked by mylonites with fluidal textures. It is of course quite possible that some of the planar microgranites and even some of the pegmatites here are not intrusive dykes at all, but shear zones along which the granite has been granulated, perhaps subjected to "metamorphic" differentiation, and then variably recrystallized (cf. Reitan, 1965; San Miguel, 1969; Shaw, 1969). These situations make it difficult to establish a chronology of dyke emplacement; in one outcrop, for example, a pegmatite dyke cuts a dark microgranite dyke in one place and is cut itself and offset by an identical microgranite a few feet away. Nevertheless, apart from such cases of possible dyke "reactivation," there does appear to be a succession of dyke phases in these marginal zones broadly similar to that established within the main body of the pluton—thus, highly deformed early pegmatite dykes and segregations of coarse pegmatitic granite are cut by a suite of microgranites, some of which may cross-cut others. These are followed in turn by pegmatites, rare felsites, and occasional microgranites, and finally by late pegmatites and quartzofeldspathic veins.

As we stated earlier, the mineral alignment in the marginal zones is generally $L > S$ fabric with steeply dipping planar and gently plunging linear components, though most of the shear zones contain dominantly planar fabrics (i.e., $L \leqslant S$), lying parallel or subparallel to their walls (see Fig. 11-8c). This alignment is expressed by the orientation of biotite, both in single crystals and aggregates, by elongate lenses and discontinuous bands of quartz, and to a lesser extent by smeared out feldspars.

In many places throughout the marginal zones, the granite, its dykes and sheets are cut at low angles (10° to 30°) by a series of discrete narrow shear zones or cleavages, distinct in both character and attitude to the mineral alignment just described (Fig. 11-8e, 10d). These *cross-cleavages* are generally restricted to marginal zones, though they may be more widespread in the northeastern part of the pluton, and their occurrence in the adjacent schists provides another link between structural events in the pluton and the envelope (see p. 260). The cross-cleavages clearly deflect the mineral alignment and lie with steep dips in a NNE–SSW direction, though a conjugate set is sometimes also present. Their surfaces are strongly slickensided and coated with smeared out micas, and they invariably exhibit a horizontal or shallow plunging grooving which, since it is generally parallel to the linear component of the mineral alignment, adds greatly to the apparent intensity of the latter. However, hand specimens usually break along these cleavages, and this fact together with their distinct character permits their recognition as separate structures.

In summary then, we have evidence in these marginal zones of a powerful penetrative granulation and a later more discrete shearing along obliquely oriented cleavages. That at least a part of this cataclasis postdates the deformation within the main body of the pluton is clear from the distortion in marginal areas of late microgranites, felsites, and pegmatites, which are elsewhere largely posttectonic. It is clear that some marginal dykes and sheets are deformed both internally and externally more than others, and while we are aware that such differences may reflect differences in ductility or in the intensity or rate of strain, there is evidence from the geometry of the deformed dykes (see p. 251) that some were emplaced during the later stages of the deformation which had already affected earlier dykes. In addition, as Pitcher and Read (1959) demonstrated and as we shall describe in the following section in more detail, the abundant textural relationships indicating recrystallization and grain growth in the marginal granite show that this deformation was not a matter of cold cataclasis, except perhaps at the stage at which the cross-cleavages were produced. We therefore consider that the marginal deformation continued during the last stages of consolidation of the pluton, well beyond the time when the late felsites and pegmatites were being emplaced in central zones of the pluton. Before proceeding to a discussion of the kinematic aspects of the Main Granite deformation, we turn briefly then to a consideration of the mineralogy and textural relationships of this pluton.

6. Petrography: the Importance of Late Mineralogical Changes

All the components of the Main Donegal Pluton are characterized by the simple mineral assemblage: plagioclase, microcline, quartz, biotite, muscovite, epidote, apatite, ores, and zircon, in approximate descending order of abundance. There are, of course, many variations, such as those involved in the regular-banding (see p. 217), but details of these and other aspects of the mineralogy are given in Berger (1967T), and we wish only to comment here on some of the salient features.

While most of the other minerals appear, as we shall see, to have been active at relatively late stages in the evolution of the mineral fabric, *plagioclase* and the accessories seem to have been relatively stable since their initial crystallization. For example, bent or broken plagioclases are sometimes seen in contact with quartz and microcline which show no signs of strain (Fig. 11-9c, i). An important exception to this generalization is provided by evidence of internal transformations within plagioclases, as witnessed by what appears to be the obliteration of twin lamellae (Fig. 11-9f), and especially by the veining of central, euhedral, oscillatorily zoned plagioclase cores by offshoots from the comparatively sodic anhedral mantle (Fig. 11-10a, b). This is a common enough feature of granitic plagioclases and appears to represent an early stage in the homogenization and "decalcification" of plagioclase, leading to the formation of patchy zoning (*e.g.,* Fig. 11-10c), in response, we think, to lower temperatures or deformation or both (*cf.* Hutchinson, 1956; Schermerhorn, 1960; Cannon, 1966; and see p. 334), though others have stressed the importance of magmatic resorption (Vance, 1965), exsolution (Taubeneck, 1967a), or even partial melting (Raase and Morteani, 1968; Raase, 1969).

Microcline, as we stated earlier (p. 218), appears to be relatively late in origin, enclosing quartz, biotite, and plagioclase crystals, occasionally in zones parallel to its outer edges (see p. 332 for comment), and veining and replacing plagioclase (Fig. 11-9a). It is commonly microperthitic with

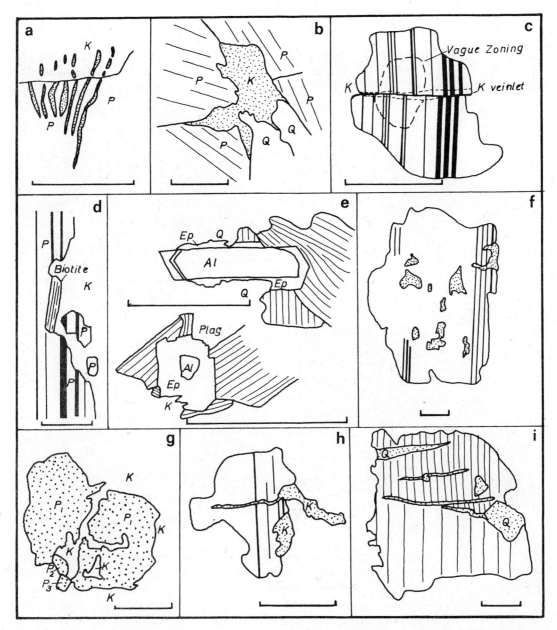

FIGURE 11-9. Main Granite textures: (a) myrmekite invaded by microcline (K) which replaces only the plagioclase (P) leaving the quartz rods (stipple) undisturbed; (b) microcline (K) growing along crystal boundaries; plagioclase (P), quartz (Q); (c) albitic "vein" along fracture in plagioclase. Note offshoot of microcline (K) along fracture; (d) microcline replacing plagioclase. Note subhedral shape of plagioclase inclusions, which are coaxial with the large plagioclase; (e) epidote–allanite in biotite (lined). Epidote corroded by quartz (Q), plagioclase, and microcline (K); (f) antiperthite: patches of microcline (stippled) coaxial with each other but not with the plagioclase host; (g) irregular microcline (K) vein in plagioclase. P_1, P_2, P_3 are differently oriented plagioclase grains; (h) microcline (K) veining plagioclase; (i) quartz veins (stippled) in plagioclase. Note deflection of twin lamellae on either side of veins. Scale = 0.5 mm.

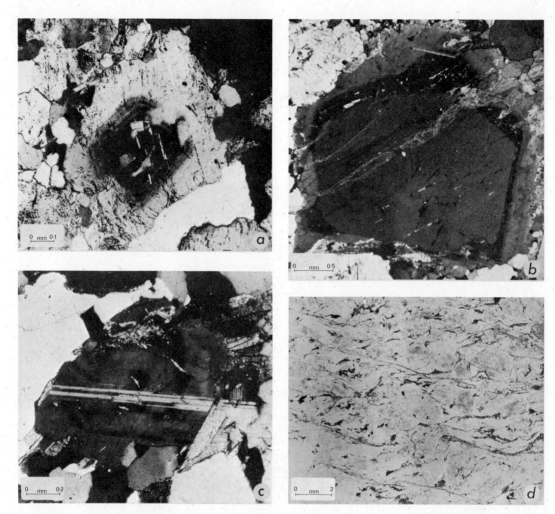

FIGURE 11-10. Photomicrographs of textural relationships. (a) Plagioclase with oscillatory zoning cut by broad sodic mantle. Note muscovite laths along fractures. Crossed Nicols. (b) Veins of sodic mantle cutting vaguely-zoned core of plagioclase. Crossed Nicols. (c) Vague relicts of oscillatory zoning in plagioclase. Crossed Nicols. (d) Cross-cleavages (marked by granulated micas) cutting a weakly-foliated granite sheet from Glenkeo. Plane Light.

patchy grid twinning, and as usual, is separated from adjacent plagioclases by myrmekitic intergrowths or albite rims or both. Coarse, often antiperthitic, patches of microcline in plagioclase (Fig. 11-9f) seem likely to be the result of replacement of the latter by the former.

Biotite, according to the standard interpretation of its textural relationships, would be of early origin (pre- or syn-plagioclase), but we note that it is part of a late, penetrative, and discordant mineral alignment (p. 219) and must therefore have acquired this preferred orientation at a correspondingly late stage (see p. 331 for a general comment on discordant mineral fabrics). The mechanism of this realignment is unknown, but evidence that it has been operative may be provided by the preferential concentration of presumably early epidote and zircon in the immediately surrounding crystals as well as in the biotites themselves. Likewise, the ragged outline of otherwise euhedral epidote crystals, where they emerge from their normal biotite host (Fig. 11-9e), seems to indicate "movement" of biotite by corrosion or by other means (see p. 194).

Quartz was clearly active at late stages, as indicated by its frequent veining of plagioclase and microcline (Fig. 11-8e and 11-9i), by the occasional presence of serrated edges suggesting grain-boundary migration, and by the widespread occurrence of undulose and patchy extinction (also indicative of lattice transformations, *cf.* Spry, 1969) and of healed planes of fluid inclusions parallel to joints (see Berger and Pitcher, 1970).

Further evidence of late mineralogical changes is provided by the sericitization of plagioclase, the occurrence of muscovite along cleavage planes in microcline, and the apparent local resorption of myrmekite in microcline (Fig. 11-9a). An intriguing feature is the sporadic appearance in thin sections of *microveins* which cut the mineral alignment. Their mineralogy is controlled largely by that of the host grain, but along some, a migration of quartz and potash feldspar can be detected; comparable veins have been recorded by Cheng (1944, p. 114), Harry (1953, p. 294), Taubeneck (1967), and especially by Carstens (1966).

In contrast to the noncataclastic central portions of the Main Granite, mechanical granulation is widespread in its *marginal zones,* concentrated mainly along thin discontinuous bands which wind around resistant feldspar crystals. These microscopic shear zones are marked chiefly by quartz and by minute laths of mica, some of which are clearly shreds torn off larger crystals. The feldspar augen commonly have tails or shadow zones, occupied by small granules of the larger crystal. Although the mineral alignment here is thus dominantly a mechanical effect, there is much evidence of "constructive" mineral activity even at this stage in the evolution of the Pluton. Thus, many of the aligned micas are fresh and undeformed, the microclines are more perthitic and exhibit better grid twinning than in central areas of the Granite, and myrmekite and coarse antiperthite are more prevalent here. The acceleration by deformation of such thermally activated transformations in feldspars is, of course, well known (see p. 334). Small elongate grains of microcline are common along microshear zones, and their highly irregular edges, which interlock with other minerals, again suggest grain-boundary migration, possibly stress-induced (Spry, *loc. cit.*). Of particular importance is the virtual absence of deformation lamellae in, and extensive granulation of, quartz; undulose extinction is widespread, but there appears to be a complete gradation from large quartz grains with patchy extinction and incipient subgrains to mosaics of polygonal crystals (*cf.* Phillips, 1965).

In thin section, the *cross-cleavages* of the marginal zones so often merge with the main cataclastic foliation that they are not easily delineated. Where they can be distinguished microscopically (Fig. 11-10d), they are marked by extremely narrow zones of much more intense granulation, which affects even muscovite and quartz, and they clearly formed at a stage when the minerals of the Main Granite were virtually incapable of strain healing.

In summary then, the mineralogy of the Main Donegal Pluton confirms our deductions from the field relationships that deformation and crystallization-recrystallization continued over an appreciable period of time. Our model for the production of the regular-banding by the potash-feldspathization of a trondhjemitic parent implies that the primary crystallization of microcline was later than that of most of the plagioclase, biotite, and quartz. The mineral alignment which postdates the banding must have involved the realignment of these minerals in response to deformation. The corrosion by quartz and feldspar of euhedral epidotes previously enclosed in biotite, some of the homogenization of plagioclase, and perhaps some of the replacive growth of microcline and the veining of feldspars by quartz and mica may well have taken place at this time. The strain involved in the marginal cataclasis was largely relieved by the granulation of feldspar, by the smearing out and recrystallization of biotite, by the growth of some new muscovite, the recrystallization and perhaps renewed exsolution of microcline, and by the continued recrystallization of quartz. The last event in the evolution of the

Main Granite mineral fabric, the formation of the cross-cleavage, appears to have been almost entirely retrogressive. Thus, the overall picture is one of early crystallization overlapped by later recrystallization, accompanied by falling temperatures and decreasing amounts of deformation.

SECTION C. THE GRANITE DEFORMATION: ITS GEOMETRY AND PATTERNS OF STRESS AND STRAIN

1. Introduction

We have established so far a chronology of events within the Main Granite and in its marginal zones. Although the full evidence cannot be stated until we have dealt with the structures in the envelope, we must emphasize here that the bulk of this deformation, as the penetrative nature of the mineral alignment suggests, took place *in situ* but before the complete consolidation of the pluton. We now proceed to examine the nature of this deformation through an analysis of the geometrical relationships of the mineral alignment, the dykes and their internal fabrics, the inclusions, and certain features in marginal zones, parameters which provide much information about the conditions of stress and strain. There are many details to be considered, and some of our conclusions concerning the deformation raise questions of more general interest, for example, relating to the mode of emplacement of the pluton. However, rather than break the continuity of our discussion we shall postpone these matters until the summary section of the chapter.

2. The Mineral Alignment: Kinematic Significance of the L-S Fabric System

We may recall that the penetrative alignment of minerals in the Main Donegal Pluton is a fabric superimposed on the banding and certain early dykes. The predominance of $L < S$ and S fabrics throughout central positions of the pluton (see p. 220) suggests, as we showed earlier (p. 94), a flattening deformation ($0 < k < 1$), while in marginal zones the $L > S$ to $L = S$ alignments indicate a greater component of constriction ($1 < k < \infty$). That the linear component of the fabric in the latter areas does in fact represent the local direction of extension is confirmed by the occasional presence of veins of quartz in broken feldspar porphyroclasts along fractures normal to the mineral lineation (see Fig. 11-8e).

It has been shown that lineations lying along surfaces oblique to the principal planes of the ellipsoid are frequently controlled by the direction of maximum elongation within that particular plane and may have no relationship to the direction of bulk extension, that is, to the X axis of the deformation ellipsoid (cf. Ramsay 1967, p. 160). However, we are dealing here not with a lineation controlled by discrete surfaces within the Main Granite, but with the linear component of a completely penetrative three-dimensional fabric. Furthermore, as we shall see in the following pages, the deformation ellipsoids constructed from other parameters are generally in agreement with those suggested by the mineral alignment. We turn now to the most important of these, the inclusions which abound throughout the pluton.

3. The Inclusions: Important Parameters for the Analysis of Strain

(A) Preamble

The Main Donegal Pluton contains numerous inclusions of varied lithologies, which are all aligned with their longest axes in the plane of the host foliation. As Pitcher and Read demonstrated, they may occur as isolated individuals or in long narrow zones consisting of one or more lithological types.

These zones or raft trains together constitute a pronounced ghost stratigraphy whose significance we shall discuss after dealing with the structures of individual inclusions.

The majority of the inclusions are clearly recognizable as relicts of the Dalradian metasediments or of the regional metadolerite swarm, but pods of the Thorr Pluton are common in the west and northwest, as are inclusions of the Ardara Pluton and its associated appinite suite in the southwest. The degree of alteration in these xenoliths and of contamination in the immediately adjacent host granite varies considerably, depending largely on the lithology involved. Inclusions of igneous derivation are usually little affected, quartzites are likewise fairly distinct, though thoroughly recrystallized, and limestones and other calcareous xenoliths develop skarns at their edges and calc-silicate minerals throughout (Gindy, 1953a). Pelitic and semipelitic inclusions are of course the most altered and are frequently rich in muscovite, fibrolite, and potash feldspar. They are often partly digested and may remain only as vague ghosts or as mafic schlieren in a contaminated granite host. As in the case of the Thorr Pluton, it is quite clear that exchange reactions between inclusions and host granite have been operative to varying degrees. However, despite the evidence for assimilation and contamination on a local scale, we concur fully with Pitcher and Read's conclusion that replacement of country rocks by the Main Granite was very restricted in extent.

In addition to these lithologies, there are also many inclusions of relatively fine-grained granitic material which have no counterparts outside the pluton. Some of these inclusions appear to be derived from the dark regular-bands or from the early microgranites by disruption or feldspathization or both; others may be representatives of earlier crystallized portions of the pluton (Pitcher and Read, 1959, p. 271), but whatever their precise origin, the petrographic affinity of all these pods with the host granite indicates a close genetic connection.

(B) Shape and Orientation

Most inclusions are roughly lenticular both in plan and section, though their shape varies considerably depending partly on their lithology and partly on their position within the pluton. The shape of the metasedimentary xenoliths is determined largely by their internal structure, as shown by the tabular form of quartzite and some pelitic inclusions, which are easily penetrated by the host granite along their bedding or schistosity planes (Fig. 11-11). Inclusions of igneous origin (Thorr, Ardara appinites, dolerites, and granitic pods), on the other hand, are dominantly ellipsoidal or somewhat diamond-shaped (Fig. 11-12).

All but a very few of the inclusions lie with their long axes parallel to the host granite foliation, and even those which lie slightly athwart this direction may prove in depth to have a concordant attitude as far as the bulk of the inclusion is concerned. One often "senses" that inclusions have preferred orientation within the host foliation plane, defining thus a subhorizontal lineation (*e.g.,* Pitcher and Read, 1959, p. 287), but this may be largely an effect of the predominance of horizontal exposures; certainly it is very difficult to detect such an alignment where both horizontal and vertical surfaces are available.

(C) Internal Structures

Most of the inclusions exhibit an internal mineral alignment which lies parallel to their long dimensions and thus to the host foliation. Indeed in many cases, especially those involving the granitic pods, the host mineral alignment crosses directly into the inclusions, and is thus, at least in part, superimposed. In other situations, the schistosity in metasedimentary inclusions is axial plane to tight internal folds, and though it has certainly been accentuated during the granite deformation, it is probably equivalent to the regional, pre-granite, S_{2-3} schistosity described earlier in Chapter 3.

FIGURE 11-11. Quartzite xenoliths at Glenleheen.

Figure 11-13 illustrates the orientation of these and other internal structures of xenoliths from the Doochary and Glenleheen areas.

Perhaps the most spectacular evidence of the granite deformation comes from the folding and boudinage of microgranite and pegmatite dykes where they cross certain rafts. The behavior of these dykes is best seen on the roadside at Glenleheen, where large limestone enclaves have been partly quarried out, thus providing magnificent exposures of their internal structure (Fig. 11-14). Other splendid examples can be found in the large limestone enclave on Gubbin Hill and in several of the calcareous rafts around Derryloaghan.

There is a consistent pattern in the deformation of these dykes and of certain apophyses from the host granite. Where they can be seen cutting the granite they are relatively undeformed and exhibit only those features described earlier (p. 220), but where they cross enclaves, those which lie at high angles to the maximum dimensions of the rafts are buckled, while those dykes, or portions of dykes, which lie close to this direction are necked down and often boudinaged. Figure 11-15b illustrates an example from Glenleheen where what appears to be a planar offshoot from the enclosing granite is tightly folded and thinned out as it crosses into the enclave, the limbs of the fold exhibiting pinch-and-swell structure on several axes. Directly above this structure is one of the later microgranite dykes, which though quite planar where it cuts the granite, is bent into an open fold where it cuts the enclave (Fig. 11-15a); its normally weak mineral alignment (in the granite) becomes much stronger

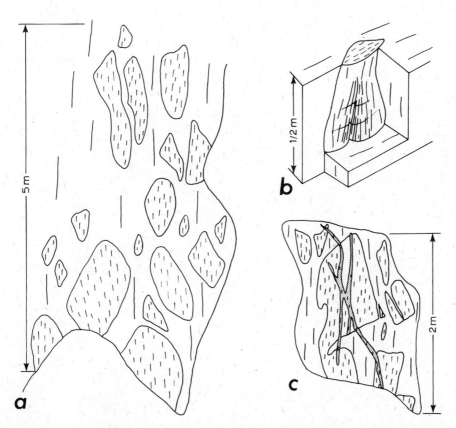

FIGURE 11-12. Inclusions of igneous origin. (a) Lenses of Thorr tonalite, Derryveagh raft train. (b) Metadolerite, Doochary. (c) Enclaves of Thorr tonalite at Brockagh, cut by microgranite (stipple) dykes.

where it is folded in the limestone, exhibiting a marked lineation parallel to the fold axis. Earlier quartzofeldspathic veins in this enclave are similarly deformed (*ibid.*) and, like the walls of the raft itself, are affected by several directions of necking. The succession of dyke phases here again clearly demonstrates the overlap in time of dyke emplacement and deformation.

In the Gubbin Hill enclave (Fig. 11-16a), a whole succession of microgranite and pegmatite dykes is similarly deformed. Here some dykes which lie close to or along the limestone foliation maintain a constant width and exhibit a powerful mineral lineation which lies parallel to their walls, while other dykes in similar positions have weaker lineations and are necked down. This is best illustrated by the steep limb of a large folded pegmatite dyke which shows two directions of necking, a stronger one normal to the mineral lineation and to the fold axis and a weaker one subnormal to it, thus producing elliptical or pillow-shaped "swells" (Fig. 11-16b).

In these, as in other cases throughout the granite, the axes to the folded veins lie within the plane containing the long axes of the host enclave, and thus usually parallel to its schistosity (Fig. 11-13, 11-15, 11-17). In the Glenleheen and Gubbin Hill rafts, the fold axes have generally shallow plunges, as do the mineral lineations and the minor necked-down zones in the latter enclave (Fig. 11-17).

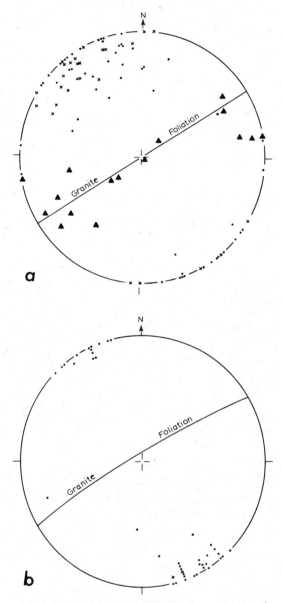

FIGURE 11-13. Stereoplots of structures in enclaves at Doochary (a) and Glenleheen (b): ×, poles to bedding or schistosity, or both; •, poles to foliation in inclusions of igneous origin; ▲, axes to folded internal veins.

A more complex example from Derryloaghan is shown in Figure 11-18, where a microgranite sill is highly contorted, with fold axes varying greatly in plunge and to lesser degree in trend, though all trend northeast and southwest. This is one of the few rafts in which the internal bedding in the limestone lies at high angles to the granite foliation; in one place, this bedding is folded to form a marked eyed structure. It is worthwhile noting that in another part of this enclave a steep NE–SW trending dyke, which thus lies at high angles to the bedding, is boudinaged on shallow axes. This relationship between dyke deformation and enclave structure is, of course, the reverse of the usual

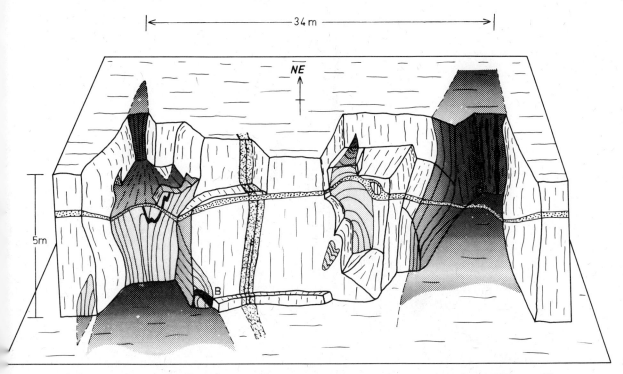

FIGURE 11-14. General view of partly quarried calcareous enclaves along roadside at Glenleheen. Stipple of varying density indicates limestone and calc-silicate enclaves. Solid lines indicate bedding. Irregular close stipple indicates pelitic xenolith. Close regular stipple indicates late microgranite dyke; the southeastern extension of this is shown in Fig. 11-5a. Solid black indicates quartzofeldspathic and granitic veins and dykes. *Note*: Dashed lines on upper surface of diagram and on vertical walls receding from viewer are construction lines and have no geological significance. Other dashed lines indicate lithological boundaries or mineral alignments.

case (see Fig. 11-17), and provides a nice confirmation that the deformation of the dykes is related to the host granite structures and not to the internal structures of the enclaves.

(D) Interpretation of Internal Structures: Ptygmatic Folds?

Folded veins in xenoliths in other plutonic bodies seem to have received little attention in the literature, though their occurrence is by no means restricted to Donegal (*e.g.*, Ramsay, 1963; for general review, see Berger and Pitcher, 1970). Similarly deformed veins in metamorphic terrains have, however, been the subject of much discussion, and a good deal of confusion has arisen by grouping these veins under the general term "ptygmatic." Although Sederholm, who coined this term in 1907, used it to refer to the tortuous quartzofeldspathic veins common in migmatitic rocks, Ramberg in his analysis of progressive deformation (1959) extended it to describe virtually any folded competent layers embedded in an incompetent host. Kuenen, in his recent review of the problem, has proposed the definition that "ptygmatic features are multiple tortuosities of a sheet of rock embedded in a host showing a simpler pattern (or none at all)" (1968, p. 145; see also Mehnert, 1968), and Ramsay has pointed out that their form resembles the curve known as an elastica, which is a type of parallel fold resulting from compressive strains greater than 36% (the theoretical maximum for true parallel folds) in a layer whose initial wavelength to thickness ratio is large (1967, p. 387). To minimize confusion, it seems best to restrict this term further to refer to dykes

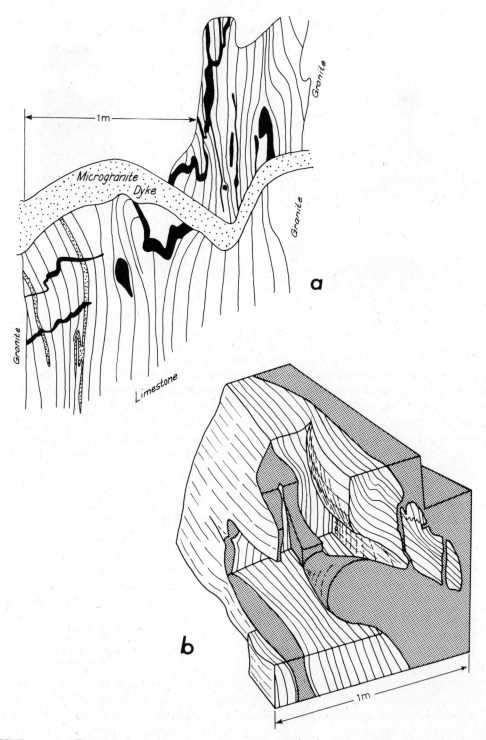

FIGURE 11-15. Details of northwesternmost enclave shown in Fig. 11-14. (a) Openly folded microgranite dyke (open stipple) post-dating folds in earlier quartzofeldspathic veins (black). Note the early isocline in the bedding (close stipple). (b) Three-dimensional view of deformed apophysis from host granite in lower right-hand corner of enclave (B on Fig. 11-14). Note association of folding with pinch-and-swell on both shallow and steep axes. Short dashed lines have no geological significance.

FIGURE 11-16. (a) General view of large quarried-out limestone raft on Gubbin Hill, looking eastward. Vertical white line is 4 meters long. (b) View along large granitic dyke shown in (a), showing "swollen" portions with major necking axis plunging steeply towards camera and minor necking axis plunging gently away.

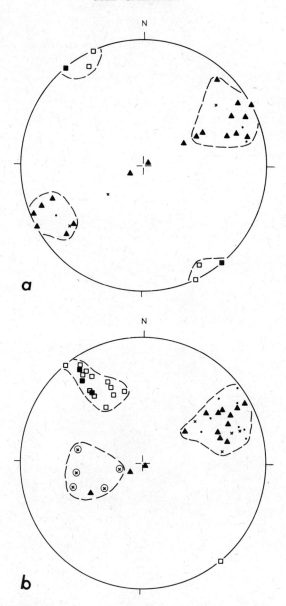

FIGURE 11-17. Stereoplots of structures in the enclaves at Glenleheen (a) and Gubbin Hill (b), illustrated in Fig. 11-14 and 11-16: □, poles to limestone foliation; ■, poles to granite foliation; ▲, axes to folded granitic veins; ●, mineral lineations in granitic veins; ×, axes to necked-down zones; [⊗, major necked-down zones in (b) and ×, minor necked-down zones in (b)].

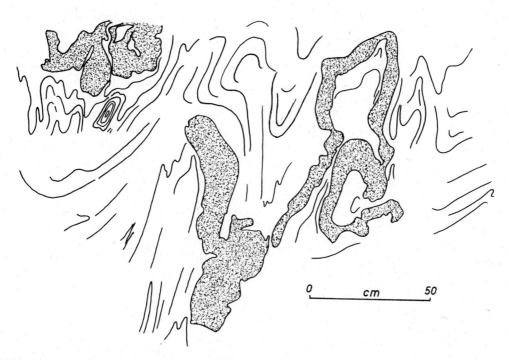

FIGURE 11-18. Highly contorted granitic vein in calcareous raft at Derryloaghan. The plunge of fold axes and necked-down zones varies greatly.

and veins without internal mineral alignments and folded on a minor scale so that their axes and axial planes have no strict preferred orientation, a situation in keeping with the structures illustrated in Sederholm's accounts. The folded dykes in the Main Granite enclaves are thus not generally ptygmatic, though on occasion they may fairly be placed into this category (*e.g.*, Fig. 11-18).

One suggestion for the origin for ptygmatic dykes and one which might, at first, seem applicable to the present situation is that their shape was acquired during intrusion either as a result of injection or metasomatic growth along tortuous fractures (Read, 1928; Shelley, 1968; Slobodskoy, 1970), or as a result of the "backing up" of a plastic intrusion against an impenetrable barrier (Wilson, 1952; Prinz, 1965). However, the consistent parallelism between fold axes, axial planes, and the walls of the enclaves, and between the internal mineral alignment and the host granite fabric, such as obtains in the Main Granite dykes, is hardly likely to arise during the course of primary injection (see also Kuenen, 1968). The obvious explanation is that these folded dykes were originally planar or nearly so and were buckled during deformation of the enclaves. The folding of enclave structures adjacent to the dykes in some cases (*e.g.*, Fig. 11-14, 11-18), and the systematic association between folding and boudinage or necking, support this view.

It is hardly necessary to review here the general principles of progressive deformation of single layers of relatively low ductility (high competency) in a highly ductile host,* for these have been adequately dealt with by Ramberg (1959), Flinn (1962), Ramsay (1967) and Talbot (1970). Suffice it to point out that the structural relationships of the deformed dykes and veins could well have arisen

---

* See p. 255 for a discussion of the relative ductility of the various components of the main granite.

from a basically irrotational deformation of the enclave host, such as illustrated schematically in Figure 11-19. During such a deformation, initially planar dykes which lie at high angles to the XY plane of the ellipsoid will buckle, and their limbs will rotate toward this plane. These limbs, together with layers which lay initially at low angles to XY, will be subject to extension, necking and boudinage, or both, once they rotate through the position of the surface of no infinitesimal longitudinal strain (Ramsay, *ibid.*, Chapter 4).

The orientation of fold axes and of directions of necking and boudinage depends upon the kind of deformation and upon the initial orientation of the layers with respect to the principal strain axes. The facts that the fold axes all lie in a plane parallel to the walls of the enclave, and that several directions of boudinage are present, indicate clearly that we are dealing with a flattening deformation of the type $1 \geqslant k \geqslant 0$ (*cf.* Fig. 11-17 with Flinn, 1962, Fig. 6; see also Ramsay, *loc. cit.*, pp. 154–62). That extension did not, however, take place equally in all directions within the XY plane is suggested by the strong shallow-plunging mineral lineations in many of the deformed dykes and by the fact that where several directions of necking are present, the major directions are normal to this lineation. Thus, the deformation ellipsoid lies somewhere within the field $1 > k > 0$, the plane of flattening is vertical with a NW–SW trend, and the direction of maximum extension (X), at least in the Glenleheen and Gubbin Hill rafts, plunges gently to the northeast. That is, of course, in general agreement with the conclusions obtained earlier from the mineral alignment. Again, the complexly folded dyke in the Derryloaghan raft (Fig. 11-18) suggests a constriction with $k > 1$, a conclusion which fits in nicely with the strong linear mineral alignment in this marginal zone, though the boudinage of a vertical NE–SW trending dyke provides evidence of a degree of NW–SE flattening even here.

(E) Structures in Inclusions as a Guide to Bulk Deformation of the Pluton

We have seen that, on a small scale, the orientation and general shape of the deformation ellipsoid derived from the geometry of the internal structures in the inclusions is in full agreement with that obtained from the geometry of the mineral alignment. But the xenoliths also provide us with information about the bulk deformation of the Pluton, for the situation is clearly that of a ductile medium (the granite) in which are embedded particles (the inclusions) whose viscosity may be greater or less than that of their host. The behavior of such systems has long been the subject of study by physicists and engineers, but has only recently been applied to geological systems, chiefly by Ramberg (1959), Flinn (1962), Ramsay (1967), Gay (1968*b, c*) and Elliott (1970). Nevertheless, the special case of rigid particles in fluid media—a situation of special importance in the study of sedimentary fabrics— has received rather more attention, often with conflicting conclusions (Rusnak, 1957; Glen and others, 1957; Johansson, 1965; Rees, 1968; Gay, 1968*a*; Lindsay, 1969).

In a general irrotational strain such as that resulting from the flattening of a consolidating magma already in place, all inclusions, unless equidimensional, would with continued deformation rotate into the XY plane, while those already in this position would remain so. In addition, unless the inclusions were rigid, and the internal structures just described show clearly that this was not the case, they would themselves be flattened regardless of their attitude, those lying parallel to XY being the most deformed. However, tabular inclusions originally lying at high angles to XY would be foreshortened or buckled or both, though with a sufficiently intense deformation, they might well unfold and begin to suffer extension. Since no buckled inclusions were seen, apart from occasional folded schlieren, the inclusions could hardly have had a random orientation prior to deformation.

This latter conclusion is, of course, supported by the general parallelism between the NE–SW trend of the inclusions and that of the envelope structures and stratigraphy. In the northeast end of the pluton around Crockmore and south of Glen Lough, the country rock schists can be followed directly along strike into zones of aligned enclaves separated by massive sheets of granite (see Pitcher and

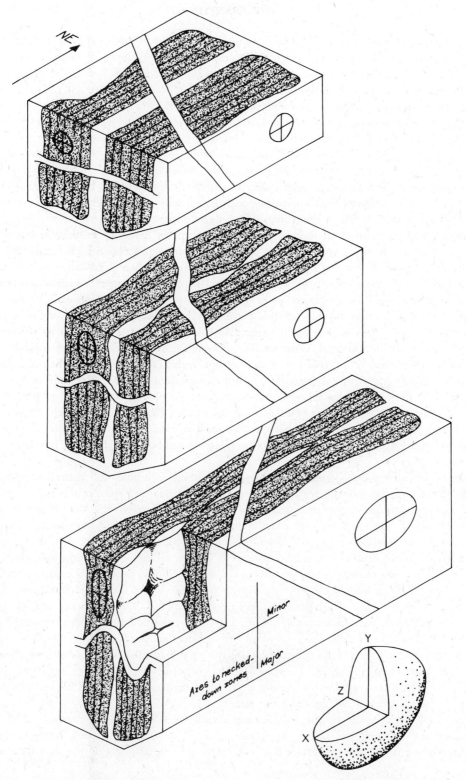

FIGURE 11-19. Schematic diagram showing effects of progressive deformation on enclaves in the Main Granite, and the relationship of their internal structures to the deformation ellipsoid.

Read, 1959). The crude parallelism between the long dimensions of tabular xenoliths, their internal schistosity, and the corresponding structures in the adjacent envelope, is nicely illustrated on Crockmore and Binaniller hills. An even more remarkable concordance between enclaves and the envelope structures can be seen in the Derryloaghan rafts described earlier (p. 238). These comprise the only group of metasedimentary inclusions observed within the Main Granite in which the internal bedding and schistosity is markedly discordant to the steeply dipping host granite foliation, here rather weakly developed. These xenolithic elements dip to the northwest at shallow to moderate angles in precisely the same attitude as the corresponding structures in the adjacent Falcarragh Limestone. Again we conclude that the inclusions in the Main Granite had at least a crude preferred orientation prior to deformation, though clearly this original alignment has been accentuated by the flattening. There is nothing here incompatible with irrotational strain, and we can now utilize the shape of the inclusions to tell us something of the magnitude of this strain.

Although the determination of finite strain from the shape of deformed particles is fraught with difficulties (see *e.g.,* Ramsay, 1967, Chapter 5), we can nevertheless obtain a crude estimate of the intensity of deformation suffered by the inclusions by measuring their axial ratios (*cf.* Watterson, 1965; Ramsay and Graham, 1970). However, since the shape of many metasedimentary enclaves was undoubtedly determined by the injection of granitic sheets along planes of ease of penetration (see p. 235), they are less suitable for this purpose than inclusions of igneous origin, which we think were originally more or less equidimensional. This assumption is supported by the relatively angular shape of the inclusions of Thorr tonalite forming the swarm of unoriented enclaves around Glen Lough (see Map 1 in folder). Recalling our earlier conclusion that the inclusions are essentially oblate (k approximately equal to 0), then measurements of their axial ratios on any section normal to XY, the plane of the host granite foliation, would provide an estimate of the parameter $b = Y/Z$ which relates the intensity of a flattening deformation (see Chapter 1). Field determinations of the length/width ratio of these inclusions (Thorr and Ardara types, metadolerites, appinites, and granitic pods) throughout the central parts of the pluton varied between 6:1 and 1:1 with an average of about 3:1.*

If we could determine the "viscosity" ratio between these inclusions and their host granite, we could produce a similar estimate for the latter (*cf.* Gay, 1968c), but since this is not yet possible, the 3:1 ratio provides a minimum or a maximum for the strain suffered by the Main Pluton during the period in which the xenoliths were deformed, depending upon whether they were on the whole more or less ductile than their host (see p. 255 for further discussion).

One feature worthy of comment is the occasional *en echelon* arrangement of inclusions which occurs where the raft trains run somewhat oblique to the main NE–SW trend of the granite structures. This is seen chiefly in the Glenleheen and Kingarrow areas where individual inclusions maintain their usual parallelism, but where the raft zones trend more ENE–WSW (Pitcher and Read, 1959, p. 287–294). Pitcher and Shackleton (1966, p. 150) suggested that this pattern originated by the intrusion of granite sheets along the axial plane cleavage of plunging folds (Fig. 11-20a). However, although this mechanism may have operated locally, the ghost folds, like their country rock equivalents around the Main Donegal Pluton, have no more than very shallow plunges (see Fig. 3-2). Furthermore, this explanation requires that the long axes of individual rafts should be inclined to their internal bedding, which is not generally the case.

If the pluton was undergoing an irrotational strain, then both passive elements, such as the boundaries of raft trains, and active components, such as individual inclusions, rotate towards the XY plane of the deformation ellipsoid. But if the rafts are generally less ductile than their host, as

---

* No consistent differences in shape were noted between inclusions of different lithology in any one area nor between inclusions of similar lithology from one area to another, though more careful studies might well establish such variations.

seems to be the case here (p. 255), they lag behind the rotation of the passive elements (*cf.* Ramsay, 1967, p. 109, Gay, 1968*b*; see, however, Collette, 1959, Fig. 5); this is the reverse of the actual situation (Fig. 11-20b*ii*). If, however, the central portion of the pluton is more ductile than marginal areas, that is, if some sort of boundary layer effect is operative, then individual inclusions will be subject to a rotational component which will tend to speed up the rotation of the inclusions (see

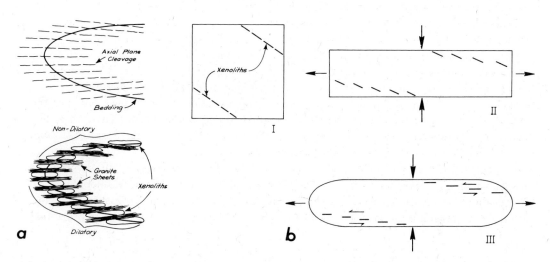

FIGURE 11-20. Origin of *en echelon* pattern of rafts in the Kingarrow and Glenleheen areas (see Pitcher and Read, 1959, Fig. 10). (a) Sheeting of plunging regional folds along their axial plane cleavage, as suggested by Pitcher and Shackleton, 1966. (b) Origin by deformation of raft trains originally oblique to the plane of flattening: i, before deformation, rafts parallel to the sides of the raft train; ii, after homogeneous pure shear, rotation of individual rafts lags behind that of the raft train; iii, after inhomogeneous flattening with boundary-layer effects.

Fig. 11-20b*iii*, and Gay, *loc. cit.*). Although the internal structures of the inclusions and the deformation of certain dykes (see below) suggest a predominantly irrotational flattening, we are dealing here with a vastly different scale, and the degree of angular shear strain is hard to estimate.

We have now established that the deformation of the Main Granite away from the marginal zones was essentially a flattening along the path $1 > k > 0$, and that the intensity of strain registered by the inclusions of igneous origin was, in terms of the *b* ratio, approximately equal to 3. Before turning to a consideration of the marginal zones, we draw attention to a feature of the later microgranite and felsite dykes which provides us with information about the pattern of stresses in these central areas.

4. Late Dykes and the Regional Stress Pattern in Central Areas of the Pluton

It will be recalled that many of the felsite and later microgranite dykes are characterized by the presence of internal mineral foliations which lie oblique to both the dyke walls and the host granite alignment. These oblique foliations are the result of lateral movements of the dyke walls in such a way that the acute angle between the walls and the internal fabric points in the direction toward which the walls are moving (see p. 223 and Fig. 11-6).

The problem of the kinematic significance of the mineral alignment is an involved one (Watterson, 1968*a*; Berger, 1971*a*), but we believe that we are dealing here with a general strain, and we interpret the dykes as shear zones in a larger body (the whole pluton) under stress. Although it is difficult to prove, we think that away from the marginal zones of the pluton, the host granite at this time was capable of ductile flow only in narrow zones adjacent to some of the dykes (*cf.* Fig. 11-6b, c). There is no sign of any external deformation of these dykes, and the mineral fabric of the host granite shows no evidence of having undergone appreciable readjustment at this stage; indeed, in some cases the sharp margin of the dykes slices through virtually unhealed feldspar crystals in the wall rock. The relatively small amount of offset along the walls, together with the fact that oblique foliations can be formed even where the dykes pinch out or end abruptly (Fig. 11-6a), agrees nicely with the model of narrow ductile zones in a much larger and relatively rigid host.

If this is indeed the situation, we are dealing with amounts of strain exceedingly small compared to the total volume under stress, that is, with a case approaching that of incremental or infinitesimal deformation. If this is so, then as Berger (*loc. cit.*) has shown, all we need to locate the direction of maximum compression (σ_1) is a number of dykes of different attitudes and known sense of displacement, which of course is readily given by the pattern of the internal foliation. For such a dynamic analysis, the assumption is also made that the direction of slip along the walls is normal to the line of intersection between the internal foliation and the dyke walls.

Fig. 11-21a is a stereoplot of normals to walls and to internal foliations in all felsite and later microgranite dykes in the Doochary and Glenleheen areas in which oblique foliations were noted. Many of the dykes are clustered into the three areas shown on this diagram by dashed lines, and the three dykes representing the average positions of each of these groups are shown schematically in

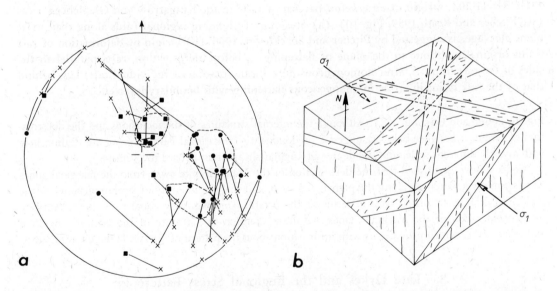

FIGURE 11-21. Dynamic analysis of structures in felsite dykes: (a) Stereoplots of poles to 19 felsite dykes at Doochary (●) and 13 dykes at Glenleheen (■) with poles to internal foliation (×). (b) Schematic diagram showing three felsite dykes representing the three clusters outlined by dashed lines in (a). Sense of movement along walls given by attitude of internal foliation indicates that the direction of maximum compression (σ_1) came from the northwest and southeast quadrants.

Fig. 11-21b. From this, it can readily be seen that the shear couples implied by the orientation of the internal foliations indicate a subhorizontal compression somewhere from the northwest and southeast quadrants. Berger (*loc. cit.*) has shown how the principal axes of stress can be located more precisely from this data, and we need only record here his conclusion that σ_1 at this stage in the deformation was horizontal and oriented in the direction 153°, a result in harmony with the other deformation parameters discussed so far. There are, of course, a number of problems inherent in any such analysis (*ibid.*), but we are confident that further studies of dykes with internal oblique foliations will increase their utility in structural investigations.

5. The Deformation in Marginal Areas

We return now to a consideration of the deformation pattern in the marginal zones of the Main Donegal Pluton. We have seen earlier (p. 227) that these areas are characterized by the presence of numerous highly deformed dykes representing all of the suites found in other parts of the pluton, by shear zones on several scales, by a particularly strong mineral alignment, and by a general cataclasis.

It is unfortunately not possible to use the axial ratios of the inclusions in these marginal zones to estimate the relative amount of strain, for most of these appear to have been "slabbed off" the envelope by sheeting, and the shapes of others were determined in part by the intersections of the numerous marginal dykes (see Fig. 11-8, 11-11, 11-12). However, there are several means of determining the orientation of the principal axes of stress and strain here.

The displacement of host elements along the shear zones and cross-cleavages (Fig. 11-8c, e) is consistent with a flattening from the northwest and southeast quadrants, and the predominance of the NNE–SSW set of cross-cleavages over the E–W set again suggests a maximum compression coming from north-northwest and south-southeast, since in this case the E–W cleavage would be under much greater normal stress and would be harder to develop than the other cleavage. It is interesting, in this connection, to recall the experiments of Means and Paterson (1966, Fig. 5), in which aggregates of platy minerals with their schistosity oriented at high angles to the maximum compressive stress developed shear zones not unlike the incipient cross-cleavage mentioned above.

The geometry of the $L \geqslant S$ mineral alignment (see p. 234) also suggests flattening from the northwest and southeast, and the shallow plunge of the linear component of this fabric suggests that the direction of maximum extension (X of the deformation ellipsoid), at the time when the mineral orientation was formed, was subhorizontal, plunging gently to the southwest in the southwest and to the northeast in the northeast. The orientation of the cross-cleavages and of the generally horizontal striae on their slickensided surfaces (Fig. 11-8e) indicates a similar strain pattern in the last stages of the deformation.

The geometrical pattern of dykes in marginal zones of the pluton and in the adjacent metasediments is much the same as that of dykes cutting the limestone rafts in the Glenleheen district (see p. 236), with dykes at high angles to the granite foliation being folded and those at low angles subject to uniform extension, necking, or boudinage (Fig. 11-22). Although Pitcher and Read emphasized boudinage on steep axes (1959, Fig. 8), there are many examples of pinch-and-swell on shallow axes too, producing, as in the small roadside quarry at Lackagh Bridge, ellipsoidal or even spherical "swells" and boudins.

It is theoretically possible to use the geometry of these deformed dykes to determine both the orientation of the axes of the deformation ellipsoid and their lengths relative to one another for the period during which the dykes were deformed (Ramberg, 1959; Flinn, 1962; Ramsay, 1967). As planes at high angles to the XY plane rotate during a progressive flattening, they pass at some point

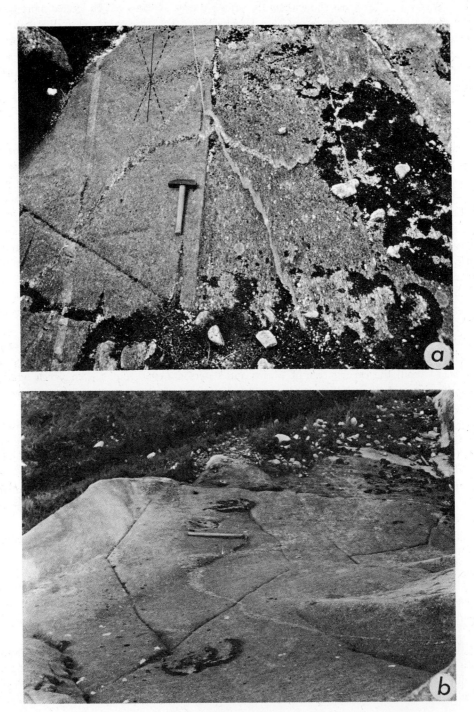

FIGURE 11-22. Deformation of microgranite and pegmatite dykes in granite. (a) Relative deformation of marginal dykes at Dunlewy varies with angular relationship to host foliation (parallel to hammer). Insert shows direction of lines of no finite longitudinal strain (dashed) and trace of XY plane of deformation ellipsoid. (b) Folded early pegmatite dyke near Doochary. Hammer is parallel to mineral alignment in both host rock and pegmatite.

from an attitude where their "stratigraphic" length, measured along their enveloping surface, is less than their original length to an attitude where it is greater, this position being that of no finite longitudinal strain (NFLS). The position of the lines of NFLS in two dimensions is readily located by separating dykes which have been shortened from those which have been extended or which have suffered no change in length (Fig. 11-22a). By measuring the attitude of a large number of dykes of varying trend and distinguishing whether they are folded, boudinaged, or extended (apparently) uniformly, the three-dimensional surface of NFLS can be determined. The angular relationships of this surface to the axes of the deformation ellipsoid are directly related to the k value and to the axial ratios, thus providing in theory an excellent method for a complete strain analysis (Talbot, 1970).

Application of this method to the deformed dykes in several parts of the Main Donegal Granite indicated that the Z axis of the ellipsoid was horizontal with trends between 142° and 149°, but unfortunately there was too great an overlap between the plotted positions of shortened and extended dykes to define the surface of NFLS and thus the position of the X and Y axes and the axial ratios. Such an overlap suggests that some dykes were being emplaced during deformation, so that their deformation would be controlled by a new NFLS surface which lags behind the earlier ones in its rotation towards the XY plane (Ramsay, *loc. cit.*); syntectonic emplacement is, of course, also indicated by other lines of reasoning (p. 226). However, despite these difficulties, the geometry of the marginal dykes confirms that at least the later increments of strain were the result of a horizontal flattening from the NNW–SSE.

SECTION D. A DISCUSSION SUMMING UP THE DEFORMATION MODEL

In the foregoing pages, we have interpreted the internal features of the Main Granite in terms of a pronounced deformation, acting on the pluton before it had completely consolidated. Our detailed studies support Pitcher and Read's original emphasis on the overall unity of the Granite, and it now remains for us to bring together the various observations discussed in this chapter.

In *terms of chronology*, the first structure we can recognize is the regular-banding, which we consider to be the result of the segregation of alkalis into planes of differential movement before and during the crystallization of potash feldspar (see Fig. 11-7, 11-23). Since the banding is generally parallel to the main NE–SW trend of the pluton, it provides little information about the pattern of strain at this early stage (T_1 on Fig. 11-23). Neither is it possible to determine the amount of movement necessary to segregate the alkalis, but since the host bands often show no perceptible offset across the cross-bands (see Fig. 11-4), we think that the banding process could have taken place along incipient "fractures" in the crystallizing granite which were generated perhaps by inhomogeneities during deformation. Of course, there is also the possibility that the bands represent planes of shear generated during the actual emplacement of the pluton; we return to this suggestion later.

Following the formation of the regular-bands, a number of early pegmatite dykes and segregations appeared and were subsequently folded or boudinaged (Fig. 11-22b). Cutting these and all the bands is the marked mineral alignment, which away from the marginal areas is an L < S fabric. The orientation of this and of the magnetic fabric suggests that the direction of maximum flattening (Z of the deformation ellipsoid) was subhorizontal and normal to the NE–SW foliation plane, in which the direction of maximum extension (X) plunged gently to the northeast in the northeast and to the southwest in the southwest. Sometime during the later stages of formation of the mineral alignment, a suite of sporadic early microgranite dykes was emplaced, and their subsequent remobilization indicates that the host granite was still ductile enough to deform homogeneously at this stage (T_2), at least on a local scale. However, by the time the next group of dykes, the later microgranites, the intermediate pegmatites and the felsites, were intruded, movements in the interior of the pluton were restricted, so that the dykes sometimes acted as zones of slip, a process which resulted in the formation of oblique internal foliations in many of them. The geometry of these dykes and their internal structures indicate that the direction of maximum compressive stress at this stage was horizontal and from the north-northwest and south-southeast. The deformation of these late dykes apparently marks the last stage (T_4) in the deformation of the central portions of the Main Granite although there may have been later minor shear zones along which small quartz veins were localized.

The wide variety of inclusions also exhibit various features due to strain, though these are more difficult to fit into our chronology. The penetration of the granitic pods by the host mineral alignment suggests that they were being deformed at the time this alignment was produced, and the folding of occasional late microgranite dykes where they cross certain enclaves (Fig. 11-14) indicates that the latter were still being somewhat deformed at this stage (T_3). However, the deformation of the inclusions and their internal veins provides a valuable parameter for strain analyses, confirming the orientation of the principal axes of finite strain as stated above and indicating again an essentially oblate deformation ellipsoid ($1 > k \geqslant 0$), with the Y/Z ratio (b), on the basis of the more homogeneous inclusions of igneous origin, approximately equal to 3.

The marginal zones of the pluton were able to respond to deformation after the central portions had become stable, for late microgranite and felsite dykes, together with all the associated pegmatites and aplites, are folded, boudinaged, or extended without disruption, depending on their attitude, and virtually all are penetrated by the intense $L \geqslant S$ host mineral alignment. Other unique features in the marginal zones include the displacement of host elements along oblique healed shears and certain microgranite dykes, and as the last stage in the deformation, the formation of a series of small-scale cataclastic shears, the cross-cleavages, at low angles to the host foliation. Despite the differences in chronology, the kinematic situation here is similar to that in the interior of the Granite, with the exception that the stronger linear component of the mineral alignment in the marginal areas indicates a greater component of stretching ($\infty > k \geqslant 1$), with the X axis of the deformation ellipsoid still subhorizontal. However, the sense of displacement along the shear zones again indicates a subhorizontal compression from the northwest and southeast, and suggests that the ellipsoid, during the last stages of the marginal deformation, was once again nearly oblate ($1 > k \geqslant 0$).

Our analysis of this deformation is of course, based on information from a number of different lithologies in various areas throughout the Pluton which acquired their structural character at different times during the deformation, and it is not always an easy matter to relate the behavior of these individuals to that of the whole Pluton. Nevertheless, we are confident that the bulk strain is adequately represented by an oblate, or in the margins, a slightly prolate ellipsoid.

The intensity of bulk deformation as expressed by the axial ratios a and b of the deformation ellipsoid is not, however, so easily obtained. The relationship between the degree of deformation of the xenoliths and dykes and that of the host granite involves a consideration of their relative ductilities (or apparent viscosities), which we can estimate for a number of situations in central portions of the pluton (see Fig. 11-24). Taking the ductility of the regular-bands as representative of the granite matrix, their deflection around both country rock xenoliths and the granite inclusions (see Fig. 11-12a, 11-23) indicates that at the time this deflection took place (T_1 on Fig. 11-23), the inclusions were less ductile than their host, as indeed one would expect. Presumably, this situation would have existed

FIGURE 11-23. Schematic representation of major events in the Main Donegal Pluton. Position of horizontal bars with regard to horizontal axis gives the relative chronology; their position with regard to the vertical axis is arbitrary. The several "time zones," T_1, T_2, etc. so defined are discussed in the text. Width of solid bars indicates in a general way the relative abundance of the various dyke suites. Above are two curves representing the relative intensity (and the k-values) of deformation throughout this sequence of events, the "main phase" referring to events away from the marginal zones. Dashed lines are conjectural.

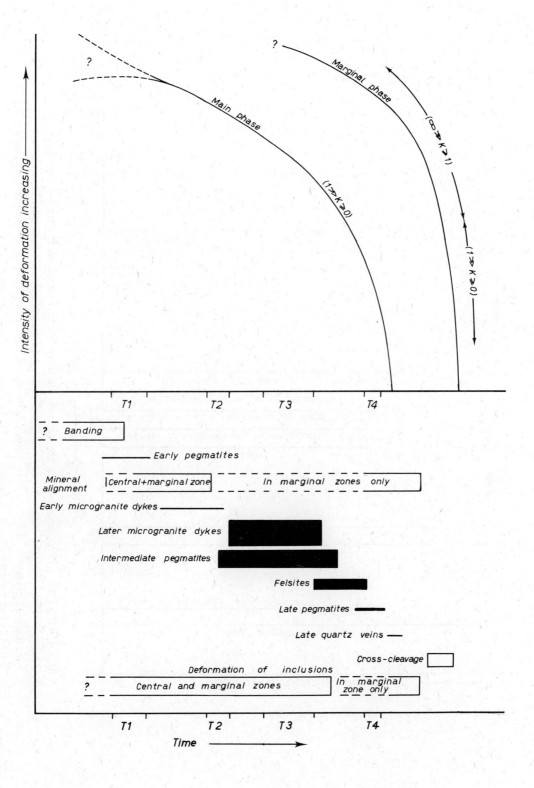

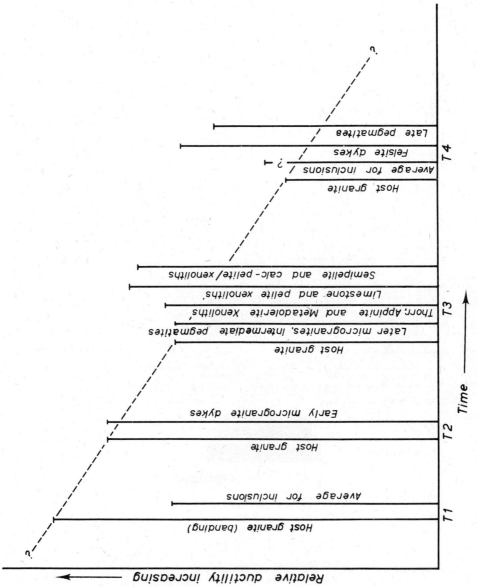

FIGURE 11-24. Diagram showing the relative ductility of the various components of the Main Donegal Pluton at the stages in its evolution defined in Fig. 11-23. Comparison between the hierarchies given for different time stages is based on the assumption of a steadily decreasing ductility for the host granite (dashed line). See text for further comment.

from the time the xenoliths were incorporated into the intruding granite, which would itself become less and less ductile as crystallization proceeded. The reactivation of the early microgranite dykes at time T_2 suggests a low ductility contrast between them and their host. But the lateral movement of the walls of the felsite dykes, and in particular the fact that certain late pegmatite dykes are folded where they cut the felsites but not where they cut the host granite, indicate that the latter was, by time T_4, considerably less ductile than the dykes, though the contrast was probably minimal immediately adjacent to the dyke wall (see Fig. 11-6b, c).

The various inclusions are very useful in this respect, and the different behavior of dykes, chiefly the later microgranites and intermediate pegmatites, which cut them, suggests a hierarchy of relative ductilities during time T_3. In the first place, the general absence of external deformation features where these dykes cut the Granite indicates that they have similar ductilities. Xenoliths of metadolerite, appinite, and Thorr tonalite rarely show more than a slight folding of these internal dykes, but the latter are markedly affected where they cut limestone and pelite rafts, and to a lesser extent, where they cross enclaves of semipelite and calcareous flags.* Unfortunately, only one or two examples of felsite dykes cutting xenoliths were observed, and if we are to judge by these cases the similar behavior of the dykes, where they cut both the host granite and the xenoliths, indicates a low contrast between the latter two and a considerably higher ductility for the felsite. Although we have thus some idea of the relative ductilities of the various components of the Main Granite, Fig. 11-24 reinforces our earlier conclusion that our estimate of an axial ratio of $b = 3$ for the inclusions of igneous origin merely provides a minimum for the bulk strain or the granitic host during the period T_1 to T_3 and probably tells us nothing of the intensity of any earlier stages of deformation.

We have now detailed many of the heterogeneities of the Granite deformation related both to variations in ductility at any one time and to changes throughout time in the physical state of the whole pluton and its components. Nevertheless, the only departures from homogeneity on a large scale concern the marginal deformation, the unusual situation regarding the Thorr xenoliths in the Glen Lough area (p. 246),** and the possible NW–SE increase in ductility from margins to the center of the pluton at about time T_1 to T_2, as suggested by the shear couples implied by the *en echelon* pattern of inclusions in the Kingarrow and Glenleheen areas (*ibid.*).

Turning now to *general considerations* of the deformation, we must emphasize that it affected xenoliths derived from the present envelope of the pluton and not brought up from depth, that the marginal sheets and the dykes which cut the Granite and its envelope are deformed together, and that the mineral alignment crosses from the Granite into the adjacent country rocks. These relationships clearly indicate that we are dealing from time T_1 onward (see Fig. 11-23), not with strain accumulated during intrusion of the pluton but with a superimposed deformation, whatever its cause. The marked banding in marginal areas (*not* the regular-banding) and the irregular "banding" which occurs throughout the pluton (see p. 215), we view as the result of the flattening of textural and compositional inhomogeneities during the granite deformation.

It is possible, however, that the regular-banding formed during the emplacement of the Granite rather than as an early phase of the *in situ* deformation. Indeed, the complex relationships between banding and areas of nonbanded xenolithic granite at Doochary (see Berger, 1971*b*) indicate considerable mobility in this central area, as does the thorough mixing of xenolith types here and elsewhere (see Fig. 11-1). In the final analysis, however, we can see no way of deciding firmly whether the regular-banding formed (1) *in situ* during an early, highly mobile stage of the deformation or (2) during emplacement as a result, for example, of differences in the rate of flow into place of a crystal mush.

---

* We are ignoring here variations in ductility of these dykes from positions where they cut the host granite to those where they cross the enclaves; in fact mineralogical changes, especially in dykes cutting calcareous lithologies, do suggest the possibility of such variations.
** For details see Berger, 1967T.

Whether or not the deformation began during emplacement or was entirely superimposed, we are fairly confident from the ghost stratigraphy and structure, and from the sheeted margins, that the Main Granite got into place by the wedgelike intrusion of a plexus of granite sheets, as Pitcher and Read (1959) originally suggested. However, we think now more in terms of vertical emplacement and can see no convincing evidence for the lateral influx of granitic material from the northeast envisaged by Pitcher and Read; their "intrusion-lineation" is clearly a result of the *in situ* deformation, and the presence of raft zones in the southwest end of the pluton is hard to reconcile with intrusion from the northeast. Examination of Map 1 in folder suggests that the splaying out of raft trains in the northeastern end of the pluton may be due largely to the fact that the major train here, that of the Thorr Pluton xenoliths, cuts at a low angle across the trend of the other zones and of the adjacent envelope stratigraphy (see also Fig. 11-1). We think now that this pattern may in fact be a pre-Main Granite relict, indicating that the intrusion of this northeastern lobe of the Thorr Pluton (see p. 97) caused a perceptible deflection of its own country rocks, now preserved in the Main Granite ghost stratigraphy. Nevertheless, there is still some angular divergence between the raft trains in the Crockmore–Leahanmore area and the adjacent envelope, which does suggest a wedging apart of the country rocks, especially when it is realized that all such angles must have been greater prior to the deformation. In other words, the opening up of the regional stratigraphy by the intruding Main Granite must originally have been *more pronounced and the whole pluton wider and less elongated than at present.*

Having discussed the internal features of the Main Donegal Pluton in terms of deformation, it is natural to attempt to deal in the same manner with the intensely developed structures in the surrounding country rocks and with their complex metamorphic assemblage. Although much attention has been focused on the envelope rocks (*e.g.,* Pitcher and Read, 1960*b,* 1963; McCall, 1954; Iyengar *et al.,* 1954; Rickard, 1962; and theses by Pande, Tozer, Cheesman, Berger, Edmunds, and Naggar; see Bibliography II), our understanding of them is still far from complete. And we must frankly warn the reader that what we do know about these rocks poses a major dilemma in the chronology of structural and metamorphic events in the aureole as related to those in the pluton. Even the discussion of the problem creates serious difficulties, so complex is the geological history of these rocks. It is thus with considerable caution that we proceed to a discussion of the envelope of the Main Donegal Pluton.

CHAPTER 12

THE ENVELOPE OF THE MAIN DONEGAL PLUTON: A DEBATE ON CHRONOLOGY

CONTENTS

1. Introduction: the Previous Model 258
2. A Structural Veneer: the Later Events, DMG_2 and DMG_3 260
   (A) Preamble 260
   (B) The DMG_3 Structures 260
   (C) The DMG_2 Structures 261
   (D) Summary of Structures in the Envelope Schists 266
3. The Metamorphism of the Metasediments in the Envelope: an Early Event, DMG_1 268
   (A) Preamble 268
   (B) The Microstructures 270
   (C) The Porphyroblasts and the Early DMG_1 Event 272
   (D) The Late Growth of Phyllosilicates 277
   (E) Other Late Minerals 279
   (F) The Late Retrogression 279
   (G) A Discussion of the Metamorphic Chronology 280
4. The Metamorphism and Deformation of the Preexisting Igneous Rocks 281
   (A) The Metabasites 281
   (B) The Lamprophyres and Felsites 281
   (C) The Earlier Members of the Donegal Granite Complex 282
   (D) A Further Discussion of the Metamorphic and Structural Chronology 283
5. The Special Problem of the Mullions 283
6. The Correlation of Deformative Events in Envelope and Pluton: the Deformation of the Marginal Sheets 286
7. The Origin of the Main Donegal Pluton and its Complex Aureole 288
   (A) A Summary of the Sequence of Events 288
   (B) The Gneiss-Dome Hypothesis 290
   (C) Superimposed Deformation versus Forceful Emplacement 291
   (D) The Regional "Hot Spot" Model: an Alternative for the Development of the Envelope 293

1. Introduction: the Previous Model

In Chapter 11, we interpreted the internal features of the Main Donegal Pluton as the result of a powerful deformation, largely superimposed on the Granite after its emplacement but prior to its complete cooling. We turn now to a discussion of the structural and mineralogical changes brought about in the country rocks as a result of these forces.

It has, of course, long been known that the intensity of deformation and metamorphism in the Dalradian increases rapidly as the margin of the Main Donegal Pluton is approached. In places immediately adjacent to the contact, coarse schists and even gneisses are developed, mullion structures appear, and everywhere folds are tight and cleavages intense. Structural effects extend outward in places as much as 5 kilometers from the contact (Fig. 12-1), but mineralogical changes are generally confined to a zone 2–3 kilometers wide bordering the Pluton. In particular, Pitcher and Read (1960b) recorded the development of a new crenulation cleavage trending subparallel to the contact and especially well exhibited by the Ards Pelites on the northwest side of the Main Granite (see also

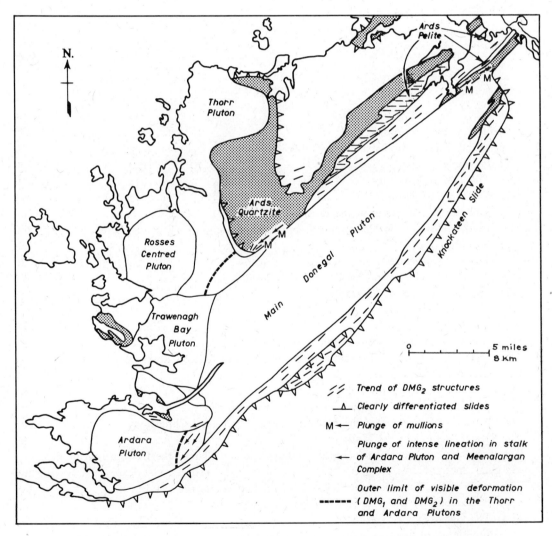

FIGURE 12-1. Distribution of "New" Structures around the Main Donegal Pluton.

Rickard, 1961, 1962). According to them, this cleavage becomes intensified toward the pluton, and accompanied by the tightening of folds, develops into a new penetrative schistosity (and lineation), albeit often following the older structural trends. This new cleavage was attributed to the forceful emplacement of the pluton, and the pronounced recrystallization and the growth of porphyroblasts of andalusite, sillimanite, kyanite, staurolite, oligoclase, and garnet were interpreted as syntectonic growths resulting from the high PT conditions induced by this forceful intrusion. Unfortunately for the ease of presentation, detailed studies by the junior author (Berger, 1967T) in a number of key areas in the envelope have shown that the structural and metamorphic history of these rocks is far more complex than earlier workers recognized, and the correlation of these events with those inside the Pluton has created some major problems.

The envelope of the Main Granite consists, for the most part, of metasediments belonging to the Creeslough Succession, and since the evidence for the polyphase history is best seen in pelitic and semipelitic rocks, we shall restrict ourselves largely to a discussion of these lithologies. As regards the contacts against other kinds of rocks, we shall need to discuss later the nature of the boundary between the Main Donegal and Trawenagh Bay Plutons (Chapter 13), but immediately important is the Thorr quartz diorite which forms the country rock in the west, and also provides a more or less continuous marginal strip from Crockator to Cock's Heath Hill in the northeast (see Map 1 in folder). Although the Thorr Pluton must have had here, as elsewhere, its own aureole prior to the intrusion of the Main Granite, we can find little trace of it now along this northwestern margin, and we assume that the metamorphic events associated with the later pluton have resulted in the obliteration of the earlier aureole (*cf.* Pitcher and Read, 1963).

One of the most intriguing and controversial features of the Main Donegal Pluton is the special metamorphic character of its envelope rocks, especially the pelites (see p. 309, and Pitcher and Read, 1963; Naggar, Atherton, & Pitcher, 1970; Naggar and Atherton, 1970). The coexistence in the aureole of the three aluminosilicates together with staurolite, garnet, biotite, chlorite, and more rarely, chloritoid and cordierite, makes this a unique situation that has few counterparts, if any, among granite aureoles presently described. Another complexity concerns the textural relationships of these minerals to each other and to the various minor structures, and it is perhaps not surprising to find that these rocks provide evidence of several distinct metamorphic episodes.

Since our main purpose in this chapter is to describe the various structural and metamorphic events in the envelope and to relate them as far as possible to events recognized in the Main Donegal Pluton itself,* we take this opportunity to review in Table 12-1 the chronology of regional, pregranite events that we discussed in detail in Chapter 3. The remarkable ghost stratigraphy in the Thorr and Main Donegal Plutons, it will be remembered, suggests the presence of the relicts of several major folds, including some of D_4 age, and this situation, together with the fact that the Main Granite cuts across both limbs of the F_4 Mulnamin Anticline in the Maas area, clearly indicates that the regional D_4 movements preceded the emplacement of the Donegal Granite Complex (see p. 61 for a discussion of the time relations between granite emplacement and the regional metamorphic events).

The new cleavages and folds described by Pitcher and Read are clearly superimposed on D_{2-3} structures, and since the former also deform the granite sheets, they must, of course, postdate the late regional D_4 event. The recent work has shown, moreover, that these late structures are the result of *two distinct tectonic events,* which we detail below. In addition, there is much indirect evidence that in the interval between these and the regional D_4 event there was at least *one other phase* of deformation, this time possibly directly associated with the intrusion of the Main Donegal Pluton. However, most important is the realization that these various post-D_4 structures, except possibly the latest, were formed while the pluton was hot.

---

* Discussion of various geochemical and petrogenetic aspects of the Main Granite aureole can be found in Chapter 14.

TABLE 12-1. Regional Chronology

		Summary of Regional Tectonic and Metamorphic Chronology (from Table 3-1)
D_1		S_1 bedding schistosity. Rare F_1 minor folds. Early tectonic slides.
	MS_1	First growth of phyllosilicates and quartz.
D_2		S_2 crenulation cleavage. F_2 recumbent folds.
	MP_2	Peak of regional metamorphism, growth of garnet and plagioclase porphyroblasts.
D_3		S_3 crenulation cleavage coaxial and coplanar with D_2 structures
	RMP_3	Retrogression
D_4		Upright open major folds and unrelated minor crenulations.
	MP_4	Phyllosilicate porphyroblastesis
	RMP_4	Retrogression

Although there is some reason to suspect that some of these events were of regional significance, we prefer not to elevate them to the status of regional episodes, since their structures are localized near the Main Granite. We shall thus, for the sake of clarity, refer to these "aureole" deformations as belonging to DMG_1, DMG_2, and DMG_3 events, and in what follows we describe their effects in reverse order, starting by consideration of the latter two phases, since it is only by "removing" the complication they add to the tectonic history that we can begin to discuss the earlier and more cryptic event (or events!).

2. A Structural Veneer: the Later Events, DMG_2, and DMG_3

(A) Preamble

The structures produced during these movements are essentially those described briefly by Pitcher and Read, though there is now the complication afforded by the need to separate DMG_2 from DMG_3 events. Since the geometry of these structures varies somewhat from area to area, we are forced to deal separately with their occurrence along the northwest, southeast, and southwest margins, restricting ourselves mainly to the metasediments and commenting only where necessary on nonpelitic lithologies.

(B) The DMG_3 Structures

The contact schists near the Main Donegal Pluton frequently exhibit a series of late cleavages which are best developed in a number of *shear zones* running parallel to the margin of the pluton. These belts of shear are generally less than a hundred meters wide and occur sporadically throughout the envelope, especially in the outer Creeslough Formation and in the Upper Falcarragh Pelites, southwest from Lough Greenan (Pitcher and Read 1960b, p. 25). In them, the rocks are markedly cataclastic and often acquire a strong platiness; chloritic and hematitic material is common. The country rock schistosity is smeared out in these zones, and any folds present are generally disrupted.

The associated *late cleavage* consists of a series of irregularly spaced, small-scale fractures or shear zones which lie at low angles (25°–30°) to the major schistosity. These shears are identical in style and orientation to the late *cross-cleavage* in the marginal zones of the pluton (p. 229), and we take them to be the result of the same movements. As in the marginal granite, they trend generally

NNE–SSW, though in some places a conjugate set is present (Fig. 12-3c, d). Their surfaces are often slickensided with subhorizontal striae, and their intersection with the host schistosity often produces a marked scalloping. Indeed, in many places the host fabric is bent by open ruckles, which, though not accompanied by discrete shears, have the same asymmetry as, and appear to represent, an incipient stage in the development of the cross-cleavage.

The cross-cleavage, which is probably equivalent to the late shear-fracture of McCall (1954, p. 172), occurs throughout the envelope wherever schistose lithologies are present but becomes more prominent toward the margin of the pluton. In the shear belts, the cross-cleavage is especially prominent and may even be axial-plane to local crenulations and minor flexural-slip folds. Thus we view the shear zones as an intense development of the cross-cleavage, and it is these structures which we now distinguish as *the DMG_3 episode*. For the sake of convenience, we also include in this structural phase certain very late kink bands and fractures which are best developed in the shear belts and which may be synchronous with, or later than, the SMG_3 cross-cleavage.

(C) The DMG_2 Structures

Initial stages in the development of the earlier DMG_2 structures are best seen to the *northwest of the Main Donegal Pluton* in the southernmost outcrops of the Sessiagh–Clonmass Formation west of Errigal (Rickard, 1961, 1962), and also in the Ards Pelites between Muckish and Sheephaven. In both these situations, the normally shallow S_{2-3} regional cleavage is warped about gently to open minor folds and minute crinkles (Figs. 3-4b, 12-2a, b; 12-3a; also Rickard, 1961, Fig. 2).* These structures develop rapidly southeastward toward the contact into close, flexural-slip to chevron folds (Fig. 12-4a) with an obvious axial-plane crenulation cleavage (Pitcher and Read, 1960b, Fig. 3; Rickard, 1962, Fig. 4h), the transition sometimes occurring within a distance of about 200 meters as, for example, on Crockatee and along the beach at Ards Friary. Since the bedding in the Ards Pelites is frequently disrupted by the intense S_{2-3} cleavage, the latter is the major surface which is folded; thus refolded F_{2-3} folds are seen only on occasion (Fig. 12-2a, b). The trend in these FMG_2 folds throughout this part of the Ards Pelites outcrop is generally ENE–WSW, and thus lies at low angles to the trend of the F_{2-3} and F_4 folds (Fig. 12-1; 12-3a, b).

It is significant that the intensity of the FMG_2 folds is related to the attitude of the S_{2-3} cleavage, with the folds rarely occurring where the latter dips steeply. Despite Pitcher and Read's contention (1960b, p. 11) that the DMG_2 structures become intensified southeastward toward the contact, to become the dominant folds and cleavages in the Creeslough Formation, we agree with Rickard (*loc. cit.*, p. 231) that the latter schists rarely exhibit FMG_2 folds "because they already dipped steeply and could not be refolded by horizontal compression." We therefore regard the main folds and axial plane fabrics of the Creeslough Formation as pre-DMG_2 structures, and the variation in sheet dip of S_{2-3} from subhorizontal on top of Crockatee (Fig. 12-3a), through moderate southeasterly on the southeast flanks of this hill (Fig. 12-3b), to steep southeasterly in the Creeslough Formation (Fig. 12-3c), as being largely a result of the D_4 Creeslough Downwarp referred to in Chapter 3 (see Fig. 3-2, 12-5f).

The polyphase nature of the thoroughgoing schistosity in the Creeslough Formation is difficult to recognize; in most places, it seems to be a perfectly penetrative fabric, which most workers on the Scottish Dalradian would have no hesitation in labeling an S_1 structure. Nevertheless, there are scattered examples of earlier isoclinal folds (*e.g.*, Fig. 12-2d, f), and the microscopic textures, as

---

* Between Errigal and Muckish, part of the area described by Rickard (1962) as the "Devlin Belt of Superimposed Folds," early stages in the development of DMG_2 structures are not generally well shown, for the massive Ards Quartzite here seems to have acted as a buttress against which DMG_2 folds and cleavages in the Ards Pelites are particularly strongly developed.

we shall see later, also attest to a polyphase development of this fabric, which we view as an intensified S_{2-3} structure, as at Ardara (see p. 178). Further, this main schistosity is occasionally folded coaxially about axial planes whose dip varies from shallow northwesterly to vertical (Fig. 12-2e, f, 12-3c, and 12-4a); these new folds are so similar in style and orientation to the FMG_2 folds in the adjacent Ards Pelites—as described above—that we have little hesitation in making a positive correlation. The cleavage which accompanies these FMG_2 folds is variably developed, but within about 500 meters of the margin of the Main Donegal Pluton, it becomes a strong, steeply dipping mineral fabric and may well form the dominant schistosity in places; this is best seen in the Crocknawama–Croagh area (Fig. 12-5f).

The problem in the Creeslough Formation, as in the narrow, tectonically thinned strip of Ards Pelite to the southeast, is to separate the regional D_{2-3} from other deformative events which evidently intervened between this and the DMG_2 event just described. The main evidence, as we shall see later, comes from a study of the microstructures, but in the absence of distinct macroscopic representatives, we can only conclude that these intervening movements had as their main effect the disruption and tightening of F_{2-3} folds (cf. Fig. 12-2c) and the intensification of the S_{2-3} cleavage, though perhaps the development of the well-developed mineral lineation (parallel to F_{2-3}) and widespread boudinage (Fig. 12-5d) can also be included here.

The structural situation on the *southeastern side of the Pluton* is somewhat different from that just described. As we discussed in Chapter 3, the major folds here, which are overturned to the northwest, appear to be F_{2-3} structures to which the main schistosity, an intense S_{2-3} cleavage, is axial plane. The second-phase character of this fabric, which dips at low angles to the southeast in the northeast but which becomes steeper to the southwest, is rarely obvious, but it can be seen in places to be a strong crenulation cleavage (Fig. 12-4b) folding a penetrative fabric (S_1) which cuts at low angles across bedding, though no F_1 folds have been definitely recognized.

Inital stages in the formation of DMG_2 structures are not well shown here, particularly as the massive Slieve Tooey Quartzite seems to have acted as a barrier. Even within the belt between this quartzite and the contact of the Pluton, there is difficulty in separating D_{2-3} from DMG_2 structures, as is best illustrated in the *Lough Salt–Crockmore area*. Here, in both the Sessiagh–Clonmass Formation and the Upper Falcarragh Pelites, the minor folds seem at first glance to represent a single generation of generally northward-overturned structures with a strong S_{2-3} axial plane cleavage. However, on closer examination, this fabric can frequently be seen to be strongly crinkled about a new crenulation cleavage often lying at relatively low angles to the earlier cleavage (Fig. 12-2g, i);

FIGURE 12-2. Minor structures in envelope schists. (a) Tectonic inclusions of semipelitic and psammitic beds in Ards Pelites, folded about regional S_{2-3} cleavage which is in turn crinkled by minute FMG_2 crenulations. Top of Crockatee. (b) F_{2-3} folds bent by open FMG_2 folds. Southeast flank of Crockatee. Ards Pelites. (c) Profiles of flattened F_{2-3} folds in pelites and semipelites of Creeslough Formation, 400 meters northeast of Lough Aleane. (d) F_1 fold cut by S_{2-3} schistosity (or F_2 cut by S_3?) Creeslough Formation. Northeastern shore of Lackagh River, northeast of Bishop's Island. (e) Profile of F_{2-3} and FMG_2 folds in cliff exposure of Creeslough Formation along bog road just north of Crocknawama. Insets show minute F_{2-3} folds refolded by FMG_2. (f) Three generations of folds in Creeslough Formation. Northeastern shore of Lackagh River at Creevagh Point. FMG_2 on F_{2-3} on F_1. Quartz segregations black. (g) FMG_2 folds superimposed on F_{2-3}. Upper Falcarragh Pelites (Lough Greenan Schists). Top of Binnadoo, southeast of Crockmore. (h) F_{2-3} folds in limestone post-dated by FMG_2 crinkles in adjacent (Lough Greenan) schists. Near Malt Kiln west of Lough Reelan. (i) Two sets of FMG_2 crenulations, sketched from a hand specimen from Crockmore Septum. (j) Disharmonic (quasi-flexural) FMG_2 crenulations. Sketched from thin section from Crockmore Septum.

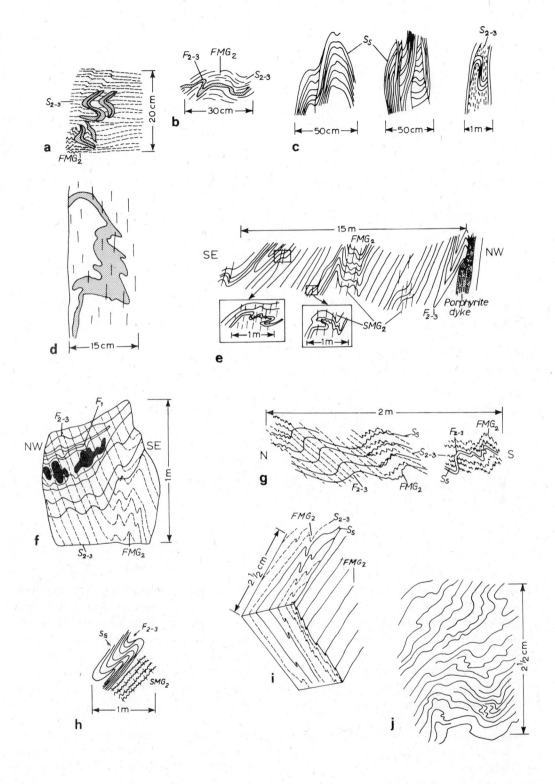

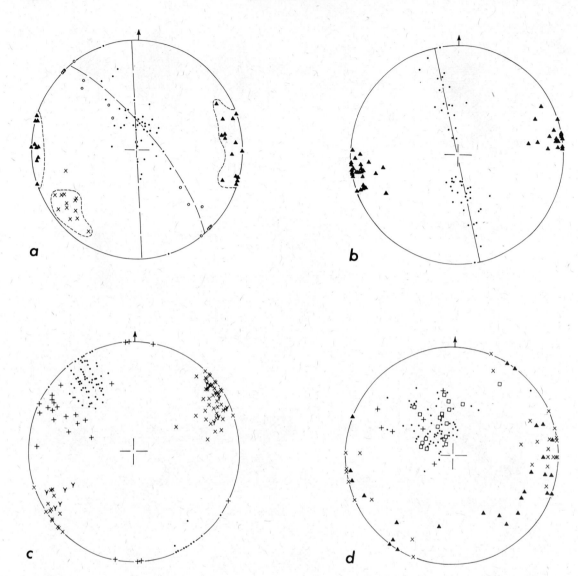

FIGURE 12-3. Stereoplots of minor structures from the Main Granite envelope. (a) Ards Pelites on top of Crockatee, northwest of Dunlewy Fault. (b) Ards Pelites on southeastern flank of Crockatee, southeast side of Dunlewy Fault. (c) Creeslough Formation in Crockatee-Crocknawama area. (d) Crockmore-Binmore area; mainly Upper Falcarragh Pelites and Sessiagh-Clonmass Formation: ○, poles to Ss (in a only); ●, poles to S_{2-3} (and to Ss in c and d); ×, axes to F_{2-3} folds (and to FMG_2 in c); Y, L_{2-3} mineral lineation (in c only); ▲, axes to FMG_2 folds and crenulations (in a, b, and d); □, poles to SMG_2 (in d only); +, poles to SMG_3 (the cross-cleavage, in c and d).

nevertheless, polished surfaces or thin sections are often necessary to convince the sceptic of the polyphase character of these schists (Fig. 12-4b). We are content to include this late structure within the DMG_2 episode, since it is similar in style to the FMG_2 folds on the northwest side of the Pluton, and since, as we shall see later, it also affects granite sheets.

Changes in intensity of both generations of structures as one approaches the pluton are not as well shown as to the northwest, but in general terms, rather angular folds of both generations in the

FIGURE 12-4. Plates of minor structures. (a) FMG_2 chevron folds in Creeslough Formation, Faymore River northeast of Lough Aleane. Note minute F_{2-3} fold in light band in upper part of photo. (b) Photograph of a thin section of Upper Falcarragh Pelites from Binnadoo showing minute F_{2-3} folds with a strong axial-plane cleavage-banding (S_{2-3}) folded by open FMG_2 microfolds along which there has been no growth of minerals. Plane light. (c) Granite sheet necked on two axes to form pillow-shaped boudins. Quarry at Carbat Gap on Fintown-Doochary road. (d) (i)—Folded pegmatite vein in Crockmore Septum, with strong axial-plane SMG_2 cleavage in host pelites. Axial-plane foliation in pegmatite marked chiefly by books of muscovite. (ii)—Folded pegmatite vein in Crockmore Septum. Note marked SMG_2 crenulation cleavage in host schists.

Sessiagh–Clonmass Formation give way northwestward to tighter folds with stronger cleavages in the Upper Falcarragh Pelites (on Binnadoo). The associated cleavage (SMG_2) likewise varies from a moderately developed crenulation cleavage to what is, in many places, a marked schistosity whose secondary nature is sometimes hard to establish.

In the Crockmore septum, there are also situations in which several distinct sets of crenulations and minor folds are superimposed on the S_{2-3} fabric. In some cases, these may form conjugate structures; in others, there is no clear distinction between them (Fig. 12-2i, j). Whatever their precise origin, these small-scale folds all postdate the S_{2-3} schistosity, and since there appears to be no consistent age difference between them, we group them all as DMG_2 structures. It is interesting to note in this connection that some of the recent experiments of Weiss and Paterson (1968) and Weiss (1969) have produced markedly disharmonic and interfering crenulations and small-scale flexural-slip folds which are strongly reminiscent of the present situation, thus strengthening our opinion that all these minor structures are the result of the one episode.

The *area from Crockmore to Fintown* has not been reexamined in the light of the chronological analyses presented herein, but earlier work (Pitcher and Read, 1960b; Pulvertaft, 1961) suggests that the main schistosity is again an S_{2-3} fabric, and although FMG_2 folds have not been recognized as such, we suspect that many of the E–W to NW–SE trending minor folds recorded by Pulvertaft belong to this generation.

Certainly in the *Fintown area,* the situation is again one where a strong S_{2-3} schistosity, lying axial plane to occasional tight to isoclinal minor folds, is folded about generally upright NE–SW to E–W trending FMG_2 folds (Fig. 12-5a, b), these having an axial-plane crenulation cleavage which appears to become more intense toward the Pluton. Several sets of crinkles and folds with similar geometric and metamorphic styles are common and are all superimposed on the S_{2-3} fabric as in the Crockmore area, though they are much more prominent here. Their attitude is variable and one set seldom folds another, so that we again attach them all to the DMG_2 episode.

The southwestern envelope is something of a disappointment as far as the record of the DMG_2 event is concerned (Fig. 12-1). Although the presence of several major folds has been well established between Maas and Cleengort Hill, namely the F_{2-3} Maas Syncline and the F_4 Mulnamin Anticline (see Chapter 3), minor DMG_2 structures are difficult to recognize near the margin of the Main Granite. However, in the Lower Falcarragh Pelites which occupy the core of the Maas Syncline, the steep, generally penetrative S_{2-3} axial-plane cleavage is crinkled about steeply westward-plunging crenulations (Iyengar et al., 1954) which are accompanied in places by minor flexural-slip folds with near-vertical axial planes. Since these structures are markedly incongruous with the Mulnamin Anticline, and since they have similar metamorphic and geometric styles to DMG_2 structures elsewhere, we attribute them to these late movements.

(D) Summary of Structures in the Envelope Schists

In summary, then, the schistose metasediments surrounding the Main Donegal Pluton are characterized by a south to southeast dipping schistosity, which is axial plane to major and minor folds overturned to the north and northwest (Fig. 12-5f); these we consider to be tightened D_{2-3} structures. During the DMG_2 movements, these structures were deformed about minor structures, which, at least in the northwest and southeast, increase in intensity toward the margin of the Pluton, though this progression is rarely as straightforward as Pitcher and Read envisaged. Thus, a normally open crenulation cleavage (SMG_2) becomes, along parts of the margin, a new schistosity, while along others it is merely a more pronounced cleavage with little growth of minerals parallel to it. Super-

---

FIGURE 12-5. Structures in the envelope. (a) F_{2-3} folds folded by FMG_2; Upper Falcarragh Pelites, Kingarrow. (b) FMG_2 crenulations on F_{2-3} folds in Glencolumbkille Pelites at lime kiln on saddle between Scraigs and Aghla. Note late [?DMG_3] kink band and dying out of FMG_2 crinkles as they approach a lamprophyre dyke (stippled). (c) FMG_2 folds superimposed on D_{2-3} structures. Lower Falcarragh Pelites on Carrickfad Point northeast of Portnoo. (d) Marked boudinage of metadolerite sill (stippled) in pelites on upstream side of Lackagh Bridge. Quartz (black) fills the gaps between boudins. (e) Folding of pegmatite vein (π) cutting large pelite raft in Thorr Pluton, 600 meters from the margin of Main Granite at Brockagh. (f) Schematic cross-section of Creeslough Succession from Binmore to Crockatee, showing major and minor structures in envelope, and ghost folds. Not to scale.

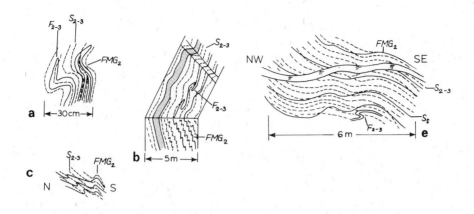
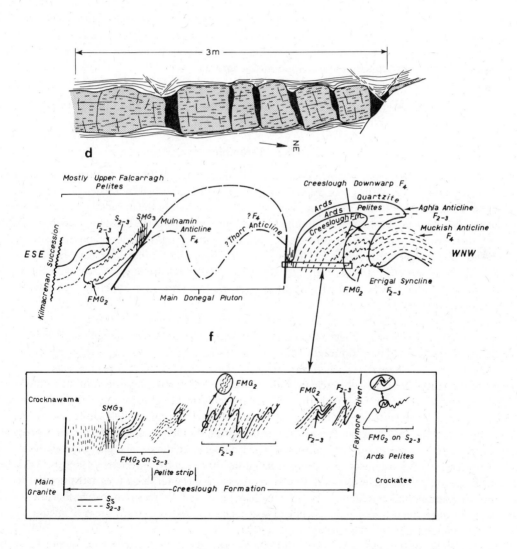

imposed on the regional S_{2-3} (and local SMG_2) fabric are the shear belts and cross-cleavage, which we view as closely related DMG_3 structures.

In addition to the D_{2-3}, DMG_2, and DMG_3 movements, consideration of the major structures in the envelope (see Chapter 3), of the textural relationships of the envelope minerals, and of the deformation of the marginal Granite (see p. 286) all indicate that also represented in the envelope are at least two more, separate, movement phases (D_4, DMG_1), occurring in the interval between the D_3 and DMG_2 movements. However, we have not been able to recognize separate minor structures associated with these two additional events, and we are thus forced to conclude that their main effect, particularly of DMG_1, has been to further tighten and intensify the D_{2-3} structures near the Main Granite. Indeed, the widespread necking and boudinage of relatively competent lithologies, the common disruption of F_{2-3} folds, and the ubiquitous mineral lineation (and pronounced mullioning in the marginal quartzites—see p. 283) parallel to the F_{2-3} axes, are more readily understood in the light of this situation. Thus, what appears in the field to be a simple structural situation with all fold axes parallel and a single planar mineral alignment is, in reality, the result of a coaxial and generally coplanar superposition of several deformation phases, and it is only through the infrequent examples of refolded folds and cleavages that the true polyphase natures of these structures can be recognized in the field. For evidence of the more cryptic post-D_4–*pre*-DMG_2 events we now turn to a consideration of the metamorphic character of the envelope metasediments.

3. The Metamorphism of the Metasediments in the Envelope: an Early Event, DMG_1

(A) Preamble

The distribution of the major minerals around the Main Donegal Pluton is shown on Fig. 14-1, and the progressive increase in grade toward the Pluton, well established by earlier workers, is given in another form in Table 12-2. This change is perhaps best followed in the metadolerites, which change isochemically (Kirwan, 1965T) from highly altered, fine-grained, often massive, saussurite-uralite-bearing rocks away from the Pluton to fresh, medium-grained, schistose amphibolites with robust hornblende and sphene at the contact. However, as details of their mineralogy have been given by Pitcher and Read (1960b) and Kirwan (*loc. cit.*), we shall concentrate our attention here on the pelites, for they exhibit both a varied mineralogy and the microstructures necessary for establishing a metamorphic chronology.

In some places, it is possible to make a broad distinction between an outer and an inner aureole, as for example, at Fintown, where andalusite is restricted to the Glencolumbkille Pelites to the southeast and staurolite to the Falcarragh Pelites to the northwest (see however, Naggar and Atherton, 1970), or in the Crockmore area where the Binadoo Shear Belt (DMG_3) separates chlorite-bearing schists to the southeast from biotite schists with andalusite, sillimanite, kyanite, and staurolite to the northwest. However, for the most part, the only zonal arrangement of major minerals is the tendency for andalusite to be restricted to areas well outside the sheeted margin, while kyanite and sillimanite (including fibrolite) tend to occur closer to the main contact. As Pitcher and Read (*loc. cit.*) predicted, there appears to be a compositional control exercised on some of these minerals, particularly kyanite (see Chapter 14). We may note in passing that Turner has recently suggested (1968, p. 223) that the inner aureole of the Main Donegal Pluton represents a metamorphic facies transitional between hornblende hornfels and amphibolite facies. This may be so, but we must emphasize that these rocks are not hornfelses, but schists identical in hand specimen to rocks from many areas of medium-grade regional metamorphism (*cf.* Reverdatto et al., 1970).

Detailed studies of thin sections of pelitic lithologies throughout the envelope have revealed a variety of textural relationships involving microstructures associated with the various deformational events.

TABLE 12-2. Variations in Mineralogy of Important Envelope Rocks Toward the Main Donegal Pluton

	Regional	Aureole (2-3 km) Outer	Aureole (1 km ±) Inner	Sheeted Zone	Rafts in the Pluton
Metadolerites	Labradorite (An_{54}—variably saussuritized)	Quartz	Oligoclase—Andesine (An_{28-35})		
	Augite → Uralite		robust Hornblende		
			(blue-green)	(brown)	
	Ilmenite → Sphene		Sphene		
			brown Biotite		
	Orthoclase				
	Chlorite				
Calc-Pelites					Diopside
			Tremolite (Actinolite)		
		Wollastonite		Wollastonite	
				Grossular	
		Quartz and Calcite			
				Microcline	
		Chlorite	Phlogopite—Biotite		
Pelites	Muscovite (g),	Plagioclase (g), and		Quartz (g)	
	green Biotite (g)		red-brown Biotite (Pg)		
	Chlorite (P), largely retrograde from Biotite				
				Microcline (g)	
		Plagioclase (P)			
	Garnet (P), progressively less chloritized				
				new Garnet (g)	
	Chloritoid				
			Staurolite (Pg)		
		Andalusite (P)		Andalusite (P)	
			Kyanite (Pg)		
				Sillimanite (Pg)	
	Muscovite (P)			Muscovite (P)	
				Cordierite (g)	

(P) as porphyroblasts.
(g) in groundmass.
(In part, modified from Pitcher and Read, 1960b, Figure 9; Kirwan, 1965T).

In these rocks, andalusite, sillimanite, kyanite, staurolite, garnet, oligoclase, chlorite, biotite, and muscovite occur as porphyroblasts, and with the exception of andalusite, they also occur as groundmass minerals. Of the minerals in general, the phyllosilicates and quartz form one more or less coeval suite clearly postdating the others, which vary somewhat in their time relations to one another.* However, as a necessary framework for the discussion of these important metamorphic minerals, we comment first upon the general metamorphic character of the several microstructures.

(B) The Microstructures

The S_1 schistosity is occasionally preserved as inclusion trails in porphyroblasts (the Si fabric), or rarely, as an independent external (Se) fabric in certain psammitic or semipelitic lithologies. In areas well away from the Main Granite (see p. 58), it is marked by graphite and ores, and by minute laths of muscovite, chlorite, chloritized biotite, and rare tourmaline, together with crudely aligned crystals of plagioclase and quartz.

As we have said, the dominant fabric S_{2-3} is often parallel to S_1 and thus forms a strong penetrative schistosity, but its second-phase nature can be ascertained in the cores of folds, where it is clearly an intense crenulation cleavage along which there has been much segregation of minerals to form a prominent banding (Fig. 12-4b). The preservation of this crenulation in the porphyroblasts, together with the difference in grain size between the Si minerals and their Se counterparts, and coupled with the deflection of the schistosity around the same porphyroblasts, indicate a composite origin for this fabric, which we comment on below.

The progressive nature of the formation of the new cleavage, SMG_2, can easily be demonstrated. In outer parts of the aureole, as on Crockatee, Binmore, and near Maas, the groundmass micas are sharply bent by the FMG_2 crenulations (Fig. 12-6a), accessory tourmaline prisms are broken, and quartz grains are highly strained, sometimes developing marked Boehm lamellae. As the Pluton is approached, bending of individual crystals becomes less prominent and the micas tend to form polygonal or herringbone arcs, suggesting increasing syntectonic crystallization or subsequent strain healing (Fig. 12-6b; see also Harte and Johnson, 1969; Misch, 1969). Indeed, as we saw earlier, close to or within the sheeted margin, a new growth of micas parallel to SMG_2 is often seen.

In contrast, the SMG_3 cross-cleavage is *everywhere* cataclastic, bending micas and straining quartz, and this episode, as we shall see later, appears to be associated with a general retrogression of many minerals throughout the aureole.

FIGURE 12-6. Photomicrographs of aureole rocks. (a) FMG_2 crenulations in Ards Pelites on Crockatee. Note retrogressive style, exemplified by kinking of large white mica lath aligned parallel to S_{2-3}. Plane light. (b) FMG_2 microfold from inclusion of Upper Falcarragh Pelites in granite sheet at Barnesbeg (Fig. 12-12d). Micas form polygonal arc and show no signs of strain. No growth of minerals along SMG_2. Plane light. (c) Plagioclase porphyroblast from Upper Falcarragh Pelites at Lough Greenan, preserving a crenulation cleavage. Note deflection of Se around the porphyroblast, and the contrast in size between Si and Se minerals. May be an increase in size of Si inclusions near margins of plagioclase. Crossed nicols. (d) Staurolite porphyroblasts in Ards Pelites at Glenkeo. Si trails in largest staurolite appear to indicate syntectonic growth. Note increase in Si toward the margin of this crystal, discordance between Si and Se and deflection of Se around porphyroblasts. Plane light.

---

* Since this account was written, Naggar and Atherton (1970) have published a rather different metamorphic chronology, with which we do not entirely agree.

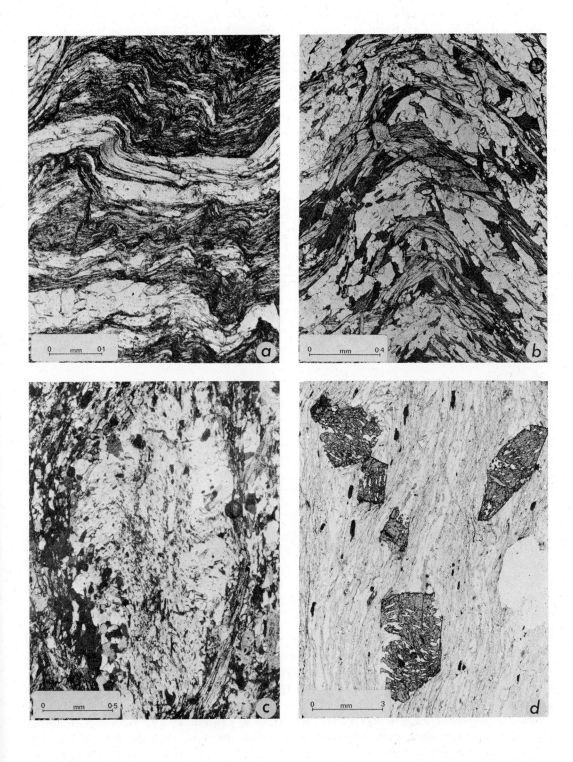

(C) The Porphyroblasts and the Early DMG₁ Event

As we stated above, andalusite, sillimanite, kyanite, staurolite, garnet, and oligoclase, all appear to have grown relatively early in the paragenesis (Table 12-3). These earlier porphyroblasts are generally small, rarely exceeding 2–3 centimeters in size, though andalusite frequently exceeds the others in its dimensions. The aluminosilicates in particular are best developed in scattered "sweat-outs" aligned generally parallel to the main schistosity (Pitcher and Read, 1960b).

TABLE 12-3. Chronology of Major Porphyroblasts within the Envelope of Main Granite

	Early porphyroblasts	Period of flattening and matrix coarsening, DMG_1	Late porphyroblasts	"New" structures of Pitcher and Read. DMG_2 of this book	Late cleavage and shear zones. DMG_3 of this book
Sillimanite	– – –		f	f	
Kyanite		───		═══	
Staurolite		───			
Andalusite		───			
Garnet		───	g	– – –	
Plagioclase		───			
Biotite			───	═══	
Chlorite			───	═══	
Muscovite			───	– – –	
Cordierite			───		
Chloritoid			───		

———, "definite;" – – – –, approximate; ═══, extensive strain-healing and recrystallisation; f, fibrolite; g, new garnet in the groundmass.

Kyanite, staurolite and oligoclase. As regards the porphyroblasts of kyanite, staurolite, and oligoclase, these have much in common (Fig. 12-6, 12-7, 12-8a). Oligoclase is generally anhedral, but regular shapes are common among the others, though intergrowth with quartz frequently leads to partly irregular and even "hornfelsic" forms (Fig. 12-7b). Where elongated, these minerals generally lie in or close to the main schistosity (S_{2-3}), and linear alignments of kyanite can sometimes be seen oriented parallel to F_{2-3} axes.

FIGURE 12-7. Textural relationships of some of the earlier porphyroblasts. (a) Plagioclase (P) and staurolite (S) with Si discordant to Se. Plagioclase preserves F_{2-3} crenulations: Garnets, black; kyanite, K. Ards Pelites at Glenkeo. (b) Irregular staurolite (heavy outline) embayed by quartz in a "sweat-out." Ards Pelites at Glenkeo. (c) Staurolite (S) and Plagioclase (P) with irregularly curving Si trails, Upper Falcarragh Pelites at Crockmore. (d) Staurolite elongated parallel to main schistosity, with slightly curving Si trails. Ilmenite, black. Same locality as (c). (e) Staurolite controlling FMG_2 microfold. Same locality as (c). (f) Idealized relationships of major porphyroblasts to external structures. Insets show contrast in grain size between Si and Se minerals. See text for further explanation.

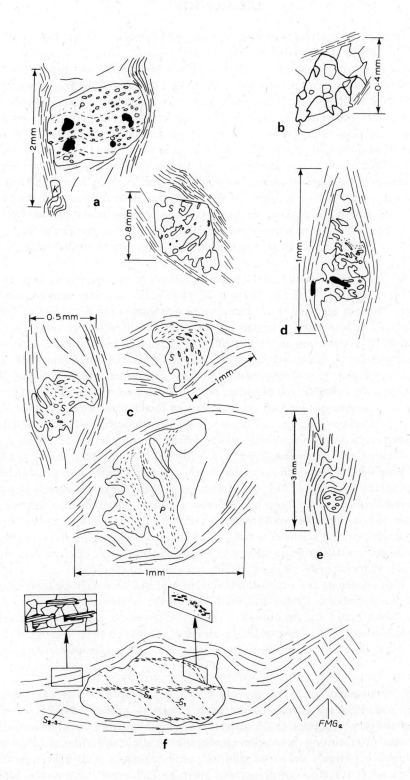

The key to the chronological significance of these porphyroblasts comes from their textural relationships. The size of the included Si grains (chiefly quartz, micas, and graphite) is almost invariably smaller than that of their Se counterparts (Fig. 12-6c; Pitcher and Read, 1960b, p. 14), and only occasionally can a progressive increase in size toward the edge of the host crystal be detected in them. The Si trails may be straight, but in many cases they are curved in such a way as to suggest the preservation of earlier open microfolds; indeed, in a few cases involving oligoclase, more than one microfold is preserved (Fig. 12-6c; Pitcher and Read, 1960b, Plate II, Fig. 5). Despite the conclusion of Pitcher and Read (1963, p. 286) that "the restriction of a single sigmoid to each of the porphyroblasts without regard to size strengthens our opinion that the porphyroblasts were rotated during the formation of the microstructures," we now think that, as in the case of the regional MP_2 porphyroblasts, the general lack of symmetry of the Si patterns about the center of the host crystal argues against their syntectonic growth. Nevertheless, there are rare indications of syntectonic Si patterns (Fig. 12-6d), and slight increases in the curvature of Si near the margins of some porphyroblasts (Fig. 12-7c), which may indicate that their later stages of growth took place during movements in the matrix.

A characteristic feature of virtually all these porphyroblasts is the discordance between their Si trails and the Se fabric (Fig. 12-6c, d; 12-7a, c, d; 12-8a). In some thin sections, the inclusion trails of all porphyroblasts are parallel, or nearly so, to one another but discordant to the external schistosity; in others, trails in adjacent crystals may be sharply inclined to each other (e.g., 12-6d), especially where FMG_2 crenulations are present (Fig. 12-8a, and Pitcher and Read, 1963, Fig. 17).

These relationships are, of course, strikingly similar to those involving the regional MP_2 porphyroblasts discussed in Chapter 3, and we advance a similar explanation for their origin. Thus, we consider that the kyanite, staurolite, and oligoclase porphyroblasts overprinted relatively open crenulations which were subsequently tightened around them to produce the strong external schistosity, whose origin as a crenulation cleavage can often be seen only through the helicitic Si fabric (Fig. 12-7f). The difference in size between the Si and Se fabrics is readily accounted for by a period of general recrystallization and "matrix coarsening" (cf. Harte and Johnson, 1969) during this flattening. The frequent "sweat-outs" of quartz and alumino-silicates, together with the common skeletal margins to many porphyroblasts (e.g., Fig. 12-7b), we attribute to resorption or to equilibrium growth during this general recrystallization event. Naggar and Atherton (1970, p. 559) claim that these sweats show "clear hornfelsic texture," but we would point out that similar augen are common in regional metamorphic terrains involving amphibolite facies or higher grades. Our conclusion is that the growth of the porphyroblasts was thus largely a static one, though the later stages of crystallization, in some cases, appears to have overlapped with the flattening and general coarsening.

The timing of this flattening and recrystallization relative to the chronology we have established is obviously of prime importance. Pitcher and Read were of the opinion that the "rotation" of the porphyroblasts, which caused the discordance between the Si and Se fabrics, was a result of the deformation which produced the "new strain-slip cleavage" (1963, p. 286); that is, of the DMG_2 movements of this study. However, the discordance between Si and Se is marked *even where DMG_2*

FIGURE 12-8. Photomicrographs of aureole rocks. (a) Porphyroblasts of staurolite and kyanite (K) in Ards Pelites at Glenkeo. Note small garnets (G), and open FMG_2 crenulations which post-date the porphyroblasts. (b) Andalusite porphyroblast from Upper Falcarragh Pelites at Lough Greenan. Overprinting crenulation cleavage. Note slight deflection of Se around andalusite. Plane Light. (c) Highly deformed chlorite porphyroblast in Ards Pelites on Crockatee. Note divergence of (001) cleavage planes and their general parallelism to the SMG_2 axial planes. (d) Small chlorite porphyroblast in Upper Falcarragh Pelites on Binnadoo. Note bending of (001) cleavage and crenulations on Si trails. Crenulations in matrix are of DMG_2 age.

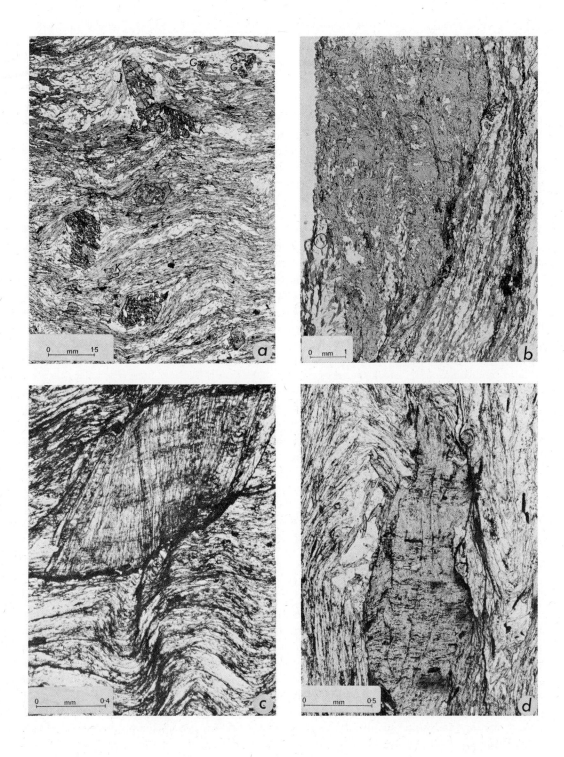

microstructures are absent (Fig. 12-6c, d; 12-7a, c, d); indeed, in many cases one of the main effects of these movements has been to destroy the alignment of Si trails from one porphyroblast to another. The metamorphic "style" of the FMG$_2$ crenulations presents another objection to Pitcher and Read's interpretations, for as we stated earlier, it is only in certain places immediately adjacent to the margin of the Pluton that there is any evidence of a general recrystallization and new growth of minerals during DMG$_2$, whereas we have to account for the widespread difference in size between the Si and Se fabrics which arose during this flattening. Furthermore, in some places, particularly at Glenkeo, kyanite prisms are strongly bent and kinked by the DMG$_2$ microstructures (Pitcher and Read, 1960b, Plate III, Fig. 3; Dewey, 1965, Fig. 29). We have little doubt, therefore, that the flattening which caused the bulk of the discordance between the Si and Se fabrics, together with the general matrix coarsening, took place *prior to* the DMG$_2$ movements, though the latter, like the DMG$_3$ microstructures, certainly accentuated the geometrical contrasts. This episode of flattening and syntectonic recrystallization, together with the growth of the earlier porphyroblasts, we term the DMG$_1$ *event*. Later, we shall include certain other events under this designation, but first we must deal with the rest of the aureole minerals.

Andalusite. This mineral occurs as porphyroblasts which are generally much larger than those of the minerals we have just discussed; indeed, it may on occasion contain them as inclusions (Pitcher and Read, 1963, Fig. 18). In some cases, there is little difference between the relationship of andalusite to the various microstructures and that involving the other porphyroblasts, but in many places, the angular discordance between Si and Se is considerably less than for the others (Pulvertaft, 1961, p. 271; Pitcher and Read, *loc. cit.*, p. 287). This situation appears to be largely the result of the andalusite having overgrown several microlithons (parallel to which it is generally elongated, Fig. 12-8b), in contrast to the other more equant porphyroblasts, which seem, for the most part, to have grown in crestal regions of existing crenulations (Fig. 12-7a, c). Close examination shows that the included crenulations in the andalusite are again more open than their Se counterparts; indeed, in some cases, the external fabric is effectively a penetrative schistosity (Fig. 12-8b). Where staurolite, kyanite, and garnet inclusions in andalusite possess Si trails, these are generally similar in size to, and continuous with, the Si fabric of the host porphyroblast (Pitcher and Read, *loc. cit.*, Fig. 18), and we can find no convincing evidence that the growth of andalusite was separated in time from that of these included minerals by movements in the matrix; neither have we seen any evidence of andalusite having grown syntectonically. Nevertheless, the inclusion of kyanite, staurolite, and garnet in andalusite, together with the observation that there is a lesser contrast in size between the included and groundmass minerals in the andalusite than in the other porphyroblasts, suggests that the growth of the andalusite may well have outlasted that of these minerals, and by implication, that of plagioclase. FMG$_2$ crenulations, on the other hand, clearly postdated even the andalusite.

Sillimanite. The timing of sillimanite poses another problem. Robust sillimanite prisms are only present at the immediate contact and in rafts within the Main Granite, but fibrolite is far more widespread, especially within and just outside the sheeted zone. Where sillimanite prisms and mats of fibrolite show a preferred orientation, this invariably lies in the main (S$_{2-3}$) schistosity, and may, on occasion, be parallel to the F$_{2-3}$ crenulations (Fig. 12-9c). There are many examples of fibrolite mats being folded by FMG$_2$ crenulations, but individual needles are seldom bent, and sometimes fibrolite lies along the new SMG$_2$ cleavage. Thus it seems that sillimanite was present prior to the DMG$_2$ movements, but, like the micas, healed itself of strain accumulated during this episode near the contact of the pluton. However, the occasional fibrolite fringes to andalusite crystals, and the common association with late muscovite which we comment on below, suggest that some sillimanite, at least, formed later than andalusite, and thus later than kyanite, staurolite, garnet, and plagioclase (*cf.* Tozer, 1955; Pulvertaft, 1961; Pitcher and Read, 1960b, plate III, Fig. 1 and 2; 1963, Fig. 19).

Garnet. Another problematic mineral is garnet, which occurs both as variably altered porphyroblasts and as relatively fresh smaller grains. In most places, the altered porphyroblasts exhibit the same textural relationships to the microstructures as the kyanite, staurolite, and oligoclase (*e.g.*, Pitcher and Read, 1960*b*, Plate II, Fig. 3), but at Fintown, in several localities in the Glencolumbkille Pelites, there are some which are enclosed in andalusite and plagioclase porphyroblasts which themselves overprint crenulations bent around the garnets. In these cases, there is thus no doubt that these particular garnets predate the other porphyroblasts, and it seems reasonable to view them as regional (MP_2) relicts chloritized during RMP_4. Elsewhere, however, smaller grains of garnet seem likely to be coeval with the kyanite, staurolite, and plagioclase (see below, p. 279).

Among the earlier porphyroblasts there is thus no unequivocal evidence of replacement, one by the other (or even absorbtion), and each mineral appears to have grown independently from the groundmass, as Pitcher and Read concluded. Some garnet may be earlier, but we can see little evidence from the textures that kyanite, staurolite, and oligoclase porphyroblasts grew at different times; all overgrew early crenulations and ceased their growth before or during the early stages of the period of flattening, which was accompanied by a general coarsening of the matrix. The growth of andalusite and sillimanite appears to have outlasted that of the other minerals, but all predate the DMG_2 movements during which fibrolite was the only mineral of this group to have undergone appreciable readjustment.

(D) The Late Growth of Phyllosilicates

In addition to the groundmass chlorite, biotite, and muscovite—which define the major schistosity S_{2-3} and which have occasionally lined themselves up with the SMG_2 cleavage in areas near the contact—there are, in many places, porphyroblasts of these minerals which overprint the main schistosity and which have complex relationships to the DMG_2 microstructures (Fig. 12-8c, d; 12-9a, b, c). Of these, muscovite occurs chiefly near the contact, chlorite in the outer aureole, and biotite in the inner aureole. Biotite is, however, commonly replaced to varying degrees by chlorite, and even many of the apparently fresh chlorites contain pleochroic halos and patches of higher birefringence. The problem here is whether all the chlorite is retrograde from biotite, this retrogression being more effective in the outer than inner aureole, or whether chlorite is being progressively "upgraded" to biotite near the contact.

Phyllosilicate porphyroblasts clearly overprint the main schistosity which is deflected around them only where strong FMG_2 crenulations are present (Fig. 12-8d, 12-9b); they appear, thus, to be later than the porphyroblasts discussed above. Throughout the outer, and parts of the inner, aureoles, groundmass micas and chlorites, together with those phyllosilicate porphyroblasts which lie close to the main schistosity, are bent and kinked by the FMG_2 crenulations (Fig. 12-6a), but there are also many biotite and chlorite porphyroblasts here which appear at first sight to overprint these crenulations (Fig. 12-8c; 12-9a, b; and Pitcher and Read, 1960*b*, p. 14; Rickard, 1964, p. 683). These contain *Si* trails of elongate ores outlining minute folds which are invariably more open than the external FMG_2 crenulations, and close examination shows that the porphyroblasts are often severely deformed with kinked and crushed cleavages. It might seem then that they, like the earlier porphyroblasts, grew over early open microfolds which were subsequently tightened around them (*cf.* Harte and Johnson, *op. cit.*, Plate I). However, further studies show that the (001) cleavages of the phyllosilicates are almost always parallel or subparallel to the axial planes of the *Si* microfolds (Fig. 12-8c, d; 12-9b), and we favor the explanation that these porphyroblasts wholly predate the DMG_2 movements which deformed certain crystals by slip along (001) to form *genuine shear folds* in their previously straight inclusion trails. The presence of many porphyroblasts with straight trails, even where FMG_2 crenulations are marked, and the severe deformation of those crystals which are elongated subparallel to the main schistosity (Fig. 12-6a), support this explanation. Indeed, in certain areas, as for example in the Glencolumbkille Pelites at Fintown, small phyllosilicate porphyroblasts grow so that (001)

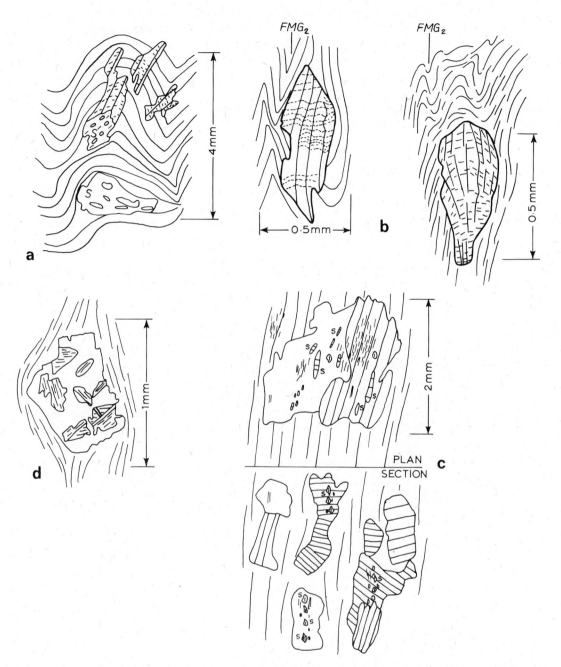

FIGURE 12-9. Sketches of textures involving late phyllosilicates and other minerals. (a) Chlorite (after biotite) apparently overprinting FMG_2 microfolds which postdate staurolite (s). Upper Falcarragh Pelites at Binnadoo. (b) Chlorites with crenulated Si trails. Upper Falcarragh Pelites on Binnadoo. (c) Random muscovite porphyroblasts overprinting main fabric which contains a linear (sub-horizontal) alignment of sillimanite prisms: fibrolite, short dashes; robust sillimanite, s. Pelitic inclusion in strip of Thorr Pluton adjacent to Main Granite at Lackagh Bridge. (d) Garnet with random inclusions of biotite and muscovite. Upper Falcarragh Pelites at Barnesbeg.

makes a high angle—generally 90°—with the main schistosity, even where FMG_2 folds are absent or very open (see also McCallien, 1935, Fig. 2). We suspect that either the preexisting schistosity exercised some kind of control on the orientation of the phyllosilicates, as in the case of certain garnets (Powell, 1965) and plagioclase porphyroblasts (Kennedy, 1969) elsewhere (see also Turner and Weiss, 1963, p. 442), or that they grew under the influence of an early or pre-DMG_2 stress system with the same orientation as that which caused the later DMG_2 movements. The occasional elongation of the phyllosilicates and chlorite knots after garnet along the FMG_2 cleavage supports the latter alternative. In any case, it seems clear that the chlorite, biotite, and muscovite porphyroblasts in the outer, and much of the inner, aureoles postdate the flattening of the schistosity around the earlier porphyroblasts, and that they predate at least the bulk of the DMG_2 episode.

In summary then, the phyllosilicate porphyroblasts show clear evidence of having overgrown the main (S_{2-3}) schistosity, and are thus younger than garnet, kyanite, staurolite, andalusite, and oligoclase. Since they also overprint folded fibrolite mats, they are generally later than this mineral as well. In most places, FMG_2 crenulations deform the micas and chlorite, though they may in some cases have still been growing after the onset of the DMG_2 movements, but in parts of the inner aureole, especially within the sheeted margin, the phyllosilicate porphyroblasts, like most of their Se counterparts, appear to have undergone considerable readjustment during the DMG_2 episode. The SMG_3 cross-cleavage is, however, clearly later.

(E) Other Late Minerals

In addition to the major minerals just described, there are also several other mineral phases worthy of comment. One occurrence of *chloritoid* and one of *cordierite* have been found in the Fintown area, and although their exact timing is not known, the former, like the other phyllosilicates, apparently overprints the main schistosity, and the latter also seems to be relatively late (Pitcher and Read, 1960b, p. 24; Naggar and Atherton, 1970).

In many pelitic lithologies in the inner aureole, there are small euhedral to anhedral *garnets* which may occur as inclusions in plagioclase, andalusite, or staurolite, or as separate groundmass grains (Fig. 12-7a; 12-8a; 12-9d). In distinction to the generally dirty and chloritized garnet porphyroblasts described above, these are usually clear, though on occasion the groundmass grains may exhibit dusty patches, especially in their core regions where straight or curved inclusion trails may also be seen. These patches tend to disappear near the contact, and it seems that we may be dealing with a regeneration of the earlier garnets near the margin, though in this case the garnets enclosed in the earlier porphyroblasts are either of an earlier generation as Naggar and Atherton (*loc. cit.*) claim, or they have grown inside existing crystals. In places in the inner aureole especially, these small garnets enclose euhedral plates of biotite, chlorite, and muscovite which, in contrast to their external counterparts, show no preferred alignment (Fig. 12-9d); indeed, at Lough Finn, such garnets form complete or partially breached "atolls" around the micas. The origin of these atoll forms is not well understood (*cf.* Atherton and Edmunds, 1966; Edmunds and Atherton, 1971), but the included phyllosilicates seem to have had a growth history different from their Se cousins (*cf.* Hess, 1971, and see also Chapter 14).

(F) The Late Retrogression

Finally, we comment on the general retrogression prevalent in most parts of the envelope but more pronounced in the outer than inner aureoles. We have already pointed out the frequent chloritization of biotite, and andalusite is commonly partly or even completely replaced by fibrous white mica (the shimmer aggregate of Pitcher and Read, 1960b, esp. Plate III). Kyanite and staurolite may also be partly altered to muscovite, especially along cracks; plagioclase may be variably sericitized, and garnet is generally more or less altered to chlorite, though this applies less to the small garnets just mentioned. Some of this general retrogression is clearly associated with the late DMG_3 movements, as can be seen by the concentration of chlorite along the cross-cleavages (see also Pulvertaft, 1961, p. 271; McCall, 1954, p. 172; Pitcher and Read, 1960b, p. 26), or even later, as suggested by the occasional presence of undeformed quartz and chlorite veinlets. However, there is also evidence that some of the retrogression away from the Pluton was associated with the DMG_2 episode, as shown for example by the occasional alignment of chlorite knots after garnet along the axial planes of FMG_2 crenulations. The only signs of the earlier regional retrogression (RMP_3 or RMP_4, see Chapter 3) in the aureole is, as we saw earlier, the inclusion of chlorite and biotite knots after garnet in andalusite porphyroblasts at Fintown.

(G) A Discussion of the Metamorphic Chronology

The sequence of mineral growth in the pelites of the Main Donegal Pluton envelope is shown schematically in Table 12-3. The general increase in metamorphic grade toward the Pluton, as illustrated in Table 12-1, involves each of these episodes.

There is no difficulty in attributing the late retrogression to the DMG_3 and perhaps to the latter part of the DMG_2 events (Table 12-4), but prior to this, there were two phases of metamorphic reconstruction, separate in time. The later event, during which the phyllosilicate porphyroblasts grew, took place mainly prior to the formation of the DMG_2 structures, though there may have been an overlap in places. These porphyroblasts were separated in time from the earlier, generally static,

TABLE 12-4. Correlation of Metamorphic and Tectonic Events in the Envelope Pelites of the Main Donegal Pluton

Tectonic Episodes	Metamorphic Episodes (recorded in pelitic lithologies)
DMG_3 movements: production of shear belts and SMG_3 cross-cleavage	*Late retrogression:* chloritization of biotite, shimmerization of aluminosilicates, etc.
DMG_2 movements: production of "new" folds and cleavages which become the dominant structures only in a few localities in the inner envelope	Late fibrolite [?], phyllosilicates
	Static growth of phyllosilicate porphyroblasts, late garnet [?]
DMG_1 movements*: coaxial and coplanar (with earlier structures) flattening of regional schistosity around early porphyroblasts	General recrystallization and *matrix coarsening*
	Growth (dominantly static) of *early porphyroblasts* of kyanite, staurolite, garnet, plagioclase, sillimanite [?]; andalusite slightly later.
D_1 to D_4 ———Regional pre-Granite events——— MS_1 to RMP_4	

* These are the *late* DMG_1 movements of Table 12-5.

porphyroblasts of kyanite, staurolite, oligoclase, garnet, sillimanite, and andalusite by a phase of flattening of the schistosity around these early minerals (to account for the discordance between Si and Se fabrics). This was accompanied by a general recrystallization of the matrix (to account for the contrast in size between the Si minerals and their Se counterparts). If the earlier minerals are to be attached to the Main Granite as aureole growths, as the spatial connection suggests, then their growth and the main flattening of the matrix around them must lie in the interval between the regional D_4 and the later, local, deformative events, DMG_2: we have here defined this flattening and matrix coarsening as due to the *latter part* of the DMG_1 event (Table 12-4), of which the growth of the early porphyroblasts represents an earlier, generally static part. The DMG_1 event was in turn followed by the growth, largely static, of the phyllosilicate porphyroblasts, and the whole assemblage was deformed by DMG_2 and DMG_3 movements. Despite this complex history, we emphasize that *during all these events, with the possible exception of the last, the Main Granite appears to have been a source of heat,* as attested especially by the change in metamorphic style of the DMG_2 structures toward the Pluton (*cf.* Heitanen, 1968; Kays, 1970).

If this sequence of events is correct, it again means that what appears in the field to be a simple schistosity and a mineral lineation is in fact a composite D_2–D_3–D_4–DMG_1 structure further intensified by DMG_2 and DMG_3 movements. It also implies that, apart perhaps from the early garnet porphyroblasts at Fintown (see p. 277), either the MP_2 porphyroblasts never grew in this area, or more likely, that all trace of this regional metamorphism has been obliterated in the Main Granite envelope. There is, however, another interpretation of the structural and metamorphic chronology which reduces the need to appeal to such a repeated flattening of the marginal schistosity, but which carries the rather unhappy implication that the earlier porphyroblastesis in the envelope took place *prior* to the emplacement of the Donegal Granite Complex, and that the real aureole of the Main Donegal Pluton is only represented by the phyllosilicates. This possibility we return to later, after detailing the effects of this complex structural history on rocks other than pelites in the aureole.

4. The Metamorphism and Deformation of the Preexisting Igneous Rocks

(A) The Metabasites

We deal first with the structural changes brought about in the regional metadolerite sills (see Chapter 2) as they approach the Main Granite. As these enter the area affected by DMG_2 structures (see Fig. 12-1), they begin to lose their normal massive appearance and take on a foliation which is strongest at and parallel to their margins, this change being best shown in the Creeslough Formation farthest from the contact. Nearer the edge of the Pluton, the metadolerites exhibit a strong alignment of hornblende (an L to L > S fabric; see Pitcher and Read, 1960*b*, Fig. 5), which is generally parallel to the mineral alignment in the host schists. Despite many variations in texture, there is a clear progression from massive, generally unfoliated sills at distances over 2–3 kilometers from the contact to strongly foliated and lineated sills within the sheeted marginal zone. This change is accompanied by increased amounts of necking and boudinage, especially in the metadolerites of the Creeslough Formation (Fig. 12-5d, Map 1 in folder), and by the folding or disruption of internal quartzofeldspathic veins. In a few localities near the Pluton, the metadolerite foliation is bent into small-scale, relatively open folds (lacking an axial-plane fabric) which are sometimes associated with necking and sometimes with late kink bands or the cross-cleavage.

(B) The Lamprophyres and Felsites

Those regional lamprophyre and felsite dykes which occur near the Main Donegal Pluton are likewise deformed. Since they usually lie parallel to S_{2-3} (Fig. 12-5b), they exhibit a strong internal foliation with a linear component of variable intensity, and necking and boudinage are common.

Cross-cutting quartzofeldspathic veins are generally folded by what appear to be D_{2-3} structures (*cf.* Fig. 11-8b), and apart from certain of the porphyritic felsite dykes of the Meenalargan Hill area (Iyengar *et al.,* 1954, p. 227), all of these dykes and veins show the effects of the DMG_2 and DMG_3 movements (Fig. 12-5b).

(C) The Earlier Members of the Donegal Granite Complex

Structures developed in the *Meenalargan Appinitic Complex* and in the Stalk of the *Ardara Pluton* to the southwest of the Main Granite are also instructive. Although an ENE–WSW trend is common throughout the whole of the Meenalargan Complex, near the Main Granite it exhibits an unusually strong mineral lineation which occasionally becomes a kind of mullioning (Iyengar *et al.,* 1954), plunging gently to the southwest and decreasing in intensity away from the Pluton. The neck of the Ardara Pluton likewise acquires a mineral alignment which dies out progressively from a strong southwest-plunging L > S fabric near the Main Granite to a weak L fabric at the Maas–Glenties road, beyond which the normal concentric foliation of the Ardara Pluton predominates (Akaad, 1956*a*; and Chapter 8). In addition, the frequent mafic aggregates in the Ardara monzodiorite become more flattened and drawn out, so as to lie parallel to this mineral alignment.* Both the Meenalargan Complex and the Ardara Pluton are cut by numerous granitic dykes and sills, many of which appear to be genetically related to their host rocks (Chapter 7, 8), while others are undoubtedly Main Granite sheets. Those dykes which occur closer to the Maas–Glenties road show no signs of internal or external deformation, but those near the Main Granite are folded, necked down or disrupted (Akaad, *loc. cit.,* Fig. 6), depending on their attitude, and are invariably crossed by the host mineral alignment.

But perhaps the best area to demonstrate the increase in structural complexity toward the Main Granite is the Brockagh–Thorr district where the *Thorr Pluton* forms the immediate country rock (Pitcher, 1953*a*; see also Iyengar *et al., loc. cit.,* p. 226). The normal Thorr tonalite, it will be recalled, is a weakly foliated rock enclosing many rafts of country rock which preserve D_{2-3} structures, and which, together with their host, are cut by sporadic undeformed pegmatic and aplite dykes, many of which are probably associated with the Rosses Complex (Chapter 5). At Lough Ardmeen, for example, well over a kilometer from the contact at Brockagh, these rafts are undeformed, except for local mobilization near their borders, but toward Brockagh, at about 650 meters from the contact, the tonalite and its dykes begin to take on a moderate foliation, trending NE–SW.

Within 300 meters of the contact, the penetrative mineral alignment, now clearly an L < S fabric with L plunging gently (<30°) to the southwest, becomes quite pronounced, and dykes cutting the tonalite are folded or necked down. The bedding in the pelitic and semipelitic rafts is frequently disrupted and mobilized and, together with the S_{2-3} schistosity, is often folded into upright folds which also affect any dykes present (Fig. 12-5e). These folds trend NE–SW, but generally lack an axial-plane fabric except immediately adjacent to the contact. They have a general similarity to FMG_2 folds elsewhere in the envelope, and the SMG_3 cross-cleavage is also strongly developed in this narrow contact zone.

In summary then, the country rocks of igneous origin demonstrate even better than the metasediments the progressive increase in intensity of deformation toward the Main Donegal Pluton. The effects of this deformation are noticeable up to $1\frac{1}{2}$ to 2 kilometers from the margin in the neck of the Ardara Pluton and in the Meenalargan Complex, and perhaps for $1\frac{1}{2}$–2 kilometers from the northwest margin in the metadolerites, but in the southeastern envelope and in the Thorr–Brockagh district, they generally extend no farther than 700 meters–1 kilometer from the contact. The deforma-

---

* It is of interest to note here that Hall (1966*a*) has shown that the normally highly ordered microcline of the Ardara Pluton loses its high triclinicity in the stalk of the pluton, a situation which he attributed to *reheating* of these feldspars by the Main Granite.

tion "aureole" so defined extends generally well beyond the outer part of the sheeted zone and well beyond the zone where minerals begin to grow along the SMG_2 cleavage in the schists.

(D) A Further Discussion of the Metamorphic and Structural Chronology

We have seen in the foregoing that in the envelope rocks of igneous origin adjoining the Main Granite there are penetrative mineral alignments which originated prior to the DMG_2 event, though of course they must have been intensified to some extent by these movements. Such structures in the metadolerites are so poorly developed away from the margin of the Pluton and so progressively intensified toward it that it is logical to connect them with the emplacement of the Pluton. Of course, we could argue that this intensification in the deformation of the metadolerites is the result of pre-granite deformation being localized in areas now occupied by the Pluton (see p. 64). However, it would be difficult to deny a genetic connection between the pluton and similar structures produced in the adjoining members of the Donegal Complex.

The degree of regrowth and alignment of the hornblende in the aureole metadolerites is also completely progressive. It would seem the obvious course to link the growth of hornblende in the metadolerites with that of staurolite, kyanite, and the other early porphyroblasts in the pelites, except that the latter appear to have grown statically and the former syntectonically—unless, of course, the hornblendes aligned themselves along previously established strain directions (e.g., L_{2-3}) by mimetic growth, a situation which the senior author considers unlikely. If this distinction between static and syntectonic minerals is valid, there seem to be two possible solutions. First, it might be argued that the production of the mineral alignment in the metadolerites was later than the growth of the early porphyroblasts and a part of the same event which resulted in the syntectonic matrix coarsening. This might be reasonable except that in the pelites the general recrystallization (matrix coarsening) following the growth of the last of the porphyroblasts, andalusite, is far less intense than, and therefore unlikely to be, the temporal equivalent of the penetrative recrystallization in the metadolerites. The senior author thus prefers the view that the preferred orientation of hornblende in the metadolerites (and presumably the strengthening of the S_{2-3} cleavage in the pelites, though the effects of this have been masked by subsequent events) was followed by the growth of the early porphyroblasts in the pelites, and in turn, by flattening of the schistosity around these porphyroblasts and by the accompanying matrix coarsening. These three events we group into the DMG_1 episode, distinguishing early from late movements by the intervention of a phase of mineral growth, though of course the time involved in this static event may well be minimal compared to that involved in the two movement phases (see Table 12-5).

5. The Special Problem of the Mullions

Where the Ards Quartzite outcrops near the Main Granite in the Crocknawama–Lackagh Bridge–Cock's Heath Hill area in the northeast, and in the region around Crockator and Cor Hill in the west and southwest, it exhibits a magnificent mullion lineation (Fig. 12-10); indeed, it was in the regional memoir describing the Crockator area that the term *mullion* was first used in the geological literature (Nolan in Hull 1891, p. 53).

The precise origin of mullion structures is still the subject of some debate (see, for example, Boschma, 1963; Cloos *et al.,* 1964; Whitten, 1966; Ramsay, 1967), and, apart from stating at the outset that the mullions are always parallel to the axes of folds in the quartzite where these are present, and that they may be bounded partly by bedding surfaces, mineral foliations, or curved parting surfaces (McCall, 1954, p. 166; Pitcher and Read, 1959, p. 281; Rickard, 1962, p. 232), we shall restrict ourselves here to a discussion of their chronological significance.

In his account of the complex structure at Crockator, Rickard (1963) interpreted scattered north

FIGURE 12-10. Photo of mullion structure in Ards Quartzite adjacent to the Main Donegal Pluton, near Lackagh Bridge.

to northeast plunging minor chevron folds (his L_4) in the core of the major bend, together with gently southwest-plunging mullions (his L_5) on the east limb of the structure, as results of the forceful emplacement of the Main Granite. His descriptions suggest a correlation between these folds and those of the DMG_2 episode described above, and we would readily attribute the mullions to the same movements, were it not for certain relationships in the outcrop of Ards Quartzite around Lackagh Bridge.

In this area, the Quartzite is characterized by upright minor folds with axes horizontal or gently northeast-plunging and with a strong penetrative axial-plane mineral alignment (Fig. 12-11c, d). The mullions are invariably parallel to these fold axes. Although no refolded folds or fabrics were seen here, it seems reasonable to link these structures with the dominant folds in the adjacent Creeslough Formation, that is, as D_{2-3} structures. This conclusion is supported by the structural relationships of the many granite sheets which cut the Quartzite (see Fig. 12-11).

In a few places, granitic dykes cut across limbs of folds in the Quartzite and are either folded with more open profiles (Fig. 12-11d) or are planar and possess a strong mineral fabric, (L > S, Fig. 12-11c), indicating that they postdate early folds which have subsequently been flattened together with the dykes. It is significant that the axes to folded sheets are almost invariably parallel to the fold axes and mullions in the Quartzite. Such a parallelism can, of course, arise during a stretching deformation ($k = \infty$) in which the degree of strain is extremely large (*cf.* Watterson, 1968*b*), but it is clear from the relatively open profiles of the folded sheets, together with the strong planar component ($k < 1$) of their internal mineral alignment, that they have not been deformed to anything like this

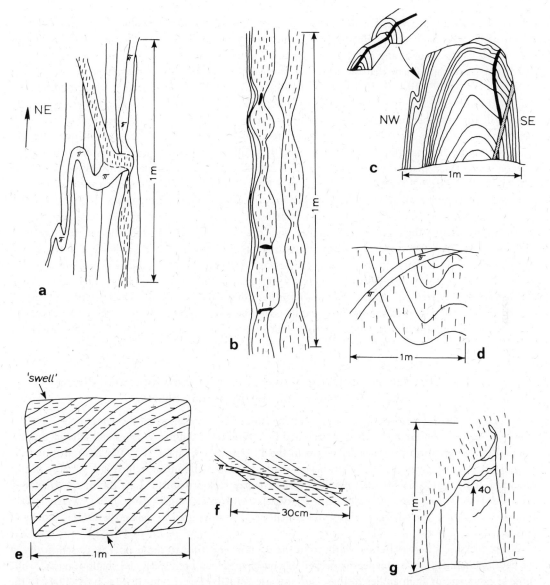

FIGURE 12-11. Structures in Ards Quartzite at Lackagh Bridge. Unless otherwise indicated solid lines denote bedding in quartzite. (a) Deformation of granitic sheets: microgranite, dashes; pegmatite, π. (b) Pinch-and-swell in microgranite sheets (dashed); quartz-filled tension gashes, black. (c) Strongly foliated microgranite dyke (black on block diagram, stippled on detailed section) cutting at low angles across both limbs of (?) F_{2-3} fold in quartzite. (d) Pegmatite dyke (π) cutting across fold in quartzite. (e) Mullions (solid lines) bent by a "swell" in the quartzite caused by necking of adjacent granite sheet which is hidden. Mullions are crossed by the quartz elongation lineation (dashed). Vertical surface looking to the southeast. (f) Mullions (solid lines) cut by quartzofeldspathic vein (π) in which internal mineral lineation parallels the quartz elongation lineation of the host quartzite. (g) Xenolith of mullioned quartzite (mullions plunging 40° to the northeast) in marginal granite sheet with horizontal mineral lineation. Plan view.

extent. The only alternative seems to be that the sheets were not initially random, and we suggest, in fact, that their intrusion was guided by preexisting passageways which were themselves controlled by the mullion structure, or at any rate, by the fold axes in the host to which the mullions are parallel.

To support this argument that the intrusion of the sheets postdates the formation of the mullion structures, we point to the situation in many localities in the Lackagh Bridge area where mullions are deformed around necked-down zones in the granite sheets (Fig. 12-11e), or where there is a slight angular discordance between the mullions and the mineral lineations in the sheets (Fig. 12-11f). Furthermore, on many outcrops here, the Quartzite exhibits a faint linear alignment of quartz grains which invariably lies parallel to the lineation in the adjoining granite sheets and which crosses the mullion structures in places (Fig. 12-11c, f). In addition, there are occasional inclusions of mullioned quartzite in the marginal sheets (Fig. 12-11g, and McCall, 1954, p. 166), in which the mullions are bent and markedly oblique to the linear component of the host mineral alignment.

It seems reasonable to view all the quartzite mullions throughout the envelope as one synchronous structure, and despite their general parallelism with the margin of the Pluton, the quartz elongation lineation in the Ards Quartzite, and with the linear component of the mineral alignment in the sheets, we place great emphasis on those localities where these structures are discordant. Thus, we conclude with McCall (1954, p. 166) that the mullions were present prior to the emplacement of the granite sheets, though they were, of course, *intensified* further during the DMG_2 movements which deformed the sheets. The formation of the mullions we therefore view as synchronous with the development of the hornblende alignment in the metadolerites, that is, as a part of the early DMG_1 event (see Table 12-5).

6. The Correlation of Deformative Events in Envelope and Pluton: the Deformation of the Marginal Sheets

A discussion of the relationship between the structural (and metamorphic) events in the envelope and those in the Main Granite is best approached by reference to the many dykes and sills of granitic composition which form the sheeted margin to the Pluton (see Chapter 11). We have already seen that all but the coarsest pegmatite sheets have a marked mineral alignment, which is a variably healed cataclastic fabric ($L \geqslant S$) oriented parallel to the country rock structures. Like the marginal granite proper, the sheets are also cut by the strongly cataclastic cross-cleavage (see p. 229), which is continuous with that in the adjacent schists and which we have now used to define the DMG_3 structural episode.

Where the sheets lie parallel or nearly so to the contact, and thus to the main (S_{2-3}) schistosity in the envelope schists, they are generally necked or boudinaged (on both steep and shallow axes), while those at moderate or high angles to these surfaces are folded (Fig. 12-4c, 12-11a, b, d, 12-12); the situation is thus the same as that involving dykes lying within xenoliths inside the pluton (see p. 236). Fold axes invariably lie within the S_{2-3} or SMG_2 surfaces but their precise attitude, of course, depends on the initial orientation of the sheet. In many places it is clear that it is the FMG_2 folds in the schists which deform the sheets, with the mineral alignment in the sheets forming an axial-plane fabric continuous with and, allowing for "refraction," parallel to the SMG_2 cleavages in the host schists (Fig. 12-4d, 12-12b, d). Those situations in which the folds in the sheets are not matched by folds in the host (Fig. 12-12a, f), are easily explained by differential responses to the flattening.

A careful search was made for evidence regarding the timing of the sheets relative to the DMG_1 movements, but the only sheets which showed signs of polyphase deformation were a group of granite sills standing well away from the main contact at Crockastoller, northeast of Fintown. These exhibit a strong cataclastic foliation which is parallel to their walls and which is folded about what otherwise seem to be normal FMG_2 folds, without however, the formation of a new axial-plane fabric (Fig. 12-12e). Despite this, the host schists show no trace of this cataclastic fabric, which must therefore

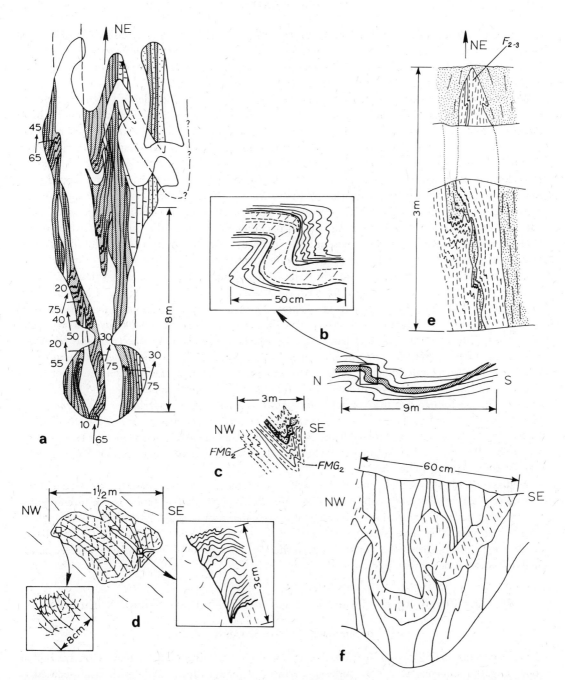

FIGURES 12-12. Structures associated with granite sheets. (a) Association of folding with pinch-and-swell structure in granite sheet (blank) cutting a pelite (close stipple)-quartzite (ribbed)- semipelite (open stipple) raft in marginal granite, Crocknawama. Arrows show trend and plunge of FMG_2 crenulations. (b) Composite aplite-pegmatite sheet in Crockmore septum, folded by FMG_2 folds. Note axial-plane mineral alignment in pegmatite portion. (c) Pegmatite vein (stippled) post-dated by what appear to be two sets of DMG_2 structures. Crockmore Septum. (d) Xenolith of Upper Falcarragh Pelites in large granite sheet at entrance to Barnesbeg Gap. FMG_2 folds die out as they approach the enclosing granite. (e) Granite sheets (stippled) with strong internal foliation (dashed lines) folded by (?) FMG_2 folds. Crockastoller near Kingarrow. (f) Folded pegmatite dyke in Ards Quartzite at Lackagh Bridge. Note fanning of axial-plane foliation.

have been preferentially accepted by the granite sheets *after* the general recrystallization. Thus, the fabric cannot be ascribed to DMG_1 movements, but is simply local evidence for two stages within the DMG_2 event itself; in this connection, we may recall that the pelites of the Fintown area are characterized by several sets of DMG_2 folds and cleavages (p. 266).

We are now in a position to enquire whether the DMG_2 structures which deform both the envelope rocks and the marginal sheets are responsible for the whole of the *in situ* deformation of the Main Donegal Pluton, or whether they account only for the later phases such as the marginal cataclasis. We start by again noting the equivalence of the cross-cleavage in the marginal Granite with the SMG_3 cleavage in the schists (Table 12-5), and also by recalling our earlier conclusion (p. 230) that this cross-cleavage postdated the marginal cataclasis in the Pluton, a process which itself overlapped upon and outlasted the main deformation phase in the internal portions of the Granite (see Fig. 11-23). It seems clear therefore that the DMG_2 structures in the envelope must have been responsible for at least part of this cataclasis, and attesting to this conclusion is the generally cataclastic nature of the (SMG_2) mineral alignment in the sheets, and the similarity between the structural styles of the dykes cutting the marginal Granite and those in the adjacent country rocks.

The real problem here is the extent to which the DMG_2 episode can be correlated with earlier events in the Pluton, and some evidence bearing on this question comes from the structures in pelitic rafts within the Main Granite. As we showed in Chapter 11, these rafts are for the most part characterized by one dominant mineral fabric, a schistosity which is axial plane to occasional folds, and which has clearly been flattened during the later part of the Granite deformation. We suspect that this schistosity predates the DMG_2 event, since in a number of xenoliths, especially in the Binaniller Hill–Crockmore area, there are small-scale chevron folds and crenulations on this surface which are identical in geometric and metamorphic style to the FMG_2 folds in the inner envelope (Fig. 12-12d). It is of great importance to note that in some rafts *these DMG_2 structures are associated with the folding of post-T_1, pre-T_2 dykes* (see Fig. 11-23) *which cross the enclaves in the same way as the FMG_2 folds in the envelope fold the marginal sheets*. Thus, we conclude that the DMG_2 episode in the envelope was responsible not only for the marginal cataclasis, but also for much of the later (post-T_2) stages of the Granite deformation (Table 12-5).

We have already shown that prior to the DMG_2 event, there were two phases of movement in the aureole (the earlier associated with the mullions and with the penetrative fabric in the metadolerites, and the later associated with the matrix flattening) separated by a static growth of porphyroblasts. These events, which we have collectively defined as the DMG_1 episode, must then be the correlatives of the pre-T_2 events within the Pluton. This conclusion we shall amplify as we now proceed to the final discussion of our chronology.

7. The Origin of the Main Donegal Pluton and its Complex Aureole

(A) A Summary of the Sequence of Events

We begin this final discussion by summarizing the geological history of the Main Donegal Pluton, combining the sequence of events outlined in the envelope (Table 12-4) with that deduced in Chapter 11 for the Main Granite itself (Table 12-5).

It is important at the outset to reemphasize the late date of the emplacement of the Main Donegal Pluton. Not only is it later than the latest regional deformative event, D_4, and the succeeding retrogression, RMP_4, but it most clearly postdates the Thorr and Ardara Plutons and representatives of the appinite suite.

From the evidence of wedging apart of the country rock stratigraphy (p. 256), the intrusion of the Main Granite is thought to have been accomplished by a forceful injection of a plexus of coalescing granite sheets. If our reading of the evidence is correct, this must have been accompanied

TABLE 12-5. History of Development of the Main Donegal Pluton

	Events in the Pluton (see Fig. 11-23)	Events in the Envelope (see Table 12-4)	
↑ Time — Marginal Cataclasis / Main Phase of Deformation / Deformation of Enclaves / Mineral Alignment / Marginal Sheets	Cross-cleavage (mainly in marginal zones)	DMG_3 shear zones and cleavages	
		Late retrogression	
	T_4—Late pegmatites and quartz veins	DMG_2 movements	
	T_3—Intermediate pegmatites and later microgranites	Phyllosilicate porphyroblasts	
	T_2—Remobilization of early microgranites		
	T_1—Deformation of regular-banding and emplacement of early pegmatites	Late DMG_1 movements: flattening of the schistosity and matrix coarsening	The DMG_1 Events
	Formation of regular-banding and widespread potash feldspathization	Early porphyroblasts (static)	
	Emplacement of Main Donegal Pluton	Early DMG_1 movements: penetrative mineral alignments in metabasites and mullions formed in quartzites	

by the first stage (early DMG_1) of the coaxial and coplanar flattening of the regional structures in the aureole pelites, and by the production of penetrative mineral fabrics in the marginal metadolerites and earlier plutons and of the mullions in the quartzite. During, or directly following, the emplacement of the Pluton there was a period of generally static nucleation and growth of porphyroblasts of kyanite, staurolite, garnet, plagioclase, and possibly sillimanite. These minerals may not have been entirely coeval, but there is no clear textural evidence for separating them in time; the growth of andalusite, however, appears to have outlasted that of the other minerals. Although we are not sure, it seems reasonable to view the formation of the regular-banding within the Pluton and the accompanying potash feldspathization (the T_1 event) as taking place at broadly the same time as this early porphyroblastesis.

In the envelope, the regional schistosity which was tightened and intensified during the granite emplacement and then overprinted by the early porphyroblasts, was again flattened, this time around the porphyroblasts, and this late DMG_1 event was accompanied by another general recrystallization (the matrix coarsening). Inside the Pluton at about this late DMG_1 time, the regular-banding was deformed, early microgranite dykes were emplaced and "remobilized," and a penetrative mineral alignment was developed. Subsequently, a series of microgranite and pegmatite dykes were injected into the cooling granite and its immediately adjacent country rocks (to form the majority of the marginal dykes and sills), and this event (T_2 to T_3) may well be correlated with the largely static, post-DMG_1, growth of phyllosilicate porphyroblasts in the aureole. These phyllosilicates, together with the marginal dykes, were then deformed by DMG_2 movements, which generated new folds and cleavages in many parts of the envelope. Even inside the Pluton, the DMG_2 movements are recorded by the deformation of the microgranite and pegmatite dykes where they cross metasedimentary enclaves. In the marginal portions of the Main Granite, the latter part of these DMG_2 movements resulted in a general cataclasis, and in the interior, in the local deformation of late felsite dykes. The final event in this long history was the production of the late cross-cleavage in both the marginal granite and the envelope rocks during the completely retrogressive DMG_3 episode, which was also responsible for the formation of the zones of intense shearing in the country rocks.

This matching of chronologies is, of course, considerably oversimplified. In the envelope, for example, we have argued the case for essentially static episodes of crystallization separated by movement events, while inside the Pluton, the evidence seemed to indicate a more or less continuous history of deformation and simultaneous crystallization, decreasing in importance with the passing of time, that is, with the cooling of the Pluton. The true picture probably lies somewhere between these two extremes, but we think that the above synthesis provides a good working model, and it seems that the episodic nature of the metamorphic events in the envelope may be due in part to the periodic dissipation of accumulated stresses which were steadily built up within the Granite.

We have now complicated the much simpler picture presented by Pitcher and Read, but in doing so, have resolved some of the problems their work raised. In particular, their conclusion that "structures in a" (the linear mineral alignments and mullions) and "structures in b" (the fold axes) were "formed simultaneously by outward push and horizontal drag of the body of consolidating granite magma" (1960b, p. 13), we reinterpret in terms of the complex superposition of coaxial and coplanar deformation phases.

We can now consider several possible explanations for the complex situation just summarized, and in particular for the connection, if any, between regional stresses and those due to forceful intrusion.

(B) The Gneiss-Dome Hypothesis

At the outset, we must reject one hypothesis which has often been proposed for this and similar situations elsewhere. This is the concept of mantled gneiss domes, outlined first in detail by Eskola (1949) and subsequently elaborated in various forms by, among others, MacGregor (1951), Preston (1954), Billings (1956), Lauerma (1964), Cloos and others (1964), Nicholson (1965), Mallick (1967), De Waard and Walton (1967), and Thompson and others (1968). According to this view, the Main Donegal Pluton would represent a diapirically emplaced body of granite which brought up with it from deeper levels an envelope of high-grade rocks. While the intense internal deformation of the Main Granite accords well with this idea, it is hardly consonant with the facts that this pluton is the latest in a series of otherwise relatively high-level granites, that it is sharply discordant in the west and southwest, and particularly that it preserves an excellent ghost stratigraphy and structure which has certainly not been brought up from deeper crustal levels. Further, and in contrast to the normal rim synclines of gneiss domes, the envelope rocks all along the northwest margin of the Main Granite are bent downward instead of upward (by the D_4 Creeslough Downwarp—see Fig. 12-5f). We are, of course, not arguing that forceful emplacement was not involved, merely that the unique structural and metamorphic character of the Main Granite and its envelope is the result of tectonic and thermal events which took place at the presently exposed crustal level.

(C) Superimposed Deformation *versus* Forceful Emplacement

It may be argued that the Main Granite is a result of the same kind of forceful emplacement which produced the Ardara Pluton. As we saw in Chapter 8, the latter body also shows evidence of internal deformation and of polyphase development in its aureole, but consideration of the data we have presented in the foregoing shows that the similarity is more apparent than real. Thus, the central portions of Ardara are massive, and the structural trends in the deformed outer parts, though locally discordant, are on the whole controlled by the contact with the country rocks. In the Main Granite, on the other hand, the mineral alignment is ubiquitous and completely penetrative, except in respect to the later dykes, and is markedly discordant to the margin of the pluton in the northeast and southwest where it is continuous with the polyphase mineral fabrics in the envelope. In fact, the most fundamental question concerns the cause of the deforming stresses, whether they were in some way related to the forceful emplacement of the Main Donegal Pluton, or were entirely superimposed from outside, albeit while the Pluton was still hot enough to respond by ductile flow (*cf.* Berger and Pitcher, 1970). Was Kilroe right after all when he stated that the emplacement of the Main Granite "marks an important epoch in the geological history of Donegal, having occurred, it would seem, between earlier and later periods of metamorphism" (Hull, Kilroe and Mitchell, 1881, p. 30)?

In an attempt to resolve this problem, a search was made for possible regional representatives of the structures in the aureole, particularly those we have labelled DMG_2. In Glinsk and on the Clonmass Peninsula in the north, in Loughros to the southwest, and along the northern flanks of the Ardara Pluton (Fig. 12-1, 12-5c), there are scattered late, generally upright, NE–SW to E–W trending, minor flexural-slip folds with associated crenulation cleavages which are virtually identical in style and trend to the FMG_2 folds in the envelope of the Main Granite. Some of these, especially those on Glinsk,* postdate late chlorite and chloritized biotite porphyroblasts in the same way as do the DMG_2 structures around the Main Granite, and they appear, from several other lines of evidence, to be later than the granitic plutons. We suspect therefore that they are also the result of the DMG_2 movements.

Further, in the Kilmacrenan Succession, Cambray (1960a) has recognized late angular folds of varying trend in the Termon Formation east of Fintown (his F_5 folds), and Howarth and others (1966) have recognized what could be similar late crenulations (their F_5 structures) in the Glencolumbkille district; again, it is tempting to link these and certain other late "brittle" folds throughout the Kilmacrenan Succession with the DMG_2 event recognized in the Creeslough sequence. However, against this particular correlation is the fact that late carbonate porphyroblasts, which overprint or are syntectonic with these late structures, are pseudomorphed in the aureoles of several appinitic complexes by thermal hornblende (see Chapter 7). Since the appinites were emplaced well before the Main Donegal Pluton, this situation prevents us from extending the DMG_2 episode over much of the Kilmacrenan Succession.

We are, thus, uncertain of the status of the DMG_2 movements as a widespread deformation phase;

---

* Although Pitcher and Read (1963, p. 274) attributed these to the emplacement of the Fanad Pluton, recent work by Edmunds (1969T) has shown that these cleavages are superimposed on the latest phases of the Fanad aureole.

certainly they do not appear to have a direct correlative elsewhere in the Dalradian (see, however, below). The spatial extent of the DMG_3 movements is likewise difficult to determine, though it is certain that their effects are much more obvious near the Pluton.

We recall now our earlier conclusion that the DMG_2 movements deformed the marginal sheets and were also responsible for the later (post-T_2) part of the deformation of the Pluton. There is, thus, no doubt that the Main Granite and the vast majority of its minor intrusive phases were in place, though by no means completely cooled, by the time the DMG_2 structures were formed, and since their geometry and attitude are hardly consonant with upward diapiric movements from below the present level of exposure, we are forced to the conclusion that the DMG_2 movements are superimposed on the Pluton as a result of regional stresses, the effects of the later being localized for some reason. No doubt a regional stress field, of a normally insufficient intensity to cause any appreciable deformation, would exert quite a profound effect in areas where the ductility was relatively high, due for example to the presence of a hot granitic body.

The same situation seems to hold for the latter part of the DMG_1 movements which caused the early flattening in the envelope and the T_1 to T_2 deformation within the Pluton (see Table 12-5), for we can see no way that the penetrative mineral alignment (which is of late DMG_1 age, see Table 12-5) could have formed throughout the Pluton and discordant to its margins as the result of movements directly related to its emplacement. We could accept diapiric action as a cause only if we could find, as at Ardara, evidence of a late body of granite whose outer margins controlled the trend of this mineral alignment (see Berger and Pitcher, 1970, for further discussion of this point). Thus, while there is little doubt that the late DMG_1 movements took place while the Pluton was still in the process of consolidation, we believe that they must also have been the result of stresses originating in the country rock. As for the early DMG_1 movements, it is these which we imagine developed as the result of a genuine diapiric action, though their effects have been nearly obliterated by the later events.

Thus, although we envisage an emplacement mechanism by sheeting, we are not sure how forceful this was. If it was a forceful wedging, as Pitcher and Read originally thought, then it is perhaps possible that the stresses superimposed by this invasion on the country rocks accumulated there until the influx of material diminished. Relaxation of these stresses might then cause the late DMG_1 deformation of the Pluton and the mobilization of deeper levels within the granite to form the microgranite and pegmatite dykes, which were then deformed by DMG_2 movements, possibly as a last phase in this reciprocal action (see also Chapter 15). If, on the other hand, the original sheeting was more or less passive, then we can think of no alternative to a genuinely regional source for the DMG_1 and DMG_2 events, though such stresses must have been insufficient to generate detectable structures elsewhere. Whatever the answer, the interplay of movement and crystallization, both in the envelope and within the Pluton, is clear: *deformation was synchronous with the granite event.*

We have looked for similar situations elsewhere, but can find no close parallels. The complex sequence of plutonic activity in the Appalachians of New England invites comparison, for there are here many partly or wholly deformed plutons (Page, 1968), and schistose rocks of regional aspect bearing evidence of polyphase development are found in intimate association with granitic rocks, but from the published accounts (see p. 293) we are not sufficiently informed as to the chronological sequence or the genetic control. There are many other areas where complex relationships between granitic intrusion, deformation, and "regional" metamorphism occur (*cf.* Read, 1957), but it seems that only in the Caledonides have field studies reached the level of sophistication to permit comparisons in this detail. Thus, there are examples in the Moine and Dalradian of Scotland of late, otherwise post-tectonic dykes which have been deformed by several sets of structures in a "private orogeny" (*e.g.,* Peach et al., 1912, p. 116–26; Johnson and Dalziel, 1966; Dearnley, 1967), although even here the regional status of these movements is not yet entirely clear (*cf.* Platten, 1968). We could, of course, follow the suggestion of Berger and Cambray (1969) that the stresses which deformed the Main Donegal Pluton (DMG_2 in particular) were "foreign" representatives of the end-Silurian movements active mainly in Wales and southeast and western Ireland (the Cymrian orogenic episode of Rast and Crimes, 1968), especially as it was these late events which could have given rise to the K-Ar ages of 372–394 million years obtained on the Donegal Granite Complex by Brown and others (1968). Both suggestions, however, require a very long cooling period for the Main Granite, which was emplaced at about 470 million years (see p. 90).

(D) The Regional "Hot Spot" Model: an Alternative for the Metamorphic Development of the Envelope

In attaching the growth of kyanite, staurolite, andalusite, oligoclase, and garnet porphyroblasts chronologically and spatially to the Main Donegal Pluton, we have assumed that any earlier porphyroblasts resulting from regional events (particularly MP_2) had either been obliterated by subsequent events in the envelope or had never been formed in this area; also, that the regional S_{2-3} schistosity was overprinted by these Main Granite porphyroblasts, only to be tightened around them in precisely the same manner as the S_2 cleavage was flattened about the regional MP_2 porphyroblasts by the D_3 movements. However, it is worth explaining that the junior author began with the much simpler hypothesis that the porphyroblasts in the envelope were, in fact, of the same age as those which formed during the regional MP_2 event (*i.e.*, that the late DMG_1 movements were the representatives of the D_3 regional phase), and that the Main Granite was thus neatly emplaced into a "hot spot" or "root zone" in the regional MP_2 isotherms.

With this "hot spot" model (in contradistinction to the one where deformation and recrystallization were essentially synchronous with the granite event), the presence of the penetrative mineral alignments and the mullions in the envelope present no special problem, for they would simply be another reflection of the increased intensity of the regional D_{2-3} movements and intervening MP_2 metamorphism in this area. The single schistosity surface in the envelope would thus be a tight S_{2-3} structure which had simply been intensified by the local coaxial movements, DMG_2. On the basis of this concept, the deformation of the Pluton itself must be ascribed to the DMG_2 movements, the obvious contrast between the crenulation cleavage (SMG_2) in the envelope and the penetrative mineral alignment in the granite being a result of a marked difference in ductility between the relatively cool country rocks and the relatively hot intrusive (*cf.* Fig. 11–24). According to this model, the Main Granite was emplaced by more or less *passive* sheet intrusion with an aureole represented by a local regrowth of the phyllosilicates and by the production of cordierite, late garnet, andalusite, and fibrolite (see Table 12-3). This model emphasizes the separation in time between these latter minerals and the earlier kyanite, staurolite, garnet, and plagioclase (*cf.* Naggar and Atherton, 1970). The absence of more profound contact effects would not be surprising in view of the common observation that plutonic bodies emplaced into existing high-grade rocks seldom cause more than minor thermal effects in their envelope (*e.g.*, Sabine, 1963; Hart, 1964; Soper, 1963; Mercy, 1965; White and others, 1967; Page, 1968).

The idea of a granitic body rising into a restricted zone of highly deformed and metamorphosed rocks is, of course, not a new one. Many workers have implied or directly invoked the idea of a vanguard of plutonic or metamorphic activity or both advancing above a rising magma (*e.g.*, Barrell, 1921; Read, 1957; Heitanen, 1961, 1968; Soper, 1963; Binns, 1966; Dewey, 1967). The New England Appalachians, in particular, provide many excellent examples of granitic plutons surrounded more or less concentrically by metamorphic zones of regional aspect which are often very narrow and which they may cut in places (*e.g.*, Billings, 1956; Murthy, 1957; Green, 1964; Hamilton and Myers, 1967; Thompson and Norton, 1968; Guidotti, 1970; Page, 1968; Kays, 1970: see also Jenness, 1963, and Williams, 1964, 1968; Gabrielse and Reesor, 1964, for comparable situations in Newfoundland and British Columbia). Though the time relationships between the formation of these zones and the emplacement of the granites is still a matter under discussion, it seems that the zones are not due to simple contemporaneous contact metamorphism. Indeed, a common opinion is that the granitic and migmatitic rocks in the higher-grade zones are not, as Barrow originally claimed (1893), the cause but the end product (through metasomatism or anatexis) of this metamorphic activity (Harry, 1958; Johnson, 1965; Thompson and Norton, *loc. cit.;* Turner, 1968, p. 378).

When looking for a reason why such a spatial association should exist between early metamorphic "hot spots" and later intrusions, it seems worth considering the possibility that such "hot spots" retain their heat and hence their relatively high ductility for long periods of time so as to provide more attractive sites for upward migrating magmas than adjacent cooler environments. In this regard, it is interesting to recall Watson's (1964) conclusion that large areas of orogenic belts, such as the Caledonides and the Alps, seem to be capable of retaining their metamorphic heat pattern throughout succeeding generations of plutonic events (*cf.* Sutton, 1965).

One difficulty in invoking a simple causal connection in the present situation, *i.e.,* between an early-formed "hot spot" and the relatively late Main Granite, is that we have recorded a relatively "brittle" phase of regional folding (D_4) acting prior to the emplacement and following the regional MP_2 metamorphism.

Another argument against the "hot spot" model is provided by the deformation recorded in the igneous rocks of the Meenalargan Complex, the Ardara and the Thorr Plutons. Where these intrusives lie adjacent to the Main Granite, there is a strong contrast in metamorphic style between the structures in their igneous rocks and those in the metasediments surrounding or included in them (p. 282). Unless we have exaggerated this contrast or unless these igneous rocks were considerably more ductile than the metasediments, then it seems that the structures in them demand a late and more intense deformation than we have been led to expect from the examination of the DMG_2 structures in the aureole schists, and such as might well have been provided during the actual emplacement (as DMG_1).

Yet another cogent objection to the "hot spot" or "root zone" model is presented by the close correspondence between the increase in metamorphic and structural intensity and the margin of the Pluton, so exact as to mean that, apart from a possible "outlier" in southern Rosguill where the structural intensity and metamorphic grade rise locally (Knill and Knill, 1958, 1961), the Main Donegal Pluton rose with such precision that it nowhere transected the boundary of the presumed "hot spot."

Here the authors find themselves faced with a dilemma, for neither of our models is completely satisfactory. First, the model associating deformation with granite emplacement and cooling does not easily account for the complete continuity in the field between the regional S_{2-3} cleavage and the schistosity in the envelope. Second, there is the apparent absence of minor structures associated with the early stage of forceful emplacement, the DMG_1 episode, implying that the sole effect of these movements was the coplanar flattening of the S_{2-3} cleavage, resulting in a composite fabric which the field geologist would have little hesitation in labeling an early penetrative schistosity! Third, Berger is puzzled by the absence from much of the aureole (though not all of it) of representatives of the earlier MP_2 garnets and plagioclases, and finally, there is no detectable difference in metamorphic style between what must be regional MP_2 garnets in the outer Creeslough Formation and in the Upper Falcarragh Pelites southwest of Glenties and the garnets in the Main Granite envelope. Berger would emphasize that all of these factors accord well with the "hot spot" concept, but Pitcher thinks that the evidence for a late and intense deformation revealed by the structures in the Ardara and Thorr Plutons, together with the perfect correspondence between the proposed "hot spot" and the margins of the Pluton, particularly the excellent gradation in intensity of metamorphism toward the contacts, provides sufficient reason to reject this model and embrace the concept of a "deformation aureole." At this point, we are at an impasse, and although the balance of evidence may favor the "deformation aureole" or "DMG_1 model," the junior author prefers not to reject completely the possibility of a "hot spot" situation.

Here we leave the Main Donegal Pluton. Whether its unique character as an intensely deformed and yet high-level pluton which is post-tectonic with respect to the main regional movements is the result of unusual (for the metamorphic Caledonides) regional stresses, or whether it results from some kind of accumulation of stresses due to diapiric intrusion, we simply do not know. It may, however, be of some consolation to the reader to learn that despite these problems, our new interpretation of the Main Granite structures has helped to resolve certain difficulties in previous models for the Trawenagh Bay Pluton which forms part of the western boundary to the former; this Pluton we deal with in the following chapter.

CHAPTER 13

THE TRAWENAGH BAY PLUTON: AN ENIGMA

CONTENTS

1. Introduction 296
2. The Margins 297
3. The Magmas 298
   (A) The Rock Types 298
   (B) The Enclosures in the Magmas 298
   (C) The Fabric 299
4. The Relationship between the Trawenagh Bay and Main Donegal Plutons 299
5. Discussion: Mode and Time of Emplacement 300

1. Introduction

Throughout our discussion of the Main Donegal Pluton, we referred only briefly to the area of granitic rocks which appears to form a bulbous protrusion to the west. The relationship between this Trawenagh Bay Pluton and the Main Granite has always been something of an Achilles' heel in Donegal geology, but we can now clarify to some extent the link between their two histories. Before we do this, however, we need to outline the general character of this particular pluton.

The Trawenagh Bay Pluton (Fig. 13-1) is a more or less rectangular, sharply bounded body of biotite granite some 50 square kilometers in area. It was first clearly distinguished from the Thorr Pluton by Gindy (1953b), who mapped it in its entirety as his "Newer Granite," and from whose account much of the following is taken. The junction with the G_1 member of the Rosses complex in the north is also sharp, though not always easy to find as it lies between two rather similar granites. Indeed, were it not for the fact that several porphyrite dykes belonging to the Rosses swarm (see Map 1 in folder) are cut off abruptly by the marginal rocks of the Trawenagh Bay Pluton, it would be hard to ascertain the time relationships between the two granites.

As regards the relationship with the Main Donegal Pluton, we can note that in the townland Meenatotan, granite sheets which cut the Thorr Pluton—and which Pitcher and Read (1959) connected with the Main Granite—are truncated by the Trawenagh Bay Pluton; nevertheless, the junction

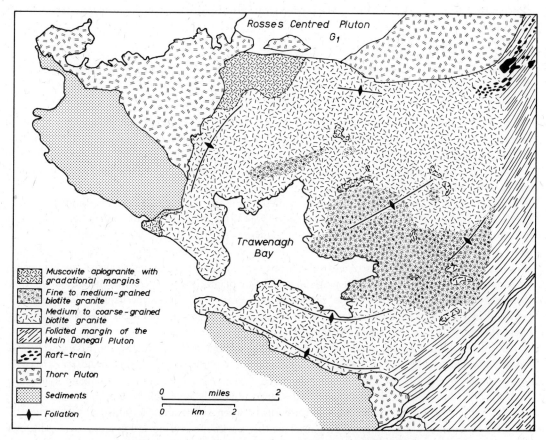

FIGURE 13-1. Outline map of the Trawenagh Bay Pluton.

between the latter two bodies appears to be transitional, the characteristic structures of the Main Granite dying out westward into the Trawenagh Bay Pluton. Pitcher and Read interpreted the rather equivocal evidence as indicating the younger age of the latter, the pegmatitic zone lying between the two plutons representing "an elongated protrusion from the roof forming a partition dying out in depth" (*loc. cit.*, p. 298). This partition shielded the Trawenagh Bay Pluton from the effects of deformation, which they considered to be associated with the forceful emplacement of the Main Granite.

2. The Margins

Except against the Main Granite, the outer edge of the Trawenagh Bay Pluton is knife-sharp, steeply dipping and, over considerable lengths of the contact, extraordinarily planar. Apophyses from it are rare. There are no structures in the adjacent country rocks which can be ascribed to pressures exerted during emplacement, nor, remarkably, are there more than slight mineralogical changes in the envelope. Thus, no contact effects have been detected in the Rosses complex, and in the Thorr Pluton there is only a local change in the color of biotite (*pace* Gindy, *loc. cit.*). The high-grade metasediments flanking the southern and western contacts are virtually unaffected, though sometimes, at the immediate contacts, there is local recrystallization, and random porphyroblasts of muscovite may appear, clearly postdating both the robust sillimanite and the fibrolite mats of the Thorr aureole

(Pitcher and Read, 1963). A few minor changes can also be detected in the adjacent metadolerites and marbles.

3. The Magmas

(A) The Rock Types

The Trawenagh Bay Pluton is for the most part a rather homogeneous, equigranular, medium to coarse-grained, biotite granite with diffuse areas of somewhat finer-grain size in its central and southeastern portions (see Fig. 13-1). The most obvious variant is a garnetiferous, muscovite-rich, aplitic to pegmatitic granite, devoid of biotite, which occurs in the northwestern and western corners and as scattered patches in central areas.

The biotite granites are modally similar, regardless of differences in grain size. They consist of sodic plagioclase (An_{10-15}) with occasional oscillatory zones (average proportion by volume 31%), a microperthitic microcline (31%), quartz (27%), myrmekite (5%), dark green biotite (4.3%, often altered to chlorite or muscovite), and with epidote (0.2%) as a common accessory. Replacive textures are rare, though veinlets of quartz and potash feldspar along fractures and cleavages in plagioclase are sometimes seen, and small, nonperthitic inclusions of plagioclase in microcline are common. Apart from the usual undulatory extinction in quartz, the only evidence of mechanical deformation is provided by the slight bending or kinking of plagioclase twin lamellae, though adjacent quartz and microcline crystals show no signs of strain.

The normal biotite granite grades into the marginal aplogranitic variant by the disappearance of biotite, accompanied by the incoming first of muscovite and then of garnet (Gindy, *loc. cit.*), the former finally averaging 12% by volume, the latter 0.3%. Aplitic textures predominate, but coarse pegmatitic patches also occur. Muscovite varies antipathetically with microcline and is occasionally in radiating or plumose form. Delicate intergrowths between quartz and the feldspars or between the feldspars themselves are common. The garnet in this muscovite-bearing aplogranite is a red-brown almandine-spessartite with a small proportion of the grossularite end member, and is similar in composition to garnets from the Rosses and Main Donegal Plutons (Hall, 1965*c*). Gindy (1956*b*) considered that it crystallized, together with muscovite, from a volatile-rich magma and as an alternative to biotite, but Hall pointed out that even if the components of the garnet and muscovite were to have been combined to form a mica, this would still have had the composition of a muscovite rather than a biotite, and he concluded that the garnet appeared because of the enrichment of the magma in manganese and the reluctance of this element to enter the muscovite lattice.

This garnetiferous variant has every appearance of being a marginal rock; it was preferentially collected in the angular reentrants, and its special mineralogy clearly suggests that it represents crystallization under the influence of accumulations of volatiles. If so, those areas of this rock type lying well away from the margins may represent remnants of a roof phase, implying that the pluton had a flattish roof. Then as regards the slightly finer-grained variety of the center, we are of the opinion that it represents a separate (second ?) surge of magma.

The variation in chemical composition of the Trawenagh Bay Pluton is very like that of the Rosses and overlaps that of the Main Donegal (see p. 346), so that we ought not to be surprised at the close connection between these plutons as revealed by the field evidence: indeed, we shall later consider that they were all three derived from the same source.

(B) The Enclosures in the Magmas

For the most part, the Trawenagh Bay Pluton is free of inclusions, though irregular biotite schlieren are occasionally seen which have been interpreted by Gindy (1956*a*) as inhomogeneities within the magma. Although it is difficult to rule out the possibility that these represent the end stages of digestion of mafic-rich xenoliths—as we have previously argued in the case of the Thorr, Rosses, and Main Donegal Plutons—it must be pointed out that recognizable inclusions of country rock are rare and confined solely to the outer portions of the Pluton, where they are all sharply bounded and appear to have been incorporated purely mechanically into the granite.

In just a few localities, there are angular inclusions of a granodiorite within the normal biotite granite. These are thought to be relicts of a possible early pulse which has now been totally disrupted; though outside the pluton itself, it is still represented in outcrop by certain early granodioritic sheets within the country rocks of the Crohy Hills (*pace* Gindy, *loc. cit.,* p. 406).

(C) The Fabric

For the most part, the rocks of the Pluton appear to be structureless, but faint mineral alignments can sometimes be seen in marginal portions, where they lie generally parallel to the contacts. Generally, it is rather difficult to determine whether these are linear, planar, or L-S fabrics, but in the south, Gindy (1953b) recorded scattered steep to vertical planar alignments. In central zones, a faint alignment trending generally west-southwestward can occasionally be seen, and this increases rapidly in intensity near the edge of the Main Donegal Pluton. In just one locality, near the Dunglow–Meenacross road junction, we have observed a vague banding reminiscent of that in this latter pluton.

Most interesting is the fact that recent analyses of the anisotropy of magnetic susceptibility in several locations throughout central areas of the Trawenagh Bay Pluton have demonstrated the existence of weak, gently southwest plunging lineations, even where none was visible in the field (King, 1966). Thus it appears that, in distinction to those fabrics which are clearly related to the nearby contacts in the north, west, and south, those in the eastern and central areas are parallel to the penetrative fabrics in the southwest end of the Main Granite, and they are therefore of the greatest importance in establishing the relationship between these two plutons.

4. The Relationship Between the Trawenagh Bay and Main Donegal Plutons

Pitcher and Read expended much effort in trying to map a contact between the Trawenagh Bay Pluton and the pegmatitic margin of the Main Granite, but as a traverse from one to the other in any area will show, there is no marked change in mineralogy, and the cataclastic mineral alignment and cross-cleavage in the latter fades out rapidly westward along strike, with only a slight divergence in trend in this direction (see Map 1 in folder). Indeed, as we have already said, it is natural to view the west- to southwest-trending alignments around Trawenagh Bay as the westerly extension of the Main Granite fabric.

Perhaps the most significant clue comes from examination of the raft train of inclusions of pelite, semipelite, and Thorr granodiorite, which enters the Trawenagh Bay Pluton in the extreme northeast (see Fig. 13-1). Whereas in the Main Granite these are aligned parallel to the host foliation, they become rapidly disoriented as they enter the Trawenagh Bay body, and the group as a whole swings round into a position of crude concordance with the nearby junction with the Thorr Pluton. Moreover, the signs of internal deformation in these xenoliths, which are ubiquitous where they occur in the Main Granite (see p. 235-6), are absent where the Trawenagh Bay granite forms the host; this is perhaps best exemplified by the undeformed nature of the internal veins in the inclusions in the latter. These relationships clearly indicate that the Trawenagh Bay Pluton was intruded prior to the bulk of the Main Granite deformation.

Nevertheless, the Pluton was evidently later than the first stages of emplacement of the Main Granite, as is shown by both the truncation of sheets belonging to the latter at Meenatotan and also

300   THE GRANITES

by the fact that the raft train mentioned above passes from the Main Granite into the Pluton. The Pluton thus seems to have been intruded at a late stage in the emplacement history of the major body, so that the problem arises how it is possible to explain both its comparative lack of deformation structures and the contrasted mode of emplacement.

5. Discussion: Mode and Time of Emplacement

As Pitcher and Read pointed out (1959), there is not very much evidence to indicate the mechanism by which the Trawenagh Bay Pluton was emplaced, for it would seem to belong to that numerous group of plutons in which little evidence of their emplacement history is preserved. However, even in this comparatively homogeneous body, there is some evidence of different pulses or surges of magma, represented now by the fragments of the early granodiorite and the two grainsize variants of the biotite granite. Then, as regards the *mis-en-place,* the lack of accommodation structures, the sharp margins, and lack of a preferred marginal fabric certainly argue against diapirism. On the other hand, the remarkably planar nature of the contacts suggests to us that material was broken away along early formed fractures, just as in the case of the Rosses Pluton. Further, if the isolated patches of garnetiferous aplogranite represent outliers of a roof phase, as we have suggested above, then a

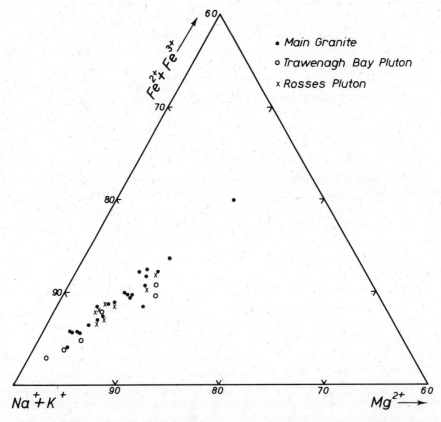

FIGURE 13-2. Plot of alkalis, total iron, and magnesium in the rocks of the Main Donegal, Trawenagh Bay, and Rosses Plutons. Illustrates the general overlap of chemical composition.

flat planar roof may have existed, probably with contacts like those of the walls. Thus, a stoping mechanism seems called for, though whether major or piecemeal we cannot decide. It would, of course, be tempting to compare the Trawenagh Pluton with cauldrons with flat roofs, like the Mourne Mountains Complex, for example, but this seems to be an unwarranted extension of the evidence.

Whatever its mode of intrusion, however, the Trawenagh Bay Pluton must have been in place before at least the bulk of the Main Granite deformation, and yet have escaped nearly all of its effects. This might have been due to the Trawenagh Bay Pluton or its margins being rigid enough to act as a buttress against which the Main Granite was deformed. However, this explanation implies a considerable time lapse between the emplacement and the deformation of the Main Granite, a period long enough to permit the intrusion and consolidation of the Trawenagh Bay mass, a proposition which we cannot accept.

The opposite view that the Trawenagh Bay Pluton remained liquid during the deformation and that the absence of strain in the xenoliths results from a very high ductility contrast between a liquid host and "solid" inclusions is also untenable, for it requires that the Pluton maintained a near-liquidus temperature while the adjacent Main Granite cooled and solidified.

Another possibility worth considering is that the two granites were emplaced at much the same time and perhaps even as one unit, just as the similarity in their bulk chemistry suggests, particularly insofar as the Mg/Fe ratios are concerned (Fig. 13-2). The lack of deformation in the Trawenagh Bay Pluton might then be a result of the locus of the deforming stresses having somehow avoided this body, while imparting to the Main Granite its strong structural grain.

Whatever the answer, we seem to be left with a model involving a remarkable combination of an entirely passive, permitted intrusion in close association, in space and time, with an active, forcibly emplaced body. It also remains a puzzle why the latest stages of deformation (Chapter 12) are so restricted to the latter. We do not, therefore, feel entirely satisfied that we have interpreted the evidence aright, and we still regard this particular problem as one of the most intractable in Donegal.

CHAPTER 14

THE CONTROLS OF CONTACT METAMORPHISM

CONTENTS

1. General Statement 302
2. The Chronology of Reaction and Recrystallization in the Aureoles 303
3. The Possibility of Changes in Bulk Composition 312
4. The Effect of Host Rock Composition on Mineral Reactions 314
5. The Physico-chemical Controls 317
6. Mineral Paragenesis in Relation to the Physico-chemical Controls: the Importance of Rate Processes 319
   (A) Garnet 319
   (B) Cordierite and Staurolite 321
   (C) Staurolite and Andalusite 324
   (D) The Aluminosilicates 324
7. Some Further Conclusions 326

1. General Statement

We know of nowhere else where there is so varied a display of contact metamorphic phenomena as in Donegal, a variety which Pitcher and Read (1963) believed was the result of the different modes of emplacement. According to them, reactive stoping at Thorr and Fanad was associated in both cases with a static recrystallization, as a result of which andalusite hornfelses were produced in wide zones, overlapped near the plutons by metasomatic, pneumatolytic, fibrolite-bearing assemblages. On the other hand, forceful emplacement involved a synkinematic recrystallization and produced carapaces of contact schists. Thus, during the movements produced by the radial distension of the Ardara Pluton, Pitcher and Read reported that andalusite grew as augen knots in the outer part of the aureole (along with staurolite, and, as is now known, with kyanite), while in the inner zone the crystallization of andalusite was very limited. Where emplacement was accomplished by lateral wedging combined with horizontal stretching, as with the Main Donegal Granite, contact schists were also produced, but here an even more thoroughgoing deformation was recorded by the development of very strongly preferred mineral orientations in staurolite-garnet-kyanite schists of regional aspect. In these rocks, andalusite was added to the assemblage during a late static event, in which the trespass of fibrolite was confined to a very narrow zone at the contact. Permitted intrusion, best illustrated by the Barnesmore Complex, produced very subdued, static aureoles with perfect hornfels textures and bearing andalusite and cordierite; the occurrence of fibrolite is here restricted and sporadic.

In what follows below, we examine specific points of these proposals in some detail, commenting on the chronology of deformation and growth of the various contact minerals, the possibilities of bulk chemical changes, the control of original composition on mineral growth, and the nature of the physical controls. Finally, we reconsider the proposed connection between the manner of emplacement and the type of aureole produced.

2. The Chronology of Reaction and Recrystallization in the Aureoles

In the accounts of the various plutons, we have shown that emplacement in all cases postdated both the latest regional deformation, D_4 and the retrograde regional metamorphic event, RMP_4. This relative dating naturally applies to the associated aureoles, and although we have found it necessary to discuss at length the special problems met with in the envelope of the Main Donegal Granite, the final conclusion was that, except in Central Donegal, a low-grade assemblage of rocks was presented to post-tectonic granites.

It is necessary to note, at this juncture, that there is some evidence of an increase in the regional retrograde effects toward the aureoles of the Fanad and Ardara plutons (Edmunds and Atherton, 1971). We cannot, however, decide whether this indicates a connection in time between RMP_4 and the intrusive event, or represents merely a natural reinforcement of the regional retrogressive effects in the outermost part of the aureoles—a matter we return to later (p. 314).

As we have seen, the country rocks show a most varied response to contact metamorphism (Fig. 14-1). There are, first, those plutons which produce no structural effects whatsoever in their envelopes; these are the Rosses, Trawenagh Bay, and Barnesmore. The former two were emplaced, mostly, into older granites and their associated high-grade aureoles, so that any purely thermal effects would be difficult to recognize; certainly, there is little to see. The contact effects within the *Barnesmore aureole* were limited to the production of simple hornfelses which carry andalusite and cordierite; these are accompanied by a little fibrolite near the contacts, the mineral growing quite independently of the andalusite (Fig. 14-2). There is here some slight evidence of polyphasal development: the fact that andalusite shows two habits of growth suggested to Smart (1962) that there had been two accessions of heat, corresponding to the two main pulses of granite intrusion. Otherwise, there is nothing in the least complicated about the contact effect of these three plutons, which, at the most, produce a simple static recrystallization completely lacking in metasomatic activity.

In contrast, the *Fanad aureole* (see p. 134) presents a much more thoroughgoing recrystallization. The earliest stages of this thermal metamorphism were formerly thought to have overlapped with movement (*cf.* Knill and Knill, 1961, p. 289; Pitcher and Read, 1963, p. 274), but we would now claim that the contact effects are entirely static and that any alignments of contact minerals are of mimetic origin. Usually, in fact, the new minerals, especially the andalusite, are quite disoriented, not even taking much advantage of the old schistosity.

The rapid disappearance of chlorite and the gradual reduction of muscovite are clearly connected with the production of aureole biotite, but as we have seen, there is some considerable variation in the rest of the mineralogy, apparently depending on the original composition. Thus, in the semipelites of Rosguill (Fig. 6-1, p. 133), cordierite is the most characteristic mineral, yet in nearby Fanad the pelites show a neat sequence of mineral zones (Fig. 14-3)—biotite, new garnet, andalusite, cordierite,

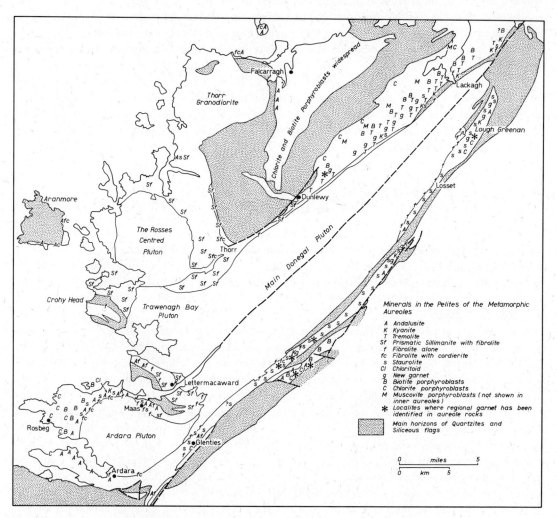

FIGURE 14-1. The distribution of the metamorphic minerals associated with the Donegal Granite Complex. Plagioclase porphyroblasts omitted.

fibrolite, robust sillimanite—which also represent a time order in mineral development according to Naggar and Atherton, 1970:

New Minerals	Time	Space
Biotite	earliest	outer aureole
Garnet		
(Staurolite)		
Andalusite	↓	↓
Cordierite		
K-feldspar and fibrolite		
Sillimanite (robust prisms)		
New muscovite	latest	inner aureole

In detail, we note that staurolite, occurring rarely in this Fanad aureole, predates the andalusite

FIGURE 14-2. Pelitic hornfels. Andalusite (*stipple*), biotite (*line*), muscovite (*pecked line*), cordierite, and some quartz as background. Bundles of fibrolite needles are shown diagrammatically. Southern contact of Barnesmore Pluton. After Smart (1962).

in a manner which we shall see is general in Donegal (as it is, indeed, in many metamorphic situations the world over, according to Schermerhorn, 1959). Further, with respect to the aluminosilicates, we interpret the textural evidence in the inner aureole as indicating that the mats of fibrolite were of wholly later growth than the andalusite and *grew independently from groundmass material* and in close association with cordierite: robust sillimanite developed from these mats and the groundmass, only rarely growing epitaxially on the andalusite. The clear diminution in muscovite as the contact is approached shows that this mineral is one term in the aluminosilicate reaction: a new K-feldspar is, however, not abundant. To complicate matters, a *new* muscovite appears along with the sillimanite, and in such contact situations, tourmaline is a notable accessory.

In Fanad, despite this rather complete progression of mineral zones, the textural evidence indicates two distinct, though overlapping stages of mineral growth. Further, the way in which the fibrolite zone encroaches onto an established andalusite zone indicates that at least two separate reactions are represented in the aureole, implying not just a mere increase in temperature but a distinct change of conditions. Indeed, we suspect that the fibrolite, sillimanite, the new muscovite, and tourmaline grew under the influence of a developing *hydrous phase,* a proposition which we now follow up by reference to the contact rocks of the Thorr Pluton.

The *Thorr aureole* represents an even more thoroughgoing contact metamorphism, in which the metasomatic processes have played an important role. In its description, we emphasized that the structural effects attributable to emplacement, particularly a new crenulation cleavage, were clearly overprinted by the new minerals, showing that the recrystallization was entirely static in character. Thus, the quartzes show the familiar triple junctions of equilibrium growth, and such preferred mineral orientations as there are, are entirely due to mimetic growth along the crenulations of this cleavage; Fig. 14-4 shows a beautiful example of andalusite overgrowing this structure by a combi-

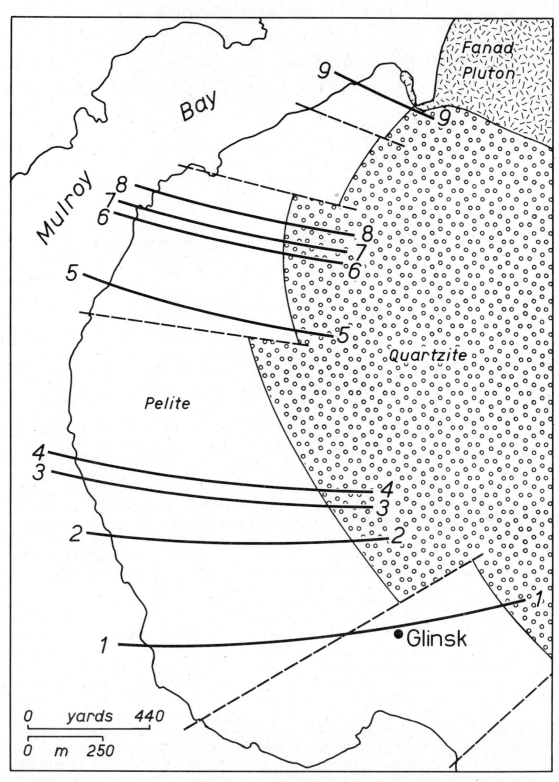

FIGURE 14-3. The metamorphic zonation of that part of the Fanad aureole around the hamlet of Glinsk, Fanad. After Edmunds (1969T): 1, outer limit of new biotite growth; 2, south of this line new biotite is restricted to knots of chlorite; 3, new garnet appears; 4, andalusite appears—chlorite disappears; 5, cordierite appears; 6, K-feldspar appears; 7, fibrolite appears; muscovite disappears; 8, andalusite becomes pleochroic; 9, sillimanite appears.

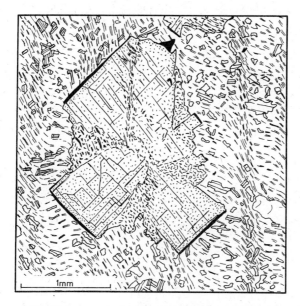

FIGURE 14-4. Andalusite porphyroblast overprinted on an early crenulation cleavage. Thorr aureole. Inishbofin. After Pitcher and Read (1963).

nation of dendritic and layeritic growth (according to Rast, 1965, p. 86), yet with the c-axis aligned along the pucker.

At Thorr, particular interest again attaches to the relationship between the fibrolite and sillimanite (again with cordierite) in the inner aureole and the andalusite in the outer. Where andalusite is the more important, as in northern outcrops, fibrolite is weakly developed, while the reverse is true in the southern outcrops. In fact, andalusite was formerly present in all parts of the aureole, but in the south has been largely replaced by white mica and fibrolite, the intimate association of which (within the pseudomorphs after andalusite) indicates contemporaneity of growth (according to Pitcher and Read, *loc. cit.*). Another interpretation is that the muscovite nucleated onto fibrolite (Fig. 14-5) and is thus sensibly later, but the closeness of the spatial connection in the field between the two minerals convinces us that any time difference must be of small degree.

A particularly complete example of this two stage reconstruction, where one aluminosilicate bearing assemblage is replaced by another, is seen in the townland of Lettermacaward (Fig. 5-3). Here, the *metasomatic* zone, with its fibrolite-muscovite-tourmaline association, progressively encroaches onto the purely *metamorphic* zone with andalusite as its characteristic aluminosilicate, finally producing a reconstruction so complete as to all but obliterate the early stage in the inner aureole.

This late-stage encroachment of a metasomatic zone seems to be a general feature of the Donegal aureoles, and we shall need to discuss its cause. For the moment, however, we turn our attention to the contact rocks of the Ardara Pluton and the Main Donegal Granite, aureoles in the development of which contemporaneous deformation had an important part to play.

We have seen how the regional structures were deformed in such a way as to concordantly wrap around the Ardara Pluton. In this position, the old structures, S_1, S_2, S_3, perhaps even S_4, were squeezed together and reinforced by further flattening, so as to nearly obliterate the early record of regional polyphase metamorphism. In fact, the only traces of the early fabric are the augen-like units of quartz and crenulated biotite which represent the microlithons originally produced by the intersection of S_1 and S_{2-3} (see Fig. 8-7).

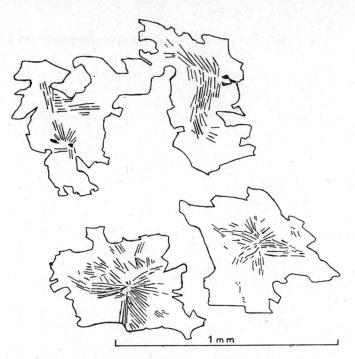

FIGURE 14-5. Muscovite porphyroblasts with centrally placed clusters of fibrolite needles. Thorr aureole, Meenatotan. After Pitcher and Read (1963).

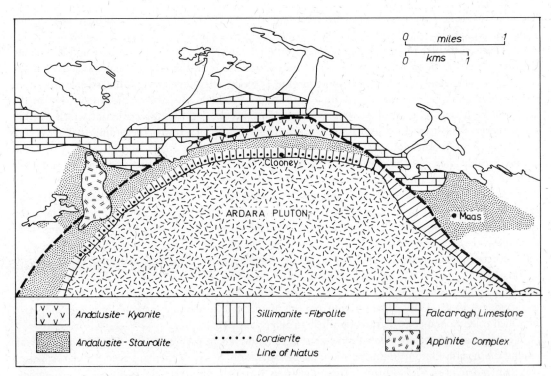

FIGURE 14-6. The northern part of the Ardara Aureole. Partly after Akaad (1956a), modified after Naggar and Atherton (1970). See also Fig. 8-6.

A more or less contemporaneous recrystallization (p. 180) led to the production of the Ardara *aureole* (Fig. 14-6), in which sillimanite, garnet, and cordierite are abundant in a narrow inner zone, and staurolite, kyanite, and andalusite in an outer. In more detail, especially within the northern part of the aureole, it is possible to separate this outer zone of the contact schists into two subzones (Naggar and Atherton, 1970), an outer characterized by kyanite and andalusite together, and an inner by andalusite alone. Later, we give an explanation of this particular zonation on the basis of a compositional control.

In these contact rocks, new porphyroblasts overprint the aligned micas which form the reinforced schistosity, even those within the fabric of the relict microlithons, yet the micas external to the larger porphyroblasts still sweep around them, showing that a further flattening must have occurred after the main period of growth (Fig. 14-7a and Fig. 8-7, p. 181). In the case of andalusite, it may even be that growth overlapped this later episode of deformation, or so we interpret the evidence as depicted in Fig. 14-7a.

In detail, as we have already seen, the kyanite and staurolite occur in small equidimensional crystals which are rather evenly scattered, suggesting, as demonstrated below, that they were produced under conditions of rapid heating. The andalusites form much larger crystals growing mimetically in the schistosity planes and often enclosing this kyanite and staurolite, never the reverse; possibly, therefore, two reactions are involved, separate in time, the earlier giving rise to the mineral pair kyanite and staurolite, and the later to andalusite. In both cases, the minerals must have *arisen directly from the groundmass, for there is no evidence of the replacement or resorbtion of one aluminosilicate in favor of the other* (see also Naggar and Atherton, *loc. cit.*).

In the inner aureole, in contrast to the outer, there is no trace of the later episode of deformation, and both andalusite and sillimanite overprint a single schistosity which we take to represent a completely reconstructed regional structure. Either recrystallization here outlasted deformation altogether or the most highly compressed inner hornfelses failed to respond to the imposition of the new structure; we have suggested the former because the cored andalusites of the inner aureole seem to represent two stages of growth (Fig. 14-7b), the earlier stage being a possible equivalent in time to the augen in the other aureole.

Where the zones of andalusite and sillimanite impinge, there is no pseudomorphous relationship, and as we have seen elsewhere, sillimanite rises independently from the groundmass. Even where fibrolite can be seen to fringe andalusite, it always forms an intergrowth with muscovite. Nevertheless, there is very little evidence of the kind of encroachment of one zone on the other which we have observed in the aureoles so far described.

The situation pertaining in the *aureole of the Main Donegal Pluton* was similar to that at Ardara, in so far as the adjacent metasediments were compressed by an early, DMG_1, phase associated with emplacement (see Chapter 12), and the older structures (S_1, S_{2-3}) were intensified and brought into such close coincidence as to be generally indistinguishable. In marked contrast to Ardara, however, is the presence of later, sometimes penetrative, deformations (DMG_2 and DMG_3), whose effects we described in detail earlier.

Kyanite, staurolite, and garnet—some of the latter apparently regenerated from the relicts of the regional garnet (p. 277), just as in the Fanad aureole—are much more abundant contact minerals here than at Ardara. Again, they typically occur as small porphyroblasts overgrowing the mica fabric, which is nevertheless also flattened around them by the DMG_2 movements (see Fig. 12-6, -7, -8). Although Naggar and Atherton (*loc. cit.*) have erected a detailed growth chronology, mainly on the basis of inclusion of one mineral by another, we think that the only really significant time difference is between the growth of the two mineral associations, kyanite-staurolite-garnet-plagioclase and fibrolite-biotite(chlorite)-muscovite (see Table 12-3). However since the former can be enclosed in andalusite (Fig. 14-8), which generally occurs as much larger porphyroblasts, around which the degree of flattening of the S_e fabric appears to be less than around the other porphyroblasts, and we

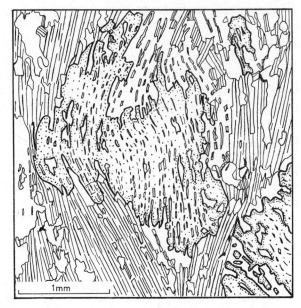

(a)

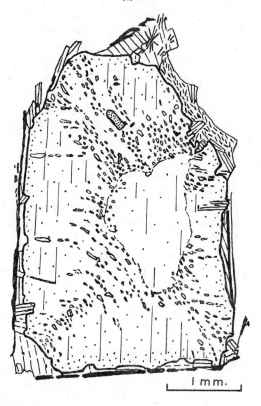

(b)

FIGURE 14-7. (a) Andalusite auge in biotite-quartz contact-schist. The schistosity is equivalent to S_{2-3}. (b) Cored andalusite porphyroblast outlined by graphite. Ardara aureole, Clooney. After Pitcher and Read (1963).

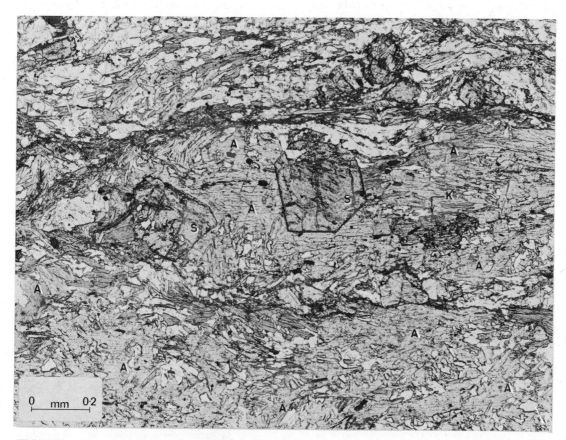

FIGURE 14-8. Inclusion of staurolite (S) and kyanite (K) in andalusite (A) porphyroblast. Aureole of Main Pluton, near Lough Greenan.

agree with Pitcher and Read (1963) that its growth outlasted that of the other minerals. Naggar and Atherton (1970) interpret the not infrequent augenlike lenticles of granular intergrowths of quartz, biotite, staurolite, or kyanite as relics of an early (DMG_1) hornfelsic texture which originally affected the whole groundmass (see Fig. 12-7b), a conclusion with which we also do not agree (see p. 274).

Near to the contacts with the Main Granite, in intimate relation with the granite sheets, robust sillimanite prisms occur sporadically and seem to be roughly coeval with the other porphyroblasts, but the abundant fibrolite here presents a special problem, for needles of this mineral are enclosed in the garnet, staurolite, and kyanite, as if of earlier growth. This is an example of what Read called the problem of the "spearing minerals," and we follow him in interpreting the evidence as whisker growth of sillimanite *into* these other minerals—implying that the fibrolite was, in reality, of later growth. Certainly the occasional fibrolite fringes around andalusite indicate later growth; further, fibrolite needles have recrystallized in places during the DMG_2 movements (see p. 276).

In summary of the main effects of contact metamorphism, we conclude that there is an order in the paragenetic sequence of porphyroblast growth in the Donegal aureoles. In general terms, there is first the production of garnet (in part regenerated from earlier garnets) together with or followed by kyanite, staurolite, plagioclase, and possibly some sillimanite; andalusite outlasted these minerals, and fibrolite, cordierite and phyllosilicate porphyroblasts were later (*e.g.* Table 12-3). At each of these stages, however, we repeat that the new minerals arise directly from the groundmass and rarely by conversion of existing porphyroblasts; indeed, *coexistence of aluminosilicate species is usual*. On the other hand, there is an important antipathy—where staurolite is abundant cordierite is uncommon and the reverse, though there are some intermediate situations, as in the Rosguill part of the Fanad aureole. Further, sillimanite is much more abundant in some aureoles than in others, and we later comment on the possible reason for this difference, after, however, briefly discussing certain of the possible controls of the metamorphism, particularly the compositional effect and heat supply.

Finally, we cannot lose sight of certain late-stage effects. In all the aureoles, there is the sporadic conversion of aluminosilicates and staurolite to shimmer-aggregates of white mica and chlorite, respectively, a process often accentuated along late shear zones. These phyllosilicates also form late porphyroblasts, especially well displayed in the Main Granite aureole, in close association with, or earlier than, the process of shimmerization. Presumably, this represents nothing more than an accommodation to cooling, marking the increase in relative stability of phyllosilicates over that of aluminosilicates in systems involving water. Of course, this is not simply a matter of hydration, for it also represents a potassium metasomatism; and though the actual amounts of potassium added were clearly not great, we are at a loss to identify the source, be it by redistribution among the phyllosilicates or by a contribution by the pluton during the late stages of its cooling.

3. The Possibility of Changes in Bulk Composition

In discussing the controls of this multi-event contact metamorphism, we first consider the possibility of bulk chemical change being involved. This is a matter which has often been discussed in general terms, though there are comparatively few studies in which the original material is compared with its metamorphosed equivalent in such a way as to properly take into account the range of original composition. Even since the introduction of rapid methods of silicate analysis, we know only of the detailed work by Engel and Engel (1958), Miyashiro (1958), Harme (1958, 1959), and Steveson (1970) in the regional situation, and by Pitcher and Sinha (1958), Bowler (1958), Floyd (1965, 1967), and Joyce (1970) in local metamorphism. This is perhaps surprising when we consider the clear guide given so early (1877) by Rosenbusch in his classic chemical study of the composition of the aureole of the Barr–Andlau Granite, but the fact is that there are very few ideal cases where single horizons can be traced into and through a prograde metamorphic area.

We are, ourselves, not able to comment objectively on the effect of the regional metamorphism on bulk composition in Donegal, except to note that all our analyses of metasediments and metadolerites show compositions within the normal range of sediments and dolerites; we suspect, therefore, that regional metamorphism, up to the energy level of the amphibole facies, is isochemical in character for all components, except perhaps hydroxyl. We have, however, a good deal of information concerning the effect of the local metamorphism, mainly because the Upper Falcarragh Pelites (locally known as the Clooney Pelitic Schists) can easily be followed into the two aureoles (Fig. 14-9) of the Ardara and Thorr Plutons. Using the then newly developed rapid methods of silicate analysis, Pitcher and Sinha (1958) were able to show that the compositions of these pelites were all but identical, horizon for horizon, both inside and outside the Ardara contact zone; this despite a considerable vertical change of composition (a negative correlation between Mg^{2+} and K^{1+}) due to an original upward facies change. The exceptions were the wholesale dehydration of the hornfelses and a very minor addition of the alkalis within a hundred meters or so of the contact. These changes resulted from dehydration reactions leading to the general growth of biotite at the expense of chlorite and sericite, coupled with the sporadic appearance of new sodic plagioclase and muscovite in the immediate contact zone.

It is of interest to note that Flinn (1959) has properly criticized the lack of statistical control of both the sampling procedure and the collation of the results in this particular study. He provides a

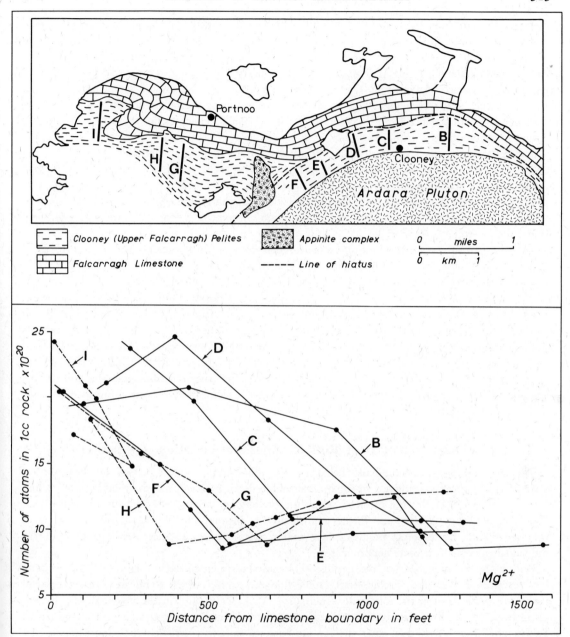

FIGURE 14-9. Sketch map (top) of the area north of the Ardara Pluton to show the sample-traverses on which are based the records of the variations in Mg^{2+} (bottom) within the horizon of the Upper Falcarragh Pelites. After Pitcher and Sinha (1958).

guide for future work of this kind by reexamining the raw data, using simple t-tests and one- and two-way analyses of variance. The result is an *objective* verification of the conclusions of Pitcher and Sinha—that the contact effects were largely isochemical except for the transfer of water.

In a less rigorous study of the composition of this same pelite formation where it enters the aureole of the Thorr Pluton on Cleengort Hill, Mercy (1960a, p. 119, 126) showed that the pelites, even here, were not chemically changed except for a relative dehydration. In yet another study, this time involving the aureole of the Fanad Pluton, Edmunds (1969T) failed to detect any change, other than dehydration in the pelitic horizons striking into this aureole at Glinsk (Fig. 14-3); indeed, his detailed work shows that diffusion must have been restricted to the volume of a single porphyroblast (*cf.* Atherton and Edmunds, 1966; Edmunds and Atherton, 1971). Finally, we note that although the general concordance of the Main Donegal Granite prevents the tracing of sedimentary horizons into its aureole, it has been possible to study the composition of the rather uniform quartz-dolerite sills in relation to distance from the contact (Kirwan, 1966T); once again, there is no significant change in composition.

The general conclusion is clearly that, over most of the aureole and with the notable exception of $(OH)^{1-}$, the contact effects were isochemical, even over small distances. The only trace of bulk metasomatism—involving the movement of alkalis (Pitcher and Sinha, *loc. cit.*, p. 403; Mercy, *loc. cit.*, p. 121)—is found in association with the fibrolite zones adjacent to the contacts, where, at a stage of near melting (as evidenced by the frequent mobilization phenomena), significant diffusion might well be expected to occur.

It is the fate of the water resulting from this dehydration which presents the main problem—whether it was absorbed into the magma or expelled outward to form a cortege of emanations preceding the pluton. If the latter, we think that heated solutions under pressure would have left ample evidence of their former presence, if only by accelerating retrograde changes in mineral assemblages. There is, indeed, some evidence of this in the case of the Fanad aureole, where Edmunds (1969T) has shown an increase in the thoroughness of the earlier retrogression toward the aureole proper—representing just the kind of hydrothermal front we have suspected. This is not an isolated finding, because around the Ardara aureole, there is also an increase in the chloritization of the regional garnets. Although the evidence available is not conclusive, we think that the bulk of the widespread RMP_4 retrogression may well be the result of a cloud of H^{1+} ions rising in the van of the granites (*cf.* Naggar and Atherton, 1970).

4. The Effect of Host Rock Composition on Mineral Reactions

The relationship between the bulk chemical composition of the contact pelites and their mineral assemblages has been investigated in some considerable detail by Naggar and Atherton (1970) and Edmunds (1969T), following the lead set by the preliminary work by Pitcher and Sinha (1958), Pitcher and Read (*loc. cit.*), and Pitcher and Smart (in Pitcher, 1965). The results, leaving aside the accounts of the technical problems involved in the sampling analysis and separation of the individual minerals, are quite simply stated.

Within the aureoles many of the pelitic lithologies show a quite narrow range of composition, but nevertheless there are specific stratigraphic horizons within which there are marked original variations, particularly in the total magnesium (1.4–6.1% MgO) and in the magnesium/iron ratio (0.34 to 0.70); we have already met with one such example in dealing with that part of the Upper Falcarragh Pelites where they enter the Ardara aureole. It is this variation in magnesium to iron which provides a particularly interesting study of the interrelation of mineral paragenesis and original composition.

A first finding is that there is a most consistent correlation in the distribution of magnesium and iron between the aureole *biotite* and its host rock (Fig. 14-10). The ratio is independent of position in each of the separate plutons, so that the control of mineral composition by the original host rock composition is particularly clear. This appears to be a general finding, for, as Atherton (1965, 1968) has pointed out, the composition of biotite is sensitive, first to the host rock composition and secondly to the coexisting phases (*cf.* also Butler, 1965; Albee, 1948; Hess, 1971). In the present case, the garnet content, being less than 2.5%, has little effect.

A second conclusion concerns the presence or absence in the aureoles of *new almandine garnet*.

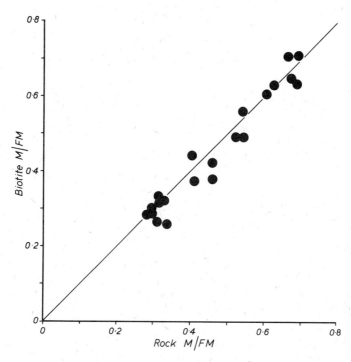

FIGURE 14-10. Plot of M/FM for biotite and host-rock in various aureoles of Donegal. According to Naggar and Atherton (1970).

Apparently this mineral, though stable enough in a purely thermal environment, only appears in rocks with appropriate iron-rich compositions. This is particularly well shown in the Fanad aureole, where the growth of garnet is favored in the relatively iron-enriched knots which represent the regional garnet; in fact, in this aureole, the growth of new garnet is entirely restricted to these sites (Edmunds and Atherton, 1971). This clearly illustrates the principle that during polymetamorphism, a rock may often be considered as consisting of two systems—the porphyroblast and the groundmass. In the more general case, Edmunds (1969T) finds that, other factors being equal, the amount of garnet produced is a function of Mn^{2+} content of the rock, *i.e.*, spessartitic garnet nucleates at lower temperatures than the less manganese-rich varieties (*cf.* Green, 1963; Albee, 1965).

Compositional restrictions to the growth of *staurolite* have also been suggested, but in the Donegal aureoles, it appears in rocks with a wide range of compositions, with the exception of those with especially high magnesium/iron ratios. In other situations, Juurinen (1956) also finds this wide latitude in the compositional control of staurolite growth, which also seems to apply even where the state of oxidation of the rock is considered, despite the predictions of Ganguly (1968).

Perhaps the most important conclusion, certainly the most novel, is the restriction of the growth of *kyanite* to aureole rocks with especially high magnesium/iron ratios (Naggar, 1968T; Naggar *et al.*, 1970; Naggar and Atherton, 1970). This result is undeniable and unequivocal (Fig. 14-11), being based on the analyses of some 80 samples from a large number of different situations. It is clearly shown in the Main Granite and Ardara aureoles, especially the latter, where kyanite is completely restricted to that part of the pelitic hornfelses—here a particular stratigraphic horizon—which is rich in magnesium relative to iron: the close association of kyanite-bearing and kyanite-free rocks denies the possibility of any differences in the physical conditions controlling growth. Concerning the kyanite-free aureoles of Thorr, Fanad, and Barnesmore, the composition of the contact rocks falls outside the kyanite field as defined by rocks from the Ardara and Main Donegal aureoles (Fig. 14-12). Thus, one important reason for the absence of kyanite is the lack of an appropriate rock composition, and in iron-rich pelites, staurolite should occur in its stead. That staurolite appears

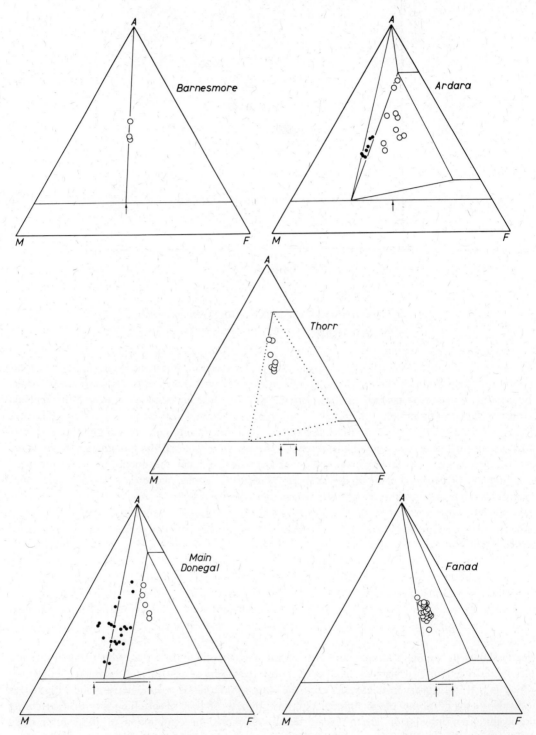

FIGURE 14-11. A–M–F plots for Donegal aureoles showing the compositional restriction of kyanite-bearing rocks (●) kyanite-free rocks (○). After Naggar and Atherton (1970). Spacing between arrows represents range of M/FM in coexisting biotites.

as a natural alternative to kyanite is certainly true in the case of the Ardara and Main Donegal aureoles; this is not to say that the two minerals do not occur together. However, neither mineral occurs in any quantity in the Thorr aureole, and both are completely lacking at Barnesmore, despite the fact that staurolite might be expected in pelites of appropriate composition. There must be a further reason for the restriction of this mineral pair, as we shall discover below.

As regards *andalusite* and *sillimanite,* on the other hand, Naggar and Atherton find that both minerals develop in rocks of both high and low magnesium/iron ratios, and further, there is no obvious difference between the bulk composition of andalusite and sillimanite-bearing rocks. Concerning the latter point, we suggest that the fluorine content, as yet not systematically determined, might yet prove to be an important factor in this respect (see Pitcher, 1965).

During our discussion of the chronology of mineral growth in the aureoles of the Ardara and Main Donegal Granite, we came to the conclusion that garnet, kyanite, and staurolite appeared, and sometimes even completed their growth, before that of andalusite and sillimanite. It seems, therefore, that two series of reactions might have been involved (Naggar and Atherton, 1970), as follows:

(1) A relatively early, *first reaction,* producing any of the following assemblages, depending on the magnesium/iron ratio of the rock, thus:
High ratios: kyanite + biotite + muscovite + garnet
Intermediate ratios: kyanite + biotite + muscovite + staurolite + garnet
Lower ratios: biotite + muscovite + staurolite + garnet
(2) A relatively later, *second reaction,* producing andalusite, irrespective of the magnesium/iron ratio of the rock.

We may add a *third reaction,* that which produces abundant sillimanite and fibrolite. It is, like (2) above, also independent of the magnesium/iron ratio, though we have suggested a measure of compositional control, in the sense that the reaction seems to be facilitated by the presence of "volatiles."

This chronological aspect of the reactions provides a valuable clue to the restriction of the pair, kyanite-staurolite. Perhaps the first reaction, which leads to the formation of these two minerals, is slow in comparison to the second, which produces andalusite, thus providing an opportunity for one reaction to overstep the other; differences in the rate of heating could produce much the same effect.

Before we follow up this matter of the control by rate of reaction, we much emphasize that, in our opinion, there was no great hiatus in the course of metamorphism. Neither do we envisage a later reaction starting only after the cessation of an earlier, for the textural evidence seems to us to suggest considerable overlap. We are not dealing with the superposition of metamorphisms but with phases of a single event.

5. The Physico-chemical Controls

The fact that there is a clear mineral zoning in the aureoles obviously suggests a relationship to a falling temperature gradient away from the intrusion. There is, however, little we can say here about the form of the heat front emanating from the various intrusions. We are not qualified to attempt the kind of calculations made by Lovering (1935, 1955), Larsen (1945), Jaeger (1964), Turner (1968), and Reverdatto and others (1970), and we are certainly not able to take into account the effect of differences in rock conductivity, the latent heat of crystallizaton, and the thermal contributions of the metamorphic reactions, particularly those involving dehydration.

There are, however, one or two field observations of interest in this respect, the first being that the outer margins of the aureoles, particularly of Ardara, are relatively sharply demarcated: one wonders if this may not be a common feature in aureoles in general. It is as if the thermal gradient, falling slowly throughout the distance of the aureole (*cf.* Turner, 1968, p. 251), had finally dropped very rapidly, and from theoretical considerations (see Jaeger, 1964, p. 449), this would seem to imply that the *initial* heating was of relatively short duration. However, once set up, this sharp gradient would apparently be retained for long periods.

At this point, we must briefly refer to the possibility that there have been several accessions of heat in certain of the aureoles. Thus, for example, the two-stage growth of andalusite at Ardara could be correlated with the two pulses of intrusion, and a similar suggestion has been made in the case of Barnesmore. However, it is even more likely that, at Ardara, the two stages of heating are represented by the two mineral associations, staurolite-kyanite and andalusite-sillimanite (see p. 309). Further, the clear encroachment of the fibrolite zone onto the andalusite zone in the Thorr and Fanad aureoles might also be interpreted as being due to a second surge of heat, though in neither case is there evidence of a discontinuity in the history of emplacement. Nevertheless, these both represent kinematic situations where magma is being continually added as the intrusion stopes upward, and the idea of zonal trespass in aureoles as magma surges upward, either continuously or in pulses, is an attractive one (see also Naggar and Atherton, 1970).

Another point concerns the variation in the width of the aureoles which is apparent after taking into account the differences in the lithology of the envelope. It is most obvious that the cauldrons have the least effect on their country rocks. This does not seem to be simply due to the special lithology of their envelopes, and it may be that the mechanism involving a sinking of a major block requires that much less magma be emplaced into the crust than in the case of plutons which extend to depth: any substantial replenishment of the heat loss is prevented by the sinking block acting as a barrier to heat flow. Despite the fact that the Trawenagh Pluton was emplaced in such high-grade rocks that its contact effects might be difficult to observe, we are of the opinion that its thermal contribution to the envelope was also slight. If this pluton were to have been emplaced by the stoping of great blocks, the situation might approach that of the true cauldrons and the same explanation might apply. Doubtless, however, this general notion is far too oversimplified.

Concerning the other plutons, we can merely observe that the width of their aureoles is very crudely appropriate to the size of their outcrop (assuming that the Fanad exposures form but a fragment of a much larger body); *viz.,* Main Donegal–2400 meters, Thorr–1800 meters, Fanad–1750 meters, Ardara–800 meters. We do not, however, consider size to be any more than one among several factors (*cf.* Jaeger, *loc. cit.*; Reverdatto *et al., loc. cit.*) controlling the intensity of the contact effect.

A further observation concerns the possible effect of differences in thermal conductivity. The differing degree of contact effect around the Thorr Pluton might find an explanation in the fact that quartzite is both a better conductor than pelite and the dominant rock of the least altered parts of the envelope. Heat might be expected to drain away more rapidly in the relatively good conductor, with a corresponding lowering of the temperature near the contacts; thus the contrast in reaction. There is also some possibility that the Falcarragh Limestone which intervenes in the northern part of the Ardara aureole may have acted as a heat sink, therefore creating an effective barrier to the contact effects.

Finally, there is the possibility of a relationship between the magma composition and the temperature of metamorphism (*cf.* Naggar, Atherton and Pitcher, 1970; Reverdatto *et al., loc. cit.*). Thus, the variation in the importance of sillimanite and cordierite—whatever the complications introduced by the hydrothermal conditions we have proposed—may indicate that either high temperatures were held for varying periods, or that different maximum temperatures were reached in the several aureoles, or both. On the basis of differing temperatures, we might expect that these were relatively highest in the Fanad aureole where sillimanite is joined (rarely) by corundum, not much less in Thorr, then less in Ardara, the Main Donegal Granite, Barnesmore, and Rosses, in this order. To a considerable extent, this parallels the mineralogy of the plutons—Fanad is dominantly quartz-dioritic, Thorr is tonalitic-quartz-dioritic where the sillimanite zone is widest, Ardara has a relatively narrow monzotonalitic periphery, and the Main Donegal is variably granitic and granodioritic; the other plutons are wholly granitic in composition. There does, therefore, seem to be a crude correlation between magma type and the character of the metamorphism, but we doubt that it is as direct as saying that the more basic magmas were the hotter, for in our opinion these more basic rocks, in some cases at least, were produced by the contamination of granite and could not, therefore, be hotter.

In spite of all the complications involved in the mode of emplacement and environment, it would seem obvious that heat was the most important agent of metamorphism, but we cannot altogether dismiss the possibility that pressure differences are also concerned. While it may be difficult to envisage a significant pressure gradient being established at any one time across the comparatively narrow zone of an aureole, it is still likely that pressure conditions might change with time, and after all, our problem is largely one of explaining a chronological sequence of mineral reactions. Even if a pressure gradient is not directly involved, there is the known involvement of deformation at particular stages in the development of the Ardara and Main Granite aureoles, which are just those situations in which the "special" assemblage, garnet-kyanite-staurolite, is important. With these matters in mind, we can now discuss the varied mineral paragenesis in the Donegal aureoles.

6. Mineral Paragenesis in Relation to the Physico-chemical Controls: the Importance of Rate Processes

Bearing in mind the time-independence of the several reactions and the fact that, in Donegal, K-feldspar is not at all a common component in the reactions involving the production of aluminosilicates, we could write a whole series of model dehydration reactions involving chlorite and muscovite as reactants—and biotite, quartz, and staurolite, or cordierite, or almandine, or one or other of the aluminosilicates, as the main products. Some of these types of reactions have been reproduced in the laboratory (for the staurolite reaction, see Hoschek, 1969; for garnet, see Hsu, 1968; for cordierite, see Hirschberg and Winkler, 1968; and for the breakdown of muscovite, see Evans, 1965), so that it is even possible to give some estimate of the ranges of temperature and pressure involved (Fig. 14-12). Like all dehydration reactions, those we have mentioned become largely temperature dependent at raised pressures, with the clear implication that the natural zonal order of their production is simply a response to rising temperatures.

In fact, it is possible to explain most of the established relationships on the basis of the experimental findings. Selecting a suitable isobaric line (X–Y in Fig. 14-12), we can easily see how, with rising temperature, kyanite first develops from pyrophyllite and quartz, and how the stability fields of andalusite and sillimanite are progressively entered: also, the manner in which staurolite, cordierite, potash feldspar, and even corundum (present only in xenoliths) can appear in temperature order. It seems that, in principle, this is the explanation, but unfortunately, the pressures concerned seem impossibly high for this simple explanation to apply in the case of the "high-level" plutons of Donegal. In fact, we would be content to follow Turner (1968, p. 249) in his estimate—based on the sillimanite-andalusite reaction—of one kbar for the pressure pertaining in the Ardara aureole!

Be that as it may, a number of very interesting problems arise. Why are the assemblages, and therefore the reactions, so exclusive and the products of earlier reactions so completely metastable, and why is the high-pressure polymorph, kyanite, produced in a high-level contact environment? Are rate controlled processes important in this respect also? We begin a discussion of these matters by briefly examining, along with allied matters, the status of garnet, cordierite, and staurolite, turning later to the problem of the interrelationship of the aluminosilicates.

(A) Garnet

Here we have the unusual case, for Donegal, of the resorbtion of one porphyroblast to provide the material for another. As we have seen, the partially chloritized regional garnet is progressively

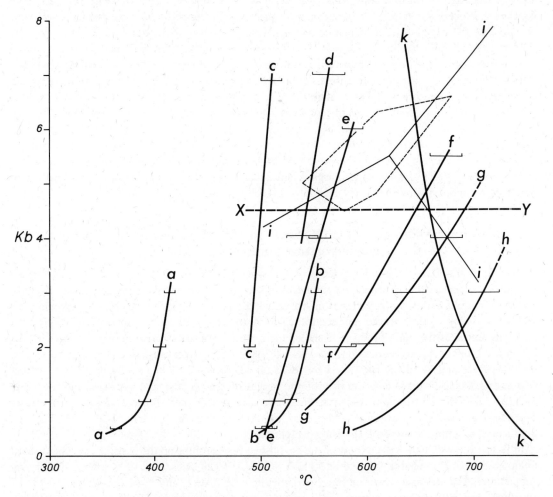

FIGURE 14-12. Experimental equilibria for various metamorphic minerals; After Naggar and Atherton (1970). (a) Mn-chlorite + quartz + fluid ⇌ spessartite + fluid (Hsu, 1968). (b) Fe-chlorite + quartz + fluid ⇌ almandine + fluid (Hsu, 1968). (c) Pyrophyllite ⇌ andalusite (kyanite) + quartz + H_2O (Althaus, 1969). (d) Chlorite + muscovite ⇌ staurolite + biotite quartz + vapor (Hoschek, 1969). (e) Chlorite + muscovite ⇌ cordierite + biotite + Al_2SiO_5 (Hirschberg and Winkler, 1968). Curve about 4 kb refers to the reaction involving ripidolite only *i.e.*, relatively Mg-rich chlorite. (f) Staurolite + muscovite + quartz ⇌ Al_2SiO_5 + biotite + vapor (Hoschek, 1969). (g) Muscovite + quartz ⇌ K-feldspar + Al_2SiO_5 + H_2O (Evans 1965). (h) Muscovite ⇌ K-feldspar + corundum + H_2O (Evans, 1965). (i) Aluminium-silicate curves and triple point location with uncertainty area (Richardson, Gilbert, and Bell, 1969). Reactions d–i, use natural starting materials. Error bars are those of the respective authors. (k) Minimum melting point curve (Luth *et al.*, 1964). (x–y) Possible crystallization path in the aureoles of Ardara and Main Donegal Plutons.

converted to chlorite-biotite knots, on which a new thermal garnet grows inside the pseudomorph of the old garnet (Fig. 14-13). The new garnet has a very different composition from the regional mineral: the latter is zoned, with a calcium and manganese-rich core and an iron-rich rim, while the new garnet is richer in both iron and manganese and poorer in calcium (Atherton and Edmunds, 1966; Edmunds and Atherton, 1971). Evidently, in the thermally controlled reactions which involve the conversion of chlorite to biotite and garnet, the iron and manganese had a stronger affinity for the new garnet; the fact that the regional almanditic garnet disappears in favor of this new Mn-rich garnet is possibly due (see Chinner, 1960) to the narrowing of the almandite-bearing rock composition field by competition from cordierite-bearing assemblages, and not, as might perhaps be expected, to an increase in the stability field for manganese-rich garnets.

In important contributions, Jones and Galwey (1964, 1966) have discussed what at first sight may be considered rather a different matter from the topic of this discussion, that is the growth kinetics of garnet in different metamorphic environments. Two contrasted situations were selected by these workers, *viz.*, the narrow contact zones adjacent to small basic intrusions and the wide aureoles of granite plutons. The two examples from Donegal were located, respectively, close to an appinite body near the town of Portnoo and in the inner aureole of the Ardara Pluton. The detailed study of the crystal size distribution of the garnets in these different contact rocks shows that this distribution is not controlled by the composition of the reactant matrix. Neither, it appears, is the determined pattern explicable on the basis that a random nucleation is followed by growth of nuclei, each competing for the available reactant. However, a most significant finding is that the size distribution seems to be simply related to the metamorphic environment. Within very different rocks, the distribution patterns were similar in each of the two contrasted environments, suggesting that the rate of heating was the single determining factor in grain size distribution.

The authors point out the probability that very small nuclei are unstable and must reach a critical size before they grow with reduction of the free energy of the system: nucleation may thus require a higher energy of activation than the growth process (Fig. 14-4). Where small basic intrusions are concerned, large rates of temperature change may be expected in the contact rocks—as a result, nucleation is accelerated in relation to rate of growth. It is thus likely that nucleation occurs relatively rapidly, ensuring that crystals would not grow large before nucleation was completed and with the result that they are more or less the same size at the completion of crystallization. On the other hand, in granite aureoles, where the rates of temperature change might well be relatively lower, nucleation rate is not as accelerated in relation to growth rate, and crystals nucleated early in the reaction may reach an appreciable size before nucleation is complete: the scatter of sizes is therefore likely to be considerable. The data fit this model so well as to confirm the thesis of Jones and Galwey *that size distribution is a function of the rate of heating of the host rock during metamorphism.* The important point is that we have again brought *rate* into the discussion, though perhaps we should express our surprise at the signal importance of this one factor, in view of the probable complexity of the controls of metamorphic textures (*cf.* Spry, 1969; Kretz, 1969).

(B) Cordierite and Staurolite

We turn now to the paragenesis of cordierite and staurolite. Cordierite occurs throughout each of the static aureoles, in contrast to the moderately deformed aureole of Ardara, which only carries this mineral in its innermost zone, and the strongly deformed rocks of the Main Donegal Granite in which cordierite is virtually absent: in every case, quite the reverse is true of staurolite. It is, therefore, tempting to invoke differences in pressure as a controlling factor (see Pitcher and Read, 1963), but the relevant experimental work (Hoschek, 1969; Hirschberg and Winkler, 1968) seems to indicate that the P/T controls of the staurolite and cordierite reactions (and that of garnet, for that matter)

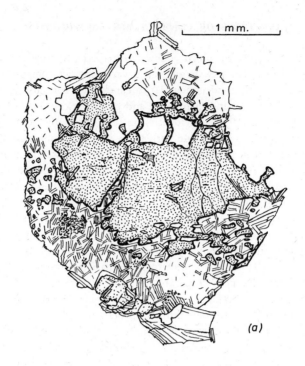

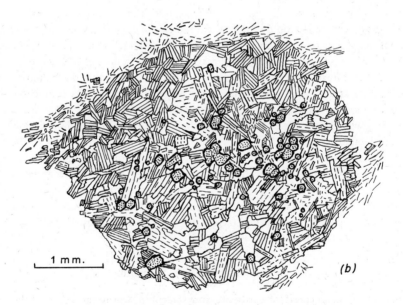

FIGURE 14-13. (a) Garnet porphyroblast (close stipple) partially replaced by a chlorite aggregate (open peck) which is itself replaced by biotite (close peck), iron oxide, and andalusite (open stipple). (b) A final stage in the conversion: biotite-muscovite knot formed in place of original garnet. Note new garnet and fringe of fibrolite. The Fanad aureole, Glinsk. After Pitcher and Read (1963).

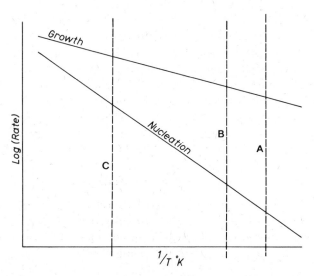

FIGURE 14-14. Diagrammatic sketch of Arrhenius plot of growth and nucleation steps. The formation of garnet in a system in which the rate of increase of temperature is such that a significant reaction begins at point A and is completed by the time position B is reached. Nucleation increases in rate as the reaction proceeds but growth is a rapid process and crystals nucleated early in the reaction reach an appreciable size before nucleation is complete. Thus there is a relatively large scatter of diameters about the average diameter, which is quantitatively measured as a relatively low value of K (a measure of the crystal size distribution). This contrasts with a reaction occurring in a system which is heated more rapidly so that the temperature has reached position C before the reaction is complete. In this case nucleation is accelerated in relation to growth, and completion of nucleation occurs more rapidly, so that crystals have not grown large before nucleation is complete. This may be expected to result in a larger value of K. After Jones and Galwey (1966).

are much the same. Certainly they overlap, though it may be that the production of cordierite is "favored" at lower pressures.

We are, nevertheless, attracted to this notion that the facilitation of one reaction rather than the other involved differences in the rates of reaction. Thus, Compton (1960) has suggested that in contact metamorphism, the combination of rapid heating and slow reaction rates can mean that reaction thresholds are overstepped. As a result, products of early reactions persist metastably in the presence of new components. Such differences in the rate of heating might be produced merely by one magma being hotter than another. On the other hand, if reaction rates are capable of being catalyzed, then overstep may be prevented and different reaction products result. Compton applies this reasoning to explain the separate appearance of cordierite and staurolite contact rocks that do not show any significant composition differences; apparently, cordierite grows where reaction rates are slow, staurolite when they have been accelerated. Compton, in fact, invoked water as the catalytic agent, leading to the view that so-called wet aureoles preferentially develop staurolite, but if we suppose that deformation will also promote reactions (by the mechanical generation of dislocations [?]), then we have one possible explanation for the antipathetic relationship of the two minerals in the Donegal aureoles (see Naggar and Atherton, 1970, for another explanation).

(C) Staurolite and Andalusite

There is no spatial antipathy between staurolite and andalusite, which occur together in the aureoles. This is not an unusual finding (Rosenbusch, 1877; Brindley, 1957; Compton, *ibid.*), and staurolite is clearly a mineral to be expected in contact environments. Indeed, we are of the opinion that the two minerals can grow separately or together, depending simply on the original rock composition (see Pitcher and Read, 1963, and Pitcher, 1965, Fig. 2, p. 335). Nevertheless, the fact that, in Donegal, the staurolite appears before the andalusite—also a general finding (*cf.* Schermerhorn, 1959)—suggests the possibility that the growth of the former may have been influenced by the special controls envisaged above, the staurolite appearing only where the reactions have not been overstepped at an early stage in the recrystallization. Its presence then effectively removes from the system just those materials which would provide cordierite.

(D) The Aluminosilicates

The relationships of the aluminosilicates need to be discussed in some detail. We first restate, even at the risk of repetition, that, in the Donegal aureoles, each species of aluminium silicate grew independently from the groundmass; there is little evidence of straight polymorphic transition and no obvious resorbtion of one mineral in favor of the other. Thus, even where sillimanite or fibrolite overgrows andalusite, it is equally important in the groundmass. A second point, which we have already referred to, is that the usually accepted andalusite-type reaction,

$$\text{Muscovite} + \text{Quartz} \rightleftharpoons \text{Andalusite} + \text{K-feldspar} + \text{Water},$$

cannot be important in the Donegal aureoles because potassium feldspar is quite rare in aluminosilicate-bearing rocks. Rather, the potassium necessarily released in the formation of these minerals must find its way into the new biotite and new muscovite.

Concerning particularly the status of fibrolite and sillimanite, we consider that they form essentially at the expense of the biotite—they do not grow merely in epitaxial relation to the mica (see Tozer, 1955; Chinner, 1961; Pitcher, 1965, p. 332). Further, the veinlike mode of occurrence, the common association with white mica and tourmaline, and the varying trespass of this assemblage upon pre-existing zones, together suggest a reconstitution under the influence of fluids moving out from the igneous bodies. Such a metasomatic late-stage and partly independent growth of sillimanite has been favored by a number of modern studies (Watson, 1949; Tozer, 1955; Autran and Guitard, 1957; Brindley, 1957; Zwart, 1958; Schermerhorn, 1959; Pitcher, 1965), and the conditions envisaged seem to be very different from those normally promoting the growth of andalusite and kyanite. The common coexistence of sillimanite with the latter minerals in the Donegal aureoles seems to show that it *nucleated and grew more rapidly than a surviving polymorph could react to the new environment,* suggesting, again, that the reactions which produced it were accelerated, possibly by the presence of H^{1+} and F^{1-}. But whatever the mechanism of the growth of sillimanite and fibrolite, there is a definite zoning in all the aureoles, with sillimanite occurring in an inner zone and andalusite in an outer—although with much overlap—which is a clear result of temperature control. It seems that at the higher temperatures near the contacts, possibly reinforced by the release of a second wave of heat and volatiles by crystallization of the magma, the early stages of a partial melting process might lead to greater mobility of ions, especially H^{1+} and F^{1-}, and so provide greater opportunities for metasomatic reactions. All this is in accord with the experimental work (Fig. 14-12).

One main problem highlighted by the contact rocks of Donegal is the relationship between kyanite and andalusite in the contact schists adjacent to the Main Donegal and Ardara Plutons. We may note first that the occurrence of kyanite in aureoles may be thought to be unusual, but there are other similar records in northwest Spain (Sergiades, 1962), northern New England (Albee, 1968; Thompson and Norton, 1968), and southwest Ghana (Lobjoit, 1964), and we suspect that the reexamination of contact schists elsewhere will provide further examples. In Donegal, the possibility of a compositional control on the growth of kyanite is now clear: it will only appear in rocks of appropriate composition, regardless of the P/T conditions. In addition, we have already shown that there is a degree of separation in time between kyanite and andalusite-bearing assemblages, and have suggested that the reaction producing the latter mineral is capable of overstepping the other: again, we have a case where the new mineral grew more rapidly than the surviving polymorph could be resorbed.

We might ask how the kyanite survives. It is perhaps just possible that under the particular P/T conditions pertaining near the boundary between the kyanite and andalusite stability fields, a magnesian kyanite of ferroan andalusite could both be stable, depending simply on the composition of the particular host rock. That aluminium silicate structures might be stabilized, or metastable nucleation enhanced, by the presence of small amounts of foreign ions has been discussed in some detail by Strens (1968), Chinner et al. (1969), and by Okrusch and Evans (1970) among others. However, Chinner and his coworkers, who utilized some material from the Donegal aureoles in their researches, conclude that the partition of iron between two polymorphs is small and the possibility of divariant equilibrium therefore negligible (see also Okrusch and Evans, *loc. cit.*). Further, it is their opinion that, modest though it is, foreign-ion stabilization would more usually lead to the coexistence of two or more polymorphs rather than the promotion of one at the expense of the other.

If foreign-ion stabilization is not the explanation, then we can only suggest that the occurrence together of these polymorphs in the Donegal aureoles implies a high degree of metastable persistence of these minerals in nature. This might well be especially true in thermal aureoles where the various steps of a reaction series would be quickly "ascended," so quickly in fact that, as we have seen, some are likely to be completely by-passed.

There is, of course, a simple enough explanation of the production of andalusite following that of kyanite. As we have already pointed out in reference to Fig. 14-13 (p. 319), we can, by choosing a suitable isobaric line, explain the development of each mineral assemblage on the simple basis of rising temperature. The problem here is that the pressures required by reference to the experimental work seem impossibly high to us. It is highly unlikely that the aureoles of the Ardara Pluton and the Main Donegal Granite were produced at great depths in the crust—even as much as 18 kilometers, according to the work of Althaus (1969, p. 114)—where regional pressures might be high enough for the primary kyanite reaction to occur. We are so strongly of this opinion because the emplacement of the Main Granite was the latest of the plutonic phenomena in Donegal, later, indeed, than certain relatively simple thermal aureoles and even the formation of a cauldron. It is perhaps tempting to suggest that diapiric intrusion may well have provided the necessary overpressures, but we then have the question as to whether such considerable pressures could be maintained locally in the higher parts of the crust (*cf.* Clarke, 1941; Rutland, 1965). Obviously, something remains unexplained, either in the laboratory or in the field.

One of us was sufficiently stimulated by this work on the Donegal aureoles to earlier review the natural paragenesis of alumino-silicates and make comparisons with the results of synthesis experiments (Pitcher, 1965). A conclusion was reached that the stability relationship of the naturally occurring aluminosilicates had not yet been closely enough simulated in the laboratory for estimates to be made of the P/T conditions of metamorphic processes. However, there has since been much new work and discussion (Newton, 1966; Fyfe, 1967; Pugin and Khitarov, 1968; Strens, 1968; Richardson, *et al.*, 1969; Zen, 1969), and the whole system has been treated in great detail by Althaus (1969), so that it seems that the form of the P/T diagram for the experimental system Al_2O_3–SiO_2–H_2O is now well enough established (see Fig. 14-12 and Fig. 8 in Althaus, *ibid.*). We would still enquire, however, if it is proper to compare natural reactions, separately producing aluminosilicates from the groundmass under *different conditions,* with a phase relation produced at one time and under the *same conditions*. Further, we are still of the opinion that the experimental "triple-point" is too high in relation to the natural situation (*cf*. Turner, 1968, Fig. 4-1).

In searching for further solutions to this complex problem, we think that it is significant that kyanite occurs only in those aureoles which have suffered deformation at the time of recrystallization —a rather outdated view! We are, of course, well aware that deformation will not, theoretically, alter the P/T conditions of polymorphic transition (*cf*. Verhoogen, 1951; MacDonald, 1957), but as we have pointed out several times above, we are not dealing, in the Donegal aureoles, *with inversions but with independent reactions,* the rates of which are subject to "external" influences of various kinds. Indeed, as Dachille and Roy (1960, 1964; see also Turner, 1968, p. 61) have shown, the effect of shearing stress is to increase the reaction rates leading toward the formation of stable parageneses, and we suggest that shearing stress could exert different effects upon different reactions. We are also led to speculate whether it is possible for the reaction producing kyanite from phyllosilicates to be so accelerated as to first appear at substantially lower pressures than at present envisaged. This is not perhaps a view which is thermodynamically tenable, but it will serve to underline the problem.

7. Some Further Conclusions

At first sight, the aureole rocks of the Main Donegal Granite have a distinctly regional aspect, so much so as to have led to opinions that these high-grade rocks represent a narrow belt of regionally metamorphosed rocks, formed before the emplacement of the granite. However, we have given clear evidence of superposition of contact metamorphism on that of the regional recrystallization in the country rocks, and indeed there are some significant textural differences from such regional situations —in so far as the characteristics of the latter can be generalized.

The most obvious of these is the generally small size and evenly scattered nature of the new garnet, staurolite, and kyanite, a feature which fits well with the expectation that rapid heating will be more of a feature of thermal aureoles than regional environments (see Jones and Galwey, 1966; Edmunds and Atherton, 1971). It could also be argued (*ibid.*) that the independent growth of each species of porphyroblast favors an hypothesis of rapid heating; this on the basis that reaction, nucleation, and growth in the phyllosilicate groundmass might more rapidly produce new aluminosilicates than cause polymorphic inversions, which are known to be very sluggish at all but high temperatures and pressures. The important point is that on rapid heating, we might well expect rate processes to be important in deciding which reaction shall predominate, or which substance shall nucleate earlier or grow preferentially: thus, the overstep of reactions might be more of a feature of aureoles than regional situations. We suspect also that deformation may also have some catalytic effect on rates of reactions. Certainly, the evidence in Donegal indicates that some importance be attributed to the intrusion process in this respect, probably simply by determining the time interval over which the several controls operated.

The differences between the several types of contact metamorphism themselves are probably the result of a combination of several factors. We note here that Reverdatto and his coworkers (1970) have recently proposed six different types of contact metamorphism, three of which, they state, are exemplified in Donegal, including one which they term the "Main Donegal type," these types being determined largely by the temperature and composition of the magma, the regional temperature of the country rocks, and most important, the depth of intrusion. Although there can be little doubt, as we have shown, that the first three factors have exercised some control in Donegal, depth can have had no importance here, since the Donegal granites were all emplaced within a relatively short time interval at a similar level in the crust.

It seems especially significant whether the magma was permissively or forcibly emplaced. During forceful intrusion, the creation of overpressures and the effect of deformation must influence not only the structural evolution of the contact rocks but the metamorphic reactions taking place within them. In permissive emplacement, on the other hand, such influences are at a minimum, and the reactions are largely controlled by temperature. Even so, the contact effects are still possibly subject to structural controls, for the penetration of the heat front into the envelope is likely to be limited where, as in cauldrons, a single pulse of magma is quickly emplaced and isolated in the upper crust as a result of the sinking of a great block along a single, permanent contact. How much greater must be the penetration of heat when a rooted intrusion gradually works upward continuously by the piecemeal stoping of a receding contact. Perhaps the late-stage overprinting of the metasomatic zone of fibrolite-muscovite over an early established metamorphic assemblage is due to the further accessions of heat, consequent on a continued upward penetration of the intrusive: the release of heat and volatiles during the final crystallization of such a considerable volume of magma would add its quota. These are all complex matters where such generalization may be unwarranted, but we conclude that the nature of emplacement mechanism does influence the chemical controls of metamorphism.

CHAPTER 15

THE FABRIC OF GRANITIC ROCKS: COMMENTS ON GRANITE TECTONICS, THE INTERPRETATION OF TEXTURES, AND ON GHOST STRATIGRAPHY

CONTENTS

1. Preamble: Primary versus Secondary Structures 328
2. The Rheology of Mobile Granitic Material 329
3. The Mineral Fabric of Granites: Magmatic Indicators and Mechanisms of Flow 332
   (A) Preamble 332
   (B) Comments on some Commonly Used Petrogenetic Indicators 332
   (C) Possible Mechanisms of Flow 333
4. Ghost Stratigraphy and Structure in Plutonic Rocks 335

1. Preamble: Primary versus Secondary Structures

Throughout this book, we have emphasized the structural character of the Donegal granites, and for the Ardara and Main Donegal Plutons we have argued the case for deformation occurring during the cooling stages of these bodies. We now want to discuss certain aspects of these conclusions in the light of contemporary ideas, though much of what follows is a condensation of our more detailed work (Berger and Pitcher, 1970), in which we made certain inferences about the physical behavior of mobile granitic material.

As we promised in Chapter 4, we have assiduously avoided use of the terms "primary" and "secondary" which, since the pioneering work of Cloos and Balk, have been used to describe the structures in igneous rocks. According to Balk (1937), primary structures were those that "developed during the time of consolidation" (p. 7), and were to be distinguished from secondary structures such as those in metamorphic rocks which were the result of "deformation subsequent to their solidification and deposition" (p. 132). In classic granite tectonics, primary fabrics were regarded as alignments of particles which were generally parallel to the walls of plutonic bodies and which originated by the rotation of previously random particles into positions of dynamic stability, according to the laws of fluid flow. As Mayo argues (in Nevin's well-known text on structural geology, 1949), since the minerals in granitic rocks show in general "very little evidence of strain," they must have been "embedded in a liquid that completed its crystallization after flow had ceased" (p. 187; see also Balk, *loc. cit.*, p. 10). Mayo goes on to conclude that "on purely structural evidence then the granite was a liquid magma of high viscosity, replete with solid suspensions" which were "oriented by the differential flow" of magma past its wall rocks. It seems to us that these are the fundamental assumptions upon which the vast majority of studies of granitic structures have been based.

However, it is clear from modern studies that many of the features used to distinguish "primary structures" can be matched in the metamorphic complexes of the basement, or even in areas of polymetamorphosed sediments (see Balk, *loc. cit.*, pp. 151–2, for a list of such criteria; also Billings, 1954; Taubeneck, 1964). Further, there is not the slightest doubt that essentially solid rocks can flow during deformation, as shown for example by the folding of sediments without megascopically visible fracture, or by the movement of salt diapirs. We see no great difficulty, then, in discussing structures in granitic rocks—even in igneous rocks in general—in terms of deformation, by which we mean simply change in shape (see p. 94). Indeed, we think that the terms "primary" and "secondary," as normally applied to granitic structures, are misleading and fulfil no useful function (see Berger and Pitcher, *loc. cit.*).

2. The Rheology of Mobile Granitic Material

There is still much debate about the physical state of granitic material during its movement into place, largely because of a confusion over the definition of the term "magma" and the degree of "liquidity" involved. We have elsewhere given a review of this matter (Berger and Pitcher, *loc. cit.*), in which we consider granitic material which is mobile under geological conditions to be made up very largely of solid components. Some of our reasons for this are summarized in the following. We are, of course, aware of the difficulty of defining precisely the terms "solid" and "liquid," but our concern here is to contrast the structural behavior of mobile plutonic rocks at either end of the spectrum and to argue against analogies with the behavior of liquids.

Though we know little about the actual rheological behavior of granite magmas below their liquidus temperatures, recent experimental work has shown that comparison with the behavior of liquids is inappropriate. Thus, Shaw (1969) and his colleagues (1968) have shown that when significant amounts of crystals (or bubbles) appear, magmas acquire yield values and behave as pseudoplastics or Bingham bodies. Further, recent model experiments using plastic materials (by Ramberg, 1963, 1967, 1970) have indicated that diapiric plutons cannot be emplaced unless the "magma" is either completely crystalline or consists of a mush with so little interstitial material that its strength and bulk viscosity approaches that of the surrounding rocks.

The more familiar approach to the problem is, of course, through field studies of granitic rocks, and in this respect the Donegal granites can provide much material of interest. In particular, many granites exhibit mineral alignments which are so like those of metamorphic rocks in expressing an L–S fabric that we have argued in this book and elsewhere (Berger and Pitcher, *loc. cit.*) against making a distinction between granitic and metamorphic rocks merely on grounds of differences in their rheological behavior. We believe that the kinematic significance of a planar fabric which results from flow of granites under stress has much in common with that involving metamorphic fabrics.

A particular clue is provided by the abundant evidence for the existence of fractures and other sharp discontinuities in mobile granitic material prior to its complete consolidation. Examples are provided by the regular-banding and cross-bands of the Main Donegal Pluton (Berger, 1971*b*) and by the presence in it of early healed shear zones (Figs. 11-3, -4), a structure often recorded in the literature (see Berger and Pitcher, *loc. cit.*). A special case is that of the synplutonic dykes, which were emplaced at a very early stage in the consolidation history of plutons (see Fig. 11-5) and show, therefore, that their host rocks were "solid enough to maintain cracks for the intrusion of dykes but still capable of slow fluid motion" (Krauskopf, 1941, p. 16). Such early fractures are more likely to occur, we believe, in the medium of a dense crystal mush than in a viscous liquid (for comments on joints in granitic rocks, see Berger and Pitcher, *loc. cit.*).

A most instructive approach to the problem of the ductility of mobile granites at the level of intrusion can be made by reference to the deformation of inclusions in them (*cf.* Chapters 8 and 11). Commonly, as we have seen, the three-dimensional form of the "basic autoliths" so clearly mirrors that of the local deformation ellipsoid that the inclusions are easily recognized as having been shaped by the application of "unequal" stress (*cf.* Watterson, 1965; Kretz, 1967; Ramsay and Graham, 1970). The extent to which the inclusions are so deformed depends, it seems, largely on the ductility contrast between them and their host rock—the lower the contrast, the greater the deformation (see Ramsay, 1967). Since deformed xenoliths which preserve sedimentary structures were clearly never in a highly fluid condition, this puts a similar limit on the ductility of the host granite (*cf.* Fig. 11-24).

Further evidence that strong structural grains can develop in granitic rocks below their liquidus temperatures, but before they have cooled completely, is provided by the "synplutonic" deformation of internal dykes and marginal apophyses to plutons. We saw in Chapters 11 and 12 how both the marginal sheets to the Main Donegal Pluton and the several generations of internal dykes (where the latter cut enclaves) were folded and boudinaged in a regular way prior to the complete consolidation and cooling of the Pluton. The geometry of these deformed minor intrusions leaves no doubt that they were more competent (less ductile) than their hosts at the time of deformation (Fig. 11-24). Where the host is a metasediment, the dykes and sheets could thus not have been anything like liquid when they were deformed, and since the same dykes also cut the Main Granite proper, this must also have been relatively solid, for it seems extremely unlikely that such late dykes could be less "liquid" (less ductile) than their genetically related host pluton.

The frequent occurrence of alignments of particles discordant to internal contacts within plutons, and to their margins, is of prime importance in discussing the rheological behavior of mobile granitic material (see Berger and Pitcher, *loc. cit.*, for full discussion). In many plutons, mineral alignments can be seen to cross schlieren, banding, walls to internal dykes, and junctions between different portions of the pluton (see Map 1 in folder; Fig. 5-3, 5-4, 8-3, 8-5, 9-1, 11-5, 11-7, 11-8b, 11-22b, and Chapters 5, 6, 8, and 11). The simplest view is that these alignments are later than the features which they cross and that they result from uniform deformation of the rocks on both sides of the discordant junction, which is thus a passive feature. In these cases of discordant fabrics, comparison with fluid behavior is again inappropriate, for it is unlikely that such lithological boundaries could form in a fluid and remain regular in form throughout a subsequent deformation in which the maximum compressive stress is at low angles to them.* Clearly, therefore, such mineral alignments formed when the host was in a solid or near solid state.

Even more important in this respect are the many examples of plutons whose structural grain is discordant to their outer margins. We can distinguish two cases here: those where discordances are local within a pluton whose structural grain is on the whole concordant with its walls, and a second category of those plutons where the internal structures cut straight across the walls and are continuous with those in the country rocks.

* We are here simply following the modern consensus that mineral alignments form parallel to the directions of maximum extension in a deformed body (*e.g.*, Ramsay, 1967; and see Chapter 11).

Many workers have pointed out that the discordances in the first category (*e.g.,* Fig. 5-1, 5-3, 8-5 and Chapter 8) originate as a result of stresses exerted on early-emplaced marginal portions of plutons by new diapiric accessions of material in central zones (Berger and Pitcher, *loc. cit.*). But what is of great importance is the role of the ductility of the marginal portions relative to that of the adjoining country rocks. If the ductility of the pluton is considerably higher than that of the adjoining envelope, as is normally considered to be the case, a unit volume in the outer part of the former will be much more flattened than one in the latter during a distension of the pluton, and the resultant fabric will be parallel or at low angles to the wall (*ibid.,* Fig. 2). On the other hand, unless there is a complete loss of cohesion (*i.e.,* slip) along the contact where the discordance occurs—and the general welding of junctions, together with the frequent presence of irregularities such as apophyses, argues against this—then a cross-cutting fabric which lies at high angles to the contact implies a low ductility contrast, and hence a more nearly similar physical behavior, between the marginal pluton and its wall rocks.

In the second case, the structural grain is thoroughly penetrative and everywhere parallel or nearly so to that in the envelope rocks, particularly where the margin of the pluton cross-cuts the structural trend of the latter. The Main Donegal Pluton provides a clear example of this type (see Map 1 in folder), and we have elsewhere given reference to others (Berger and Pitcher, *loc. cit.*; see also Cannon, 1970). Although many workers have used such relationships to argue or even prove a replacement mechanism for such cases (*i.e.,* "ghost structures," see Cole, 1902, for the Main Granite, and see below), we prefer the alternative view that they result from the deformation of the whole pluton and its envelope in a regional stress field not controlled directly by diapiric intrusion (see below)—that is, that the plutons are syntectonic in the strictest sense. Again, a low ductility contrast between the pluton and its country rocks is indicated, otherwise the more "fluid" pluton would be deformed more than the country rocks, and the resulting structural grain would be nearly like that in the first category.

It may be relevant, in this connection, to consider whether the regional stresses themselves can be related in any way to those generated by diapiric intrusion, as in the case of the Ardara and Main Donegal Plutons (Chapters 8 and 11). It seems possible that when a forcefully emplaced magma, driven by buoyancy differences in the crust (Ramberg, 1967), reaches a tectonic level at which its driving force is no longer able to move it upward (or sideways), it may in turn be subjected to inwardly directed stresses originating in the country rocks which it has just heaved aside. The operation of such a "reciprocal" model would clearly depend upon the rate of emplacement, the crustal level involved, and upon the ductility contrast between the pluton and its country rocks. We would expect such a process to be most effective at relatively high levels in the crust, where the country rocks, away from the aureole, are able to store up strain energy without suffering appreciable deformation while the pluton was being emplaced (*cf.* Price, 1966). In the final analysis, however, it would probably be impossible to distinguish between this model and that of a regional stress field whose source was unrelated to the rise of the pluton; in both cases, increased ductility due to higher temperatures near the pluton would lead to the bulk of the strain being taken up by deformation in the immediate envelope and the pluton itself (for further discussion, see Chapter 12).

In summary, then, we have shown that the ductility of mobile granitic material on *final* emplacement is often little different from that of its country rocks, and that flow can take place at quite a late stage in the crystallization of a plutonic unit. Further, magmas are clearly capable of fracture under certain conditions, while still able to flow under others. We are thus inclined to agree with Mehnert (1968, p. 213) that there are no convincing structural criteria capable of proving the presence of a liquid phase, and we are convinced that the "liquidity" of mobile granitic material has generally been overemphasized.

3. The Mineral Fabric of Granites: Magmatic Indicators and Mechanisms of Flow

(A) Preamble

If we view the flow structures in granitic rocks as the result of ductile deformation of essentially solid material, then we must comment on the many aspects of their mineralogy which traditionally have been regarded as the products of crystallization from fluid melts. Can these be relicts from early high-temperature liquid phases preserved throughout later ductile flow, or are they, at least in part, the result of reactions in the solid state? And if we deny fluid behavior in at least some granites, then by what mechanisms do they flow? It is to these two problems that we now turn our attention, taking the opportunity to stress again the frequent ambiguity of textural relationships.

At the outset, we must emphasize that the granitic rocks which we have used in the preceding pages to argue flow in the solid state are, in general, no different in chemistry, mineralogy, and textural relationships from many other relatively high-level granites. The same arguments used in other situations to indicate magmatic descent can be applied to the Donegal granites; indeed, in thin section, the Main Donegal Pluton (p. 230) is in no way distinct from many granites whose fluid behavior might never be questioned. There are, of course, the marginal areas of this pluton, and of Ardara, which provide evidence for cataclastic flow, but we are concerned here with the bulk of these units which show little or no signs of granulation or of mechanical breakdown of minerals.

(B) Comments on Some Commonly Used Petrogenetic Indicators

In general, we do not think that textural relationships between minerals in granitic rocks can properly be used to establish a detailed paragenetic sequence from which to argue crystallization from a magma. In the following, we consider briefly several individual textures involving feldspar which have often been used to indicate magmatic (fluid) origin; in doing so, we are really stressing the importance of *convergence phenomena* (Read, 1957; Mehnert, 1968).

One common feature of potash feldspar—the zonal distribution of inclusions of other minerals— has long been considered as a reliable criterion of magmatic (fluid) parentage. Here, inclusions (generally quartz, plagioclase, or biotite) which are oriented with their longest dimensions parallel to crystal faces in their host have been ascribed to the incorporation of preexisting crystals into potash feldspar growing in a fluid medium (*e.g.*, Oen, 1960; Bateman *et al.*, 1963; Hibbard, 1965; Mehnert, *loc. cit.*; Kerrick, 1969). However, this feature, which is found in both the Thorr and Main Donegal Plutons, has also been recorded in potash feldspar porphyroblasts in metamorphic rocks (Drescher-Kaden, 1948, p. 225; Smithson, 1963, 1965; Sylvester, 1964; Ohta, 1969), and, very significantly, in demonstrably late feldspars which cross dyke walls (Booth, 1968; Dickson, 1969; Oen, *loc. cit.*). There must clearly be other (solid-state) mechanisms such, perhaps, as exsolution (Dickson, *ibid.*), capable of producing this relationship.

The presence of oscillatory zoning in plagioclase has, of course, long been used in petrogenetic discussions, there being a weighty mass of opinion that it provides a strong indication or even proof of a magmatic origin (*cf.* Mehnert, *loc. cit.*).* But there is an increasing number of records of similar plagioclases in a wide variety of metamorphic rocks; there are examples in the Falcarragh Pelites of Donegal (see p. 114), and we can refer to many others (Runner, 1943, p. 453; Emmons, 1953, p. 113; Goodspeed, 1959, p. 244; Hooper, 1962, p. 41; Smithson, 1963, p. 131; Ketskhoveli and Shengelia, 1966; Van Diver, 1969; Spry, 1969, p. 163). Rarely is the oscillatory zoning in these rocks as common or as delicate as in unequivocally igneous rocks, neither is the contrast between

---

* Indeed, there have been several recent attempts to utilize zonal sequences in plagioclase as a guide to physiochemical history of magmatic evolution (*e.g.*, Wiebe, 1968; Hutchinson, 1970).

adjacent zones as great, since plagioclases in most metamorphic rocks of granitic composition are relatively sodic, but until precise limits on these parameters can be set, we prefer to keep an open mind.

Various characteristics of twinning in feldspars are also commonly used in petrogenetic discussions. The relative frequence of twin laws has been used in many attempts to distinguish between igneous and metamorphic rocks (*e.g.,* Smith, 1962; Felici, 1964; Mehnert, *loc. cit.*), and although the predominance of complex laws in volcanic feldspars stands in contrast to the simpler forms in metamorphic rocks, overlaps arise where high-grade metamorphics and migmatites are concerned (Vernon, 1965; Cannon, 1966; Vogel and Spence, 1969). Likewise, use of the presence of growth twins to indicate a magmatic origin, under the assumption that they result from rapid crystal growth in a supersaturated (fluid) environment (Marfunin, 1963; Seifert, 1964), has become equivocal with the realization that other mechanisms can produce similar twins (Vernon, *loc. cit.;* Spry, *loc. cit.;* see also Donnelly, 1967). Indeed, we have seen porphyroblasts of plagioclase with typical growth twins in pelitic rocks along the Lackagh River (Berger, 1967T, Fig. A18).

Twinning due to synneusis (agglutination, combination) is certainly more useful in this regard, if it results from two or more partly grown crystals which come into contact in a mobile (magmatic) environment and then grow as one unit (Seifert, *loc. cit.*), but we note that Spry (*loc. cit.*) has suggested a mechanism for the formation of such units in metamorphic rocks (for examples from Donegal granites, see Hall, 1966*b,* Fig. 2; Berger, *loc. cit.,* Fig. A15E). Much more common, especially in the Thorr Pluton, are aggregates of plagioclase (and of quartz, or potash feldspar), which share neither zones nor structural linkages but which appear to fall into the category of synneusis features in the extended model of Vance and Gilreath, who interpreted the situation as "clear petrological evidence of igneous origin" (1967, p. 529). We suspect, however, that similar results might be obtained from textures resulting from polygonization and subsequent recrystallization of larger crystals (*cf.* Voll, 1960; Rast, 1965; Phillips, 1965; Spry, 1969); further studies are clearly needed (*cf.* Whitfield *et al.,* 1959; Mahan and Rogers, 1968; Vistelius, 1966, 1967).

In disputing the use of these features as petrogenetic indicators, we are not claiming that the Donegal granites, and granites in general, did not crystallize from magmas; there are of course many arguments based, for example, on chemical composition (*cf.* Mehnert, *loc. cit.*), on isotope ratios (*e.g.,* Faure and Hurley, 1963; Kolbe and Taylor, 1966; White *et al.,* 1967), and on zircon morphology (Mehnert, *loc. cit.,* but see also Saxena, 1966, 1968, and Veniale *et al.,* 1968) that may be perfectly valid in claiming a magmatic genesis. But we are concerned here with the minerals that make up the bulk of the rocks, since these must have taken part in important flow mechanisms.

(C) Possible Mechanisms of Flow

At the outset, we recall the importance in producing the granitic fabric of late, subsolidus (deuteric) reactions, which represent changes from higher to lower temperatures (see p. 93). These reactions include, to one degree or another, the late growth of potash feldspar, its exsolution, and the formation of albite rims and myrmekitic intergrowths on adjoining plagioclases, the deformation twinning and alteration of plagioclase, the chloritization of biotite, and the recrystallization and straining of quartz.

First, there are the important contributions to the granitic fabric made by the alkali feldspars, whose late origin in a large proportion of granitic rocks is, of course, well known (Eskola, 1956; Read, 1957; Mehnert, 1968). This is shown, first, by clear textural evidence (see p. 105 and p. 218; Schermerhorn, 1956*a, b;* Smithson, 1963; Watt, 1965; Mehnert, *loc. cit.*); second, by the presence of individual potash feldspars crossing dyke walls (the "constipated veinlets" of Hopson, *in* Cloos *et al.,* 1964; see also our Fig. 5-4, and San Miguel, 1955; Oen, 1960; Booth, 1968; Dickson, 1969) and junctions between other internal components of a pluton (*e.g.,* The Rosses, p. 190; Mehnert, *loc. cit.*); third, by its concentration in zones of differential shear and fracture (*e.g.,* the regular-banding of the Main Granite; also Preston, 1954; Marmo, 1962; Gore, 1968; Burwash and Krupicka, 1969), and fourth, by its parallelism with late mineral alignments which cross external junctions in plutons (*e.g.,* the Main Granite; Martin, 1953; Schermerhorn, 1956a, 1962; Oen, 1960; Smithson, *loc. cit.*).

We have nothing to add to the detailed discussions of the source of the material necessary to form the potash feldspar and of its method of transfer given by Boone (1962), Orville, (1963), Marmo (1967), and Mehnert (*loc. cit.*), among others, and we wish merely to point out that its crystallization is a complex process involving exsolution, replacement, and frequently inversion, which take place at a late stage in the evolution of the granitic fabric.

The complex aspects of the crystallization of potash feldspar, and indeed of plagioclase, are well illustrated by the continuing controversy surrounding perthitic and antiperthitic intergrowths, myrmekite, and albite rims on plagioclase crystals adjacent to potash feldspars (Voll, 1960; Rogers, 1961; Schermerhorn, 1961; Castle, 1966; Hubbard, 1967; Carstens, 1967a, b; Mehnert, *loc. cit.*; Griffin, 1969; Barth, 1969; Widenfalk, 1969; Peng, 1970; Vogel, 1970; Shelley, 1970). These relationships, which are widespread in the Donegal granites, have been variously attributed to exsolution, simultaneous crystallization, or replacement; but whatever their precise mechanism of formation, there is little doubt that they result from late reactions of considerable importance in making up the granitic fabric. What is particularly significant is the common observation that they, like certain other thermally activated transformations in feldspars (*e.g.,* orthoclase to microcline, high- to low-albite), are accelerated by deformation, as is especially well shown by the prevalence of grid twinning in potash feldspar, coarse perthite, and myrmekite in the marginal parts of the Main Donegal and Ardara Plutons (p. 170 and p. 233; see also Binns, 1966; Marfunin, 1963; Watt, 1965; Shelley, 1964; Ramberg, 1961b; Spry, *loc. cit.*; Burwash and Krupicka, *loc. cit.*). We mention also, in this connection, the homogenization of zonal differences in plagioclase sometimes recorded from deformed granitic rocks (see Fig. 11-10; and Ishi *et al.,* 1960; Sendo, 1958; Mueller, 1963; Fraser, 1966; Vogel and Spence, 1969).

Another feature of possible importance is secondary twinning, especially in plagioclase, due to deformation or to structural transformation (Seifert, 1964; Smith, 1962; Vernon, 1965; Spry, *loc. cit.*), for these generally result in some changes in the shape of the host crystal, if only on an atomic scale, and may thus be of some importance in solid-state flow. However, proper quantitative studies in granitic rocks have only recently begun (Laurence, 1970; for a general review of methods available, see Carter and Raleigh, 1969).

Probably the most important granitic mineral in this regard is quartz, for the ease with which it recrystallizes in response to strain in geological environments is well known (*e.g.,* p. 233, and Carter *et al.,* 1964; Augustithis, 1966; Carstens, 1966; Spry, *loc. cit.*). Serrated or scalloped mutual boundaries are occasionally seen and attest to the migration of boundaries in response to stress. The widespread occurrence of undulatory and patchy extinction resulting from lattice movements clearly indicates that quartz can respond to stress at late stages in the evolution of the granitic fabric (*cf.* Chayes, 1952; Blatt and Christie, 1960; DeHills and Corvalan, 1964; Phillips, 1965). Indeed, we can do no better than to recall the suggestion made by Balk near the end of a lifetime's research on plutonic rocks, that during the intrusion of more or less solid granitic material, quartz grains "easily deformed, but also easily recrystallizing, might be the carriers of the movement" (1953, p. 2473).

Thus, we think that a late, discordant alignment of quartz or potash feldspar, or both, does not present a major problem if the former recrystallizes with ease and the latter crystallizes late (from interstitial melt or by replacement) under the influence of directed stresses. However, the general consensus, borne out by textural relationships, experiment and theory, is that plagioclase and the mafic minerals are generally of early origin. Yet, where these minerals contribute to discordant fabrics, it is necessary to find some model explaining their reorientation during the flow of essentially solid bodies; here, analogies with metamorphic fabrics may prove useful (*cf.* McKenzie, 1968; Holland and Lambert, 1969; Spry, 1969).

What is clearly needed is a kinetic interpretation of the granitic fabric, and detailed studies of the importance of late deuteric activity in permitting solid-state flow and in producing preferred orientations. We suggest that the crystallization-recrystallization of quartz and potash feldspar, aided by the structural rearrangements attendant, for example, upon exsolution, inversion, the chloritization of biotite, the homogenization of plagioclase, and upon the production of deformation twinning, provide a collective mechanism by which granitic rocks can flow when largely or even wholly crystalline. Far more work will certainly be necessary before such a claim can be substantiated; useful progress might be made, for example, by studies of grain shapes, contacts, and interfacial angles (*cf.* Kretz, 1966; Vernon, 1968; Katz, 1968; Spry, 1969). Likewise, it might prove instructive to investigate the kinematic significance of intergranular flow in the light of recent work (*e.g.*, Hsu, 1969; Brace, 1969), which shows the dependence of the strength of geological materials upon the presence of minor amounts of pore fluids (*e.g.*, interstitial melt, late volatiles, etc.). The current emphasis on solid-state flow in the emplacement and deformation of basic and ultrabasic rocks (*e.g.*, Talbot *et al.* 1963; Raleigh, Lappin, and Ragan *in* Wyllie, 1967; Burch, 1968; Brothers and Rodgers, 1969) strengthens our conviction that many of the aspects of granitic textures will be better understood when they are viewed in terms of the same kinds of reactions and transformations involved in the production of a metamorphic fabric.

As a final comment on structures in granitic rocks, we turn now to ghost patterns in plutonic rocks, taking this opportunity to review the problem in the light of such structures in the Thorr, Fanad, and Main Donegal Plutons and to stress much of what the senior author has stated elsewhere (Pitcher, 1970).

4. Ghost Stratigraphy and Structure in Plutonic Rocks

It has long been recognized that there are areas where the stratigraphy or structure of the country rocks can be traced across a plutonic unit. A *ghost stratigraphy* is usually expressed by a visible pattern of xenoliths (the overt relicts of Chapter 5), but there are also examples of cryptic ghost patterns, exhibited by variations in chemical or modal composition, and best recognized through the application of modern statistical techniques (Ianello, 1971; and see review of work by Whitten and others in Chapter 5), and in one example, by the use of magnetic surveys (Tuominen, 1961, 1966). This ghost stratigraphy is often accompanied by a *ghost structure* in which country rock elements, such as schistosity, lineations, or even folds themselves, can be followed without deviation into the pluton itself (*e.g.*, Fig. 3-2). Indeed, in areas where stratigraphic sequences in the host rocks are not easily established, as for example, in basement complexes or areas of monotonous lithology, only a ghost structure can be determined.

It is not always appreciated that ghost patterns can form in a number of ways (see Pitcher, 1970), for to judge by the literature, many workers use these relicts as major arguments for, or even absolute proof of, a replacement mechanism, echoing Read's one-time statement (in discussion of Pitcher, 1953a) that they provide "the only valid demonstrations of granitization" (*e.g.*, Buddington, 1959; Misch, 1949; Compton, 1955; Ambrose and Burns, 1956; Wynne-Edwards, 1957; Crowder, 1959; Walton, 1960; Roubault and de la Roche, 1965; Slobodskoy, 1966; Niyogi, 1966; Holtropp, 1969; and reviews by Termier and Termier, 1956; Read, 1957; Raguin, 1965; Mehnert, 1968).

One obvious kind of ghost pattern results from erosion revealing irregular projections from the roof or the floor of an intrusive unit; indeed, to Daly (1933, p. 122) all unrotated inclusions in intrusive bodies were roof pendants. Nevertheless, there are surprisingly few accounts of situations in which the attachment of inclusions to the surrounding country rocks can be unequivocally demonstrated; the country rock "outliers" of the Barnesmore Pluton provide one such example (see Walker and Leedal, 1954), and others are exhibited by the Beinn an Dubhaich Granite of Scotland (see Pitcher, 1970), the Nunarssuit Complex of South Greenland (Harry and Pulvertaft, 1963), and the Hamar Gabbro of Somalia (Daniels *et al.*, 1965). Often quoted are the so-called "roof pendants" of the American cordillera (see Bateman *et al.*, 1963; Kistler, 1966; Kistler and Bateman, 1966; Erikson, 1969), but it seems from the published observations that the term is applied here chiefly because the enclaves themselves preserve ghost structures and stratigraphy.

Ghost patterns can also be exhibited by true xenoliths, which originated as blocks separated and passively stoped from the roof, so as to maintain an orderly arrangement, though individual inclusions, especially smaller ones, are often rotated relative to their neighbors. This mechanism is, of course, merely an extension of the classic model of stoping developed by Daly (1933), who concluded that stoped blocks were often fixed not far from the roof in a congealing magma. We have argued this mechanism in some detail for the ghost patterns of the Thorr and Fanad Plutons (Chapters 5 and 6), and Cobbing and Pitcher (in press) have recorded a spectacular example from the Coastal Batholith of Peru.

A very similar situation often results when an intrusive body has wedged open the country rocks, leaving *in situ* partially disrupted septa or screens, separated by coalescing intrusive sheets, as in the case of the Main Donegal Pluton. Multiple intrusions can, of course, lead to the same result, especially where a plexus of dykes and sills is involved (*e.g.,* Daly, 1933, pp. 77–104). Such a mechanism has often been invoked (*e.g.,* Mayo, 1935, 1941; Runner, 1943; Gault, 1945; Den Tex, 1956; Buddington, 1959; Raguin, 1965, p. 37; Brindley, 1969, Fig. 9), and there are several tantalizingly brief descriptions throughout the literature of situations which could easily be interpreted in such a way (*e.g.,* Oen, 1960, Fig. 3 and 7; Gevers, 1963, pp. 215–6; Gansser, 1964, especially Fig. 141; Escher, 1966, pp. 68–72).

Various combinations of the sheeting and passive stoping models have also been invoked, as for example, by Barker (1964), who, though not stressing its crude ghost stratigraphy, suggested a mode of emplacement for the Hallowell Granite of Maine by repeated concordant injection followed by stoping of the intervening screens of metasediments. Similar explanations could perhaps be advanced for the crude ghost patterns in the Graven Complex of the Shetland Isles (see Miller and Flinn, 1966, p. 108), and indeed, for many of the "roof pendants" or septa of the Sierra Nevada and Coast Range Batholiths (*e.g.,* Phemister, 1945; Mayo, 1941; Bateman *et al.* 1963; Kistler, *loc. cit.*).

We would emphasize further that the near-perfection of the ghost stratigraphy and structure in some intrusive plutons is the result of a deformation either during or subsequent to emplacement, which results in the alignment of previously random or crudely oriented inclusions (*cf.* Grout, 1941, p. 1529, 1546). This is the case with the highly developed ghost stratigraphy of the Main Donegal Pluton, as we have argued at length in Chapter 11, and has clearly been of local importance in the Sierra Nevada (Kistler and Bateman, 1966, p. 1312).

We may mention in passing a situation which some workers have interpreted as a kind of ghost structure—the "Sederholm Effect," referred to earlier (p. 222 and p. 330), in which dykes or dyke-like bodies, cutting plutonic units, are themselves disrupted within their host and cut by it. One explanation was that such bodies were the remnants of preexisting dykes preserved while their original host rocks were replaced or granitized under more or less static conditions (*e.g.,* Callaghan, 1935); these are the "relict dykes" of Goodspeed (1955), of which Figure 10-4 provides an example. Others have argued that they represent similarly resistant tabular xenoliths of preexisting country rock (the "pseudo-dykes" of Miller, 1945, and the "relict pseudo-dykes" of Goodspeed, *ibid.*; Weaver 1958). However, there now appears to be a consensus that such dykes represent, in the main, true intrusive bodies broken up by localized movements in a still mobile host (the "synplutonic" dykes of Roddick and Armstrong, 1959; Pitcher and Read, 1960*a*; Lipman, 1963; Davis, 1963) or during a later reactivation of the enclosing rocks (Watterson, 1965; Wegmann, 1963; and see this book, p. 222, for further references and discussion).

Most of the examples we have quoted so far are from relatively "low energy" environments (the mesozonal and epizonal categories of Buddington, 1959), where the rocks which exhibit relict patterns are probably all intrusive. But the majority of recorded accounts of ghost stratigraphy and structure come from higher energy (catazonal) levels where metasomatic replacement, granitization, and anatexis are the rule, (*e.g.,* Cushing, 1910; Hewitt, 1956; Wynne-Edwards, 1957, 1967; Harpum, 1961; Engel and Engel, 1963; Cannon, 1970; and examples quoted by Read, 1957; Buddington, *loc. cit.;* Badgley, 1965). Although we suspect that some of these situations are the result of the mechanisms we have just dealt with, there are undoubtedly situations where the ghost patterns involve static relicts which have resisted some kind of replacement mechanism (the "skialiths" of Goodspeed, 1948). We may quote, for example, the phacoliths of the Adirondacks which, according to Engel and Engel (*loc. cit.,* p. 351), exhibit "relict stratigraphic sequences whose form, complexity, and continuity are totally inconsistent with magmatic intrusion no matter how subtle, multiple, selective, or passive," though even here, there is dissension (see Buddington, 1963). In terms of Read's Granite Series, these are the autochthonous granites, while parautochthonous members are represented by ghost patterns which have begun to break up as the host rock becomes "unstuck" from its birthplace (*e.g.,* Allaart, 1958). Pitcher (1970) has discussed these high energy situations, and we need only emphasize once again the importance of deformation in producing the ghost alignments.

Cryptic ghost patterns, where a structure is revealed by local variations in modal or chemical composition, have been given a special credence in the argument for granitization processes (*cf.* Tuominen, 1961, 1966; Jahns, 1948, p. 94). However, as we have argued in the case of the Thorr Pluton (Chapter 5), we believe that such patterns can also arise during contamination of a magma by included blocks of country rock.

In summary then, we conclude that ghost patterns have an important bearing on the mode of emplacement of the rocks that exhibit them, but that they do not provide *prima facie* evidence of replacement.

CHAPTER 16

THE FORM OF THE DONEGAL GRANITES AND THE ORIGIN OF THE MAGMAS

CONTENTS

1. General 338
2. The Form of the Plutons at Depth 338
3. The Time of Emplacement Relative to the Regional Thermal Event 341
4. The Chemical Variation within the Plutons at the Present Level 342
5. The Bulk Compositions 344
6. The Source of the Granite Magmas 345
   (A) The Geological Reasoning 345
   (B) The Geochemical Reasoning 347
   (C) The Significance of the Initial Sr Isotope Ratio 349
   (D) The Contribution of the Appinitic Suite to the General Hypothesis . 349
7. Conclusion: a Final Model 350

1. General

We have now discussed in some detail the origins of the compositional variation and the mode of emplacement of each of the plutons of Donegal. Necessarily, our conclusions are limited in application to the present level of erosion, and unfortunately, it is only too probable that mechanisms will change with depth. Nevertheless, with all the information at our disposal, we ought to be able to usefully comment on the extension of granite downward and the possible origin of the magmas.

2. The Form of the Plutons at Depth

We have already commented on the results of the gravity investigations of Cook and Murphy (1952), and we can now add to these certain more detailed records obtained by Riddihough (1969) and Young (1969), despite the fact that these are not yet fully analyzed.

Even without these calculations, however, mere inspection of the gravity anomaly map (Fig. 16-1) shows that a likely model for the isolated pluton at Barnesmore is that of a cylindrical body, though

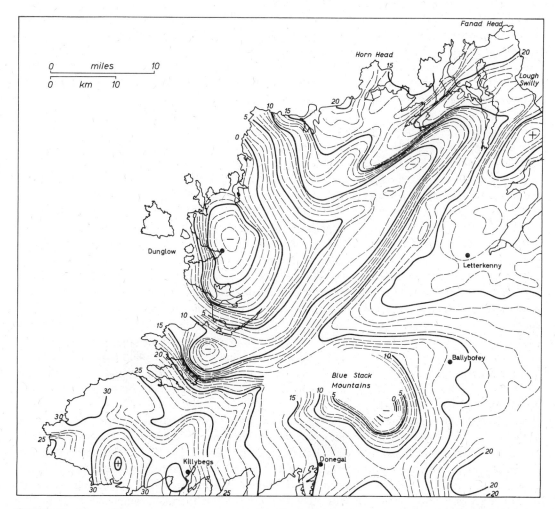

FIGURE 16-1. Bouguer gravity map of Co. Donegal with contour intervals of 1 milligal. After Riddihough (1969).

we would need detailed profiles before we could confirm our belief that the walls are steeply outward-dipping to depth. Thus, Barnesmore apparently has the shape of many so-called, post-kinematic granite plutons (Bott, 1956; Bott and Smithson, 1967), but the calculated thickness, assuming a density contrast of 0.15, is here only 5.5 kilometers (Cook and Murphy *loc. cit.*, p. 14): a possible explanation for this being the presence of a stoped block of country rock (or granodiorite) at depth (see p. 204).

The gravity data associated with the nested plutons making up the Donegal Granite Complex, again unanalyzed though they are, provide certain valuable clues to the overall three-dimensional shapes of the constituent members. Thus, steep contacts to depth appear to be the rule along the northwesterly margin of the Main Granite, but along the opposite margin, it seems likely that there is an outward sloping contact in the Barnesbeg area, which becomes steeper, however, to the southwest.

Concerning a model three-dimensional shape for the Main Granite, this can only really be determined in the northwest and outside the influence of other bodies. Thus, Dr. Young allows us to report that, along the Barnesbeg section, the final computed model possesses an apparent thickness of 3.2

kilometers beneath the area of granite producing the minimum Bouguer anomaly, decreasing to 1.5 kilometers under the southeastern flank, and indicating a quite thin wedge-shaped structure with an arched floor. Young's findings further suggest that a relatively shallow southeastern contact extends outward at depth as the upper surface of a thin tongue—which probably represents a complex of sheets like that at Crockmore. Most intriguing is the presence of an increasing negative anomaly, southwestward and *along* the long axis of this Main Granite, which may indicate the presence of a southwest-sloping "floor" to the intrusion as a whole. This concept is greatly complicated by the knowledge that a basic part of the Thorr Pluton may exist below the Main Granite, but nevertheless, one likely model shape, based on these preliminary results and our knowledge of the sheeted nature of the body, is of a great tongue of granite thickening southwestwards and lying more or less horizontally in the crust. This hypothesis—that the Main Granite is partly floored—is attractive to only one of us (W. S. P.), who is probably influenced by the fact that such an interpretation could support the view of Pitcher and Read (*loc. cit.*) that intrusion of magma was more lateral than vertical!

The data pertaining to the Ardara Pluton (Fig. 16-1) suggests that the deep form of the latter is cylindrical, almost vertical, yet, again according to Young, not very deep reaching. We have suggested to him, however, that a possible model is turnip-shaped, with the basic material of the exposed periphery closing inward at depth and filling the root. Perhaps the Toories is rather similar to this.

It is the western and greater part of the Thorr Pluton, in combination with the Plutons of the Rosses and Trawenagh Bay, which provides the most considerable Bouguer anomaly in Donegal—a negative anomaly 30 milligals in amplitude. We think that the distribution of the isogals suggests an oval cupola-shaped body, with a steeper contact in the west than in the east (Fig. 16-2). Concerning thickness, we mention that the density contrast seems to even out at the modest depth of nearly 10 kilometers. Unfortunately, the presence of quartzite as country rock will complicate interpretations, and further, the lithological similarity, and therefore the densities, of the several units within the Complex itself make it unlikely that gravity data will ever differentiate between their separate contributions to the total effect.

We have too little geophysical data to advance farther into an objective discussion of the overall shape at depth of the Donegal Complex, but we are prepared to make some guesses in order to stimulate further investigation. It seems likely that all the six plutons of the main cluster, despite the cross-cutting relationships between them at the present level, probably represent closely time-related pulses of magma which emanated from a limited part of the middle crust lying somewhere beneath the western seaboard of Donegal. The possible form of the multi-component body so built up in the higher crust is illustrated in Fig. 16-2.

Summarizing the structural situation by reference to this figure, clearly the Thorr Pluton represents the core of the Donegal Complex. From it, at depth, emerge the nearly upright [?] conical bodies of Toories and Ardara, around which the crust has been stretched. Possibly at Ardara, the outer monzodiorite represents a skin of mobilized basic facies of the Thorr Pluton, brought up by the ascending granite diapir and also left trailing behind as a kind of root or tail. From this Thorr center protrudes, laterally, the serrated tongue of the Main Donegal Granite; this represents magma forcefully sheeted along the grain of the country rocks. It might be suggested that the marginal strip of tonalite also represents Thorr material which was mobilized, smeared along one wall of this body, and finally partially disrupted during this intrusive event, though, as we have argued earlier, the evidence indicates that a considerable body of the Thorr rocks was in place here prior to the influx of the Main Donegal Pluton. Then, within this framework of more or less diapiric plutons lie the permissively emplaced bodies of Trawenagh and the Rosses.

What we are most uncertain about is the possible depths which could be assigned to the parts of the Complex, particularly to its core. We are fully aware, of course, that various kinds of models can be constructed from the geophysical evidence, but the relatively shallow depths at which the density contrasts even out below the present erosion surface are puzzling. Whatever the shape at

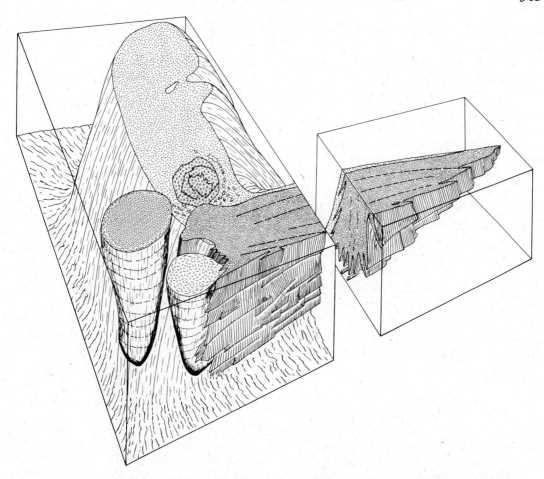

FIGURE 16-2. Schematic block diagram to show a possible three-dimensional shape of the Donegal Granite Complex.

depth, however, this is so far a rather static model we have built up, and what we have to do now is consider the relative times of emplacement and bring this model to life.

3. The Time of Emplacement Relative to the Regional Thermal Event

Unfortunately, we have as yet very little exact evidence by which to measure the difference in time, either between the emplacement of the several plutons, or between emplacement in general and the peak of regional metamorphism. Leggo's (1969) Rb/Sr based determinations of 470 ± 1 million years for both the Rosses and Trawenagh Bay Plutons do, in fact, indicate comparable times for an early stage in the crystallization history of these particular members.* On the other hand, his finding that certain of the comagmatic minor intrusions (Rosses aplites) provide dates of 430 million years suggests a longer history of emplacement than is compatible with our ideas on the formation of such intrusions. However, we are not satisfied that late-stage dykes, with all the opportunities for late-stage enrichment which exist in connection with them, may not yield spurious results. What is required

---

* We have already noted that the revision of the ^{87}Rb decay constant yields an older age of 498 ± 5 million years. The date of the main regional metamorphic event (Caledonian I) on this same basis is 520 ± 30 million years in Connemara (Leggo and Pidgeon, 1970).

here is surely the separate determination of the ages of early stage of crystallization for all the plutons by the application, for example, of the Th-Pb method on a specific fraction of the zircons (*cf.* Silver and Deutsch, 1963). The available K-Ar dates cannot do more than give a date to the late stages of a cooling (or deformation) history which is probably common to the complex as a whole; this appears to be in the range 372 ± 6 to 394 ± 8 million years (Table 4-1).

The earlier age of 470 million years may not, of course, represent the time of emplacement, but only a particular point in the early stages of consolidation. We certainly have no idea how much earlier was the actual generation or how long the magmas took to migrate upward through the crust, but the assignment of any reasonable span of time to these processes brings the generation of granite at depth nearer to the recorded "peak period" of metamorphism (520 ± 30 million years). Anyway, there is now much evidence that wholesale cooling of the Caledonian orogen must have occupied a very considerable period (Watson, 1964; Sutton, 1965; but see Fitch *et al.*, 1969), probably of the order of 100 million years, which is but another way of stating that mobile belts remain regions of high heat flow for a long time. Indeed, the middle crustal temperatures in the central zone of the fold belt may have remained sufficiently high during this period so that even a relatively slight increase in heat flow, or a relatively rapid drop in pressure, could have induced remelting and the generation of minimum melting-point fluids. On this view, acid magmas might well appear throughout the development of an orogen; and each of the several phases of deformation and uplift during the Lower Palaeozoic, whatever their cause, might well have provided the necessary changes in physical conditions leading to the production of granitic magmas. We shortly return to a discussion of this possibility, but before doing so, we must comment on the nature and possible source of the material produced at this time or times.

4. The Chemical Variation within the Plutons at the Present Level

We have seen that in Donegal there is a marked contrast between the chemical variation such as is found within the plutons and that existing within the small appinitic bodies. The appinitic rocks manifestly belong to a true differentiation series, however complex this may be, but the situation as regards the granitic rocks is even more complex and needs discussion.

The kind of variation within the Donegal Granite Complex has been determined in some detail by Mercy (1963), who also discusses these findings in the light of compositional changes within the Caledonian granites as a whole. He finds a remarkable uniformity in the chemical trends, and this is shown by the several plots of Figure 16-3 (see also Fig. 13-2). Not only do all the granites of Caledonia show a like differentiation trend, but in addition, they all have similar relative proportions of the elements Mg^{2+}, Fe^{2+}, Na^{1+}, K^{1+}, and Ca^{2+}. Further, these granites show a great regularity in the variations of the concentrations of the major and minor elements, variations which are often related in an orderly manner to geographical position within a complex or to the time order of emplacement of the units. This is, of course, a general confirmation of the earlier findings of Nockolds and his collaborators (Nockolds and Mitchell, 1946; Nockolds and Allen, 1953) who, like Mercy (*loc. cit.*, p. 208), consider that such a regular pattern in Caledonian plutonic magmatism is best accounted for on the basis that crystallization differentiation was the central process involved, contamination being of secondary importance only.

However, as is well understood (*cf.* Krauskopf, 1967, p. 399), major element variation curves alone do not constitute conclusive proof that crystallization differentiation has been active, and there certainly seem to be other possibilities in Donegal. Thus, our studies of the Thorr Pluton demonstrate that such a pattern may arise as a result of the contamination of granite magma during the engulfment of country rocks of mixed lithology (p. 120). Further, in the case of the Ardara Pluton, a trend of crystallization differentiation established earlier in the appinite suite has been inherited, albeit in a diluted form, during the process of mixing of members of this suite with a separate granite

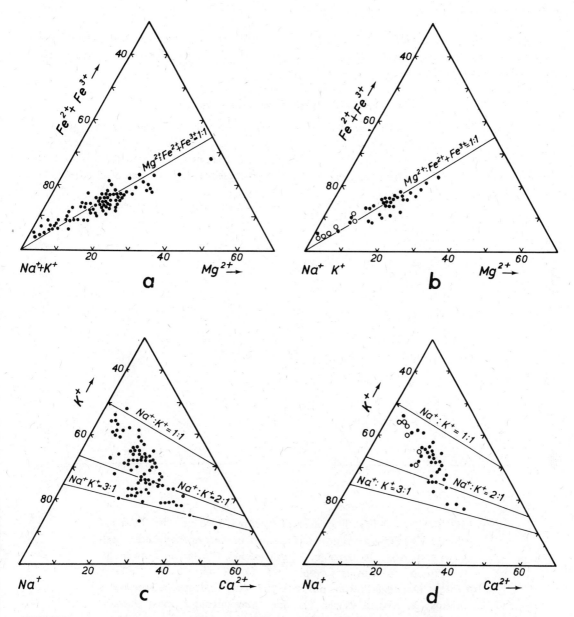

FIGURE 16-3. Relative proportions of elements in the Caledonian granites according to Mercy (1963): (a) for all Caledonian granites; (b) for the Thorr Pluton (●) and Rosses Complex (○); (c) for all Caledonian granites; (d) for the Thorr Pluton (●) and Rosses Complex (○).

magma. In addition, a scheme of magmatic evolution has been proposed for the Rosses Complex, which involves progressive partial fusion of the underlying rocks, giving rise to magma with a range of granitic compositions; this is further modified by the assimilation of low melting-temperature constitutents, derived from the material displaced during intrusion. Even in the Fanad Pluton, where it is likely that the main rock type has been derived by an essentially magmatic process, the present marginal variations are due to high level contamination. In brief, many of the variations we have been dealing with can be partly explained otherwise than on the basis of conventional differentiation theory.

Central to these explanations involving high-level contamination is the need for parent granitic magmas, the existence of which at the present level is strongly supported, as we see immediately, by a consideration of the quantities involved.

5. The Bulk Compositions

It is easy to determine the approximate surface areas covered by the different rock types in the plutons (Table 16-1), and from this calculation, it is seen that granitic types (as defined earlier, p. 91) greatly predominate.* We can also calculate the mean composition of each compositional part of these plutons on which we have sufficient analytical data: these averages are presented in Table 16-2 and Fig. 16-4a).

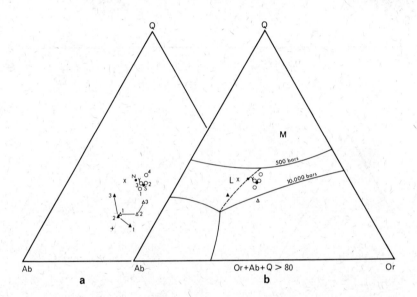

FIGURE 16-4. (a) The average compositions of the Donegal Plutons. (b) The average compositions of Donegal granites with Q + Or + Ab > 80 percent: ×, Barnesmore Pluton; Y, Trawenagh Bay Pluton; ○, Rosses Complex (1, G_1; 2, G_2; 3, G_3; 4, G_4); ●, Main Donegal Pluton, northwestern (N) and southeastern (S) variants; ▲, Ardara Pluton (1, outer component; 2, most contaminated inner component; 3, least contaminated inner component); △, Thorr Pluton (1, Dunlewy Type; 2, Thorr Type; 3, Gola Type); +, Fanad monzotonalite. Also average compositions of (L) Lewisian (Laxford type) rocks and (M) the Moine quartzo-feldspathic schists of Lough Derg; the former from Lambert (personal communication), the latter from Pitcher (unpublished work). Data concerning the experimental system $NaAlSi_3O_8$–$KAlSi_3O_8$–SiO_2–H_2O from Luth et al. (1964). Dashed line is locus of isobaric minima at water pressures intermediate between 500 and 10,000 bars.

---

* A reservation is introduced by the impossibility of knowing the former extension of the most basic variety of Thorr rocks.

TABLE 16-1. Areas Covered by Main Rock Types

	Granite (*sensu stricto*)	Tonalite and granodiorite	Quartz diorite
All Plutons	86.6%	7.6%	5.8%
Excluding Barnesmore	85.8%	8.1%	6.1%

Of course, the use of simple averages in this way is open to criticism, a matter which has been discussed in some detail by Whitten (1962). Though we must accept that the sampling on which the averages are based is not truly statistical, we would, nevertheless, point out that the result of averaging the Thorr analyses of Mercy is in no way different from the statistical mean calculated by Whitten from a proper target population: even the use of the trend surface seems unable to change the results (Whitten, 1962, taking into account the *errata*).

Another and more geological criticism is that, as one would expect from theoretical considerations, basic rocks are more likely to reach the surface than granites (Harris *et al.*, 1970), with the result that calculations of magma quantities based on plutonics alone might be quite incorrect. Any such volcanic rocks have long since been removed from Donegal, except perhaps for the traces represented by the boulders of andesitic lavas in the Old Red Sandstone outlier of Ballymastocker (p. 365), but the objection remains.

However, even with these reservations, we have to deal with the data presented to us which show, with the exception of Fanad, that the most abundant rock type in each individual pluton of the Donegal Complex is granitic in composition, a finding which suggests to us that such rocks are parental. Even so, there is a clear separation into three compositional groups of granites (Fig. 16-4a) embracing (1) the Main Donegal, Trawenagh, and Rosses Plutons, (2) the Ardara Pluton (3) the Thorr Pluton, in addition to the dioritic rocks represented by the Fanad Pluton; unfortunately we have, as yet, insufficient data to include the Toories and Barnesmore in this collation.

We very much doubt that these groups can be considered as merely terms in a steadily differentiating parent magma; rather they are of separate generation and it seems as if either the source or the physical conditions pertaining at the time of origin were different for each group.

6. The Source of the Granite Magmas

(A) The Geological Reasoning

When we approach the problem of the deep source of the Donegal magmas, we necessarily run out of direct evidence, and we do not intend to debate at length the origin of granites in general because this has been done elsewhere (*e.g.*, Read, 1957; Mehnert, 1959, 1968; Winkler, 1967). All we know is that, in Donegal, very considerable volumes of granitic magmas, derived from depth, arrived in the upper crust in a number of pulses. There is no convincing field evidence that such magmas originated by crystal differentiation, for there is nothing like the proportion of intermediate rocks we might then expect, nor any trace of a positive Bouguer anomaly, which might suggest a basic accumulate at intermediate depth in the crust. In fact, the evening out of the gravity contrast at moderate depths (Riddihough, 1969) might indicate that the source lies somewhere in the sialic crust, and we may conjecture what kind of material is present at such depths in the region of Donegal.

On the basis of the tectonic model we have provided earlier (Chapter 3), we can be reasonably sure that, in the area of the Donegal Complex, Moine and Lewisian lithologies lie deep beneath the exposed Dalradian. Additional support for this view comes from the fact that both outcrop in Donegal, and even more suggestive is the way in which the Moine, which normally lies to the north

TABLE 16-2. Average compositions of constituent parts of the Donegal Plutons*

		Rosses Complex				Trawenagh Bay Pluton		Main Donegal Pluton			Thorr Pluton			Ardara Pluton				Fanad Pluton	Barnesmore Pluton
		G_1-G_{1A}	G_2	G_3	G_4	Normal	marginal	NW	SE	"Strip"	Thorr	Gola	Contact	1	2A	2B			
Wt. %	SiO_2	73.1	74.0	74.4	76.4	72.9	75.3	71.9	71.6	63.1	65.4	69.6	65.8	62.9	65.1	70.0		60.5	71.1
	TiO_2	0.21	0.19	0.20	0.09	0.23	0.8	0.27	0.32	0.74	0.64	0.42	0.64	0.88	0.60	0.31		0.74	0.18
	Al_2O_3	14.2	13.8	13.6	13.2	14.7	13.8	15.0	15.0	17.1	16.6	15.0	17.0	17.0	16.6	15.7		18.82	15.3
	Fe_2O_3	0.50	0.56	0.50	0.42	0.57	0.14	0.47	0.52	0.99	0.94	0.96	0.71	1.33	1.12	0.60		1.56	0.40
	FeO	0.93	0.79	0.70	0.22	1.02	0.41	1.27	1.37	2.97	2.50	1.35	2.7	2.98	2.35	1.25		3.15	1.89
	MnO	0.03	0.04	0.03	0.04	0.03	0.01	0.03	0.03	0.06	0.05	0.05	0.05	0.08	0.04	0.04		0.09	0.05
	MgO	0.62	0.57	0.52	0.28	0.56	—	0.61	0.64	2.41	1.73	1.32	1.8	1.92	2.07	1.23		1.95	0.92
	CaO	1.24	0.97	0.91	0.39	1.52	0.76	1.95	2.12	3.70	2.77	1.81	2.8	3.48	2.98	1.70		3.65	1.76
	Na_2O	4.05	3.91	4.04	3.82	3.85	3.60	3.74	3.87	4.20	4.14	4.02	3.8	4.41	4.77	5:02		4.71	4.32
	K_2O	4.60	4.66	4.62	4.82	4.10	5.00	4.31	3.80	3.24	4.37	4.89	3.5	4.21	3.47	3.15		3.19	3.12
	H_2O	0.73	0.67	0.70	0.68	0.67	0.02	0.53	0.57	0.85	0.76	0.60	0.77	0.66	0.72	0.62		0.84	0.72
	P_2O_5	0.08	0.06	0.07	0.05	0.07	0.27	0.09	0.10	0.26	0.24	0.14	0.19	0.26	0.17	0.09		0.34	0.06
n		21	11	5	5	8	1	11	6	9	10	17	7	9	6	11		5	1
σ	SiO_2	1.35	1.38	0.47	0.53	2.2		0.98	1.34	4.24	1.65	3.07	1.69	1.75	3.06	1.72		2.9	
	TiO_2	0.05	0.04	0.04	0.03	0.07		0.06	0.07	0.19	0.07	0.14	0.08	0.11	0.17	0.07		0.19	
	Al_2O_3	0.67	0.75	0.76	0.77	0.84		0.33	0.82	1.12	0.42	1.05	0.47	1.03	0.43	1.05		0.88	
	Fe_2O_3	0.17	0.14	0.06	0.13	0.09		0.27	0.08	0.39	0.28	0.15	0.28	0.23	0.30	0.19		0.29	
	FeO	0.26	0.23	0.07	0.10	0.32		0.23	0.23	0.74	0.32	0.61	0.33	0.36	0.65	0.25		0.68	
	MnO	0.01	0.02	0.01	0.04	0.01		0.01	0.01	0.02	0.01	0.02	0.01	0.06	0.05	0.04		0.01	
	MgO	0.16	0.13	0.08	0.12	0.18		0.11	0.13	0.75	0.33	0.68	0.29	0.31	0.56	0.39		0.35	
	CaO	0.47	0.26	0.18	0.12	0.37		0.46	0.37	1.05	0.46	0.65	0.38	0.37	0.84	0.42		0.65	
	Na_2O'	0.14	0.30	0.26	0.35	0.21		0.37	0.31	0.27	0.29	0.36	0.30	0.29	0.32	0.20		0.15	
	K_2O	0.48	0.28	0.19	0.27	0.22		0.37	0.62	0.59	0.52	0.50	0.48	0.13	0.12	0.33		0.39	
	H_2O	0.12	0.12	0.12	0.19	0.18		0.10	0.09	0.13	0.15	0.19	0.31	0.08	0.08	0.19		0.14	
	P_2O_5	0.03	0.02	0.03	0.03	0.03		0.03	0.03	0.08	0.07	0.06	0.07	0.05	0.05	0.02		0.08	
Ab/An		6.5	7.0	9.7	2.3	4.8		3.5	3.3	2.0	2.7	4.2	2.5	3.3	2.9	5.7		2.5	4.3
F/M		2.3	2.3	2.3	2.3	2.8		2.9	3.0	1.6	2.0	1.75	1.5	2.2	1.7	1.5		2.4	2.5
Larsen Index		89.8	91.2	93.0	95.3	87.1		85.3	83.7	67.9	76.2	84.4	74.1	73.0	74.6	84.1		68.0	82.6

* Analyses of rocks from Rosses and Thorr Plutons by Mercy (1960a, b), from the Ardara Pluton by Hall (1966d), from the Main Donegal and Trawenagh Bay Plutons by P. Curtis (unpublished work), from the Fanad Pluton by M. Brotherton (unpublished work), and from the Barnesmore Pluton as published by Walker and Leedal (1954). Ardara: 1-outer monzodiorite; 2A-monzotonalite (outer part of central component); 2B-granodiorite (inner part of central component). σ—Standard deviation of n samples.

of the Dalradian outcrop, appears again in Donegal to the south of this outcrop. Indeed, in the area of the Complex, the Moine may lie but a few kilometers deep, that is, if we can judge from the close stratigraphic relationship between the Moine and the Ballachulish (Creeslough) rocks in Scotland.

In the south of Donegal, at Lough Derg, the Moine consists largely of semipelitic and psammitic gneisses, the chemical composition of which, despite the obvious variations and the inadequacy of sampling, is reasonably represented by the average, "M," plotted on Fig. 16-4b. In contrast, we can only guess at the possible bulk composition of the Lewisian. However, from the observations of Bowes and others (1968) on the Lewisian rocks on Inishtrahull, an island lying off the north coast of Donegal, it seems that the quartzofeldspathic gneisses may be generally representative of the main outcrop, so that we may accept, simply for the purpose of discussion, a central composition ("L" on Fig. 16-4b) given by Lambert (personal communication).

It is now interesting to consider whether, on remelting, such rocks could theoretically yield granitic fluids of the composition of the several granitic magmas of Donegal.

(B) The Geochemical Reasoning

In discussing the probability of remelting, we have the advantage of reference to a considerable literature on the general problem of the production of granites by experimental anatexis (*e.g.,* Tuttle and Bowen, 1958; Winkler and von Platen, 1961, 1965; von Platen, 1965; Kleeman, 1965; Mehnert, 1968; Piwinskii and Wyllie, 1968; James and Hamilton, 1969; Merrill *et al.,* 1970). Plotting the compositional data as in Fig. 16-4b, and following the experimental deductions—keeping in mind that the cotectic lines will be shifted both by changes in the anorthite content and the water vapor pressure—we can reach certain tentative conclusions.

It would seem that the psammitic gneisses of the Moine (represented by "M" on Fig. 16-4b), being quartz-rich and with potash feldspar as the dominant feldspar, would be relatively refractory. Such early mobilisates as there were would be, therefore, relatively small in volume and much richer in the components of potash feldspar and quartz than any Donegal granite at the present crustal level (*cf.* Fig. 16-3). Further, this would remain true for a considerable part of the remelting process.

On the other hand, the Lewisian ("L" on Fig. 16-4b), which in bulk composition is so near to a possible eutectic, could have easily supplied in quantity mobilisates approaching the required compositions. Thus, the Main Donegal, Trawenagh Bay, and Rosses Plutons, which have bulk compositions very near a possible ternary minimum, might well represent such mobilisates derived, it seems, at comparatively moderate temperatures and pressures. This explanation would also hold even for what are, according to us, the parent rocks of the Ardara and Thorr Plutons, provided that much higher water pressures were envisaged. Clearly, one of the factors controlling the variation in the composition of the anatexitic magmas could have been changing water pressures at the times of generation.

This explanation, involving changes in pressure, has previously been utilized by Hall (1969a) to explain what he interprets as a systematic change of composition of Caledonian granites in *space* (see, however, Chayes, 1970): apparently the magmas originated under water pressures that increased toward the center of the orogenic belt. What we suggest here is that this kind of explanation may also hold in respect of *time.* Water pressures might well fall off quite rapidly after a thermal event, and in this situation, early magmas, produced at relatively high vapor pressures, would necessarily be more intermediate in composition than magmas produced when the pressure had fallen. Perhaps the Thorr and Ardara Plutons (even the Toories) were generated under conditions of higher water pressure than the later members, *viz.,* the Main Donegal, Trawenagh Bay, and Rosses Plutons. This thesis is at least in accord with the established order of emplacement, and if acceptable, could be extended to include the monzodiorite of the Fanad Pluton by assuming that this represents the earliest of all the mobilisates and the one formed at the highest water pressure—perhaps even at the highest temperatures. Of course, depth might be just as relevant as time in explaining these differences in initial pressures, but we cannot tell which is the most important: whether the different magmas were generated at different levels of the crust or at different times.

The calculation of actual depths to the zones of remelting involves many assumptions, including the composition and concentrations of the volatile fraction and especially that $P_{H_2O} = P_{Load}$, and is probably naive in the extreme. However, if we keep in mind that in the remelting processes, the effect of increasing the An-content of the source material (von Platen, 1965) is opposite to that of increasing the pressure (Luth et al., 1964), we may expect that those pressures required for minimum melting (3–10 kbars) according to the simple system $KAlSi_3O_8$–$NaAlSi_3O_8$–SiO_2–H_2O, will be rather too high in actuality, as will be the approximately equivalent depths of 12 and 35 kilometers. Probably the range of depths concerned for the *generation* of the parental granites of Donegal is of the order of 10–30 kilometers, which is a reasonable estimate for middle crustal positions at the time of formation of the Caledonian trough.

We have already suggested that the temperatures in the middle crust may have continued high enough for remelting to remain a possibility over a considerable span of time, so that a quite modest increase in heat flow, or a fall in water vapor pressure, could have led to remelting. It seems likely to us that phases of deformation are connected with increased heat flow—perhaps deformation belts provide channels for the rise of "thermal domes" (see p. 294); and, further, we would suggest that the succeeding relaxation and uplift would be accompanied by a relatively rapid fall of vapor pressure at depth—relative, that is, to the rate of cooling. It is possible, in fact, that a series of magmas might be produced, progressively changing in composition in response to these changing conditions, and in such a manner as to explain the differing compositions of the Donegal Granites with time.

There were, according to Rast and Crimes (1968), three such periods of deformation followed by uplift in the Caledonides of the British Isles: the late Cambrian *Grampian Event* at 530–510 million years, the mid-Ordovician *Lakelandian Event,* not yet precisely dated, and the end-Silurian *Cymrian Event* at 420–390 million years. Intrusion of granite accompanied each of these events, and Bell (1968) even provides a date range of 440–460 million years (or 465–490 million years, using the lower ^{87}Rb decay constant, see p. 90), for mid-Ordovician plutonic rocks. However, the confusion concerning the significance of the recalculation of the Rb/Sr dates for the Donegal granites prevents us from definitely assigning the latter to any of these events, though formerly we were content to suppose that they were generated in the so-called Lakelandian phase.

This time model may fit with the generation of magmas at depth, but what we are certainly not clear about is the allocation of absolute times to the several phases of the peculiarly local deformation event associated with the emplacement of the Ardara and Main Donegal Plutons. It would, perhaps, be possible to include the synplutonic phases (DMG_1 and DMG_2) with the mid-Ordovician Event, and the very late cataclasis (DMG_3) with the end-Silurian Event; indeed, this allocation could probably be fitted with the record of the two kinds of age data, the 470 million years (or 498 million years) date obtained by the Rb/Sr method and the maximum of 395 million years (Table 4-1) obtained by the K-Ar method (*cf.* Berger and Cambray, 1969). However, we believe that these deformation phases were really very closely connected in time, and we hesitate to propose that the crystallization, cooling, and polyphase deformation extended over a span of 75 million years. It is, however, an interesting idea worthy of further investigation.

This linking of magma generation with orogenic events is, of course, very tenuous, both in respect to the preliminary character of the data and in the implication that such events are universal in their operation in any particular orogen. Nevertheless, whatever the complications of timing and the determination of pressures at depth, we do not think they deny the possibility that the granites of Donegal were generated largely by remelting of Lewisian-like rocks of the middle crust.

(C) The Significance of Initial Sr Isotope Ratios

Another approach to the problem of the origin of the Donegal Granites is by the interpretation of the initial $^{87}Sr/^{86}Sr$ ratio derived by the isochron method of Faure and Hurley (1963), who claim that it may be possible to distinguish between those magmas generated in the sialic crust and those derived from the mantle. The value for the Rosses and Trawenagh Bay Plutons, obtained by Leggo (1969), of 0.708 ± .002 suggests a mantle source for such rocks, which would have had a ratio of 0.704 at the time suggested for their generation (Hurley, 1968). The slightly higher value may be easily explained by the assimilation of radiogenic strontium from the country rock, just as has been shown by detailed work elsewhere by Brooks and Compston (1965). Remelting of Lewisian rocks, as suggested above, should have produced rocks with an appreciably higher ratio, such as was obtained for many of the Scottish Caledonian granites by Bell (*loc. cit.*).

The possibility of deriving granites from the mantle has been discussed by Ringwood and others (1966), and we have nothing to contribute directly to their ideas. Some intrusives, such as the Cordilleran batholiths of America, certainly do bear the stamp of derivation from basic rocks, whether it be by crystallization or remelting processes (*cf.* Piwinskii, 1968*b*); in such bodies, tonalite, diorite, and gabbro predominate over granite in the strict sense, and the presence of low strontium ratios is not at all unexpected. The derivation of predominantly granitic magmas, as in the case of the Caledonian plutons, is, we believe, a rather different matter, and we are reluctant to abandon the model involving, at least in part, remelting of the sialic crust, whether it be of metasedimentary or "primary granitic" material (*cf.* Bell, *loc. cit.*). The isotope ratios might not be so different from mantle material when we consider how much of the Lewisian is of basic parentage, either directly in the form of basic and ultrabasic intrusions, or indirectly, as greywackes derived therefrom. What we are prepared to question is whether the deep crust is as relatively enriched in ^{87}Sr as has been suggested.

Nevertheless, recent investigations of strontium and lead isotopes in granitic rocks in general indicate that mantle material is often involved in the genesis of granites (*e.g.,* Hurley *et al.,* 1965; Hurley, 1968; Doe, 1967), and a scheme of petrogenesis involving the mixing of mantle and crustal materials may afford a satisfactory model (*cf.,* Gilluly, 1963, 1965; Hamilton and Myers, 1967; Piwinskii and Wyllie, 1968). Indeed if, as is suggested by Dewey (1969), crustal material was dragged down into the mantle during the evolution of the Caledonian mobile belt, this involvement is very likely.

(D) The Contribution of the Appinitic Suite to the General Hypothesis

We have already concluded (p. 167) that the special features of the appinitic magmas—crystallization of amphibole in preference to pyroxene and the resultant effect on the course of differentiation—are due to crystallization having taken place under high vapor pressures. Hall (1967*c*) has also shown that a comparison of the compositions of those granite plutons associated in time and place with appinites, with those lacking this association (Fig. 16-5), suggests that the former also crystallized under relatively high water pressures.

Moreover, in view of their general basaltic composition, it may be assumed that the appinitic magmas are derived from the mantle. Any melting of basaltic material already in the crust, as is suggested by Joplin (1959), would seem to Hall (*loc. cit.*) to require high temperatures unlikely in a crustal environment, and certainly at about this time in the evolution of the Caledonian mobile belt basic magmas were available, as is shown by the occurrence in Aberdeenshire of gabbros with an Rb/Sr isochron age of 486 ± 17 million years (Pankhurst, 1970). Further, if crustal material is actually dragged down into the mantle during the evolution of a mobile belt, then it seems quite possible that the granitic remelts generated in this process would be accompanied by very basic magmas.

The water in the appinitic magmas could perhaps have come from the mantle (*cf.* Hamilton and Anderson, 1967; Bailey, 1970), but we favor a source from the lower parts of the crust (see, however, the reservations of Fyfe, 1970), a notion which fits well with the association of appinites with just

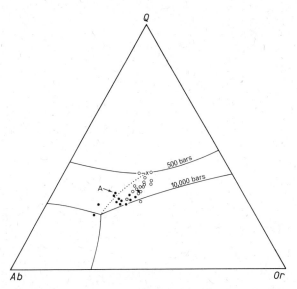

FIGURE 16-5. Q–Ab–Or plot for Caledonian granites. Those associated with appinitic rocks are indicated by (●), others by (○). The position of the granodiorite (2B, see Chapter 8) of Ardara is indicated by (A). Field boundaries in the system $NaAlSi_3O_8$–$KAlSi_3O_8$–SiO_2–H_2O are shown at 0.5 and 10 kilobars and the dotted line represents the trace of the isobaric minimum or eutectic points at intermediate water pressures. From Hall (1967c).

those granites which originated under a high water pressure. Thus, basic magma produced at the same time as these granites, and necessarily having to pass through the zone of anatexis, would be likely to absorb volatiles. On the basis of any one of the several hypotheses for the generation of granitic magmas by remelting—either by the dragging down of sialic material into the mantle or by the heating due to the ascent of basaltic asthenoliths, it would seem that the mobilization of basic rock might well be facilitated by the imbibition of water from crustal stuff.

Thus, we have a general model for the explanation of not only the special petrological features of appinites but also their close association with "wet" granite plutons and gas-drilled vents. It seems that the processes which produced granites by remelting of the sial also released basic magma from the mantle, so that we should not be surprised if mixing between these two products sometimes went further than a mere transfer of volatiles, leading to hybrid intermediate magmas, just like those which gave rise to the appinites and perhaps even the rocks of Fanad.

7. Conclusion: a Final Model

The final model, then, is one in which granitic and appinitic magmas were produced by remelting processes in the crust and mantle, respectively, during the Caledonian orogenic event. The granites were not simply produced at the peak of metamorphism and then "left" to migrate slowly into the upper crust. Rather, they were generated during each of the main deformation episodes when, in response to a slight increase in heat flow and followed by a progressive drop in pressures, a series of magma pulses of changing composition were formed.

On this general hypothesis, granitic mobilisates ought to be most abundant simply because the lowest possible temperatures and pressures are more likely to be reached; obviously, the generation of melts of intermediate composition would have required substantially higher temperatures (*cf.* Piwinskii and Wyllie, 1968).

These granitic neomagmas collected into discrete plutons which rose as a result of a buoyancy natural to their lower density (*cf.* Ramberg, 1967, 1971), albeit triggered by crustal deformation, making their way by forceful and stoping mechanisms into the upper crust and driving before them a cloud of water-rich volatiles (see p. 314).

Anatexites of this kind would probably consist of various mixtures of two materials, the granitic liquid phase of the partial melt, suspended with the "refractory" residue. It is likely that, during their ascent, they were even further diversified by the secondary processes of contamination, involving both the assimilation of solids and selective extraction of local, partial remelts. However, a limit is set to the degree of this secondary contamination by energy considerations: such remelts, suspended with residual material, could hardly carry much heat in excess of that released by crystallization. This seems to be borne out in the Donegal Complex where contamination *at the present level* is at a minimum, except in the case of the Thorr Pluton.

We must also understand that the density contrast between granites and crust is gradually reduced as crystallization takes place (*cf.* Fyfe, 1970, p. 213), though it is never lost altogether; we do not really understand the role of volatiles in this respect. It seems, however, that these "polyglot" crystal mushes and "linked-anion" liquids must often become quasi-solids in the final stages of emplacement, due to loss of volatiles in the uppermost parts of the crust. During this period and throughout the long history of cooling, a variety of solid-state reactions takes place in an attempt to redress the disequilibria necessarily induced.

The basic magmas might represent the residua from the remelting process itself, but we are of the opinion that these essentially basaltic rocks were derived directly from mantle sources. Such magmas, which were probably more mobile than the granites, were carried rapidly to higher levels, and even to the surface, by a buoyancy largely imparted by the volatile material transferred to them in their passage through the zone of palingenesis. The variation in these rocks was attained by crystal differentiation at high vapor pressures, aided by gas transfer.

In general terms, the position of the Donegal Granites in a central zone of the mobile belt presents no problem, for the generation of granite magma is most likely to occur beneath the thickest part of the sedimentary prism. Further, it seems likely that once a bubble of granite rises from depth, others will follow along the same path (*cf.* Grout, 1945; Elsasser, 1963; Fyfe, 1970).

We are not, however, clear why the granites rose into the *particular* position in the crust which they now occupy. The tight belt of folding now marking the position of the Main Pluton is largely the result of the synplutonic deformation itself, and anyway, there are too many other tight fold belts in Caledonia, most of which are quite barren of granite, for us to look for a unique, deep structural control of this kind. However, it is worth pointing to the position of the complex at the bend of the "orocline" of Western Ireland (see p. 61), a relative position it shares with the Connemara Granites much farther to the south.

We might see some significance in this location by supposing a preferred location for a deep thermal dome and a preferred path for magma at the "core" of the swing in the fold belt; at least, it is popular these days to find connections between major lineaments and earth processes!

Whatever the validity of the above model, however, the most remarkable single finding of our researches is the considerable range of contact effects at the one level in the crust. Such a variety of modes of emplacement is displayed that it seems evident that *the character of a granitic pluton, and of its relationship with its present envelope, need bear little relation to position in the crust*. The variations of contact effects are not simply records of crustal conditions changing with time, and structural classifications based on this assumption—involving epizonal, mesozonal, and katazonal granites—are evidently not valid.

PART III

LATE EVENTS

CHAPTER 17

FAULTING AND UPLIFT, EROSION, UNROOFING, AND SEDIMENTATION

CONTENTS

1. Uplift and Erosion 355
2. The Wrench Faults 357
   (A) General 357
   (B) The Age of the Faulting 359
   (C) The Leannan Fault 361
   (D) Certain Other Faults 362
   (E) Correlation with the Scottish Fault System 363
3. The Old Red Sandstone: Intermontane Molasse 363
4. The Viséan Marine Transgression 366
   (A) The Structural Setting 366
   (B) Stratigraphical Considerations 366
   (C) The Northern Delta 369
5. The Hercynian Mineralization 371

1. Uplift and Erosion

Because of the length of time involved, the history of the unroofing of the Donegal granites is probably very complex and the present degree of erosion a compound of various events. For instance, we have seen that the granites are probably pre-Middle Ordovician in emplacement, so that it is quite possible that substantial erosion of their roofs could have occurred very early, perhaps at several times during the Upper Ordovician and the Silurian. While local deposits of these ages are lacking, fragments of metasediments and granites like those of Donegal are to be found in conglomerates of Lower Silurian age in the Lesbellow inlier, only 80 kilometers to the southwest in Fermanagh (Harper and Hartley, 1938). Thus, uplift and erosion of the Donegal Highlands in Silurian times seems very likely (for recent reviews of the Silurian palaeogeography of the British Isles, see McKerrow, 1969, and Ziegler, 1970).

However, by analogy with Scotland, we can certainly expect that the main removal of material occurred during uplift at the end of the Silurian, and indeed, a trace of the resulting Lower Devonian (Old Red Sandstone) red-bed molasse is still to be found in the Fanad area of Donegal at Ballymastocker. Despite being only 11 kilometers from the nearest outcrop of the Main Donegal Granite, not a single pebble of this granite has been found within these coarse clastics. Of course, it is well known that such intermontane deposits can be entirely local in origin, deriving their material from just one valley, but the distance involved is so short that we might well otherwise have interpreted the evidence to indicate that the Donegal granite was not finally unroofed until the Devonian.

Whatever the clast content of the Ballymastocker conglomerates, their presence is clear evidence of deep erosion of the Donegal Highlands early in the Devonian. On the other hand, the K-Ar dates (Table 4-1) indicate an age for the final stage in the cooling of the granite which is well into the Devonian, a situation difficult to reconcile with the stratigraphic evidence *as it stands,* particularly the suggestion of erosion of the granites during the Silurian. One way out of the dilemma would be to treat the Ballymastocker outlier as having been downfaulted on the Leannan fracture relative to the Donegal Granite Complex, but the actual downthrow is not very great, and we would like to see the stratigraphic and radiometric evidence reexamined.

During this main period of uplift and erosion, the basement rocks of Donegal were sliced up into narrow blocks by the NE–SW trending faults which are such a special feature of the late Caledonian of Great Britian and which are clearly an integral part of the uplifting process. It may be that these faults were initiated at an even earlier date than the Devonian, and certainly, they were active at various times during both this period and the Carboniferous, and played a considerable role in controlling contemporaneous erosion and sedimentation. Thus it is likely, as we shall see, that the Carboniferous basin of Donegal is of depositional as well as structural significance and that its formation was controlled by faults.

In Lower Carboniferous times, the Devonian erosion surface was planed still further by the Viséan marine transgression, and it appears that shallow-water marine sediments covered the greater part of Donegal, if only thinly in the highlands. It seems, indeed, that the present high erosion surface represents a relict of the wave-cut platform resulting from this very transgression (Dewey and McKerrow, 1963); it is not of Cretaceous or later age, as is often assumed (see p. 3), though it may well have been modified at a later date.

There is not much direct evidence of what exactly happened in Donegal during the long interval between the Carboniferous and the Pleistocene. Nevertheless, it is likely that considerable erosion, even to the extent of stripping some of the Carboniferous cover, took place as a result of Permo-Triassic uplift. Thus, there is evidence that only a little way over the borders of the county, in Derry, red-bed conglomerates of this general age, with clasts of schist and granite, rest unconformably on the eroded Lower Carboniferous of the Lough Foyle basin (Charlesworth, 1963). We can reasonably surmise, then, that there were western highlands in New Red Sandstone times, and that in these ranges the sub-Carboniferous platform was being actively exhumed.

How far the major transgression of the Mesozoic entered the area of Donegal we can never know. The superimposed nature of the drainage, as we have previously noted (p. 3), certainly suggests the former presence of a cover, but this is just as likely to have been of Carboniferous as of Mesozoic or Tertiary age. All we can say is that there is some evidence, at least, for the former presence of a considerable extension of the plateau basalts of Antrim (p. 375), though whether they were poured out onto plain or mountain we do not know. We suspect the former, which implies that the sculpturing of the present landscape is largely of mid-Tertiary age (*cf.* George, 1967).

In view of the possible complexities in the history of unroofing, and certain doubts about the age of the granites, any attempt to calculate the thickness of the roof at the time of emplacement can be of little value. In order to make such an estimate, we might assume that the Dalradian represented continuous sedimentation until well into Cambrian times, also that above Ballachulish (Creeslough) rocks lay a full thickness of the type Perthshire (Kilmacrenan) Dalradian. Then, leaving aside the structural complications, this could lead to estimates of thickness of metasediments, removed from above the granites, of the order of 16 kilometers of rock (see p. 10). But we might well have a situation, as we have earlier proposed, where intrusion was a consequence of uplift, implying that *erosion may have removed much of the Dalradian roof before and even during emplacement.* As a result of this, the roof of the Donegal Granite Complex remained quite thin, a fact which fits well with the character of the Rosses cauldron and the presence of diatremes, but makes the special features of the Main Donegal Granite even more remarkable.

In what follows, we shall briefly describe these late events in the geological history of Donegal, dealing first with the faulting, then with sedimentation—both subaerial and marine—remembering, of course, that these processes were intimately connected.

2. The Wrench Faults

(A) General

The important NE–SW trending faults which slice up the Dalradian rocks of northwestern Ireland are only part of a system which includes the Highland Border and Great Glen Faults of Scotland (Pitcher, 1969). In fact, the former fault enters and crosses Ireland, forming much of the southeastern boundary of the Dalradian outcrop, while the latter almost certainly lies but a little distance from the north coast of Donegal, where its position is indicated by the distinct linear anomaly on the new aeromagnetic maps (Published by the Institute of Geological Sciences; see also the Institute's "Tectonic Map of Great Britain," and Riddihough, 1968; Riddihough and Young, 1970).

Of the group of subparallel faults which lie between these two master fractures, the Leannan is the central and most important member (Pitcher *et al.*, 1964), and its location and relation to the associated structures is shown in Fig. 17-1. All these faults map as highly inclined fractures, on which both strike-slip and dip-slip movements can often be proved to have taken place.

Faults oblique to the main trend are not common, but there are some lesser oblique fractures that take up part of the movement of the main set, as is shown best by the branches of the Gweebarra fault. Also, where the main faults bend, as the Leannan itself does northeast of Killybegs (Fig. 17-2), the faults split up into a series of divergent fractures. In fact, as a whole, the pattern of the belt is one of gently curving subparallel faults isolating great slices of country.

The geological maps show clearly the considerable effect these faults have on the outcrop pattern, particularly in central Donegal. This is because complexly folded rocks with near recumbent structures, commonly showing tectonic thinning and thickening in the limbs, will naturally produce a maximum of outcrop change across quite a minor fault.

Of course, the contrast between the rocks flanking the Leannan fault is the most marked (see Fig. 17-1 and p. 74). On the northwestern side of this fracture, the outcrops of the same succession of metasediments run parallel to the fault line for 96 kilometers, and over this distance, they show little change in lithology, structure, or metamorphism. In contrast, on the other side, the geology varies considerably along the fault; the Ballybofey and other folds are truncated, and rocks different in lithological facies and metamorphic state are brought into juxtaposition. This was thought (Pitcher *et al., loc. cit.*) to be the result of a left-handed strike-slip movement of some 40 kilometers, a suggestion strongly confirmed by the later work (p. 362), and one which reinforces the opinion that the Leannan fault is probably a splay of Great Glen Fault itself (Fig. 17-4 and Riddihough and Young *loc. cit.*).

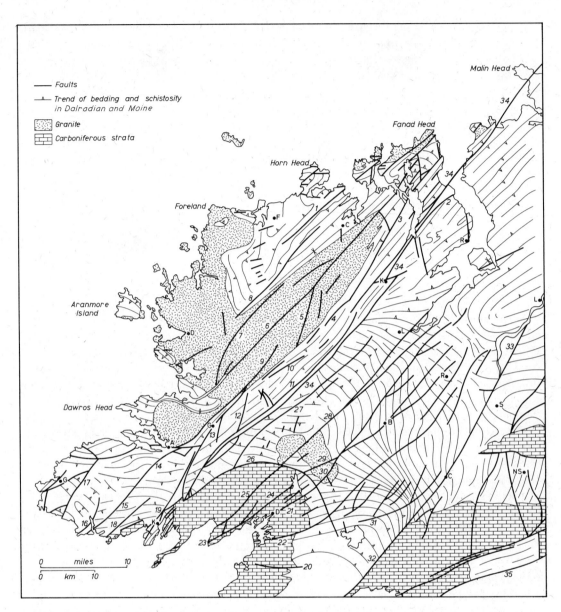

FIGURE 17-1. The fault system of northwestern Ireland (for town names see Map 2 in folder): 1, Lough Agher; 2, Knockalla; 3, Lough Salt; 4, Mossfield; 5, Stranagloch; 6, Gweebarra; 7, Owennamarve; 8, Dunlewy; 9, Errig; 10, Finn; 11, Lough Ea; 12, Aghla; 13, Carbane; 14, Glengesh; 15, Ballaghdoo; 16, Carrick; 17, Slieve League; 18, Glenaddragh; 19, Killybegs; 20, Brownhall; 21, Donegal; 22, Laghy; 23, Mountcharles; 24, Burns Mt.; 25, Eglish; 26, Boundary; 27, Carnaween; 18, Belshade; 29, Barneslough; 30, Barnesmore; 31, Derg; 32, Pettigo; 33, Lough Foyle; 34, Leannan; 35, Highland Boundary.

(B) The Age of the Faulting

The major faults may well have been located along structural lineaments early established in the crust; indeed, this is very likely to be so in the case of the Highland Boundary Fault and its extension into Ireland (*cf*. Pitcher, 1969; Rast and Crimes, 1969). Nevertheless, there is no evidence for such protofaults in Donegal itself, for it is particularly clear that the northeast-trending lateral-slip faults as a whole are later than all the granites, as well as their attendant appinitic intrusions, and lamprophyres and felsites. There is, however, a little evidence that the span of intrusion of certain late dykes overlapped with faulting: thus, a felsite occupies a NE–SW fault near Doagh in Rosguill, while another lies within the Leannan fracture zone near Kilmacrenan and seems only to be brecciated by the latest movements on that fault. Further, Leedal and Walker (1954, p. 119) hold that the important faults that cut the Barnesmore granite were initiated before the injection of the microgranite dyke swarm.

Limits to the age of the main movement can best be set by reference to those strata unconformably overlying the metamorphic rocks. There is no doubt that the Lower Devonian red beds of Ballymastocker have been faulted into their present position (see p. 363), but it is possible that they represent a valley fill along a precursor of the Leannan Fault. Evidence of multiple movement on some of the faults is provided by reference to the Carboniferous rocks of the Donegal syncline (p. 366). Many of the faults cutting these gently dipping strata have exactly the same attitude, and are even continuous with those cutting the basement. Thus, the Eglish and Burns Mountain–Mountcharles faults are represented outside the Carboniferous outcrop by the Belshade and Barneslough faults, respectively, but there is such a difference in the sense and amount of displacements in these two situations that it seems that the component recorded in the Carboniferous is the result of a posthumous movement. For example, along the Belshade line a left-lateral slip of 3.4 kilometers is compounded with a southeasterly downthrow of the order of 600 meters, but only the latter movement is recorded within the Carboniferous strata (Pitcher *et al., loc. cit.*, p. 252). Even more critical evidence of this kind is to be found where the Carboniferous boundary straddles the Leannan fault (Fig. 17-2) northeast of Killybegs; clearly, the main movement was pre-Viséan, and posthumous faulting in the Carboniferous is here of small importance.

It is interesting to follow these conclusions by reference to the sedimentation in the Carboniferous, for, as we shall show later, the presence of great changes of facies, with conglomerate wedges spreading southward from a steep littoral, strongly suggests a control by downsagging to the southeast on the very faults we are describing.

Finally, we may note that the Tertiary dykes normally cut the faults and even run along them for short distances: there are a few in this position, however, which are very closely jointed and even slightly crushed, though never displaced.

In summary, then, the main lateral movements were almost wholly later than the local Caledonian igneous activity, though there is a slight overlap with certain late dykes. Some strike-slip movements may be older than the early stages of the Old Red Sandstone deposition, though probably the most important lateral movements occurred between the Lower Old Red Sandstone and the Viséan. Significant dip-slip movement occurred on some faults in post-Lower Carboniferous times, and there were probably minor later adjustments on the same lines.

This pattern of repeated movement is generally similar to that found in the case of the NE–SW faults of Scotland, though in that situation, the chronology can be determined in somewhat greater detail. The Great Glen structure, for instance, was initiated and probably underwent its major lateral displacements at about the Lower Devonian–Middle Devonian boundary, and was reactivated, though to lesser effect, after deposition of the mid-Devonian (Pitcher, 1969). Clearly, once a grain for the relief of stress had been set up early on, it controlled movement for the rest of geological time.

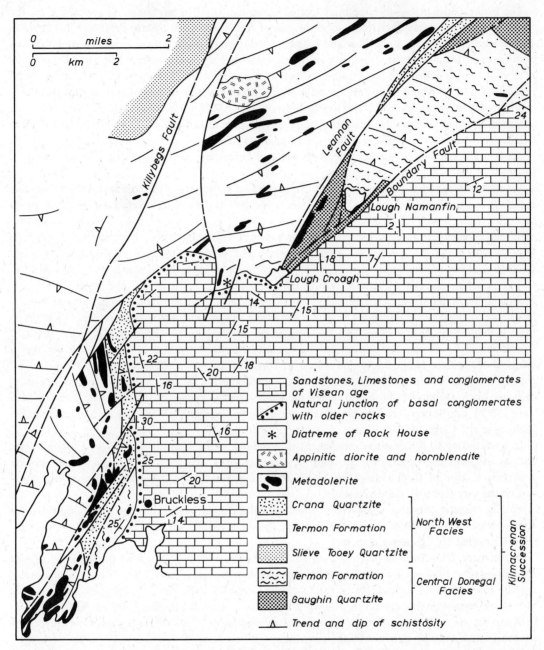

FIGURE 17-2. Sketch map of the geology around Bruckless to illustrate the time relation between faulting and the deposition of the Viséan sediments. After Pitcher *et al.* (1964).

(C) The Leannan Fault

The Leannan Fault is so important a structure in the geology of Donegal that it deserves special comment. The details of the 100 kilometer trace, shown in Fig. 17-1, are given elsewhere (Pitcher et al., 1964), and we need only to provide a summary here.

We should, perhaps, first point out that the line of this considerable fault is not expressed by a topographical feature comparable with the Great Glen of Scotland. Nevertheless, long stretches of its course are marked by a fault-line scarp on one side or the other, or by a groove, and near Kilmacrenan it runs through the broad valley of the River Leannan; in fact, in all but rare situations its precise position as a continuous line can easily be demonstrated.

However, even though the original investigators hesitated to continue the main fault directly into the Malin Head peninsula, further work does show that it passes through a valley on the southern limit of this headland. Along this particular feature, very different kinds of quartzite are brought into juxtaposition, and within the valley itself there are outcrops of lithologies quite foreign to the succession on either side. Accepting this revision, we then have the Leannan Fault striking out into the sea south of Inishtrahull, from whence its exact path is conjectural. Certain possibilities do, however, arise from the magnetic surveys of Riddihough (1968) and Riddihough and Young (1970), and one of these involves a proposal that a cross fault between Malin and Inishtrahull offsets the Leannan to the southeast.

As might be expected, the fault line is in many places a composite structure, being made up of a number of closely spaced faults, each with its own crush zone. There is, however, always one fault of far greater importance than the others, and this invariably forms the southeastern margin of the fault complex. These several faults together isolate long strips of metasediment, representing *inverted* parts of the Kilmacrenan succession (Fig. 17-3; see also p. 69 and Fig. 3-9). In the southwest, these are composed of much-broken remnants of the Slieve Tooey Quartzite and the Cranford Limestone, but in the north, it is the Upper Crana Quartzite which occupies this position, forming a broad band between the boundary faults.

Along the main fault, the zone of most intense deformation is near vertical and between 60 and 100 meters wide; the main observable effect of the movement is brecciation, and steep movement surfaces plastered with coarse breccia are much more common than smooth slickensided surfaces. The ultimate product of the deformation is a centrally placed, dark, siliceous, banded mylonite which is sometimes cut by a second generation of closely spaced, slickensided planes showing both gently plunging and steeply plunging striations, though on different surfaces. Clearly, several episodes of movement are involved; perhaps the mylonite represents intense lateral movement, and the later, striated slickensides, two periods of rejuvenation—one dip-slip, and the other strike-slip.

In view of the magnitude of the Leannan Fault and this polyphase history of movement, it is not easy to establish the kind and amount of each of the several displacements. An interpretation of predominantly lateral displacement is, however, supported by the straightness of the fault trace; by the fact that the oblique cleavage, shears, and drags within the fault zone are predominantly vertical; and by the association with parallel dislocations of proved horizontal displacement, namely the Doon and Carnaween faults.

The most obvious proof of great lateral movement is, however, provided by the structural contrasts across the fault, which are incapable of explanation by simple dip-slip displacement. It may be that differences of structure of this kind could result from different response to deformation within each of the separated blocks, especially if the present faults follow much older lineaments. However, as we showed in Chapter 3, a restoration on the basis that a left-handed strike-slip of 40 kilometers has occurred leads to such close agreement between structural and metamorphic patterns on opposite sides of the Fault that movement of this magnitude can hardly be denied.

To explain the inverted strips between the component faults of the Leannan complex, it was

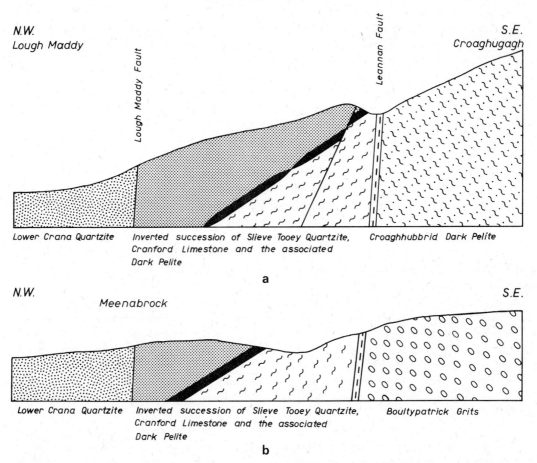

FIGURE 17-3. Sketch sections across the Leannan Fault-zone in west-central Donegal (see also Fig. 3-9). Length of sections approximately 0.8 kilometer. After Pitcher *et al.* (1964).

necessary to invoke very considerable dip-slip on the bounding faults (see p. 70), and this situation is confirmed by the downthrow of the Old Red Sandstone deposits of Ballymastocker.

Thus, it may be concluded that the Leannan Fault represents a wrench structure with a left-handed strike-slip of 40 kilometers and an apparent dip-slip of some thousands of meters to the southwest: obviously, the next stage in its investigation is to determine the geophysical contrasts across this structure and its effects on the distribution of the K-Ar cooling ages. At this moment, we remain attracted to the notion that it represents at least a splay of the Great Glen Fault.

(D) Certain Other Faults

Of the other faults of the same type in Donegal, apart from the close associates of the Leannan, we briefly draw attention to the Gweebarra, Mossfield–Lough Salt, and Barnes Lough faults (Fig. 17-1), which appear to show right-handed components of movement, and the Belshade (Leedal and Walker, 1954) and Pettigo faults, which have left-handed displacements. The details are provided elsewhere (Pitcher *et al.,* 1964), but we may mention that the probable northeasterly extension of the Belshade fault is important enough to produce a feature on the gravity anomaly map of the area around Lough Foyle (Bullerwell, 1961). Taken together, these faults indicate that approximately strike-slip displacements were the rule; the movements were, in fact, oblique-slip but more nearly lateral than up and down.

Within the blocks of country which they separate, one might expect to find minor structures such as joint systems, slickensides and striations on joint surfaces, and joint drags that might fit with the pattern of faulting. Examination of joint surfaces in quarries shows that movement did occur on all scales, and clearly part of the movement on the faults was distributed and dissipated in this way. However, we have not made enough measurements of this type to be of value, though as far as we know, the directions of the joint system bear no simple relation to the faulting. In Glencolumbkille, Howarth and others (1966, p. 146) do, however, relate some of their late-stage brittle structures to northeasterly striking faults; these are strain-slip cleavages and open folds of local importance, labeled F_5 and F_6 in their extended deformation nomenclature (see further comments in Chapter 3).

(E) Correlation with the Scottish Fault System

We have already referred to the obvious correlation of the fault systems of Scotland and northwest Ireland (see Pitcher, 1969), and therefore content ourselves by reproducing an outline map of the whole system, with certain recent amendments (Fig. 17-4). Detailed correlation between the lesser faults in each situation could obviously be made, *e.g.*, the Pettigo–Lough Foyle fault line might well line up with the Ericht–Laidon fault (see Bullerwell, 1961), and the Belshade with the Laggan, but we should perhaps await the new geophysical work being carried out in the North Channel.

In the southwesterly direction, the Leannan may pass under the Carboniferous cover of Mayo and reappear in Clew Bay (Phillips *et al.*, 1969), so approaching very close to the possible extension of the Highland Border Fault, which is here bending westwards. In fact, there appears to be a general convergence of the great faults off the west coast of Ireland, and of great interest is their possible extension to the northeasternmost Appalachians in Newfoundland (Wilson, 1962; Kay, 1967; Dewey, 1969; Pitcher, 1969); but this enters the speculative field of continental drift.

Finally, we may ask which stress system we are to identify as the cause of faulting within long term, polyphase tectonics of this kind, where the master lineaments are probably set up at an extremely early stage in the evolution of a mobile belt?

3. The Old Red Sandstone: Intermontane Molasse

What little now remains in Donegal of the debris removed by subaerial erosion of the uplifted highlands is represented by just one small patch of red beds, preserved in a narrow fault trough at the foot of the Knockalla escarpment at Ballymastocker (Fig. 17-5).* Here, 250 meters of coarse clastic deposits rest unconformably on the Upper Crana Quartzite and, together with the latter, form the local representative of the exotic fault strip, which as we have seen, nearly always lies between the two major faults of the Leannan complex. Within this strip, the actual unconformity is only very locally seen, and then only uptilted into the vertical in a separate fault sliver adjacent to the main Leannan fault; here, the surface of the underlying quartzite is brecciated and the cracks infilled with sand, as reported by Andrew (1951).

The gentle dips of the main outcrop reveal a succession of a coarse basal conglomerate, followed upward by sandstones with pebbly and silty seams, and topped by conglomerate. This latter is rather spectacular in outcrop, with boulders of mixed origin up to one meter in diameter, and in it a crude bedding, picked out by the sandy seams, dips more steeply and in a manner quite contrary to the general dip, suggesting that the conglomerate represents a fanlike torrential deposit.

---

* The only other reported outcrop (McCallien, 1937), at New Bridge Bay north of Rathmullen, is likely to be a raised beach deposit.

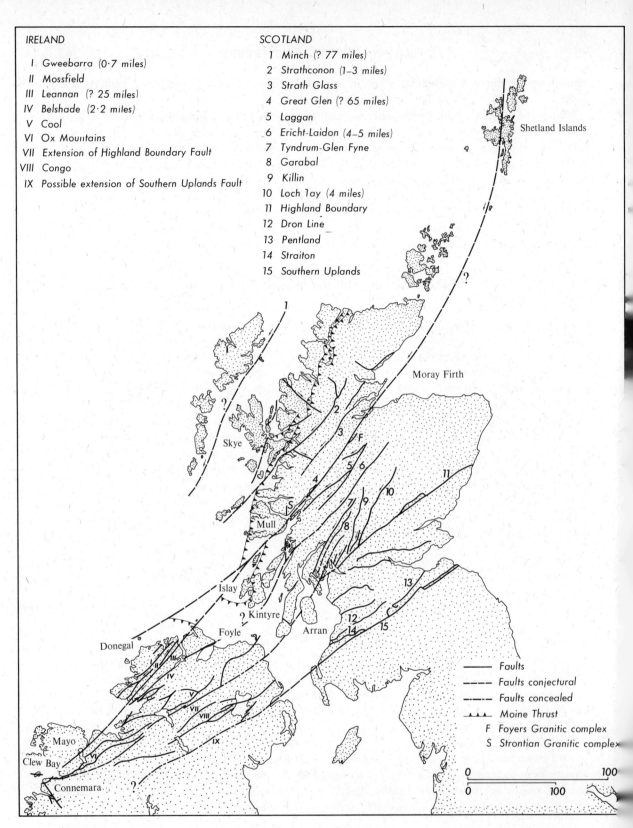

FIGURE 17-4. The faults of Scotland and Ireland. Where known, the amount of strike-slip is given in miles. After Pitcher (1969).

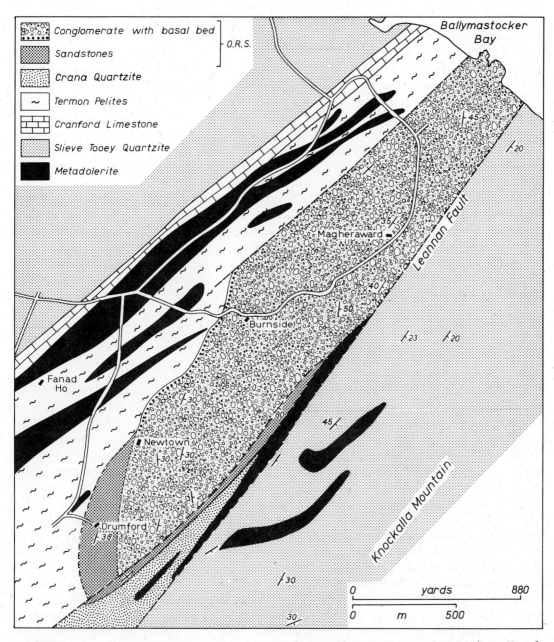

FIGURE 17-5. The Old Red Sandstone outlier of Ballymastocker, near Portsalon, Fanad, Donegal.

The boulders, which commonly have a high degree of rounding, are composed of the local basement rocks such as gritty quartzite, quartz schist, and examples of Dalradian metabasites, but there is also a fair number of clasts of a reddened porphyry of extraneous origin. Of particular interest is the nearly complete lack of boulders of the white quartzite of the adjacent Knockalla escarpment, and it seems likely that this was not exposed in the vicinity at the time of deposition of the red beds, thus confirming the very considerable movement which we have already attributed to the Leannan fault

(p. 362). Then again, though boulders of granite had been previously reported, long search has failed to reveal any during the present investigations, a point we have also previously commented on. As for the "porphyry," this is strongly reminiscent of the minor intrusions found in association with the Old Red Sandstone deposits of Antrim, and it suggests to us the former presence of Devonian igneous rocks in northwest Ireland.

The sands and sandy matrices of the conglomerates are arkoses rich in microcline; they are normally cemented by iron oxides, but chlorite is reported locally (Bishopp, 1951), which is not at all surprising, in view of the abundance of chloritized metadolerites in the adjacent Dalradian. According to Wilson (1953), the predominant heavy minerals are fresh-looking ilmenite, zircon, tourmaline, and anatase (the latter is both authigenic and detrital), while rutile and garnet are rare, and staurolite and kyanite are completely lacking. The latter observation indicates that the provenance of this assemblage was entirely located in the adjacent regional low-grade metamorphics—there is, for example, plenty of microcline in the quartzites, and tourmaline in the pelitic schists.

The mode of origin of the Ballymastocker beds is not difficult to envisage as one where intermontane conditions were coupled with rapid subaerial erosion and where deposits formed in temporary standing water were overwhelmed by torrential outwash fans. Such materials are likely to be entirely local in origin, even completely separated from similar intermontane deposits, so that detailed correlations are impossible. However, they are so like those of Lower Old Red Sandstone age in other northern areas in the British Isles (Wilson, 1953; Bluck, 1969) that their stratigraphical position is reasonably assured (see Geikie *in* Hull *et al.*, 1891; but see p. 356).

4. The Viséan Marine Transgression

(A) The Structural Setting

In northwestern Ireland, marine deposits of Lower Carboniferous age are preserved in the four shallow synclines of Ballina, Donegal, Omagh, and Lough Foyle. Of these, we shall deal in detail only with the Donegal Syncline, mainly because it alone is represented in our map area and also because it has been the subject of several recent studies (George and Oswald, 1957; George, 1958; Hubbard and Sheridan, 1965; Hubbard, 1966a, b). Nevertheless, the general setting is of great interest, as is shown by reference to an isopach map prepared by George (Fig. 17-6), for this shows dramatically the general correlation between downwarping, thick accumulation, and the structural frame for the whole area. Very clearly, these are basins of deposition as well as of structure.

In western Donegal, along this line, the Donegal Highlands still form a northern wall which can be taken to reflect the Carboniferous topography. In the sediments themselves, the very rapid facies change northwestward, from the limestones and shales of shallow seas to the conglomerates and gravels of confluent deltas, indicates very clearly the proximity of a northwestern shoreline and an extensive and topographically varied land mass beyond.

The basins are formed, then, as a response to downsagging, probably on the NE–SW trending faults. In this process, some element of tilting of the blocks of the basement must have been involved because the basal beds, and therefore presumably the unconformity itself, can dip at up to 30° away from the periphery of the basin. Strangely enough, this unconformity is often obscured by a boundary fault, and even where the basal beds are seen within a few meters of the metamorphic rocks, the contact is very rarely displayed.

(B) Stratigraphical Considerations

Not surprisingly, within so small and marginal a basin, the facies changes with these rocks are so marked, and the beds so diachronous throughout the whole of the 1300 meters of shallow-water

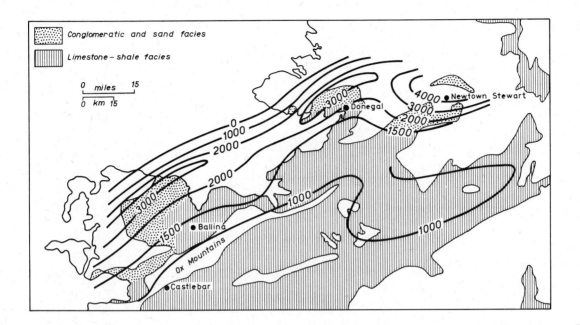

FIGURE 17-6. Map showing approximate isopachs (in feet) of the Lower Viséan rocks (Ballyshannon and Doorin-Bundoran groups) of northwestern Ireland. The correlation between downwarping and thick accumulations of deltaic sediments is very clear. After George (1958).

sediments, that it is only possible to utilize local nomenclatures for discussing the stratigraphic relationships (Fig. 17-7 and George and Oswald, *loc. cit.*, p. 140). A reference succession well within the basin is as follows:

3. Kildoney and Mountcharles Sandstones
2. Doorin–Coolmore (= Bundoran) Shales
1. Ballyshannon (= Pettigo) Limestone Group.

Particularly evident is a thickness of the conglomeratic basal beds and a concomitant replacement of calcareous by arenaceous lithologies, marking the alternation and transition between deltaic and neritic conditions of sedimentation. This is well demonstrated in the areas covered by our present study, particularly at Bruckless Harbor, by lateral facies changes within the 40 meters of calcareous shales and limestones (an extension of the Ballyshannon Limestone and Doorin shales proper), intercalated between the basal Bruckless "grits" and a thick wedge of the Mountcharles Sandstones (George and Oswald, *loc. cit.*). These calcareous beds pass quickly and laterally into an almost unbroken sequence of conglomerates and sandstones, so that the equivalent marginal deposits on the northern edge of the basin are almost entirely coarse and arenaceous. Yet even within these conglomeratic arkosic-sandstones, there are intercalations of fossiliferous shales and impure limestones such as those which occur on Bonagher Hill, also in the Clady gorge, southwest of Lough Eske and just outside our map area, on Killin Hill, and around Calhome, a locality to the north of Bruckless (for details, see George and Oswald, *loc. cit.*, Sections IIIa, IIIb, and Vc).

Because of these facies changes, the lithostratigraphic divisions are largely lost in these northern outcrops; so, too, are the rich fossil assemblages associated with the limestones. Where they do occur,

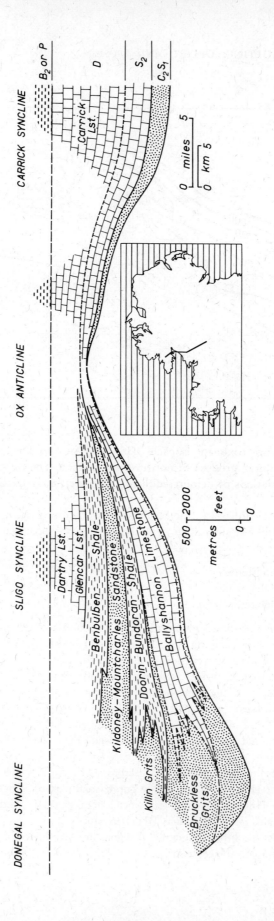

FIGURE 17-7. Sketch section across the Donegal basin and Ox Mountains ridge along the line shown in the inset. Shows the lateral changes in thickness and facies of the Viséan rocks of the northern downwarp, and the influence of contemporary shoaling along the anticlinal ridge in defining the northern basin and limiting the southward spread of terrigenous sediment. After George (1958).

they are life assemblages rich in entire specimens, and the most abundant forms are corals such as the caniniids, lithostrotionids, and auloporids, also brachiopods such as the daviesiellids; indeed, the brachiopods sometimes predominate over other forms. Crinoid debris and algal material are abundant throughout (for full faunal lists, see George and Oswald, *loc. cit.*).

The fossils are characteristically Upper Caninian in aspect, so proving the Viséan age of the local Carboniferous transgression. However, apart from assemblage differences that so clearly reflect the different depositional environments, the distribution of fossils within Donegal shows no significant change in upward succession, though over a wider area and on the basis of detailed population studies Hubbard (1966a) has detected sufficient variation to suggest that small time differences are represented. Thus, the Ballyshannon Limestone of Sligo may be slightly younger (S_2D_2) than its lithostratigraphic equivalent in Donegal (S_2); a diachronous situation which is not unexpected.

Hubbard comments in detail on the environment as determined from the life assemblages. Her conclusion is that conditions during the formation of the limestones lay between those of a dystrophic tropical belt and the shallow warm waters of an open shelf—though at any one site, a rapid alternation of environment is shown by the changes in the coral associations higher up in the section. These findings fit well with the general palaeogeographical picture provided by George (*loc. cit.*) whose account we now reproduce in large measure.

(C) The Northern Delta

Apparently the Viséan rocks of northwestern Ireland were laid down on the flanks of a considerable land mass, with high relief extending indefinitely northward, and drained by fast flowing rivers carrying much detritus southward into the Viséan sea. South and southeast of a main shoreline (Fig. 17-8), stretching southwest from northwest Donegal to Belmullet in Mayo, there were small islands, probably periodically revived by posthumous folding.

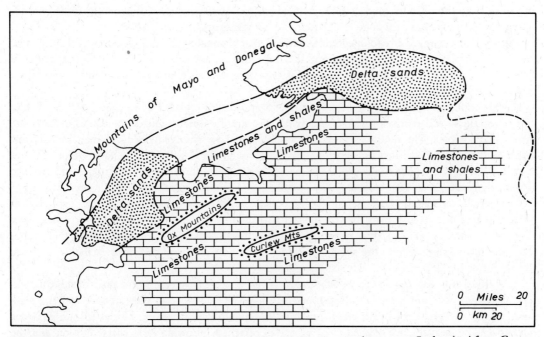

FIGURE 17-8. Paleogeographical map of the Viséan in northwestern Ireland. After George (1958).

At a few localities, this shore line may be exactly located, and elsewhere its near proximity can be safely inferred. However, the marginal sediments are not typical beach shingle but have all the characteristics of delta beds laid down where the flow of turbulent rivers is checked. They consist of boulder beds, gravels, sands, and silty shales, showing great lateral variation over short distances, and often churned up and slumped. Within a few meters, tongues of coarse conglomerates wedge out, completely massive sandstones pass into thin-bedded silts, and carbonaceous seams thicken dramatically. Occasionally, it is possible to identify gravel tongues extending seaward, suggesting the existence of powerful distributaries in a form not unlike those of a bird's foot delta. There is, in fact, ample evidence (Hubbard, *loc. cit.*) of the migration of such distributaries, *viz.*, penecontemporaneous erosion, slumping into channels, and the alternation of marine and terrestrial deposits, together with the presence of marine and brackish-water ostracod faunas.

The detritus in these largely clastic rocks consists mainly of metamorphic quartzite, strained quartz, schist, and mica, also microcline and plagioclase, all derived from obvious sources in the Donegal Highlands. Perhaps an unusual feature is the persistent arkosic composition of the beds, even where they form part of the highly calcareous sandstones of the marine intercalations. Sorting is almost invariably poor, and the details of the mineralogy confirm the macroscopic lithology in its evidence of delta sedimentation in a marine environment. Except in the calcareous layers, fossils are rare, though the carbonaceous content of many beds and the occasional sandstone casts of large lycopodian fragments hint at a land or swamp vegetation not far away.

The northerly source of the material is very well demonstrated, not only by the special distribution of the sedimentary facies, but by the directional characteristics of the sediments, such as the thinning of sand tongues, the cross-stratification, and the increase in slump structures.

Southward, and seaward in terms of Carboniferous palaeogeography, a convincing picture has been built up of a transition across a delta front. All the characteristic elements have been identified and described in terms of sedimentary facies (George and Oswald, *loc. cit.;* Hubbard, 1966) from terrestrially colonized deltaic (arkoses with rootlet horizons), transitional to perideltaic (thin-bedded sands and silts), into the offshore muds (thinly bedded calcareous shales, often abundantly fossiliferous), and finally to an offshore clear zone (cross-stratified oosparites and oomicrites).

Within these changing environments, it is natural that there should be a variation in the styles of the cross-stratification and the character of the trace fossils, an aspect which has been the special concern of Hubbard (1966a). We content ourselves here, however, in continuing this account in a general vein.

Thus, continuing "seaward," the deltaic sands and gravels are replaced by or merge into muds and limestones, the lateral changes in facies occurring very rapidly over distances sometimes of no more than a mile or two. The muds are often silty and may show slumped bedding; they contain few fossils and look like the deposits of quieter embayments or shallows within the delta environment. But more often they are of open sea origin, calcareous with many thin ribs of limestone, and may be abundantly fossiliferous with vast numbers of zaphrentoid and caninioid corals and a great variety of orthoid, strophemenoid, and spiriferoid brachiopods. They are perideltaic in character, deposited in water perhaps little deeper than that of the deltaic sands (for some of their bedding planes show abundant dessication cracks), but beyond the range of the delta distributaries. Along their inner margins, they intertongue with the pebbly quartzites and sandstones; and there was oscillatory advance and retreat of the delta front as sedimentation continued.

Still farther seaward, the shales become progressively more calcareous; but simple offshore lateral facies sequence of sandstones-shales-limestones may have its middle term unexpressed, and the delta sands, within a few miles of the inferred shoreline, merge into typical massive bioclastic Carboniferous Limestone. The local absence of shales may be due to current winnowing or to coarse erosion; but the result is a great variety of transitional rock types, from slightly calcareous sandstones (whose abundant quartz grains may be encased in oolitic skins) to the purest crinoidal and shelly calcarenites (George and Oswald, 1957, Fig. 15, p. 175), often with a sparite cement. The purity may be a measure of clear water, but it is not a measure of depth, for the bioclastic limestones are interbedded with oolites and algal calcarenites (in which aggregates of detrital grains are cemented in an algal bond) that could not have been formed on a sea floor more than a few fathoms deep. While the belt of limestone sedimentation was relatively offshore—though only a short distance offshore—in the regional environment, the rocks are strictly shallow-water neritic; the seaward slope of the floor being at most a few feet to the mile.

The reconstructed facies belts (Fig. 17-8) are discontinuously exposed at the present day, but the depositional strike leaves little doubt of their former continuity, and they follow a general Caledonoid direction from Derry into Mayo. A feature of the belts is that in a thickness of thousands of feet of sands and gravels, only comparatively minor gradations are to be recognized, either in grain size of the sediments or in loci of facies change, suggesting a progressive northward encroachment of the sea through simple subsidence. It is to be inferred that the intense erosion of the hinterland must have been maintained by continued revival of relief of the upland tracts, presumably on the fault lines we have described. There is little evidence that the early Viséan sea ever extended far beyond present outcrops, and in fact, the Donegal mountains of the present day may well be the residuals of mightier ranges of Tournasian times.

5. The Hercynian Mineralization

A late event in the geological history of Donegal is recorded by the post-Carboniferous mineralization. Unfortunately, there is no important ore deposit, but in common with the rest of Ireland (Cole, 1922; O'Brien, 1959), Donegal has its scattering of old mine workings and trial pits from which the ores of lead and zinc have occasionally been obtained. Something of the history of the discovery and removal of this ore is referred to in the relevant Memoirs of the Geological Survey of Ireland, so that we content ourselves here with providing a brief discussion of the main geological factors. Suffice it to say that, from the economic point of view, the veins are small, the disseminations local, and the likelihood of any large body of ore having remained undetected at the surface is, in our opinion, slight.

There are three principal bodies within the area covered by our detailed survey: one is located at Keeldrum, southwest of Gortahork (see Map 1 in folder), another runs within the southeastern aureole of the Main Donegal Pluton between Glenboghill (Fintown) and Glendowan, and yet another in the Faymore area in the northwestern aureole of the same Pluton. Unfortunately, there is now little to see at any of these localities.

At Keeldrum, it seems that sulphide mineralization has followed the boundary between the Slieve Tooey Quartzite and the overlying pelite-limestone assemblage of Falcarragh: here, the Horn Head Slide (p. 56) is providing the main structural control, though we also think that late faulting is concerned because brecciation has been reported in association with the actual lode.

At a number of localities between Glenboghill and Glendowan, mineralization occurs in association with a NE–SW trending fault which appears to be one of the lesser members of the system of strike-slip fractures we have previously described (see also O'Brien, *loc. cit.* p. 14). Of the mixed group of metasediments which form the country rocks, a highly silicified limestone (the local equivalent of the Falcarragh Limestone) is the most obvious host, and within it, sphalerite, galena, calcite, and very subordinate greenockite form both the cement to a fault-breccia and a filling to the tension gashes.

Obviously, we have here simple examples of faults providing the conduits for the mineralizing solutions, and carbonate horizons providing the chemical environment for the precipitation of the sulphides. Clearly, despite the proximity of granites of the Donegal Complex in each case, the late date of mineralization in the structural history denies the possibility of any connection with Lower Palaeozoic igneous activity.

In a wider context, Russell (1968) has suggested that Irish mineral deposits of this type are located at the intersection of faults with certain major north-south trending fissures which were initiated during the early stages of the continental split. Be this as it may, it is of interest to note that the Glendowan and Faymore occurrences lie on faults which would intersect the extension of one of the "fissures" identified farther south by Russell (*loc. cit.*, Fig. 1)—though there is absolutely no surface evidence of such a structure in Donegal.

The provision of an absolute age of this mineralization has, in fact, been attempted by Moorbath (1962) and Pockley (1961), using determinations of the lead isotopes from a number of Donegal localities, including those mentioned above. Unfortunately, the model ages obtained, from 340 to —80 million years, are so disparate as to defy easy explanation in geological terms: nevertheless, we are prepared to guess that the sulphides were first introduced in association with recrudescence of fault movements following the deposition of Viséan, which is in accord with general opinion (O'Brien, *loc. cit.*; Russell, *loc. cit.*).

We have now reached a final point in the history of the mobile belt which we have studied in its various stages of inception, sedimentation, deformation, uplift, erosion, and final cover. It is thus somewhat incongruous that the final episode of Donegal geology, the intrusion of a dyke swarm during the early Tertiary, forms a kind of tailpiece, in no way connected with the main history; nevertheless, it is not without interest and we present it below.

CHAPTER 18

THE TERTIARY DYKE SWARM: THE CRUST UNDER TENSION

The well-marked group of basic dykes which trends northwestward across Donegal is but part of the Donegal–Kingscourt swarm that extends almost the entire width of the north of Ireland. Despite the excellence of its representation in the present area, where dolerite dykes occur abundantly over a width of 44 kilometers, the swarm has not yet been studied in any detail, though we can draw on certain local or preliminary information (Hyland in Hull *et al.*, 1891; Walker and Leedal, 1954; Preston, 1965, 1967a, 1967b; Wilson, 1964) to supplement our own knowledge. The identification of this swarm as representing part of the Lower Tertiary igneous activity of the British Isles is not in doubt (Preston, *loc. cit.*), as can be most simply confirmed by a glance at the generalized map of the occurrence of such dykes in Ireland (Fig. 18-1).

A point to make at the outset is that a well-marked prismatic jointing normal to, or a closely spaced platy jointing parallel to, the margins renders the dykes as a whole less resistant to denudation than the host rocks: indeed, their courses are often the sites of clefts and streams, and it is likely, therefore, that many have been missed during the surveys. Despite this, however, real variations in abundance can be detected. Thus, while the apparent scarcity of dykes in the metasediments, in comparison with the granites, could perhaps be due to an overall lesser degree of exposure, we think that there are, in fact, more of these basic intrusions in the latter. Even between the Dalradian basement and the Carboniferous cover there are differences in abundance and thickness, for, as Preston (1967a) points out, the gregarious dykes of the Derryveagh and the Blue Stack Mountains are replaced by thick solitary members in the sedimentary basins; evidently, these flat-lying sediments formed the least receptive host of all.

Everywhere, the dykes fill old fractures, particularly those joints which are vertical and trend NW–SE. Such deviations as there are from the general trend, *e.g.*, to the NNW in the Rosses and Trawenagh Bay, are simply due to changes in this regional joint pattern. Occasionally, dykes side-step by running along the NE–SW trending faults, and dykes crossing the Gweebarra fault line provide good examples of this. On a regional scale, these fault lines seem to provide a barrier to dyke intrusion, as if each fault block reacted separately to the stresses leading to the opening of the old joint fractures within it.

Invariably, the joints were opened up normal to their trend, often by an average distance of 2–3 meters. However, the total amount of opening of the crust is almost impossible to calculate because, apart from the poor exposure, dykes can be multiple, can thin, thicken, bifurcate, and ramify, and as we have seen, can vary in thickness and abundance depending on the lithology of the host. Only the major dykes, of the order of 10 meters thick, can be followed continuously with any certainty, and even

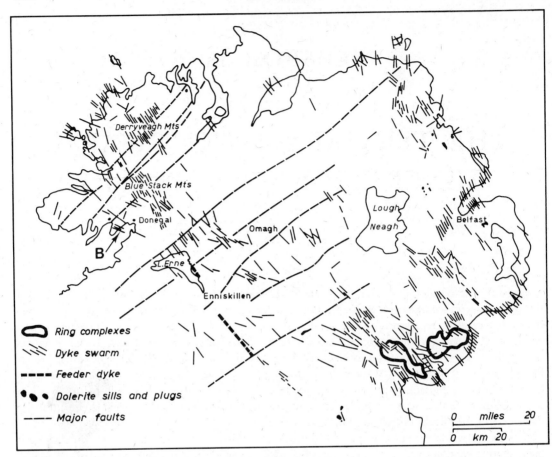

FIGURE 18-1. Diagrammatic map of the dyke swarms of the northern part of Ireland. (B) indicates position of the Blind Rock Dyke. Mainly after Preston (1967a).

these may be replaced by close-knit swarms of thinner dykes. There are so many dykes, however, that there must have been an appreciable opening of the crust. Whether this opening is of one time, representing a single flash injection of magma, is not generally answerable here, though we make certain references to this problem below; neither can we make any novel contribution to the nature of the forces involved, except to emphasize that, as far as northwest Donegal is concerned, the lines of fracture were already available; the problem is the highly specific opening of those particular master joints oriented at or near the NW–SE direction.

The majority of the dykes of this Donegal–Kingscourt swarm are simple olivine-dolerites with a single pyroxene phase; petrographically, they are tholeiites. Chilled contacts and centrally placed layers of filled or unfilled amygdales provide the only common variants. Some diversity is, however, introduced by two other types: *viz.*, a beidellite-pitchstone-felsite suite mainly restricted to the Blue Stack Mountains, and a number of so-called feeder dykes distinguished from the bulk of the olivine basalts by their considerably greater variability (Preston, 1967a).

These acid intrusions of the Blue Stack Mountains (Sollas, 1893; Walker and Leedal, 1954) are confined to an elongate zone within the Barnesmore Pluton. From a study of the alignment of their amygdales, steep and inwardly directed, it has been concluded that they originated from a deep source centered below this very zone. Though it is remotely possible that a Tertiary granite underlies the present outcrop of the Caledonian pluton, the more likely explanation of the spatial relationship is that the latter provided a conduit by which even comparatively viscous magmas could reach high levels.* Of special interest in this connection is the occasional, partial, or complete fusion of the host granite for a few centimeters from the contact of some of the basic dykes; this suggests the possibility that the contamination of basaltic magma may have occurred at depth, perhaps even by the production of small amounts of palingenetic acid melt (*cf.* Larsen and Switzer, 1939).

An even smaller proportion of the swarm shows sufficient departure from the general uniformity to suggest a different mechanism of emplacement (Preston, 1965, 1967*a,* and Wilson, 1964). Though none of these particular intrusions are present in our map area, one is nearly so—the Blind Rock Dyke of Donegal Bay (Preston, 1967*b*)—so that we will briefly refer to their special characteristics. Within these thick dykes, the presence of internal contacts have been interpreted as indicating successive surges of magma, and especially explosive activity is suggested by the abundance of xenoliths. Further, contact metamorphism is far more marked than in even thick dykes of the more usual uniform olivine-dolerite: it may reach out to 15 meters and produce contact changes comparable with those described from around the plugs in Antrim (Preston, 1967*b*). The presence of an alkaline marginal facies in the igneous rock itself is also reminiscent of this situation.

Preston concludes (*cf.* Charlesworth, 1963) that the intrusion history of these particular dykes differs considerably from that of the more usual dykes of the swarm, which he attributes to flash injection. He suggests that their magmas rose intermittently up an open fissure over an extended period of time, so that they could be regarded as feeder dykes supplying lavas at the surface. The implication is that Tertiary basaltic lavas were erupted over a much greater area of the north of Ireland than is indicated by the present outcrop in Antrim and Derry.

There are, indeed, in Donegal three pluglike masses of olivine dolerite which could represent conduits to the surface. One lies 1.5 kilometers northwest of Inver, another lies in the same geographical relation to Donegal town, and we have discovered yet another near Cranford much farther to the northwest (Fig. 18-1).

The possibility that the plateau basalt lavas formerly covered the area is bound up with the geomorphological problem of the age of the present high surface formerly alluded to (p. 356). Even allowing that this is an exhumed Carboniferous marine peneplain, this is no reason to deny that it may also represent a Tertiary surface located, in northwest Donegal at least, as a result of the superior resistance to erosion presented by the basement rocks, on removal of the cover. If so, the present outcrops of the dykes and plugs could represent intrusion of basaltic magma at a depth of only 1000 meters or less, below such a Lower Tertiary surface, and in this case, there is little doubt that magma would have reached the surface.

This is the most chastening feature of Donegal geology, the fact that so little trace is left of all that has gone on since the first unroofing of the granites, *viz.,* uplift, subaerial erosion, and the outpouring of lavas, certainly in the Devonian but possibly also at an earlier period; the partial submergence and submarine erosion in the Carboniferous; uplift and erosion in the Permo-Trias; the possible, if only partial, submergence during the Mesozoic transgressions, and finally, the blanketing by subaerial lavas in the Lower Tertiary.

---

* We are reminded of a similar situation involving the Weardale Granite of Northern England, above which the Carboniferous rests unconformably (Dunham *et al.,* 1965). In the cover immediately above the buried pluton, there are numerous mineral veins which have obviously used the Granite as a conduit.

BIBLIOGRAPHY I: GENERAL

Agrell, S. O., 1939. "The adinoles of Dinas Head, Cornwall," *Min. Mag.* **25**, 305–37.

Akaad, M. K., 1956a. "The Ardara granitic diapir of County Donegal, Ireland," *Q. J. Geol Soc. Lond.* **112**, 263–88.

―――― 1956b. "The northern aureole of the Ardara pluton of County Donegal," *Geol. Mag.* **93**, 377–92.

Albee, A. L., 1965. "Phase equilibria in three assemblages of kyanite-zone pelitic schists, Lincoln Mountain quadrangle, central Vermont," *J. Petrol.* **6**, 246–301.

―――― 1968. "Metamorphic zones in northern Vermont," in *Studies of Appalachian Geology: Northern and Maritime,* ed. Zen, E., White, W. S., Hadley, J. B., and Thompson, J. B., Jr. New York, Wiley-Interscience, 329–42.

Allaart, J. H., 1958. "The geology and petrology of the Trois Seigneurs massif, Pyrenees, France," *Leidse Geol. Meded.* **22**, 97–214.

―――― 1967. "Basic and intermediate igneous activity and its relationships to the evolution of the Julianehab Granite, South Greenland," *Meddr. Grønland* **175**, No. 1, 136p.

Alper, A. M., and Poldervaart, A., 1957. "Zircons from the Animas Stock and associated rocks, New Mexico," *Econ. Geol.* **52**, 952–71.

Althaus, E., 1969. "Das system Al_2O_3-SiO_2-H_2O. Experimentelle Untersuchungen und Folgerungen für die Petrogenese der metamorphen Gesteine," *Neues Jb. Miner.* **111**, 1, 74–110.

Ambrose, J. W and Burns, C. A., 1956. "Structures in the Clare River Syncline: a demonstration of granitization," in *The Grenville Problem,* ed. Thompson, J. E. *Spec. Publ., R. Soc. Can.* **1**, 42–52.

Anderson, E. M., 1936. "The dynamics of the formation of cone-sheets, ring-dykes, and cauldron subsidences," *Proc. R. Soc. Edinb.* **56**, 128–57.

Anderson, J. G. C., 1935. "The marginal intrusions of Ben Nevis: the Coille Lianachain complex, and the Ben Nevis dyke swarm," *Trans. Geol. Soc. Glasg.* **19**, 225-69.

―――― 1937. "The Etive granite complex," *Q. J. Geol. Soc. Lond.* **93**, 487–533.

―――― 1947. "The Kinlochlaggan syncline, southern Inverness-shire," *Trans. Geol. Soc. Glasg.* **21**, 97–115.

―――― 1948. "The occurrence of Moinian rocks in Ireland," *Q. J. Geol. Soc. Lond.* **103** (for 1947), 171–90.

―――― 1954. "The pre-Carboniferous rocks of the Slieve League promontory, County Donegal," *Q. J. Geol. Soc. Lond.* **109** (for 1953), 399–419.

―――― 1956. "The Moinian and Dalradian rocks between Glen Roy and Monadhliath Mountains, Inverness-shire," *Trans. R. Soc. Edinb.* **63**, 15-36.

―――― and Owen, T. R., 1968. *The Structure of the British Isles.* New York, Pergamon, 162p.

Andrew, G., 1928a. "Note on a sill in northwest Donegal (Irish Free State)," *Mem. Proc. Manchr. Lit. Phil. Soc.* **72**, 205–9.

―――― 1928b. "The contact relations of the Donegal granite (I.F.S.)," *Mem. Proc. Manchr. Lit. Phil. Soc.* **72**, 210–19.

―――― 1951. "Old Red Sandstone of Portsalon," *Geol. Mag.* **88**, 441–2.

Atherton, M. P., 1964. "The garnet isograd in pelitic rocks and its relation to metamorphic facies," *Am. Min.* **49**, 1331–49.

——— 1965. "The chemical significance of isograds," in *Controls of Metamorphism,* ed. Pitcher, W. S. and Flinn, G. W. London, Oliver and Boyd, 169–202.

——— 1968. "The variation in garnet, biotite, and chlorite composition in medium grade pelitic rocks from the Dalradian, Scotland, with particular reference to the zonation in garnet," *Contr. Miner. Petrol.* **18**, 347–71.

——— and Edmunds, W. M., 1966. "An electron microprobe study of some zoned garnets from metamorphic rocks," *Earth Planetary Sci. Letters* **1**, 185–93.

Aucott, J. W., 1970. "Tectonics of the Galway Granite," in *Mechanism of Igneous Intrusion,* ed. by Newall, G. and Rast, N. *Geol. J.* Spec. Issue 2, Liverpool, Gallery Press, 49–66.

Augustithis, S. S., 1966. "On the phenomenology of plastic deformation of quartz and micas in some granites," *Tectonophysics,* **2**, 455–73.

Autran, A. and Guitard, G., 1957. "Sur la signification de la sillimanite dans les Pyrénées," *C. R. Soc. Séanc. Soc. Géol. Fr.* 141.

Badgley, P. C., 1965. *Structural and Tectonic Principles,* New York, Harper and Row, 521p.

Bailey, D. K., 1970. "Volatile flux, heat-focusing, and the generation of magma," in *Mechanism of Igneous Intrusion,* ed. Newall, G. and Rast, N. *Geol. J.,* Spec. Issue 2, Liverpool, Gallery Press, 177–86.

Bailey, E. B., 1910. "Recumbent folds in the schists of the Scottish Highlands," *Q. J. Geol. Soc. Lond.* **66**, 586–620.

——— 1916. "The Islay anticline," *Q. J. Geol. Soc. Lond.* **72**, 132–59.

——— 1922. "The structure of the southwest Highlands of Scotland," *Q. J. Geol. Soc. Lond.* **78**, 82-127.

——— 1926. "Domes in Scotland and South Africa: Arran and Vredefort," *Geol. Mag.* **63**, 481–95.

——— 1934. "West Highland tectonics: Loch Leven to Glen Roy," *Q. J. Geol. Soc. Lond.* **90**, 462–523.

——— 1958. "Some aspects of igneous geology: 1908–1958," *Trans. Geol. Soc. Glasg.* **23**, 29–52.

——— 1960. "The geology of Ben Nevis and Glen Coe," *Mem. Geol. Surv. Scot.* 2nd ed., 307 p.

——— and Maufe, H. B., 1916. "The geology of Ben Nevis and Glen Coe," *Mem. Geol. Surv. Scot.,* 247 p.

——— and McCallien, W. J., 1937. "Perthshire tectonics: Schiehallion to Glen Lyon," *Trans. R. Soc. Edinb.* **59**, 79-118.

Baker, G., 1935. "The petrology of the You Yangs granite: a study of contamination," *Proc. R. Soc. Vict.* **48**, 124–58.

Balk, R., 1937. "Structural behavior of igneous rocks," *Mem. Geol. Soc. Am.* **5**, 177 p.

——— 1953. "Salt structure of Jefferson Island salt dome, Iberia and Vermillion parishes, Louisiana," *Bull. Am. Assoc. Petrol. Geol.* **37**, 2455–74.

Barker, D. S., 1964. "The Hallowell granite, south central Maine," *Am. J. Sci.* **262**, 592–613.

Barrell, J., 1921. "Relations of subjacent igneous invasion to regional metamorphism," *Am. J. Sci.* **1**, 1–19, 174–86, 255–67.

Barrow, G., 1893. "On an intrusion of muscovite-biotite gneiss in the southeastern Highlands of Scotland and its accompanying metamorphism," *Q. J. Geol. Soc. Lond.* **49**, 330-58.

Barth, T. F. W., 1969. *Feldspars.* New York, Wiley, 261 p.

Bateman, P. C., Clark, L. D., Huber, N. K., Moore, J. G., and Rinehart, C. D., 1963. "The Sierra Nevada batholith: a synthesis of recent work across the central part," *Prof. Paper U.S. Geol. Surv.* **414-D**, 46 p.

Bathurst, R. G. C., 1958. "Diagenetic fabrics in some British Dinantian limestone," *Lpool Manchr. Geol. J.* **2**, 11–36.

Bell, K., 1968. "Age relationships and provenance of the Dalradian Series of Scotland," *Bull. Geol. Soc. Am.* **79**, 1167–94.

Bennington, K. O., 1956. "Role of shearing stress and pressure in differentiation as illustrated by some mineral reactions in the system MgO-SiO_2-H_2O," *J. Geol.* **64**, 558–77.

Bennison, G. M. and Wright, A. E., 1969. *The Geological History of the British Isles.* London, Edward Arnold, 406 p.

Berger, A. R., 1971a. "Dynamic analysis using dikes with oblique internal foliations," *Bull. Geol. Soc. Am.* **82**, 781–86.

——— 1971b. "The origin of banding in the Main Donegal Granite, NW Ireland," *Geol. J.* **7**, 437–58.

——— and Cambray, F. W., 1969. "Caledonian structure and chronology in Donegal, Eire," *Spec. Paper Geol. Soc. Am.* **121** (*Abstracts* for 1968), 22.

——— and Pitcher, W. S., 1970. "Structures in granitic rocks: a commentary and a critique on granite tectonics," *Proc. Geol. Assoc.* **81**, 441–61.

Berger, J. F., 1816. "On the dykes of the north of Ireland," *Trans. Geol. Soc. Lond.* **3**, 223–35.

Bhattacharji, S., 1967. "Mechanics of flow differentiation in ultramafic and mafic sills," *J. Geol.* **75**, 101–12.

Billings, M. P., 1945. "Mechanics of igneous intrusions in New Hampshire," *Am. J. Sci.* **243**, 40–69.

——— 1950. "Stratigraphy and the study of metamorphic rocks," *Bull. Geol. Soc. Am.* **61**, 435–48.

——— 1954. *Structural Geology.* Englewood Cliffs, N.J., Prentice-Hall, 473 p.

——— 1956. *The geology of New Hampshire: Part II—Bedrock geology,* New Hampshire State Planning and Development Commission, 203 p.

Binns, R. A., 1966. "Granitic intrusions and regional metamorphic rocks of Permian age from the Wongwibinda District, northeastern New South Wales," *J. and Proc. R. Soc. N. S. W.* **99**, 5–36.

Bishop, A. C., 1964. "Petrogenesis of hornblende-mica lamprophyre dykes at South Hill, Jersey, C.I.," *Geol. Mag.* **101**, 302–13.

Bishopp, D. W., 1951. "The Moine schists," *Geol. Mag.* **88**, 371.

Blake, E. H., 1862. "On the primary rocks of Donegal," *J. Geol. Soc. Dubl.* **9**, 294–300.

Blatt, H. and Christie, J. M., 1960. "Undulatory extinction in quartz of igneous and metamorphic rocks and its significance in provenance studies of sedimentary rocks," *Bull. Geol. Soc. Am.* **71**, 1828.

Bluck, B. J., 1969. "Old Red Sandstone and other Palaeozoic conglomerates of Scotland," in *North Atlantic—Geology and Continental Drift,* ed. Kay, M. *Mem. Am. Assoc. Petrol. Geol.* **12**, 711–23.

Blyth, F. G. H., 1949. "The sheared porphyrite dykes of south Galloway," *Q. J. Geol. Soc. Lond.* **105**, 393–423.

Boone, G. M., 1962. "Potassic feldspar enrichment in magma: origin of syenite in Deboullie district, northern Maine," *Bull. Geol. Soc. Am.* **73**, 1451–76.

Booth, B., 1968. "Petrogenetic significance of alkali feldspar megacrysts and their inclusions in Cornubian granites," *Nature* **217**, 1036–38.

Boschma, D., 1963. "Successive Hercynian structures in some areas of the Central Pyrenees," *Leidse Geol. Meded.* **28**, 103–26.

Bott, M. H. P., 1956. "A geophysical study of the granite problem," *Q. J. Geol. Soc. Lond.* **112**, 45–62.

——— and Smithson, S. B., 1967. "Gravity investigations of subsurface shape and mass distributions of granite batholiths," *Bull. Geol. Soc. Am.* **78**, 859–78.

Bowen, N. L., 1920. "Differentiation by deformation," *Proc. Natl. Acad. Sci.* **6**, 159–62.

——— 1922. "The behavior of inclusions in igneous magmas," *J. Geol.* **30**, 513–70.

——— 1928. *The Evolution of the Igneous Rocks.* Princeton, N.J., Princeton University Press, 334 p.

Bowes, D. R., Hopgood, A. M., and Smart, J., 1968. "Glasgow University Exploration Society expedition to Inishtrahull," *Nature* **217**, 344–5.

——— and Jones, K. A., 1958. "Sedimentary features and tectonics in the Dalradian of Western Perthshire," *Trans. Edinb. Geol. Soc.* **17**, 133–40.

———, Kinloch, E. D., and Wright, A. E., 1964. "Rhythmic amphibole overgrowths in appinites associated with explosion-breccias in Argyll," *Min. Mag.* **33**, 963–73.

——— and Park, R. G., 1966. "Metamorphic segregation banding in the Loch Kerry basite sheet from the Lewisian of Gairloch, Ross-shire, Scotland," *J. Petrol.* **7**, 306–30.

——— and Wright, A. E., 1961. "An explosion breccia complex at Back Settlement, near Kentallen, Argyll," *Trans. Edinb. Geol. Soc.* **18**, 293–314.

———————. 1966. "The explosion-breccia pipes near Kentallen, Scotland, and their geological setting," *Trans. R. Soc. Edinb.* **67**, 109–43.

Bowler, C. M. L., 1958. "The distribution of alkalis and fluorine across some granite-killas and granite-greenstone contacts," *Abstr. Proc. 1958 Conf. Geol. Geomorph. S. W. England., R. Geol. Soc. Cornwall*, 20–21.

Brace, W. F., 1969. "The mechanical effects of pore pressure on the fracturing of rocks," in *Proc. Conf. Res. in Tectonics, Paper Geol. Surv. Can.* **68–52**, 113–23.

Bradley, J., 1965. "Intrusion of major dolerite sills," *Trans. R. Soc. N. Z.* **3**, 27–55.

Brett, R., 1964. "Experimental data from the system Cu-Fe-S and their bearing on exsolution textures in ores," *Econ. Geol.* **59**, 1241–69.

Brindley, J. C., 1957. "The aureole rocks of the Leinster Granite in South Dublin, Ireland," *Proc. R. Irish Acad.* **59B**, 1–18.

——— 1969. "Caledonian and Pre-Caledonian intrusive rocks of Ireland," in *North Atlantic—Geology and Continental Drift*, ed. Kay, M. *Mem. Am. Assoc. Petrol. Geol.* **12**, 336–53.

——— 1970. "Appinitic intrusions associated with the Leinster Granite, Ireland," *Proc. R. Irish Acad.* **70B**, 93–104.

Brooks, C. and Compston, W., 1965. "The age and initial Sr^{87}/Sr^{86} of the Heemskirk granite, western Tasmania," *J. Geophys. Res.* **70**, 6249–62.

Brothers, R. N. and Rodgers, K. A., 1969. "Petrofabric studies on ultramafic nodules from Auckland, New Zealand," *J. Geol.* **77**, 452–65.

Brown, P. E. and Miller, J. A., 1969a. "Interpretation of isotopic ages in orogenic belts," in *Time and Place in Orogeny*, ed. Kent, P. E., Satterthwaite, G. E., and Spencer, A. M. *Spec. Publ. Geol. Soc. Lond.* **3**, 137–56.

——— 1969b. "Some aspects of Palaeozoic geochronology of the British Isles," in *North Atlantic—Geology and Continental Drift*, ed. Kay, M. *Mem. Am. Assoc. Petrol. Geol.* **12**, 363–74.

——— and Grasty, R. L., 1968. "Isotopic ages of late Caledonian granitic intrusions in the British Isles," *Proc. Yorks. Geol. Soc.* **36**, 251–76.

———, Soper, N. J., and York, D., 1965. "Potassium-argon age pattern of the British Caledonides," *Proc. Yorks. Geol. Soc.* **35**, 103–38.

Bryner, L. 1961. "Breccia and pebble columns associated with epigenetic ore deposits," *Econ. Geol.* **56**, 488–508.

Buddington, A. F., 1959. "Granite emplacement with special reference to North America," *Bull. Geol. Soc. Am.* **70**, 671–747.

——— 1963. "Metasomatic origin of large parts of the Adirondack phacoliths: a discussion," *Bull. Geol. Soc. Am.* **74**, 353.

Bullerwell, W., 1961. "The gravity map of Northern Ireland," *Irish Nat.* **13**, 255–57.

Burch, S. H., 1968. "Tectonic emplacement of the Burro Mountain ultramafic body, Santa Lucia Range, California," *Bull. Geol. Soc. Am.* **79**, 527–44.

Burke, K. C., Cunningham, M. A., Gallagher, M. J., and Hawkes, J. R., 1964. "Beryl in the Rosses Granite, Northwest Ireland," *Econ. Geol.* **59**, 1539–50.

Burns, K. L. and Spry, A. H., 1969. "Analysis of the shape of deformed pebbles," *Tectonophysics* **7**, 177–96.

Burwash, R. A. and Krupicka, J., 1969. "Cratonic reactivation in the Precambrian basement of Western Canada: I. Deformation and chemistry," *Can. J. Earth Sci.* **6**, 1381–96.

Butler, B. C. M., 1965. "Compositions of micas in metamorphic rocks," in *Controls of Metamorphism*, ed. Pitcher, W. S. and Flinn, G. W. London, Oliver and Boyd, 291–98.

Callaghan, E., 1935. "Pre-granodiorite dikes in granodiorite, Paradise Range, Nevada," *Trans. Am. Geophys. Union* **1**, 302–7.

Callaway, C., 1885. "On the granitic and schistose rocks of northern Donegal," *Q. J. Geol. Soc. Lond.* **41**, 221–39.

Cambray, F. W., 1969a. "The Kilmacrenan Succession east of Glenties, County Donegal," *Proc. R. Irish Acad.* **67B**, 291–302.

——— 1969b. "The Dalradian of Donegal," in *North Atlantic—Geology and Continental Drift*, ed. Kay, M. *Mem. Am. Assoc. Petrol. Geol.* **12**, 181–93.

Cannon, R. T., 1964. "Porphyroblastic and augen gneisses in the Bartica Assemblage," *Geol. Mag.* **101**, 541–47.

——— 1966. "Plagioclase zoning and twinning in relation to the metamorphic history of some amphibolites and granulites," *Am. J. Sci.* **264**, 526–42.

Cannon, W. F., 1970. "Plutonic evolution of the Cutler area, Ontario," *Bull. Geol. Soc. Am.* **81**, 81–94.

Carey, S. W., 1955, "The orocline concept in geotectonics," *Proc. R. Soc. Tasm.* **89**, 255–88.

——— 1958. "The tectonic approach to continental drift," in *Continental Drift: a Symposium*, Dept. Geology, Univ. of Tasmania, 177–355.

Carmichael, I. S. E., 1963. "The crystallization of feldspar in volcanic acid liquids," *Q. J. Geol. Soc. Lond.* **119**, 95–131.

Carpenter, J. R., 1968. "Apparent retrograde metamorphism: another example of the influence of sructural deformation on metamorphic differentiation," *Contr. Miner. Petrol.* **17**, 173–86.

Carstens, H., 1966. "Deformation in vein genesis," *Norsk. Geol. Tiddskr.* **46**, 299–307.

——— 1967a. "Exsolution in ternary feldspars: I. On the formation of antiperthites," *Contr. Miner. Petrol.* **14**, 27–35.

——— 1967b. "Exsolution in ternary feldspars: II. Intergranular precipitation in alkali feldspar containing calcium in solid solution," *Contr. Miner. Petrol.* **14**, 316–20.

Carter, N. L., Christie, J. M., and Griggs, D. T., 1964. "Experimental deformation and recrystallization of quartz," *J. Geol.* **72**, 687–733.

———, and Raleigh, C. B., 1969. "Principal stress directions from plastic flow in crystals," *Bull. Geol. Soc. Am.* **80**, 1231–64.

Castle, R. O., 1966. "Origin of myrmekite," *Spec. Paper Geol. Soc. Am.* **87**, 198.

Chadwick, B. A., 1968. "Deformation and metamorphism in the Lukmanier region, central Switzerland," *Bull. Geol. Soc. Am.* **79**, 1123–50.

Chadwick, R. A., 1958. "Mechanisms of pegmatite emplacement," *Bull. Geol. Soc. Am.* **69**, 803–36.

Charlesworth, J. K., 1924. "Glacial geology of the northwest of Ireland," *Proc. R. Irish Acad.* **36B**, 174–314.

———1963. *Historical Geology of Ireland*. London, Oliver and Boyd, 565 p.

Chayes, F., 1952. "On the association of perthitic microline with highly undulant quartz or granular quartz in some calkalkaline granites," *Am. J. Sci.* **250**, 281–96.

——— 1957. "A provisional reclassification of granite," *Geol. Mag.* **94**, 58–68.

Chayes, F., 1962. "Numerical correlation and petrographic correlation," *J. Geol.* **70**, 44–52.

——— 1970. "On deciding whether trend surfaces of progressively higher order are meaningful," *Bull. Geol. Soc. Am.* **81**, 1273–78.

——— and Suzuki, Y., 1963. "Geological contours and trend surfaces," *J. Petrol.* **4**, 307–12.

Cheng, Y., 1944. "The migmatite area around Bettyhill, Sutherland," *Q. J. Geol. Soc. Lond.* **99** (for 1943), 107–54.

Chinner, G. A., 1960. "Pelitic gneisses with varying ferrous/ferric ratios from Glen Clova, Angus, Scotland," *J. Petrol.* **1**, 178–217.

——— 1961. "The origin of sillimanite in Glen Clova, Angus," *J. Petrol.* **2**, 312–323.

——— Knowles, C. R., and Smith, J. V., 1969. "Transition-metal contents of Al_2SiO_5 polymorphs," *Am. J. Sci.* **267A**, 96–113.

Church, W. R., 1969. "Metamorphic rocks of Burlington Peninsula and adjoining areas of Newfoundland and their bearing on continental drift in North Atlantic," in *North Atlantic—Geology and Continental Drift*, ed. Kay, M. *Mem. Am. Assoc. Petrol. Geol.* **12**, 212–33.

Clarke, S. P., 1961. "A redetermination of the equilibrium relations between kyanite and sillimanite," *Am. J. Sci.* **259**, 641–650.

Claxton, C. W., 1968. "Mineral layering in the Galway Granite, Connemara, Eire," *Geol. Mag.* **105**, 149–59.

Cloos, E., 1934. "The Loon Lake pluton, Bancroft area, Ontario," *J. Geol.* **42**, 393–99.

——— 1936. "Der Sierra-Nevada Pluton in California," *Neues Jb. Miner. Geol. Paläont.* **76B**, 355–450.

———, Hopson, C. A., Fisher, G. W., and Cleaves, E. T., 1964. *The Geology of Howard and Montgomery Counties*, Maryland Geological Survey, 373p.

Cloos, H. and Rittman, A., 1939. "Zur Einteilung und Benennung der Plutone," *Geol. Rdsch.* **30**, 600–08.

Cobbing, E. S. and Pitcher, W. S., *in press*. "The Coastal Batholith in northern part of Lima province, Peru," *Q. J. Geol. Soc. Lond.*

Cole, G. A. J., 1902. "On composite gneisses in Boylagh, west Donegal," *Proc. R. Irish Acad.* **24B**, 203–30.

——— 1906a. "On the marginal phenomena of granite-domes," *Geol. Mag.* **3** (Decade 5), 80.

——— 1906b. "On a hillside in Donegal: a glimpse into the great Earth cauldron," *Sci. Prog. Lond.* **1**, 343–60.

——— 1916. "On the mode of occurrence and origin of the orbicular granite of Mullaghderg, County Donegal," *Sci. Proc. R. Dubl. Soc.* **15**, 141–58.

——— 1922. "Memoir and map of localities of minerals of economic importance and metalliferous Mines in Ireland," *Mem. Geol. Surv. Ireland* (reprinted, 1956), 155p.

Colhoun, E. A., 1971. "The glacial stratigraphy of the Sperrin Mountains and its relation to the glacial stratigraphy of northwest Ireland," *Proc. R. Irish Acad.* **71B**, 37–52.

Collette, B. J., 1959. "On helicitic structures and the occurrence of elongate crystals in the direction of the axis of a fold," *Proc. Kon. Neder. Akad. Wetensch.* **B62**, 161–71.

Compton, R. R., 1955. "Trondhjemite batholith near Bidwell Bar, California," *Bull. Geol. Soc. Am.* **66**, 9–44.

——— 1960. "Contact metamorphism in Santa Rosa Range, Nevada," *Bull. Geol. Soc. Am.* **71**, 1383–1416.

Cook, A. H. and Murphy, T., 1952. "Measurements of gravity in Ireland: Gravity survey of Ireland north of the line Sligo-Dundalk," *Geophys. Mem. Dubl. Inst. Adv. Studies* **2**(4), 27p.

Cooray, P. G., 1961. "The mode of emplacement of some dykes and veins in the Precambrian rocks of Ceylon," *Proc. Geol. Assoc.* **72**, 73–90.

Cox, F. C., 1969. "Inclusions in garnet: discussion and suggested mechanism of growth for syntectonic garnets," *Geol. Mag.* **106**, 57–62.

Craig, G. Y., ed., 1965. *The Geology of Scotland*. London, Oliver and Boyd, 556p.

Crowder, D. F., 1959. "Granitization, migmatization, and fusion in the northern Entiat Mountains, Washington," *Bull. Geol. Soc. Am.* **70**, 827–78.

Cruse, M. J. B. and Leake, B. E., 1968. "The geology of Renvyle, Inishbofin, and Inishark, northwest Connemara, County Galway," *Proc. R. Irish Acad.* **67B**, 1–36.

Cummins, W. A. and Shackleton, R. M., 1955. "The Ben Lui recumbent syncline (SW Highlands)," *Geol. Mag.* **92**, 353–63.

Cunningham-Craig, E. H., Wright, W. B., and Bailey, E. B., 1911. "The geology of Colonsay and Oronsay with part of the Ross of Mull," *Mem. Geol. Surv. Scot.*, 109p.

Cushing, H. P., 1910. "Geology of the Thousand Islands region," *Bull. N.Y. State. Mus.*, 145p.

Dachille, F. and Roy, R., 1961. "Influence of 'displacive-shearing' stresses on the kinetics of reconstructive transformations effected by pressure in the range 0–100,000 bars," in *Reactivity of Solids*, ed., de Boer, J. H. *Proceedings of the Fourth International Symposium on the Reactivity of Solids*, Amsterdam, Elsevier, 502–511.

——————————— 1964. "Effectiveness of shearing stresses in accelerating solid phase reactions at low temperatures and high pressures," *J. Geol.* **72**, No. 2, 243–247.

Daly, R. A., 1933. *Igneous Rocks and the Depths of the Earth*. New York, McGraw-Hill, 598p.

Daniels, J. L., Skiba, W. J., and Sutton, J., 1965. "The deformation of some banded gabbros in the northern Somalia fold-belt," *Q. J. Geol. Soc. Lond.* **121**, 111–42.

Davies, R. G., 1956. "The Pen-y-gader dolerite and its metasomatic effects on the Llyn-y-gader sediments," *Geol. Mag.* **93**, 153–72.

Davis, G. A., 1963. "Structure and mode of emplacement of Caribou Mountain pluton, Klamath Mountains, California," *Bull. Geol. Soc. Am.* **74**, 331–48.

Dawson, J. B., 1962. "Basutoland kimberlites," *Bull. Geol. Soc. Am.*, **73**, 545–560.

Dawson, K. R. and Whitten, E. H. T., 1962. "The quantitative mineralogical composition and variation of the Lacorne, La Motte, and Preissac granitic complex, Quebec, Canada," *J. Petrol.* **3**, 1–37.

Dearman, W. R. 1969. "An outline of the structural geology of Cornwall," *Proc. Geol. Soc. Lond.* **1654**, 33–40.

Dearnley, R., 1967. "Metamorphism of minor intrusions associated with the Newer Granites of the western Highlands of Scotland," *Scot. J. Geol.* **3**, 449–57.

Deer, W. A., 1938. "The composition and paragenesis of the hornblendes of the Glen Tilt Complex, Perthshire," *Min. Mag.* **25**, 56–74.

——— 1950. "The diorites and associated rocks of the Glen Tilt complex, Perthshire: II. Diorites and appinites," *Geol. Mag.* **87**, 181–95.

DeHills, S. M. and Corvalan, J., 1964. "Undulatory extinction in quartz grains of some Chilean granite rocks of different ages," *Bull. Geol. Soc. Am.* **75**, 363–66.

Den Tex, E., 1956. "Studies in compatative petrofabric analysis: 1. Xeno- and autolithic fabric," *Proc. Kon. Neder. Akad. Wetensch.* **59B**, 11–22.

——— 1963. "A commentary on the correlation of metamorphism and deformation in space and time," *Geologie Mijnb.* **42**, 170–76.

De Waard, D. and Walton, M., 1967. "Precambrian geology of the Adirondack highlands; a reinterpretation," *Geol. Rdsch.* **56**, 596–629.

Dewey, H., 1915. "On spilosites and adinoles from North Cornwall," *Trans. R. Geol. Soc. Corn.* **15**, 71–84.

Dewey, J. F., 1965. "Nature and origin of kink-bands," *Tectonophysics* 1, 459–94.
——— 1967. "The structural and metamorphic history of the Lower Palaeozoic rocks of central Murrisk, County Mayo, Eire," *Q. J. Geol. Soc. Lond.* 123, 125–56.
——— 1969a. "Evolution of the Appalachian/Caledonian orogen," *Nature* 222, 124–29.
——— 1969b. "Structure and sequence in paratectonic British Caledonides," in *North Atlantic—Geology and Continental Drift,* ed. Kay, M. *Mem. Am. Assoc. Petrol. Geol.* 12, 309–35.
——— and McKerrow, W. S., 1963. "An outline of the geomorphology of Murrisk and northwest Galway," *Geol. Mag.* 100, 260–75.
———, Rickards, R. B., and Skevington, D., 1970. "New light on the age of Dalradian deformation and metamorphism in western Newfoundland," *Norsk. Geol. Tiddskr.* 50, 19–44.
Dickson, F. W., 1969. "Exsolution origin of zoned, twinned, preferentially concentrated plagioclase inclusions in K-feldspar crystals," *Spec. Paper Geol. Soc. Am.* 121, 74.
Didier, J., 1964. "Les enclaves des granites dans la littérature géologique," *Bull. B. R. G. M.* 3, 32–48.
Dietrich, R. V., 1960. "Banded gneisses," *J. Petrol.* 1, 99–120.
Dimroth, E., 1970. "Evolution of the Labrador Geosyncline," *Bull. Geol. Soc. Am.* 81, 2717–42.
Doe, B. R., 1967. "The bearing of lead isotopes on the source of granitic magma," *J. Petrol.* 8, 51–83.
Donath, F. A. and Parker, R. B., 1964. "Folds and folding," *Bull. Geol. Soc. Am.* 75, 45–62.
Donnelly, T. W., 1967. "Kinetic considerations in the genesis of growth twinning," *Am. Miner.* 52, 1–12.
Dowling, J. W. F. *et al.,* 1954. "The Great Glen fault, Donegal County," *Geol. Mag.* 91, 519–20.
Drescher-Kaden, F. K., 1948. *Die Feldspat-Quarz-Reaktionsgefüge der Granite und Gneise,* Berlin, Springer-Verlag, 259p.
Duffield, W. A., 1968. "The petrology and structure of the El Pinal tonalite, Baja California, Mexico," *Bull. Geol. Soc. Am.* 79, 1351–74.
Dunham, K. C., Dunham, A. C., Hodge, B. L., and Johnson, A. L., 1965. "Granite beneath Viséan sediments with mineralization at Rookhope, northern Pennines," *Q. J. Geol. Soc. Lond.* 121, 383–417.
Durrance, E. M., 1967. "Photoelastic stress studies and their application to a mechanical analysis of the Tertiary ring-complex of Ardnamurchan, Argyllshire," *Proc. Geol. Assoc.* 78, 289–318.
Dury, G. H., 1957. "A glacially-breached watershed in Donegal," *Irish Geog.* 3, 171–80.
——— 1959. "A contribution to the geomorphology of central Donegal," *Proc. Geol. Assoc.* 70, 1–27.
——— 1964. "Aspects of the geomorphology of Slieve League peninsula, Donegal," *Proc. Geol. Assoc.* 75, 445–59.

Edelman, N., 1968. "An attempted classification of pegmatite structures," *Geol. För. Stockh. Förh.* 90, 349–59.
Edmunds, W. M. and Thomas, P. R., 1966. "The stratigraphy and structure of the Dalradian rocks north of Recess, Connemara, County Galway," *Proc. R. Irish Acad.* 64B, 517–28.
——— and Atherton, M. P., 1971. "Polymetamorphic evolution of garnet in the Fanad aureole, Donegal, Eire," *Lithos* 4, 153–67.
Egan, F. W., Kilroe, J. R., and Mitchell, W. F., 1888. "Explanatory memoir, Sheet 24 (portions of Donegal and Tyrone)," *Mem. Geol. Surv. Ireland,* 50p.
Eggler, D. H., 1968. "Virginia Dale Precambrian ring-dyke complex, Colorado–Wyoming," *Bull. Geol. Soc. Am.* 79, 1545–64.
Elliott, D., 1970. "Determination of finite strain and initial shapes from deformed elliptical objects," *Bull. Geol. Soc. Am.* 81, 2221–36.
Elsasser, W. M., 1963. "Early history of the Earth," in *Earth Science and Meteoritics,* ed. Geiss, J. and Goldberg, E. D. Amsterdam, North Holland, 1–30.

Emeleus, C. H., 1963. "Structural and petrographic observations on layered granites from S. W. Greenland," *Spec. Paper Miner Soc. Am.* **1**, 22–29.

Emmons, R. C., 1940. "The contribution of differential pressures to magmatic differentiation," *Am. J. Sci.* **238**, 1–21.

———— 1953. "The argument," in *Selected Petrogenic Relationships of Plagioclase.*, Mem. Geol. Soc. Am. **52**, 111–17.

Engel, A. E. J. and Engel, G. C., 1958. "Progressive metamorphism and granitization of the major paraneiss, northwest Adirondack Mountain, New York: 1. Total rock," *Bull. Geol. Soc. Am.* **69**, 1369–1414.

———— 1961. "Migration of elements during metamorphism in the northwest Adirondack Mountains, New York," *Prof. Paper U. S. Geol. Surv.* **400(B)**, 465–470.

———— 1963. "Metasomatic origin of large parts of the Adirondacks phacoliths," *Bull. Geol. Soc. Am.* **74**, 349–52.

Erdmannsdörffer, O. H., 1949. "Magmatische und metasomatische Prozesse in Graniten, insbesondere Zweiglimmergraniten," *Heidelberg Beitr. Miner. Petrol.* **1**, 213–50.

———— 1950. "Die Entwicklung und jetzige Stellung des Granitproblems," *Heidelberg Beitr. Miner. Petrol.* **2**, 334–77.

Erikson, E. H., Jr., 1969. "Petrology of the composite Snoqualmie Batholith, central Cascade Mountains, Washington," *Bull. Geol. Soc. Am.* **80**, 2213–36.

Escher, A., 1966. "The deformation and granitization of Ketilidian rocks in the Nanortalik area, S. Greenland," *Meddr. Grønland* **172**, No. 9. 102p.

Eskola, P., 1949. "The problem of mantled gneiss domes," *Q. J. Geol. Soc. Lond.* **104**, 461–76.

———— 1954. "Ein Lamprophyrgang in Helsinki und die Lamprophyrprobleme," *Tscher. Min. Petr. Mitt.* **4**, 329–37.

———— 1956. "Postmagmatic potash metasomatism of granite," *C. R. Soc. Geol. Fin.* **29**, 85–100.

———— 1961. "Granitenstehung bei Orogenese und Epiorogenese," *Geol. Rdsch.* **50**, 105–23.

Evans, B. W., 1965. "Application of reaction rate method to the breakdown of muscovite and muscovite plus quartz," *Am. J. Sci.* **263**, 647–667.

Exley, C. S. and Stone, M., 1964. "The granitic rocks of southwest England," in *Present Views of Some Aspects of the Geology of Cornwall and Devon.*, R. Geol. Soc. Cornwall 131–84.

Exner, C., 1966. "Orthit in den Gesteinen der Sonnblickgruppe (Hohe Tauern)," *Tscherm. Min. Petrol. Mitt.* **11**, 358–72.

Faure, G. and Hurley, P. M., 1963. "The isotopic composition of strontium in oceanic and continental basalts—applications to the origin of igneous rocks," *J. Petrol.* **4**, 31–50.

Felici, M., 1964. "Observations on granitization and its characteristic in some rocks near Keuruu," *Bull. Comm. Geol. Fin.* **212**, 1–16.

Ferguson, J. and Pulvertaft, T. C. R., 1963. "Contrasted styles of igneous layering in the Gardar Province of southern Greenland," *Spec. Paper Miner. Soc. Am.* **1**, 10–21.

Fitch, F. J., Miller, J. A., and Mitchell, J. G., 1969. "A new approach to radio-isotopic dating in orogenic belts," in *Time and Place in Orogeny,* ed. Kent, P. E., Satterthwaite, G. E., and Spencer, A. M. *Spec. Publ. Geol. Soc. Lond.* **3**, 157–96.

Fleuty, M. J., 1964. "Tectonic slides," *Geol. Mag.* **101**, 452–56.

Flinn, D., 1956. "On the deformation of the Funzie conglomerate, Fetlar, Shetland," *J. Geol.* **64**, 480–505.

———— 1959. "An application of statistical analysis to petrochemical data," *Geochim. Cosmochim. Acta* **17**, 161–175.

———— 1962. "On folding during three-dimensional progressive deformation," *Q. J. Geol. Soc. Lond.* **118**, 385–433.

Flinn, D., 1965. "On the symmetry principle and the deformation ellipsoid," *Geol. Mag.* **102**, 36–45.

────── 1967. "The metamorphic rocks of the southern part of the mainland of Shetland," *Geol. J.* **5**, 251–90.

Floyd, P. A., 1965. "Metasomatic hornfelses of the Land's End aureole at Tater-du, Cornwall," *J. Petrol.* **6**, 223–45.

────── 1967. "The role of hydration in the spatial distribution of metasomatic cations in the Land's End aureole, Cornwall," *Chem. Geol.* **2**, 147–156.

Forbes, D., 1872. "On the geology of Donegal," *Geol. Mag.* **9**, 12–15.

Foslie, S., 1921. "Field observations in northern Norway bearing on magmatic differentiation," *J. Geol.* **29**, 701–19.

Fourmarier, P., 1959. "Le granite et les déformations mineures des roches (schistosité, microplissement, etc.)," *Mem. Acad. R. Belg.* **31**(3), 101p.

Fox, F. G., 1969. "Some principles governing interpretation of structure in the Rocky Mountains orogenic belt," in *Time and Place in Orogeny* ed. Kent, P. E., Satterthwaite, G. E., and Spencer, A. M. *Spec. Publ. Geol. Soc. Lond.* **3**, 23–42.

Fox, P. E., 1969. "Petrology of Adamant pluton, British Columbia," *Paper Geol. Surv. Can.* **67–61**, 101p.

Fraser, A .G., 1966. "Patchy zoning in plagioclase: a discussion," *J. Geol.* **74**, 514–20.

French, W. J., 1966. "Appinitic intrusions clustered around the Ardara pluton, County Donegal," *Proc. R. Irish Acad.* **64B**, 303–22.

────── and Pitcher, W. S., 1959. "The intrusion-breccia of Dunmore, County Donegal," *Geol. Mag.* **96**, 69–74.

Fyfe, W. S., 1967. "Stability of Al_2SiO_5 polymorphs," *Chem. Geol.* **2**, 67–76.

────── 1970. "Some thoughts on granitic magmas," in *Mechanism of Igneous Intrusion,* ed. Newall, G. and Rast, N. *Geol. J. Spec. Issue* **2**, Liverpool, Gallery Press, 201–216.

Gabrielse, H. and Reesor, J. E., 1964. "Geochronology of plutonic rocks in two areas of the Canadian Cordillera," in *Geochronology in Canada, Spec. Publ. R. Soc. Can.* **8**, 96–138.

Ganguly, J., 1968. "Analysis of the stabilities of chloritoid and some equilibria in the system. $FeO-Al_2O_3-SiO_2-H_2O-O_2$.," *Am. J. Sci.* **266**, 277–298.

Ganly, P., 1857. "Observations on the structure of strata," *J. Geol. Soc. Dubl.* **7**, 164–67.

Gansser, A., 1964. *Geology of the Himalayas.* New York, Wiley-Interscience, 289p.

Gates, O., 1959. "Breccia pipes in the Shoshone Range, Nevada," *Econ. Geol.* **54**, 790–815.

Gates, R. M. and Scheerer, P. E., 1963. "The petrology of the Nonewaug granite, Connecticut," *Am. Miner.* **48**, 1040–69.

Gault, H. R., 1945. "Petrography, structures, and petrofabrics of the Pinckneyville quartz diorite, Alabama," *Bull. Geol. Soc. Am.* **56**, 181–246.

Gay, N. C., 1968*a*. "The motion of rigid particles embedded in a viscous fluid during pure shear deformation of the fluid," *Tectonophysics* **5**, 81–88.

────── 1968*b*. "Pure shear and simple shear deformation of inhomogeneous viscous fluids: 1. Theory," *Tectonophysics* **5**, 211–34.

────── 1968*c*. "Pure shear and simple shear deformation of inhomogeneous viscous fluids: 2. The determination of the total finite strain in a rock from objects such as deformed pebbles," *Tectonophysics* **5**, 295–302.

Geikie, A., 1891. "The younger schists: Dalradian," *Q. J. Geol. Soc. Lond.* **47**, 48–162.

George, T. N., 1958. "Lower Carboniferous palaeogeography of the British Isles," *Proc. Yorks. Geol. Soc.* **31**, 227–318.

George, T. N., 1967. "Landform in Ulster," *Scot. J. Geol.* **3**, 413–48.

—— Harland, W. B., Ager, D. V., Ball, H. W., Blow, W. H., Casey, R., Holland, C. H., Hughes, N. F., Kellaway, G. A., Kent, P. E., Ramsbottom, W. H. C., Stubblefield, J., and Woodland, A. W., 1969. "Recommendations on stratigraphical usage," *Proc. Geol. Soc. Lond.* **1656**, 139–66.

—— and Oswald, D. H., 1957. "The Carboniferous rocks of the Donegal syncline" *Q. J. Geol. Soc. Lond.* **113**, 137–79.

Gevers, T. W., 1963. "Geology along the northwestern margin of the Khomas Highlands between Otjimbingwe-Karibib and Okahandja, South West Africa," *Trans. Geol. Soc. S. Africa* **66**, 199–258.

Giesecke, C. L., 1826. "First report of a tour through Galway, Achill and Donegal," *Proc. R. Dubl. Soc.* **2**.

Gilluly, J., ed., 1948. *Origin of Granite. Mem. Geol. Soc. Am.* **28**, 139p.

—— 1963. "The tectonic evolution of the western United States." *Q. J. Geol. Soc. Lond.* **119**, 133–74.

—— 1965. "Volcanism, tectonism, and plutonism in the western United States," *Spec. Paper Geol. Soc. Am.* **80**, 69p.

Gindy, A. R., 1951a. "A xenolithic sill at Tallabrista, County Donegal, Ireland," *Geol. Mag.* **88**, 55–59.

—— 1951b. "The production of amphibolitic and other skarn rocks from limestone at Cor, County Donegal," *Geol. Mag.* **88**, 103–12.

—— 1953a. "Progressive replacement of limestone inclusions in granite at Ballynacarrick, County Donegal, Ireland," *Geol. Mag.* **90**, 152–58.

—— 1953b. "The plutonic history of the district around Trawenagh Bay, County Donegal," *Q. J. Geol. Soc. Lond.* **108** (for 1952), 377–411.

—— 1956a. "Biotite schlieren in some intrusive granites from Egypt and Donegal, and their origin," *Bull. Inst. Désert Égypte* **6**, 143–58.

—— 1956b. "The status of garnet in some late granitic intrusions from the Eastern Desert of Egypt and in Donegal, Eire," *Bull. Inst. Désert Égypte* **6**, 159–67.

—— 1958. "The earliest pegmatites in the granitization geosynclines of Aswan (Egypt) and Donegal (Eire)," *Bull. Fac. Sci. Alexandria Univ.* **2**, 75–98.

Ginzburg, I. V., Yefremova, S. V., Volovikova, I. M. and Yeliseyeva, O. P., 1963. "The quantitative-mineral composition of granitoids and its importance for problems of petrology and nomenclature," *Int. Geol. Rev.* **5**, 1123–36.

Glen, J. W., Donner, J. J. and West, R. G., 1957. "On the mechanism by which stones in till become oriented," *Am. J. Sci.* **255**, 194–205.

Glikson, A. Y., 1970. "Geosynclinal evolution and geochemical affinities of early Precambrian systems," *Tectonophysics* **9**, 397–433.

Goodspeed, G. E., 1940. "Dilation and replacement dykes," *J. Geol.* **48**, 175–95.

—— 1948. "Xenoliths and skialiths," *Am. J. Sci.* **246**, 515–25.

—— 1955. "Relict dykes and relict pseudodykes," *Am. J. Sci.* **253**, 146–61.

—— 1959. "Some textural features of magmatic and metasomatic rocks," *Am. Miner.* **44**, 211–50.

Gore, D. J., 1968. "Potash metasomatism and granitization accomplished by boron-potassium compounds," *Rep. 23rd Intl. Geol. Cong.* (*Prague*) **4**, 275–92.

Green, A. H., 1871. Notes on the geology of part of County Donegal, Ireland," *Geol. Mag.* **8**, 553–61.

Green, J. C., 1963. "High-level metamorphism of pelitic rocks in northern New Hampshire," *Am. Miner.* **48**, 991–1023.

—— 1964. "Stratigraphy and structure of the Boundary Mountain anticlinorium in the Errol Quadrangle, New Hampshire–Maine," *Spec. Paper Geol. Soc. Am.* **77**, 75p.

Gresens, R. L., 1966. "The effect of structurally produced pressure gradients on diffusion in rocks," *J. Geol.* **74,** 307–21.

―――― 1967a. "Tectonic-hydrothermal pegmatites: 1. The model," *Contr. Miner. Petrol.* **15,** 345–55.

―――― 1967b. "Tectonic-hydrothermal pegmatites: 2. An example," *Contr. Miner. Petrol.* **16,** 1–28.

Griffin, W. L., 1969. "Replacement antiperthites of the Babbitt–Embarass area, Minnesota, U.S.A.," *Lithos* **2,** 171–86.

Grout, F. F., 1937. "Criteria of origin of inclusions in plutonic rocks," *Bull. Geol. Soc. Am.* **48,** 1521–72.

―――― 1941. "Formation of igneous-looking rocks by metasomatism: a critical review and suggested research," *Bull. Geol. Soc. Am.* **52,** 1525–76.

―――― 1945. "Scale models of structures related to batholiths," *Am. J. Sci.* **243A,** 260–284.

Guidotti, C. V., 1970. "The mineralogy and petrology of the transition from the lower to the upper sillimanite zone in the Oquossoc area, Maine," *J. Petrol.* **11,** 277–336.

Hall, A. 1965a. "The occurrence of prehnite in appinitic rocks from Donegal, Ireland," *Min. Mag.* **35,** 234–36.

―――― 1965b. "On a granite-metadolerite contact at Curran Hill, County Donegal, Ireland," *Geol. Mag.* **102,** 531–37.

―――― 1965c. "The origin of accessory garnet in the Donegal granite," *Min. Mag.* **35,** 628–33.

―――― 1966a. "The alkali feldspars of the Ardara pluton, Donegal," *Min. Mag.* **35,** 693–703.

―――― 1966b. "A petrogenetic study of the Rosses Granite Complex, Donegal," *J. Petrol.* **7,** 202–20.

―――― 1966c. "The feldspars of the Rosses Granite Complex, Donegal, Ireland," *Min. Mag.* **35,** 975–82.

―――― 1966d. "The Ardara pluton: A study of the chemistry and crystallization of a contaminated granite intrusion," *Proc. R. Irish Acad.* **65B,** 203–235.

―――― 1967a. "The variation of some trace elements in the Rosses Granite Complex, Donegal," *Geol. Mag.* **104,** 99–109.

―――― 1967b. "The distribution of some major and trace elements in feldspars from the Rosses and Ardara granitic complexes, Donegal, Ireland," *Geochim. Cosmochim. Acta* **31,** 835–48.

―――― 1967c. "The chemistry of appinitic rocks associated with the Ardara pluton, Donegal, Ireland," *Contr. Miner. and Petrol.* **16,** 156–71.

―――― 1969a. "Regional variation in the composition of the British Caledonian granites," *J. Geol.* **77,** 466–81.

―――― 1969b. "The micas of the Rosses Granite Complex," *Sci. Proc. R. Dubl. Soc.* **3,** 209–17.

Hamilton, D. L. and Anderson, G. M., 1967. "Effects of water and oxygen pressures on the crystallization of basaltic magmas," in *Basalts: The Poldervaart Treatise on Rocks of Basaltic Composition,* ed. Hess, H. H. and Poldervaart, A. New York, Wiley-Interscience, vol. 1, 445–82.

Hamilton, W. and Myers, W. B., 1967. "The nature of batholiths," *Prof. Paper U. S. Geol. Surv.* **554-C,** 29p.

Hardie, W. G., 1963. "Explosion breccias near Stob Mhic Mhartuin, Glen Coe, Argyll, and their bearing on the origin of the nearby flinty crush-rock," *Trans. Edin. Geol. Soc.* **12,** 426–37.

Harker, A., 1908. "The geology of the small isles of Inverness-shire," *Mem. Geol. Surv. Scot.,* 210p.

―――― 1909. *Natural History of Igneous Rocks.* London, Methuen, 384p.

Harkness, R., 1861. "On the rocks of portions of the Highlands of Scotland south of the Caledonian Canal, and on their equivalents in the north of Ireland," *J. Geol. Soc. Lond.* **17,** 256–71.

Harland, W. B., 1964a. "Critical evidence for a great Infra-Cambrian glaciation," *Geol. Rdsch.* **54,** 45–61.

―――― 1964b. "Evidence of late PreCambrian glaciation and its significance," in *Problems of Palaeoclimatology,* ed. Nairn, A. E. M. New York, Wiley-Interscience, 119–149, 179–84.

Harme, M., 1958. "Examples of granitization of plutonic rocks," *Bull. Comm. Geol. Fin.* **180**, 45–64.
───── 1959. "Examples of the granitization of gneisses," *Bull. Comm. Geol. Fin.* **184**, 41–58.
Harper, C. T., 1967. "The geological interpretation of potassium-argon ages of metamorphic rocks from the Scottish Caledonides," *Scot. J. Geol.* **3**, 46–66.
Harper, J. C. and Hartley, J. J., 1938. "The Silurian inlier of Lisbellaw, County Fermanagh," *Proc. R. Irish Acad.* **45B**, 73–87.
Harpum, J. R., 1960. "Granitic and metamorphic associations in Tanganyika," *Report 21st Intl. Geol. Cong. (Norden)* **26**, 42–53.
Harris, A. L., 1969. "The relationships of the Leny Limestone to the Dalradian," *Scot. J. Geol.* **5**, 187–190.
Harris, P. G., Kennedy, W. Q., and Scarfe, C. M., 1970. Volcanism versus plutonism—the effect of chemical composition, in *Mechanism of Igneous Intrusion,* ed. Newall, G., and Rast, N. *Geol. J., Spec. Issue* **2**, Liverpool, Gallery Press, 187–200.
Harry, W. T., 1953. "The composite granitic gneiss of Western Ardgour, Argyll," *Q. J. Geol. Soc. Lond.* **109**, 285–308.
───── 1958. "A reexamination of Barrow's Older Granites in Glen Clova, Angus," *Trans. R. Soc. Edinb.* **62**, 393–412.
───── and Emeleus, C. H., 1960. "Mineral layering in some granite intrusions of southwest Greenland," *Report 21st. Intl. Geol. Cong. (Norden)* **14**, 172–81.
───── and Oen, I. S., 1964. "The Precambrian basement of Alangorssuaq, South Greenland," *Meddr. Grønland* **179**, No. 1, 72 p.
───── and Pulvertaft, T. C. R., 1963. "The Nunarssuit Intrusive Complex in southern Greenland," *Meddr. Grønland* **169**, No. 1, 136 p.
───── and Richey, J. E., 1963. "Magmatic pulses in the emplacement of plutons," *Lpool. Manchr. Geol. J.* **3**, 254–68.
Hart, S. R., 1964. "The petrology and isotopic mineral age relations of a contact zone in the Front Range, Colorado," *J. Geol.* **72**, 493–525.
Harte, B. and Johnson, M. R. W., 1969. "Metamorphic history of the Dalradian rocks in Glens Clova, Esk, and Lethnot, Angus, Scotland," *Scot. J. Geol.* **5**, 54–80.
Harte, W., 1867. "Crumpled granite beds in the County of Donegal," *J. R. Geol. Soc. Ireland,* **1**, 144–45.
Hartley, J. J., 1938. "The Dalradian rocks of the Sperrin Mountains and adjacent areas in Northern Ireland," *Proc. R. Irish Acad.* **44B**, 141–71.
Hatch, F. H., 1888. "On the spheroid-bearing granite of Mullaghderg, County Donegal," *Q. J. Geol. Soc. Lond.* **44**, 548–60.
─────, Wells, A. K. and Wells, M. K., 1961. *Petrology of the Igneous Rocks.* 12th ed. London, Thomas Murby, 515 p.
Haughton, S., 1857. "On the pitchstone and pitchstone porphyry of Barnesmore and Lough Eske, County of Donegal," *J. Geol. Soc. Dubl.* **7**, 196–98.
───── 1862. "Experimental researches on the granites of Ireland: Part III. On the granites of Donegal," *Q. J. Geol. Soc. Lond.* **18**, 403–20.
───── 1864. "Experimental researches on the granites of Ireland: Part IV. On the granites and syenites of Donegal; with some remarks on those of Scotland and Sweden," *Q. J. Geol. Soc. Lond.* **10**, 268–80.
───── 1866. "On the granites of Donegal," *Trans. R. Irish Acad.* **24**, 213–45.
───── 1870. "On the constituent minerals of the granites of Scotland, as compared with those of Donegal," *J. R. Geol. Soc. Ireland* **2**, 232–35.
Heitanen, A., 1961. "Relation between deformation, metamorphism, metasomatism, and intrusion

along the northwest border zone of the Idaho Batholith, Idaho," *Prof. Paper U. S. Geol. Surv.* **424-D**, 161–64.

——— 1963. "Idaho Batholith near Pierce and Bungalow, Clearwater County, Idaho," *Prof. Paper U. S. Geol. Surv.* **344-D**, 42 p.

——— 1968. "Belt Series in the region around Snow Peak and Mallard Peak, Idaho," *U. S. Geol. Surv. Prof. Paper* **344-E**, 34 p.

Helm, D. G., and Roberts, B., 1968. "The Caledonian history of the northeastern Irish Sea region," *Scot. J. Geol.* **4**, 375–76.

Hess, P. C., 1971. "Prograde and retrograde equilibria in garnet-cordierite gneisses in south-central Massachusetts," *Contr. Miner. and Petrol.* **30**, 177–95.

Hewitt, D. F., 1956. "The Grenville region of Ontario," in *The Grenville Problem,* ed. Thompson, J. E. *Spec. Pub. R. Soc. Can.* **1**, 22–41.

Hibbard, M. J., 1964. "Development of layering in an alpine-type gabbroic intrusion as a consequence of protoclastic deformation," *Spec. Paper Geol. Soc. Am.* **76** (*Abstracts* for 1963), 205.

——— 1965. "The origin of some alkali feldspar phenocrysts and their bearing on petrogenesis," *Am. J. Sci.* **263**, 254–61.

Hickling, N. L., Phair, G., Moore, R., and Rose, H. J., 1970. "Boulder Creek Batholith, Colorado: Part I. Allanite and its bearing upon age patterns," *Bull. Geol. Soc. Am.* **81**, 1973–94.

Hill, A. R. and Prior, D. B., 1968. "Directions of ice movement in northeastern Ireland," *Proc. R. Irish Acad.* **66B**, 71–84.

Hill, J. B. and Kynaston, H., 1900. "Kentallenite and its relations to other igneous rocks in Argyllshire," *Q. J. Geol. Soc. Lond.* **56**, 531–57.

Hills, E. S., 1959. "Cauldron subsidences, granitic rocks, and crustal fracturing in southeast Australia," *Geol. Rdsch.* **47**, 543–61.

——— 1963. *Elements of Structural Geology.* London, Methuen, 483 p.

Hirschberg, A., and Winkler, H. G. F., 1968. "Stabilitats Beziehungen zwischen chlorit, cordierit and almandin bei der Metamorphose," *Contr. Miner. and Petrol.* **18**, 17–42.

Holland, J. G. and Lambert, R. St. J., 1969. "Structural regimes and metamorphic facies," *Tectonophysics* **7**, 197–217.

Holtropp, J. F., 1969. "Information on the granitoid rocks of Surinam, South America," *Bull. Geol. Soc. Am.* **80**, 2237–52.

Hooper, P. R., 1962. "The petrology of Anvers Island and adjacent islands," *Falkland Is. Dependencies Surv. Sci. Report* **34**, 69 p.

Hoschek, G., 1969. "The stability of staurolite and chloritoid and their significance in metamorphism of pelitic rocks," *Contr. Miner. and Petrol.* **22**, 208–232.

Hossack, J. R., 1968. "Pebble deformation and thrusting in the Bygdin area (southern Norway)," *Tectonophysics* **5**, 315–39.

Howarth, R. J., 1967. "Trend-surface fitting to random data—an experimental test," *Am. J. Sci.* **265**, 619–25.

——— 1970. "Principal components analysis of the geochemistry and mineralogy of the Portaskaig Tillite and Kiltyfanned Schist (Dalradian) of County of Donegal, Eire," *Math. Geol.* **2**, 285–302.

——— 1971. "The Portaskaig Tillite succession of County of Donegal," *Proc. R. Irish Acad.* **71B**, 1–35.

——— Kilburn, C. and Leake, B. E., 1966. "The Boulder Bed succession at Glencolumbkille, County Donegal," *Proc. R. Irish Acad.* **65B**, 117–56.

Hsü, J. K., 1969. "Role of cohesive strength in the mechanics of overthrust faulting and of landsliding," *Bull. Geol. Soc. Am.* **80**, 927–52.

Hsu, L. C., 1968. "Selected phase relations in the system Al-Mn-Fe-Si-O-H: a model for garnet equilibria," *J. Petrol.* **9**, 40–83.

Hubbard, F. H., 1967. "Exsolution myrmekite," *Geol. För. Stockh. Förh.* **89**, 410–22.

Hubbard, J. A. E. B., 1966a. "Population studies in the Ballyshannon Limestone, Ballina Limestone, and Renn Point Beds (Viséan) of northwest Ireland. *Palaeontology* **9**, 252–69.

——— 1966b. "Facies patterns in the Carrowmoran Sandstone (Viséan) of western County Sligo, Ireland," *Proc. Geol. Assoc.* **77**, 233–54.

Hubbard, W. F. and Sheridan, D. J. R., 1965. "The Lower Carboniferous stratigraphy of some coastal exposures in County Sligo, Ireland," *Proc. Roy. Dubl. Soc.* **A2**, 189–95.

Hull, E., 1881. "On the Laurentian beds of Donegal and of other parts of Ireland," *Geol. Mag.* **8** (Dec. 2), 506.

——— 1882. "On the Laurentian rocks of Donegal and of other parts of Ireland," *Trans. R. Dubl. Soc.* **1**, 243–56.

——— 1886. "Notes on some of the problems now being investigated by the officers of the Geological Survey in the north of Ireland, chiefly in County Donegal," *Report Brit. Assoc. Advmt. Sci.,* 660–61.

———, Kilroe, J. R. and Mitchell, W. F., 1891. "Explanatory memoir: Sheets 22, 23, 30, and 31 (part), southwest Donegal," *Mem. Geol. Surv. Ireland*.

——— Kinahan, G. H., Nolan, J., Cruise, R. J., Egan, F. W., Kilroe, J. R., Mitchell, W. F., and M'Henry, A., 1891. "Explanatory memoir to accompany Sheets 3, 4, 5 (in part), 9, 10, 11 (in part), 15 and 16 of the maps of the Geological Survey of Ireland comprising Northwest and Central Donegal," *Mem. Geol. Surv. Ireland*.

Hunt, T. S., 1864. "On the chemical and mineralogical relations of metamorphic rocks," *J. Geol. Soc. Dubl.* **10**, 85–95.

Hurley, P. M., 1968. "Absolute abundance and distribution of Rb, K, and Sr. in the Earth," *Geochim. Cosmochim. Acta* **32**, 273–83.

———, Bateman, P. C., Fairbairn, H. W., and Pinson, W. H., 1965. "Investigation of initial $^{87}Sr/^{86}Sr$ ratios in the Sierra Nevada plutonic province," *Bull. Geol. Soc. Am.* **76**, 165–74.

Hutchinson, R. M., 1956. "Structure and petrology of Enchanted Rock Batholith, Llano and Gillespie Counties, Texas," *Bull. Geol. Soc. Am.* **67**, 763–806.

Hutchinson, W. W., 1970. "Metamorphic framework and the plutonic styles in the Prince Rupert region of the Central Coast Mountains, British Columbia," *Can. J. Earth Sci.* **7**, 376-405.

Ianello, P., 1971. "The Bushveld granites around Rooiberg, Transvaal, South Africa," *Geol. Rdsch.* **60**, 630–55.

Ishi, K., Sendo, T., Ueda, Y, and Yamashita, Y., 1960. "Granitic rocks of the Hinokami-yama district, Northeastern Kitakami Massif," *Sci. Reports Tohoku Univ.* **6** (Ser. 3), 439–86.

Iyengar, S. V. P., Pitcher, W. S. and Read, H. H., 1954. "The plutonic history of the Maas area, County Donegal," *Q. J. Geol. Soc. Lond.* **110**, 203–28.

Jacobson, R. R. E., MacLeod, W. N., and Black, R., 1958. "Ring-complexes in the Younger Granitic province of northern Nigeria," *Mem. Geol. Soc. Lond.* **1**, 72 p.

Jaeger, J. C., 1964. "Thermal effects of intrusions," *Review of Geophysics* **2**, 443–66.

Jahns, R. H., 1948. "Discussion," in *Origin of Granite*, ed. Gilluly, J. *Mem. Geol. Soc. Am.* **28**, 91–98.

——— and Tuttle, O. F., 1963. "Layered pegmatite-aplite intrusives," *Spec. Paper Miner. Soc. Am.* **1**, 78–92.

James, R. S. and Hamilton, D. L., 1969. "Phase relations in the system $NaAlSi_3O_8$-$KAlSi_3O_8$-$CaAl_2Si_2O_8$-SiO_2 at 1 kilobar water vapor pressure," *Contr. Miner. and Petrol.* **21**, 111–41.

Jenness, S. E., 1963. "Terra Nova and Bonavista map-areas, Newfoundland," *Mem. Geol. Surv. Can.* **327**, 184 p.

Johannsen, A., 1931, 1932, 1937, 1938. *A Descriptive Petrography of the Igneous Rocks.* Chicago, Univ. of Chicago Press, 4 vol.

Johansson, C. E., 1965. "Structural studies of sedimentary deposits," *Geol. För. Stockh. Förh.* **87**, 3–61.

Johnson, M. R. W., 1965. "Dalradian," in *The Geology of Scotland*, ed. Craig, G. Y. London, Oliver and Boyd, 115–60.

——— 1969. "Dalradian of Scotland," in *North Atlantic—Geology and Continental Drift*, ed. Kay, M. *Mem. Am. Assoc. Petrol. Geol.* **12**, 151–8.

——— and Dalziel, I. W. D., 1966. "Metamorphosed lamprophyres and the late thermal history of the Moines," *Geol. Mag.* **103**, 240–49.

Johnstone, G. S., 1966. *British Regional Geology: the Grampian Highlands.* 3rd ed. Geol. Surv. and Mus. Edinb., 103 p.

Jones, K. A. and Galwey, A. K., 1964. "A study of possible factors concerning garnet formation in rocks from Ardara, County Donegal, Ireland," *Geol. Mag.* **101**, 76-93.

——————————— 1966. "Size distribution, composition, and growth kinetics of garnet crystals in some metamorphic rocks from the west of Ireland," *Q. J. Geol. Soc. Lond.* **122**, 29–44.

Jones, P. B., 1971. "Folded faults and sequence of thrusting in Alberta foothills," *Bull. Am. Assoc. Petrol. Geol.* **55**, 292–306.

Joplin, G. A., 1969. "On the origin and occurrence of basic bodies associated with discordant bathyliths," *Geol. Mag.* **96**, 361–73.

——— 1960. "On the tectonic environment of basic magma," *Geol. Mag.* **96**, 361.

——— 1964. *A Petrography of Australian Igneous Rocks.* Sydney, Angus and Robertson, 210 p.

Joyce, A. S., 1970. "Chemical variation in a pelitic hornfels," *Chem. Geol.* **6**, 51–8.

Jung, J. and Brousse, R., 1959. *Classification Modale des Roches Eruptives*, Paris, Masson, 122 p.

Juurinen, A., 1956. "Composition and properties of staurolite," *Annales Acad. Sci. Fen.* Ser. A, Sec. 3, No. 47, 53p.

Kaitaro, S., 1953. "Geologic structure of the late Precambrian intrusives of the Ava area, Aland Islands," *Bull. Comm. Geol. Fin.* **162**, 71 p.

——— 1956. "On central complexes with radial lamprophyric dykes," *C. R. Soc. Geol. Fin.* **29**, 55–66.

Karl, F., 1966. "Über die Zusammensetzung, Entstehung und Gesteinssystematische Stellung Tonalitisch-granitischer Gesteine," *Tscherm. Min. Petr. Mitt.* **11**, 413–38.

Katz, M. B., 1968. "The fabric of the granulites of Mont Tremblant Park, Quebec," *Can. J. Earth Sci.* **5**, 801–12.

Kay, M., 1967. "Stratigraphy and structure of northeastern Newfoundland bearing on drift in the North Atlantic," *Bull. Am. Assoc. Petrol. Geol.* **51**, 579–600.

Kays, M. A., 1970. "Mesozoic metamorphism, May Creek Schist Belt, Klamath Mountains, Oregon," *Bull. Geol. Soc. Am.* **81**, 2743–58.

Kelly, J., 1853. "On the quartz rocks of the northern part of the County of Wicklow," *J. Geol. Soc. Dubl.* **5**, 237–76.

Kennedy, G. C., 1956. "Some aspects of the role of water in rock melts," *Spec. Paper Geol. Soc. Am.* **62**, 489–504.

Kennedy, M. J., 1969. "The metamorphic history of North Achill Island, County Mayo, and the problem of the origin of albite schists," *Proc. R. Irish Acad.* **67B**, 261–80.

Kennedy, W. Q., 1935. "The influence of chemical factors on the crystallization of hornblende in igneous rocks," *Min. Mag.* **24**, 203–7.

Kerrick, D. M., 1969. "K-feldspar megacrysts from a porphyritic quartz monzonite, central Sierra Nevada, California, *Am. Miner.* **54**, 839–845.

Ketskhoveli, D. N. and Shengelia, D. M., 1966. "Metasomatic zonal plagioclase in the Tseya granitoid pluton, northern Caucasus," *Doklady Acad. Sci. USSR,* Earth Sci. Sect. **166,** 158–60.

Keyes, C. R., 1895. "Origin and relations of central Maryland granites," *U. S. Geol. Surv. 15th Annual Report,* 685–750.

Kilburn, C., Pitcher, W. S. and Shackleton, R. M., 1965. "The stratigraphy and origin of the Portaskaig Boulder Bed Series (Dalradian)," *Geol. J.* **4,** 343–60.

Kinahan, G. H., 1881. "Possible Laurentian rocks in Ireland," *Geol. Mag.* **8** (Dec. 2), 427–29.

——— 1888. "Granite, elvan, porphyry, felstone, whinstone, and metamorphic rocks of Ireland," *Sci. Proc. R. Dubl. Soc.* **6,** 169–224.

——— 1891. "A new reading of the Donegal rocks," *Proc. R. Dubl. Soc.* **7,** 14–33.

——— 1901. "Notes on the Irish Primary rocks with their associated granitic and metamorphic rocks," *Report Brit. Assoc. Advmt. Sci. (Glasgow),* 637–39.

———, Leonard, H., and Cruise, R. J., 1871. "Memoir to accompany Sheets 104, 113, with adjacent parts of 103 and 122 (Kilkieran and Aran Sheets)," *Mem. Geol. Surv. Ireland.*

King, B. C., 1947. "The textural features of the granites and invaded rocks of the Singo Batholith of Uganda, and their petrogenetic significance," *Q. J. Geol. Soc. Lond.* **103,** 37–57.

——— 1948. "The form and structural features of aplite and pegmatite dykes and veins in the Osi area of the Northern Provinces of Nigeria, and the criteria that indicate a nondilational mode of emplacement," *J. Geol.* **56,** 459–75.

King, R. F., 1966. "The magnetic fabric of some Irish granites," *Geol. J.* **5,** 43–66.

Kistler, R. W., 1966. "Structure and metamorphism in the Mono Craters Quadrangle, Sierra Nevada, California," *Bull. U. S. Geol. Surv.* **122-E,** 53p.

——— and Bateman, P. C., 1966. "Stratigraphy and structure of the Dinkey Creek roof pendant in the central Sierra Nevada, California," *Prof. Paper U.S. Geol. Surv.* **524-B,** 14p.

Kleeman, A. W., 1965. "The origin of granitic magmas," *J. Geol. Soc. Australia* **12,** 35–52.

Knill, D. C., 1959. "Metamorphic segregation structures from Rosguill, Eire," *Geol. Mag.* **96,** 374–76.

——— and Knill, J. L., 1961. "Time relations between folding, metamorphism, and emplacement of granite in Rosguill, County Donegal," *Q. J. Geol. Soc. Lond.* **117,** 273–302.

Knill, J. L., 1959. "Axial and marginal sedimentation in geosynclinal basins," *J. Sedim. Petrol.* **29,** 317–25.

——— 1960. "The tectonic pattern in the Dalradian of the Craignish–Kimelfort district, Argyllshire," *Q. J. Geol. Soc. Lond.* **115** (for 1959), 339–64.

——— 1963. "A sedimentary history of the Dalradian Series," in *The British Caledonides,* ed. Johnson, M. R. W. and Stewart, F. H. London, Oliver and Boyd, 99–121.

——— and Knill, D. C., 1958. "Some discordant fold structures from the Dalradian of Craignish, Argyll, and Rosguill, County Donegal," *Geol. Mag.* **95,** 495–510.

Knowles, R. R. and Opdyke, N. D., 1968. "Paleomagnetic results from the Mauch Chunk Formation: a test of the origin of curvature in the folded Appalachians of Pennsylvania," *J. Geophys. Res.* **73,** 6515–26.

Kolbe, P. and Taylor, S. R., 1966. "Major and trace element relationships in granodiorites and granites from Australia and South Africa," *Contr. Miner. Petrol.* **12,** 202–22.

Koschmann, A. H., 1960. "Mineral paragenesis of Precambrian rocks in the Tenmile Range, Colorado," *Bull. Geol. Soc. Am.* **71,** 1357–70.

Krauskopf, K. B., 1941. "Intrusive rocks of the Okanagan Valley and the problem of their correlation," *J. Geol.* **49,** 1–53.

——— 1967. *Introduction to Geochemistry.* New York, McGraw-Hill, 721p.

——— 1968. "A tale of ten plutons," *Bull. Geol. Soc. Am.* **79,** 1–18.

Kretz, R., 1966. "Interpretation of the shape of mineral grains in metamorphic rocks," *J. Petrol.* **7,** 68–94.

Kretz, R., 1967. "Granite and pegmatite studies at Northern Indian Lake, Manitoba," *Bull. Geol. Surv. Can.* **148**, 42p.

—— 1968. "Study of pegmatite bodies and enclosing rocks, Yellowknife–Beaulieu region, District of MacKenzie," *Bull Geol. Surv. Can.* **159**, 109p.

—— 1969. "On the spatial distribution of crystals in rocks," *Lithos* **2**, 39–66.

Kuenen, P. H., 1968. "Origin of ptygmatic features," *Tectonophysics* **6**, 143–58.

Kulish, Y. A. and Polin, Y. K., 1966. "Orbicular granites of the Burein Massif," *Doklady Acad. Sci. USSR*, Earth Sci. Sect. **169**, 173–76.

Kuno, H., 1960. "High-alumina basalt," *J. Petrol.* **1**, 121–45.

Larsen, E. S., 1945. "Time required for the crystallization of the Great Batholith of Southern and Lower California," *Am. J. Sci.* **243A**, 399–416.

—— and Schmidt, R. G., 1958. "A reconnaissance of the Idaho Batholith and comparison with the Southern California Batholith," *Bull. U. S. Geol. Surv.* **1070A**, 33p.

——, and Switzer, G., 1939. "An obsidian-like rock formed from the melting of a granodiorite," *Amer. J. Sci.* **237**, 562–568.

Larsen, L. H., and Poldervaart, A., 1961. "Petrologic study of Bald Rock Batholith, near Bidwell Bar, California," *Bull. Geol. Soc. Am.* **72**, 69–92.

Lauerma, R., 1964. "On the structure and petrography of the Ipernat Dome, western Greenland," *Bull. Comm. Geol. Fin.* **215**, 1–88.

Laurence, R. D., 1970. "Stress analysis based on albite twinning of plagioclase feldspars," *Bull. Geol. Soc. Am.* **81**, 2507–12.

Lazarenkov, V. G., 1962. "Sur les processes de l'hybidisme normal," *Lap. Vses, Miner. Obschch.* No. 1, 50–66.

Lee, J. S., 1924. "A suggestion of a new method for geological survey of igneous intrusions," *Bull. Geol. Soc. China* **3**, 123–29.

Leedal, G. P. and Walker, G. P. L., 1954. "Tear faults in the Barnesmore area, Donegal," *Geol. Mag.* **91**, 116–20.

Leggo, P. J., Compston, W., and Leake, B. E., 1966. "The geochronology of the Connemara granites and its bearing of the antiquity of the Dalradian Series," *Q. J. Geol. Soc. Lond.* **122**, 91–118.

—— and Pidgeon, R. T., 1970. "Geochronological investigations of Caledonian history in western Ireland," *Eclog. Geol. Helvetiae* **63**, 207–12.

——, Tanner, P. W. G., and Leake, B. E., 1969. "Isochron study of Donegal granite and certain Dalradian rocks of Britain," in *North Atlantic—Geology and Continental Drift*, ed. Kay, M. Mem. Am. Assoc. Petrol. Geol. **12**, 354–62.

Leveson, D. J., 1963. "Orbicular rocks of the Lonesome Mountain area, Beartooth Mountains, Montana and Wyoming," *Bull. Geol. Soc. Am.* **74**, 1015–40.

—— 1966. "Orbicular rocks: a review," *Bull. Geol. Soc. Am.* **77**, 409–26.

Lindsay, J. F., 1969. "The glacial origin of Carboniferous conglomerates west of Barraba, New South Wales: discussion," *Bull. Geol. Soc. Am.* **80**, 911–4.

Link, R. F. and Koch, G. S., Jr., 1962. "Quantitative areal modal analysis of granitic complexes: discussion," *Bull. Geol. Soc. Am.* **73**, 411–14.

Lipman, P. W., 1963. "Gibson Peak pluton: a discordant composite intrusion in the southeastern Trinity Alps, northern California," *Bull. Geol. Soc. Am.* **74**, 1259–80.

—— 1964. "Structure and origin of an ultramafic pluton in the Klamath Mountains, California," *Am. J. Sci.* **262**, 199–222.

Lobjoit, W. M., 1964. "Kyanite produced in a granite aureole," *Min. Mag.* **33**, 804–8.

Lovering, T. S., 1935. "Theory of heat conduction applied to geological problems," *Bull. Geol. Soc. Am.* **46**, 69–94.

Lovering, T. S., 1938. "Temperatures in a sinking xenolith," *Trans. Am. Geophys. Union 19th Ann. Meeting,* **1,** 274–77.

—— 1955. "Temperatures in and near intrusions," *Econ. Geol.* **50,** 249–81.

Luth, W. C., Jahns, R. H., and Tuttle, O. F., 1964. "The granite system at 4 to 10 kilobars," *J. Geophys. Res.* **64,** 759–73.

M'Adam, J., 1834. "On the geology of the district of Fannet, in the County of Donegal," *J. Geol. Soc. Dubl.* **1,** 128–39.

McBirney, A., 1959. "Factors governing emplacement of volcanic rocks," *Am. J. Sci.* **258,** 431–448.

—— 1963. "Breccia pipe near Cameron, Arizona: a discussion," *Bull. Geol. Soc. Am.* **74,** 227–232.

McCall, G. J. H., 1954. "The Dalradian geology of the Creeslough area, County Donegal," *Q. J. Geol. Soc. Lond.* **110,** 153–73.

McCallien, W. J., 1935. "The metamorphic rocks of Inishowen, County Donegal," *Proc. R. Irish Acad.* **42B,** 407–42.

—— 1936. "A note on Dalradian pillow lavas, Strabane, County Tyrone," *Proc. R. Irish Acad.* **43B,** 13–22.

—— 1937. "The geology of the Rathmullan district, County Donegal," *Proc. R. Irish Acad.* **44B,** 45–59.

MacCulloch, J., 1819. *A Description of the Western Islands of Scotland Including the Isle of Man* (3 vols.), London.

MacDonald, G. J. F., 1957. "Thermodynamics of solids under non-hydrostatic stress with geologic applications," *Am. J. Sci.* **255,** 266–281.

MacGregor, A. M., 1951. "Some milestones in the Precambrian of Southern Rhodesia," *Proc. Geol. Soc. South Africa,* **54,** xxvii–lxxi.

MacGregor, M., 1937. "The western part of the Criffel–Dalbeattie igneous complex," *Q. J. Geol. Soc. Lond.* **93,** 457–84.

McKenzie, D. P., 1968. "The geophysical importance of high-temperature creep," in *History of the Earth's Crust,* ed. Phinney, R. A. Princeton, N. J., University Press. 28–44.

MacKenzie, W. S., 1960. "Reviews of some contributions of experimental studies to petrology," *Lpool. Manchr. Geol. J.* **2,** 369–88.

McKerrow, W. S., 1969. "Silurian rocks of Ireland and a comparison with those of Newfoundland," in *North Atlantic—Geology and Continental Drift.* ed. Kay, M. *Mem. Am. Assoc. Petrol. Geol.* **12,** 284–88.

M'Parlan, J., 1802. "Statistical survey of the County of Donegal with observations on the means of improvement, drawn up in the year 1801 for the consideration and under the direction of the Dublin Society," *R. Dubl. Soc.*

Mahan, S. M. and Rogers, J. J. W., 1968. "A study of grain contacts in some high-grade metamorphic rocks," *Am. Miner.* **53,** 323–27.

Mallick, D. I. J., 1967. "The metamorphic development of the Mpande Dome in Zambia," *Geol. Rdsch.* **56,** 670–91.

Marfunin, A. S., 1963. "Some petrological aspects of order-disorder in feldspars," *Min. Mag.* **33,** 298–314.

Marmo, V., 1955. "On the classification of the Precambrian granites," *Colonial Geol. Miner. Res.* **5,** 429–37.

—— 1956. "On the emplacement of granites," *Am. J. Sci.* **254,** 479–92.

—— 1962. "On granites," *Bull. Comm. Geol. Fin.* **201,** 77p.

—— 1966. "On the petrological classification of granites," *Bull. Comm. Geol. Fin.* **222,** 69–73.

—— 1967. "On the granite problem," *Earth Sci. Review* **3,** 7–29.

Martin, N. R., 1953. "The structure of the granite massif of Flamanville, Manche, northwest France," *Q. J. Geol. Soc. Lond.* **108**, 311–42.

Mayo, E. B., 1935. "Some intrusions and their wall rocks in the Sierra Nevada," *J. Geol.* **43**, 673–89.

────── 1941. "Deformation in the interval Mt. Lyell–Mt. Whitney, California," *Bull. Geol. Soc. Am.* **52**, 1001–84.

Mead, W. J., 1925. "The geologic role of dilatancy," *J. Geol.* **33**, 685–98.

Means, W. D. and Paterson, M. S., 1966. "Experiments on preferred orientation of platy minerals," *Contr. Miner. and Petrol.* **13**, 108–33.

Mehnert, K. R., 1959. "Der Gegenwärtige Stand des Granitproblems," *Fortschr. Miner.* **37**, 117–206.

────── 1968. *Migmatites and the Origin of Granitic Rocks.* Amsterdam, Elsevier, 393p.

────── and Willgallis, A., 1961. "Die Alkaliverteilung im Malsburger Granit," *Jahresh. Geol. Landesamt Baden-Württemberg* **5**, 117–39.

Mercy, E. L. P., 1956. "The accuracy and precision of 'rapid methods' of silicate analysis," *Geochim. Cosmochim. Acta* **9**, 161–73.

────── 1960a. "The geochemistry of the older granodiorite, County Donegal, Ireland," *Trans. R. Soc. Edinb.* **64** (for 1958–59), 101–27.

────── 1960b. "The geochemistry of the Rosses ring complex, south Donegal, Ireland," *Trans. R. Soc. Edinb.* **64** (for 1958–59), 128–38.

────── 1962. "The Mullaghduff porphyry dykes," *Trans. Edinb. Geol. Soc.* **19**, 65–82.

────── 1963. "The geochemistry of some Caledonian granitic and metasedimentary rocks," in *The British Caledonides,* ed. Johnson, M. R. W. and Stewart, F. H. London, Oliver and Boyd, 188–215.

────── 1965. "Caledonian igneous activity," in *The Geology of Scotland,* ed. Craig, G. Y. London, Oliver and Boyd. 229–67.

Merrill, R. B., Robertson, J. K., and Wyllie, P. J., 1970. "Melting reactions in the system $NaAlSi_3O_8$-$KAlSi_3O_8$-SiO_2-H_2O to 20 kilobars compared with results for other feldspar-quartz-H_2O systems," *J. Geol.* **78**, 558–69.

Mikami, H. M. and Digman, R. E., 1957. "The bedrock geology of the Guilford 15-minute quadrangle and a portion of the New Haven quadrangle," *Bull. State Geol. Natural Hist. Surv. Conn.* **86**, 99p.

Miller, A. A., 1939. "River development in southern Ireland," *Proc. R. Irish Acad.* **45B**, 321–54.

Miller, J. A. and Flinn, D., 1966. "A survey of the age relations of Shetland rocks," *Geol. J.* **5**, 95–116.

Miller, W. J., 1945. "Observations on pseudodykes and foliated dykes," *J. Geol.* **53**, 175–90.

Misch, P., 1949. "Metasomatic granitization of batholithic dimensions," *Am. J. Sci.* **247**, 209–45, 372–406, 673–705.

────── 1969. "Paracrystalline microboundinage of zoned grains and other criteria for synkinematic growth of metamorphic minerals," *Am. J. Sci.* **267**, 43–63.

Miyashiro, A., 1958. "Regional metamorphism of Goscusyo–Takanuki District in the Central Abukuma Plateau," *Tokyo Univ. Fac. Sci. Jour.* **11**, Sec. 2, 219.

Moorbath, S., 1962. "Lead isotope abundance determination studies on mineral occurrences in the British Isles," *Phil. Trans. R. Soc.* **254A**, 295–360.

Mrazec, M. L., 1915. "Les plis diapirs et le diapirisme en général," *C. R. Rumania Inst. Geol.* **4**, 226–70.

Mueller, R. F., 1963. "Interaction of chemistry and mechanics in magmatism," *J. Geol.* **71**, 759–72.

Muir, I., 1953. "A local potassic modification of the Ballachulish granodiorite," *Geol. Mag.* **90**, 182–92.

Munro, M., 1965. "Some structural features of the Caledonian granite complex at Strontian, Argyllshire," *Scot. J. Geol.* **1**, 152–75.

Murthy, M. V. N., 1951. "Comptonitic dyke rocks from Inishowen, County Donegal, Ireland," *Trans. Geol. Soc. Glasg.* **21**, 205–6.

Murthy, V. R., 1957. "Bedrock geology of the East Barre area, Vermont," *Bull. Vermont Geol. Surv.* **10**, 121p.

Naggar, M. H. and Atherton, M. P., 1970. "The composition and metamorphic history of some aluminum silicate-bearing rocks from the aureoles of the Donegal granites," *J. Petrol.* **11**, 549–89.

——— Atherton, M. P. and Pitcher, W. S., 1970. "Composition of some aluminum silicate bearing contact rocks from County Donegal, Eire," *Nature* **226**, 841.

Newton, R. C., 1966. "Kyanite-sillimanite equilibrium at 750°C," *Science* **151**, 1222–25.

Nicholson, R., 1965. "The structure and metamorphism of the mantling Karagwe–Ankolean sediments of the Ntungame gneiss dome and their time-relation to the development of the dome," *Q. J. Geol. Soc. Lond.* **121**, 143–62.

Nickel, E., Kock, H. and Nungässer, W., 1967. "Modellversuche zur Fliessregelung in Graniten (Beiträge zur Tektonik von Fliessgefügen III)," *Schw. Min. Petrol. Mitt.* **47**, 399–497.

Niyogi, D., 1966. "Petrology of the alkalic rocks of Kishangarh, Rajasthan, India," *Bull. Geol. Soc. Am.* **77**, 65–82.

Nockolds, S. R., 1933. "Some theoretical aspects of contamination in acid magmas," *J. Geol.* **41**, 561–89.

——— 1934. "The contaminated tonalites of Loch Awe, Argyll," *Q. J. Geol. Soc. Lond.* **90**, 302–21.

——— 1941. "The Garabal Hill–Glen Fyne igneous complex," *Q. J. Geol. Soc. Lond.* **96** (for 1940), 451–510.

——— 1946. "The order of crystallization of the minerals in some Caledonian plutonic and hypabyssal rocks," *Geol. Mag.* **83**, 206–16.

——— 1954. "Average chemical compositions of some igneous rocks," *Bull. Geol. Soc. Am.* **65**, 1007–32.

——— and Allen, R., 1953. "The geochemistry of some igneous rocks series," *Geochim. Cosmochim. Acta* **4**, 105–42.

——— and Mitchell, R. L., 1946. "The geochemistry of some Caledonian plutonic rocks: a study in the relationship between the major and trace elements of igneous rocks and their minerals," *Trans. R. Soc. Edinb.* **61**, 533–75.

O'Brien, G. D. 1968. "Survey of diapirs and diapirism," in *Diapirism and Diapirs,* ed. Braunstein, J. and O'Brien, G. D. *Mem. Am. Assoc. Petrol. Geol.* **8**, 1–9.

O'Brien, M. V., 1959. "The future of nonferrous mining in Ireland," *Sympos. Inst. Min. Metal. Lond.,* 5–26, 35–45.

Oen, I. S., 1960. "The intrusion mechanism of the late Hercynian post-tectonic granite plutons of northern Portugal," *Geologie Mijnb.* **39**, 257–96.

Offler, R. and Fleming, P. D., 1968. "A synthesis of folding and metamorphism, Mt. Lofty Ranges, South Australia," *J. Geol. Soc. Australia* **15**, 245–66.

Ohta, Y., 1969. "On the formation of augen structure," *Lithos* **2**, 109–32.

Okrusch, M. and Evans, B. W., 1970. "Minor element relationships in coexisting andalusite and sillimanite," *Lithos* **3**, 261–68.

Orville, P. M., 1963. "Alkali ion exchange between vapor and feldspar phases," *Am. J. Sci.* **261**, 201–37.

Pabst, A., 1928. "Observations on inclusions in the granitic rocks of Sierra Nevada," *Univ. Calif. Publ. Bull. Dept. Geol. Sci.* **17**, 325–86.

Page, L. R., 1968. "Devonian plutonic rocks in New England," in *Studies of Applachian Geology:*

Northern and Maritime, ed. Zen. E-an, White, W. S., Hadley, J. B., and Thompson, J. B., Jr. New York, Wiley Interscience, 371–83.

Palmer, D. F., Bradley, J., and Prebble, W. M. 1967. "Orbicular granodiorite from Taylor Valley, South Victoria Land, Antarctica," *Bull. Geol. Soc. Am.* **78,** 1423–28.

Pankhurst, R. J., 1970. "The geochronology of the basic igneous complexes," *Scot. J. Geol.* **6,** 83–107.

Park, R. G., 1969. "Structural correlation in metamorphic belts," *Tectonophysics* **7,** 323–38.

Paterson, M. S. and Weiss, L. E., 1968. "Folding and boudinage of quartz-rich layers in experimentally deformed phyllite," *Bull. Geol. Soc. Am.* **79,** 795–812.

Peach, B. N., Gunn, W., Clough, C. T., Hinxman, L. W., Crampton, C. B., and Anderson, E. M., 1912. "The geology of Ben Wyvis, Carn Chuinneag, Inchbae and the surrounding country," *Mem. Geol. Surv. Gt. Brit.,* 189p.

Peng, C. C. J., 1970. "Intergranular albite in some granites and syenites from Hong Kong," *Am. Miner.* **55,** 270–82.

Phemister, T. C., 1945. "The Coast Range Batholith near Vancouver, British Columbia," *Q. J. Geol. Soc. Lond.* **101,** 37–88.

Phillips, W. E. A., Kennedy, M. J., and Dunlop, G. M., 1969. "Geologic comparison of western Ireland and northeastern Newfoundland," in *North Atlantic–Geology and Continental Drift,* ed. Kay, M. *Mem. Am. Assoc. Petrol. Geol.* **12,** 194–211.

Phillips, W. J., 1965. "The deformation of quartz in a granite," *Geol. J.* **4,** 391–414.

Phillips, W. S., 1956. "The minor intrusive suite associated with the Criffell–Dalbeattie Granodiorite Complex," *Proc. Geol. Assoc.* **67,** 103–21.

Pitcher, W. S., 1950. "Calc-silicate skarn veins in the limestone of Lough Anure, County Donegal," *Miner. Mag.* **29,** 126–41.

——— 1953a. "The migmatitic older granodiorite of Thorr district, County Donegal," *Q. J. Geol. Soc. Lond.* **108** (for 1952), 413–46.

——— 1953b. "The Rosses granitic ring-complex, County Donegal, Eire," *Proc. Geol. Assoc.* **64,** 153–82.

——— 1955. "Certain rapid methods of chemical analysis and their petrochemical applications," *Proc. Geol. Soc. Lond.* **1524,** 93.

——— 1965. "The aluminium silicate polymorphs," in *Controls of Metamorphism,* ed. Pitcher, W. S., and Flinn, G. W. London, Oliver and Boyd, 327–41.

——— 1969. "Northeast-trending faults of Scotland and Ireland, and chronology of displacements," in *North Atlantic–Geology and Continental Drift,* ed. Kay, M. *Mem. Am. Assoc. Petrol. Geol.* **12,** 724–33.

——— 1970. "Ghost stratigraphy in intrusive granites: a review," in *Mechanism of Igneous Intrusion,* ed. Newall, G. and Rast, N. *Geol. J.* Spec. Issue No. 2, Liverpool, Gallery Press, 123–140.

——— and Cheesman, R. L., 1954. "Summer field meeting in northwest Ireland with an introductory note on the geology," *Proc. Geol. Assoc.* **65,** 345–71.

———, Elwell, R. W. D., Tozer, C. F., and Cambray, F. W., 1964. "The Leannan fault," *Q. J. Geol. Soc. Lond.* **120,** 241–73.

——— and Read, H. H., 1952. "An appinitic intrusion-breccia at Kilkenny, Maas, County Donegal," *Geol. Mag.* **89,** 328–36.

——— 1959. "The Main Donegal granite," *Q. J. Geol. Soc. Lond.* **114** (for 1958), 259–305.

——— 1960a. "Early transverse dykes in the main Donegal granite," *Geol. Mag.* **97,** 53–61.

——— 1960b. "The aureole of the Main Donegal granite," *Q. J. Geol. Soc. Lond.* **116,** 1–36.

——— 1963. "Contact metamorphism in relation to manner of emplacement of the granites of Donegal, Ireland," *J. Geol.* **71,** 261–96.

——— and Shackleton, R. M., 1966. "On the correlation of certain Lower Dalradian successions in northwest Donegal," *Geol. J.* **5,** 149–56.

Pitcher, W. S., Shackleton, R. M., and Wood, R. S. R., 1971. "The Ballybofey Anticline: a solution of the general structure of parts of Donegal and Tyrone," *Geol. J.* **7**, 321–28.

—— and Sinha, R. C. 1958. "The petrochemistry of the Ardara aureole," *Q. J. Geol. Soc. Lond.* **113** (for 1957), 393–408.

Piwinskii, A. J., 1968 a. "Studies of batholithic feldspars: Sierra Nevada, California," *Contr. Miner. Petrol.* **17**, 204–23.

—— 1968b. "Experimental studies of igneous rocks series, central Sierra Nevada batholith, California," *J. Geol.* **76**, 548–70.

—— and Wyllie, P. J., 1968. "Experimental studies of igneous rock series: a zoned pluton in the Wallowa batholith, Oregon," *J. Geol.* **76**, 205–34.

Platten, I. M., 1968. "The metamorphism of minor intrusions associated with the Newer Granites of the Western Highlands of Scotland," *Scot. J. Geol.* **67**, 370–74.

Pockley, R. P. C., 1961. "Lead isotope and age studies of uranium and lead minerals from the British Isles and France," unpublished *Ph.D. thesis*, Oxford University.

Poole, W. H., 1967. "Tectonic evolution of the Appalachian region of Canada," in *Geology of the Atlantic Region*, ed. Neale, E. R. W. and Williams, H. *Spec. Papers Geol. Assoc. Can.* **4**, 9–51.

Powell, D., 1965. "Comparison of calc-silicate bands from the Moine Schists of Inverness-shire with similar bands from Moine-like rocks in Donegal," *Nature* **206**, 180–81.

—— 1966. "On the preferred crystallographic orientation of garnet in some metamorphic rocks," *Min. Mag.* **35**, 1094–1109.

—— and Treagus, J. E., 1970. "Rotational fabrics in metamorphic minerals," *Min. Mag.* **37**, 801–14.

Preston, J., 1954. "The geology of the Precambian rocks of the Kuopio district: Suomi Tiedeakat., Toimit," *Annales Acad. Sci. Fennica* Ser. A, Sect. 3, No. 40, 111p.

—— 1965. "Tertiary feeder dykes in the West of Ireland," *Proc. Geol. Soc. Lond.* **1626**, 149–150.

—— 1967a. "A Tertiary feeder dyke in County Fermanagh, Northern Ireland," *Sci. Proc. R. Dubl. Soc.* **3A**, 1–16.

—— 1967b. "The Blind Rock dyke, County Donegal," *Irish Nat. J.* **15**, 286–293.

Price, N. J., 1966. *Fault and Joint Development in Brittle and Semi-Brittle Rock*. New York, Pergamon, 176p.

Price, R. A. and Mountjoy, E. W., 1970. "Geologic structure of the Canadian Rocky Mountains between Bow and Athabasca Rivers—a progress report," in *Structure of the Southern Canadian Cordillera*, ed. Wheeler, J. E. *Spec. Paper Geol. Assoc. Can.* **6**, 7–26.

Pringle, J., 1940. "The discovery of Cambrian trilobites in the Highland Border rocks near Callender, Perthshire," *Report Brit. Assoc. Advmt. Sci.* **1**, 252.

Prinz, M., 1965. "Structural relationships of mafic dykes in the Beartooth Mountains of Montana–Wyoming," *J. Geol.* **73**, 165–74.

—— and Poldervaart, A., 1964. "Layered mylonite from Beartooth Mountains, Montana," *Bull. Geol. Soc. Am.* **75**, 741–44.

Pugin, V. A., and Khitarov, N. I., 1968. "The system Al_2O_3-SiO_2 at high temperatures and pressures," *Geokimia* **2**, 157–165.

Pulvertaft, T. C. R., 1961. "The Dalradian successions and their relationships in the Churchill district of County Donegal," *Proc. R. Irish. Acad.* **61B**, 255–73.

Raase, P., 1969. "über Zonarbau und Korrosionserscheinungen in Plagioclasen II," *Neues. Jb. Miner. Mh.* **5**, 189–200.

——, and Morteani, G., 1968. "über Zonarbau und basiche Kerne in Plagioclasen I," *Neues Jb. Miner. Abh.* **110**, 81–105.

Ragan, D. M., 1963. "Emplacement of the Twin Sisters dunite, Washington," *Am. J. Sci.* **261**, 549–65.

Raguin, E., 1965. *Geology of Granite.* New York, Wiley-Interscience, 314p.

Ramberg, H., 1952. *The Origin of Metamorphic and Metasomatic Rocks; a Treatise on Recrystallization and Replacement in the Earth's Crust.* Chicago, University of Chicago Press, 317p.

—— 1959. "Evolution of ptygmatic folding," *Norsk Geol. Tiddskr.* **39**, 99–152.

—— 1961a. "A study of veins in Caledonian rocks around Trondheim Fjord, Norway," *Norsk. Geol. Tiddskr.* **41**, 1–44.

—— 1961b. "Artificial and natural photoelastic effects in quartz and feldspars," *Am. Miner.* **46**, 934–51.

—— 1963. "Experimental study of gravity tectonics by means of centrifuged models," *Bull. Geol. Inst. Uppsala* **42**, 1–97.

—— 1967. *Gravity, Deformation and the Earth's Crust as Studied by Centrifuged Models.* New York, Academic Press, 214p.

—— 1970. "Model studies in relation to intrusion of plutonic bodies," in *Mechanism of Igneous Intrusion,* ed. Newall, G. and Rast, N. *Geol. J.* Spec. Issue 2, Liverpool, Gallery Press, 261–86.

Ramsay, J. G., 1962. "The geometry and mechanics of formation of 'similar' type folds," *J. Geol.* **70**, 309–27.

—— 1963. "Structural investigations in the Barberton Mountain land, eastern Transvaal," *Trans. Geol. Soc. S. Africa* **66**, 353–98.

—— 1967. *Folding and Fracturing of Rocks.* New York, McGraw-Hill, 568p.

—— and Graham, R. H., 1970. "Strain variation in shear belts," *Can. J. Earth Sci.* **7**, 786–813.

Rao, M. S., 1951. "The dyke rocks of County Donegal and the adjoining part of County Tyrone, Ireland," *Trans. Geol. Soc. Glasg.* **21**, 203–4.

Rast, N., 1963. "Structure and metamorphism of the Dalradian rocks of Scotland," in *The British Caledonides,* ed. Johnson, M. R. W. and Stewart, F. H. London, Oliver and Boyd, 123–42.

—— 1965. "Nucleation and growth of metamorphic minerals," in *Controls of Metamorphism,* ed. Pitcher, W. S. and Flinn, G. W. London, Oliver and Boyd, 73–102.

—— and Crimes, T. P., 1969. "Caledonian orogenic episodes in the British Isles and northwestern France and their tectonic and chronological interpretation," *Tectonophysics* **7**, 277–307.

—— and Litherland, M., 1970. "The correlation of the Ballachulish and Perthshire (Iltay) Dalradian successions," *Geol. Mag.* **107**, 259–72.

Read, H. H., 1928. "A note on 'ptygmatic folding' in the Sutherland granite complex," *Summ. Prog. Geol. Surv. Scot.* **1927**, 72–77.

—— 1931. "The geology of central Sutherland," *Mem. Geol. Surv. Scot.* 238p.

—— 1936. "The stratigraphical order of the Dalradian rocks of the Banffshire coast," *Geol. Mag.* **73**, 468–76.

—— 1955. "Granitization and metamorphism in Donegal, Ireland," *Jahrb. Mijnbouwkundige Varenigung te Delft,* 138–46.

—— 1957. *The Granite Controversy: Geological Addresses Illustrating the Evolution of a Disputant.* Hertsford, Thomas Murby, 430p.

—— 1958a. "A Centenary lecture: stratigraphy in metamorphism," *Proc. Geol. Assoc.* **69**, 83–102.

—— 1958b. "Donegal granite," *Sci. Prog. Lond.* **45**, 225–40.

—— 1961. "Aspects of Caledonian magmatism in Britain," *Lpool Manchr. Geol. J.* **2**, 653–83.

—— and Phemister, J., 1925. "The geology of the country around Golspie, Sutherlandshire," *Mem. Geol. Surv. Scot.* 143p.

—————— and Ross, G., 1926. "The geology of Strath Oykell and lower Loch Shin," *Mem. Geol. Surv. Scot.* 220 p.

Redden, J. A., 1963. "Geology and pegmatites of the Fourmile Quadrangle, Black Hills, South Dakota," *Prof. Paper U. S. Geol. Surv.* **299-D**, 199–291.

Rees, A. I., 1968. "The production of preferred orientation in a concentrated dispersion of elongated and flattened grains," *J. Geol.* **76**, 457–64.

Reesor, J. E., 1958. "Dewar Creek map-area, with special emphasis on the White Creek batholith, British Columbia," *Mem. Geol. Surv. Can.* **292**, 78 p.

Reffay, A., 1966. "Problèmes morphologiques dans la péninsule du sud-ouest du Donegal," *Rev. Géogr. Alpine* **54**, 287–311.

Reitan, P., 1959. "Pegmatite veins and the surrounding rocks: II. Structural control of small pegmatites in amphibolite, Rytterholmen, Kragerofjord, Norway," *Norsk. Geol. Tiddskr.* **39**, 175–95.

———— 1965. "Pegmatite veins and the surrounding rocks: V. Secondary recrystallization of aplite to form pegmatite," *Norsk. Geol. Tiddskr.* **45**, 31–40.

Reverdatto, V. V., Sharapov, V. N., and Melamed, V. G., 1970. "The controls and selected pecularities of the origin of contact metamorphic zonation," *Contr. Miner. and Petrol.* **29**, 310–37.

Reynolds, D. L., 1931. "The dykes of the Ards Peninsula, County Down," *Geol. Mag.* **68**, 97–111, 145–65.

———— 1936. "Demonstration in petrogenesis from Kiloran Bay, Colonsay," *Miner. Mag.* **24**, 237–407.

———— 1943. "The southwestern end of the Newry complex: a contribution towards the petrogenesis of the granodiorites," *Q. J. Geol. Soc. Lond.* **99**, 205–46.

———— 1946. "The sequence of geochemical changes leading to granitization," *Q. J. Geol. Soc. Lond.* **102**, 389–438.

———— 1954. "Fluidization as a geological process and its bearing on the problem of intrusive granites," *Am. J. Sci.* **252**, 577–614.

———— 1956. "Calderas and ring-complexes," *Verh. Kon. Nederl. Geol. Mijnbouwk Gen.* **16**, 355–379.

Richardson, S. W., Gilbert, M. C. and Bell, P. M., 1969. "Experimental determination of kyanite-andalusite and andalusite-sillimanite equilibria: the aluminium silicate triple point," *Am. J. Sci.* **267**, 159–292.

Richey, J. E., 1932. "The Tertiary ring complex of Slieve Gullion, Ireland," *Q. J. Geol. Soc. Lond.* **88**, 776–849.

———— 1938. "The dykes of Scotland," *Trans. Edinb. Geol. Soc.* **13**, 393–435.

———— and Thomas, H. H., 1930. "The geology of Ardnamurchan, northwest Mull, and Coll," *Mem. Geol. Surv. Great Brit.* 393 p.

Rickard, M. J., 1958. "Discussion on W. S. Pitcher and H. H. Read, "The Main Donegal Granite," *Proc. Geol. Soc. Lond.* **1557**, 40–41.

———— 1961. "A note on cleavages in crenulated rocks," *Geol. Mag.* **98**, 324–32.

———— 1962. "Stratigraphy and structure of the Errigal area, County Donegal, Ireland," *Q. J. Geol. Soc. Lond.* **118**, 207–36.

———— 1963. "Analysis of the strike swing at Crockator Mountain, County Donegal, Eire," *Geol. Mag.* **100**, 401–19.

———— 1964. "Contact metamorphism in relation to manner of emplacement of the granites of Donegal, Ireland: a discussion," *J. Geol.* **72**, 682–84.

———— 1971. "Revision of the geology of the Dunlewy and Crockator areas, County Donegal," *J. Geol. Soc. Lond.* **127**, 187–88.

Riddihough, R. P., 1968. "Magnetic surveys off the north coast of Ireland," *Proc. R. Irish Acad.* **66B**, 27–41.

———— 1969. "Magnetic map of the Ardara Granite and southern County Donegal," *Geophys. Bull. Dubl. Inst. Adv. Stud.* No. 27, 3 p.

———— and Young, D. G. G., 1970. "Gravity and magnetic surveys of Inishowen and adjoining areas off the north coast of Ireland," *Proc. Geol. Soc. Lond.* **1644**, 215–20.

Ringwood, A. E. et al., 1966. "High pressure experimental investigations into the nature of the Mohorovicic Discontinuity, the mineralogical and chemical composition of the upper mantle and the origin of basaltic and andesitic magmas," *Dept. Geophys. and Geochem. Aust. Nat. Univ.,* Publication No. 444, 251 p.

Roberts, J. L., 1966a. "Sedimentary affiliations and stratigraphic correlation of the Dalradian rocks in the southwest Highlands of Scotland," *Scot. J. Geol.* **2**, 200–23.

——— 1966b. "The formation of similar folds by inhomogeneous plastic strain, with reference to the fourth phase of deformation affecting the Dalradian rocks in the southwest Highlands of Scotland," *J. Geol.* **74**, 831–55.

——— 1970. "The intrusion of magma into brittle rocks," in *Mechanism of Igneous Intrusion,* ed. Newall, G. and Rast, N. *Geol. J.* Spec. Issue 2, Liverpool, Gallery Press, 287–338.

——— and Treagus, J. E., 1964. "A reinterpretation of the Ben Lui fold,"*Geol. Mag.* **101**, 512–16.

Robson, G. R. and Barr, K. G., 1964. "The effect of stress on faulting and minor intrusions in the vicinity of a magma body," *Bull. Volc.* **27**, 315–30.

Roddick, J. A., 1965. "Vancouver North, Coquitlam, and Pitt Lake map-areas, British Columbia, with special emphasis on the evolution of the plutonic rocks," *Mem. Geol. Surv. Can.* **335**, 276 p.

——— and Armstrong, J. E., 1959. "Relict dykes in the Coast Mountains near Vancouver, B.C.," *J. Geol.* **67**, 603–13.

Rogers, J. J. W., 1961. "Origin of albite in granitic rocks," *Am. J. Sci.* **259**, 186–93.

Ronner, F., 1963. *Systematische Klassifikation der Massengesteine.* Vienna, Springer-Verlag, 380 p.

Rosenbusch, H., 1877. "Der Steiger Schiefer und ihre Contactzone an den Graniten von Barr-Andlau und Hohwald," *Abh. Geol. Specialkarte Elsass-Lothringen,* **1**, 79–393.

Rosenfeld, J. L., 1968. "Garnet rotations due to the major Palaeozoic deformations in southeast Vermont," in *Studies of Appalachian Geology: Northern and Maritime,* ed. by Zen, E-an, White, W. S., Hadley, J. B., and Thompson, J. B., Jr. New York, Wiley-Interscience, 185–202.

Roubault, M. and de la Roche, H., 1965. "Parallele entre la géochemie des schistes Paléozoiques et celle des formations granitiques dans le Massif du Lys-Caillaouas (Pyrénées Centrales)," *Geol. Rdsch.* **55**, 301–16.

Runner, J. J., 1943. "Structure and origin of Black Hills Precambrian granite domes," *J. Geol.* **51**, 431–57.

Rusnak, G., 1957. "The orientation of sand grains under conditions of 'undirectional' fluid flow: I. Theory and experiment," *J. Geol.* **65**, 384–409.

Russell, M. J., 1968. "Structural controls of base metal mineralization in Ireland in relation to continental drift," *Trans. Instr. Min. Metal.* **77B**, 117–128.

Rust, G. W., 1937. "Preliminary notes on explosive volcanism in southeastern Missouri," *J. Geol.* **45**, 48–75.

Rutland, R. W. R., 1965. "Tectonic overpressures," in *Controls of Metamorphism,* ed. Pitcher, W. S. and Flinn, G. W. London, Oliver and Boyd, 119–39.

Sabine, P. A., 1963. "The Strontian granite complex, Argyllshire," *Bull. Geol. Surv. Gt. Brit.* **20**, 6–42.

——— and Watson, J. V., 1965. "Isotopic age determinations of rocks from the British Isles, 1955–64," *Q. J. Geol. Soc. Lond.* **121**, 477–533.

Saha, A. K., 1958. "Mineralogical and chemical variations in the Wollaston granitic pluton, Hastings County, Ontario," *Am. J. Sci.* **256**, 609–19.

——— 1959. "Emplacement of three granitic plutons in southeast Ontario, Canada." *Bull. Geol. Soc. Am.* **70**, 1293–1326.

San Miguel, A., 1955. "Les caractéristiques structurales du granite de la Costa Brava et leur signification pétrogénétique," in *Les Echanges de Matières au Cours de la Genèse des Roches Grenue Acides et Basiques.* 68th Colloques Internationaux du Centre National de la Recherche Scientifique, Nancy, 37–60.

San Miguel, A., 1969. "The aplite-pegmatite association and its petrogenetic interpretation," *Lithos* **2**, 25–38.

Savolahti, A., 1962. "The rapakivi problem and the rules of idiomorphism in minerals," *Bull. Comm. Geol. Fin.* **204**, 33–111.

Saxena, S. K., 1966. "Evolution of zircons in sedimentary and metamorphic rocks," *Sedimentology* **6**, 1–34.

―――― 1968. "The present status of zircon," *Sedimentology* **10**, 209–16.

Schermerhorn, L. J. G., 1956a. "Petrogenesis of a porphyritic granite east of Oporto (Portugal)," *Tscherm. Min. Petrol. Mitt.* **6**, 73–115.

―――― 1956b. "The granites of Trancoso (Portugal): a study in microclinization," *Am. J. Sci.* **254**, 329–48.

―――― 1959. "Igneous, metamorphic and ore geology of Castro Daire–Sao Pedro do Sul–Satao region, northern Portugal," *Comuncoes Comm. Trab. Serv. Geol. Port.* **37**, 616 p.

―――― 1960. "Telescoping of mineral facies in granites," *Bull. Comm. Geol. Fin.* **188**, 121–132.

―――― 1961. "Orthoclase, microcline, and albite in granite," *Schw. Min. Petr. Mitt.* **41**, 13–36.

―――― 1962. "The emplacement of the late Hercynian granites in Portugal: a reply," *Geologie Mijnb.* **41**, 20–25.

―――― and Stanton, W. J., 1963. "Tilloids in the West Congo geosyncline," *Q. J. Geol. Soc. Lond.* **119**, 201–34.

Sclar, C. B., 1965. "Layered mylonites and the processes of metamorphic differentiation," *Bull. Geol. Soc. Am.* **76**, 611–12.

Scott, R. H., 1862. "On the granitic rocks of the southwest of Donegal, and the minerals therewith associated," *J. Geol. Soc. Dubl.* **9**, 285–94.

―――― 1864. "On the granitic rocks of Donegal, and the minerals therewith associated," *J. Geol. Soc. Dubl.* **10**, 13–24.

Scrope, G. P., 1858. "On lamination and cleavage occasioned by the mutual friction of the particles of rocks while in irregular motion," *Q. J. Geol. Soc. Lond.* **15**, 84–86.

Sederholm, J. J., 1907. "Om granit och gneis," *Bull. Comm. Geol. Fin.* **23**, 110 p.

―――― 1916. "On synantetic minerals and related phenomena," *Bull. Comm. Geol. Fin.* **48**, 63–113.

Seifert, F., and Schreyer, W., 1968. "Fluid phases in the system K_2O-MgO-SiO_2-H_2O and their possible significance for the existence of ultramafic magmas," *Spec. Paper Geol. Soc. Am.* **101** (*Abstracts* for 1967), 197.

Seifert, K. E., 1964. "The genesis of plagioclase twinning in the Nonewaug granite," *Am. Miner.* **49**, 297–320.

Sen, S., 1956. "Structures of the porphyritic granite and associated metamorphic rocks of East Manbhum, Bihar, India," *Bull. Geol. Soc. Am.* **67**, 647–70.

Sendo, T., 1958. "On the granitic rocks of Mt. Otakine and its adjacent districts in the Abukuma Massif, Japan," *Sci. Reports Tohoku Univ.* **6**, 57–168.

Sergiades, D. A., 1962. "Geology of Palas de Rey, Provincia de Lugo, northwest Spain," unpublished *Ph.D. thesis*, Liverpool University.

Shackleton, R. M., 1958. "Downward-facing structures of the Highland Border," *Q. J. Geol. Soc. Lond.* **113**, 361–92.

Shand, S. J., 1942. "Phase petrology of the Cortlandt complex," *Bull. Geol. Soc. Am.* **53**, 409–28.

―――― 1943. *Eruptive Rocks.* 2nd ed. Hertsford, Thomas Murby, 488 p.

Shapiro, L., and Brannock, W. W., 1952. "Rapid analysis of silicate rocks," *Circ. U. S. Geol. Surv.*

―――――――――――――― 1956. "Rapid analysis of silicate rocks," *Bull. U. S. Geol. Surv.* **1036-C**, 19–56.

Shaw, H. R., 1965. "Comments on viscosity, crystal settling, and convection in granitic magmas," *Am. J. Sci.* **263**, 120–52.

—————— 1969. "The rheology of basalt in the melting range," *J. Petrol.* **10**, 510–35.

——————, Wright, T. L., Peck, D. L. and Okamura, R., 1968. "The viscosity of basaltic magma; an analysis of field measurements in Makaopuhi Lava Lake, Hawaii," *Am. J. Sci.* **266**, 225–64.

Shelley, D., 1964. "On myrmekite," *Am. Miner.* **49**, 41–52.

—————— 1968. "Ptygma-like veins in graywacke, mudstone, and low-grade schist from New Zealand," *J. Geol.* **76**, 692–701.

—————— 1970. "The origin of myrmekitic intergrowths and a comparison with rod-eutectics in metals," *Min. Mag.* **37**, 674–81.

Silver, L. T. and Deutsch, S., 1963. "Uranium-lead isotopic variations in zircons: a case study," *J. Geol.* **71**, 721–758.

Simonen, A., 1966. "Orbicular rock in Kuru, Finland," *Bull. Comm. Geol. Fin.* **222**, 93–107.

Simpson, A., 1968. "The Caledonian history of the northeastern Irish Sea region and its relation to surrounding areas," *Scot. J. Geol.* **4**, 135–63.

Simpson, I. M., 1955. "The Lower Carboniferous stratigraphy of the Omagh Syncline, Northern Ireland," *Q. J. Geol. Soc. Lond.* **110**, 391–408.

Slobodskoy, R. M., 1966. "Translucent structures in granitoids of the South Altai Narym pluton," *Doklady Acad. Sci. USSR*, Earth Sci. Sect. 168, 51–54.

—————— 1970. "Origin of ptygmatic veins in the contact aureole of granitoid batholiths," *Tectonophysics* **9**, 447–58.

Smart, T. B., 1962. "The aurole of the Barnesmore granite, County Donegal," *Irish Nat. J.* **14**, 55–59.

Smith, C. H., 1958. "Bay of Islands igneous complex, western Newfoundland," *Mem. Geol. Surv. Can.* **290**, 132 p.

Smith, H. G., 1946. "The lamprophyre problem," *Geol. Mag.* **84**, 165–71.

Smith, J. V., 1962. "Genetic aspects of twinning in feldspars," *Norsk Geol. Tiddskr.* **42**, 244–63.

Smith, R. L., Bailey, R. A. and Ross, C. S., 1961. "Structural evolution of the Valles Caldera, New Mexico, and its bearing on the emplacement of ring dykes," *Prof. Paper U. S. Geol. Surv.* **424-D**, 145–49.

Smithson, S. B., 1963. "Granite studies II: The Precambrian Fla granite, a geological and geophysical investigation," *Norges Geol. Unders.* **219**, 212 p.

—————— 1965. "Oriented plagioclase grains in K-feldspar porphyroblasts," *Contr. to Geology, Univ. Wyoming* **4**, 63–68.

Sollas, W. J., 1893. "On pitchstone and andesite from Tertiary dykes in Donegal," *Sci. Proc. R. Dubl. Soc.* **8**, 87–93.

Solomon, M., 1963. "Counting and sampling errors in modal analysis by point counter," *J. Petrol.* **4**, 367–82.

Soper, N. J., 1963. "The structure of the Rogart igneous complex," *Q. J. Geol. Soc. Lond.* **119**, 445–78.

Sørensen, H., 1969. "Rhythmic igneous layering in peralkaline intrusions: an essay review on Ilimaussaq (Greenland) and Loverzo (Kola, USSR)," *Lithos* **2**, 261–83.

Spencer, A. M., 1971, "Late Precambrian glaciation in Scotland," *Mem. Geol. Soc. Lond.* **6**, 98 p.

Spjeldnaes, N., 1959. "Traces of an Eocambrian orogeny in southern Norway," *Norsk. Geol. Tiddskr.* **39**, 83–86.

Spry, A., 1963a. "The chronological analysis of crystallization and deformation of some Tasmanian Precambrian rocks," *J. Geol. Soc. Australia* **10**, 193–208.

—————— 1963b. "Origin and significance of snowball structure in garnet," *J. Petrol.* **4**, 211–22.

Spry, A., 1963c. "Ripple marks and pseudo-ripple marks in deformed quartzite," *Am. J. Sci.* **261**, 756–66.

―――― 1969. *Metamorphic Textures.* New York. Pergamon, 350 p.

Stanton, R. L., 1964. "Mineral interfaces in stratiform ores," *Trans. Inst. Min. Metal.* **74**, 45–79.

Stauffer, M. R., 1967. "Tectonic strain in some volcanic, sedimentary, and intrusive rocks near Canberra, Australia: a comparative study of deformation fabrics," *N. Z. J. Geol. Geophys.* **10**, 1079–1108.

Stephens, N. and Synge, F. M., 1965. "Late Pleistocene shorelines and drift limits in north Donegal," *Proc. R. Irish Acad.* **64B**, 131–53.

Stephenson, P. J., 1959. "The Mt. Barney central complex, southeast Queensland," *Geol. Mag.* **96**, 125–36.

Stevens, N. C., 1959. "Ring structures of the Mt. Alford district, southeast Queensland," *J. Geol. Soc. Australia* **6**, 37–49.

Steveson, B. G., 1970. "Chemical variability in some Moine rocks of Lochailort, Inverness-shire," *Scot. J. Geol.* **7**, 51–60.

Stewart, D., 1800. "The report of Donald Stewart, Itinerant Mineralogist to the Dublin Society," *Trans. R. Dubl. Soc.* **1**.

Stone, M., 1957. "The Aberfoyle anticline, Callender, Perthshire," *Geol. Mag.* **94**, 265–76.

―――― 1969. "Nature and origin of banding in the granitic sheets of Tremearne, Porthleven, Cornwall," *Geol. Mag.* **106**, 142–58.

Streckeisen, A. L., 1967. "Classification and nomenclature of igneous rocks," *Neues Jb. Miner. Abh.* **107**, 144–240.

Strens, R. G., 1968. "Stability of Al_2SiO_5 solid solutions," *Min. Mag.* **26**, 839–49.

Sturt, B. A. and Harris, A. L., 1961. "The metamorphic history of the Loch Tummel area, central Perthshire, Scotland," *Lpool Manchr. Geol. J.* **2**, 689–711.

Sutton, J., 1965. "Some recent advances in our understanding of the controls of metamorphism," in *Controls of Metamorphism,* ed. Pitcher, W. S. and Flinn, G. W. London, Oliver and Boyd, 22–45.

―――― and Watson, J., 1951. "The Pre-Torridonian metamorphic history of the Loch Torridon and Scourie areas in the Northwest Highlands and its bearing on the chronological classification of the Lewisian," *Q. J. Geol. Soc. Lond.* **106**, 241–308.

―――――――――― 1969. "Scourian-Laxfordian relationships in the Lewisian of northwest Scotland," in *Age Relations in High Grade Metamorphic Terrains,* ed. Wynne-Edwards, H. R. *Spec. Paper Geol. Assoc. Can.* **5**, 119–28.

Sylvester, A. G., 1964. "Geology of the Vradal granite: Part III of the Precambrian rocks of the Telemark area, in south central Norway," *Norsk Geol. Tiddskr.* **44**, 445–82.

Szadeczky-Kardoss, E., 1960. "A genetical system of igneous rocks," *Report 21st. Int. Geol. Cong. (Norden)* **13**, 260–74.

Talbot, C. J., 1970. "The minimum strain ellipsoid using deformed quartz veins," *Tectonophysics* **9**, 47–76.

Talbot, J. L. and Hobbs, B. E., 1968. "The relationship of metamorphic differentiation to other structural features at three localities," *J. Geol.* **76**, 581–87.

―――――――――― , Wilshire, H. G. and Sweatman, T. R., 1963. "Xenoliths and xenocrysts from lavas of the Kerguelen Archipelago," *Am. Miner.* **48**, 159–79.

Taubeneck, W. H., 1957. "Geology of the Elkhorn Mountains, northeastern Oregon: Bald Mountain batholith," *Bull. Geol. Soc. Am.* **68**, 181–238.

―――― 1964. "Cornucopia stock, Wallowa Mountains, northeast Oregon: field relations," *Bull. Geol. Soc. Am.* **75**, 1093–1116.

Taubeneck, W. H., 1967a. "Petrology of Cornucopia tonalite unit, Cornucopia stock, Wallowa Mountains, northeastern Oregon," *Spec. Paper Geol. Soc. Am.* **91**, 56p.

—— 1967b. "Notes on the Glen Coe cauldron subsidence, Argyllshire, Scotland," *Bull. Geol. Soc. Am.* **78**, 1295–1316.

—— and Poldervaart, A., 1960. "Geology of the Elkhorn Mountains, northeastern Oregon: Part 2. Willow Lake intrusion," *Bull. Geol. Soc. Am.* **71**, 1295–1322.

Taylor, W. E. G., 1968. "The Dalradian rocks of Slieve Gamph, western Ireland," *Proc. R. Irish Acad.* **67B**, 63–82.

Tchalenko, J. S., 1968. "Evolution of kink-bands and the development of compression textures in sheared clays," *Tectonophysics* **6**, 159–74.

Termier, H. and Termier, G., 1956. *L'Évolution de la Lithosphère: I. Pétrogénèse.* Paris, Masson 653 p.

Theime, J. G., 1965. "An orbicular facies of the Leinster granite," *Proc. R. Irish Acad.* **64B**, 155–64.

Thomas, P. R. and Treagus, J. E., 1968. "The stratigraphy and structure of the Glen Orchy area, Argyllshire, Scotland," *Scot. J. Geol.* **4**, 121–34.

Thompson, J. B., Jr. and Norton, S. A., 1968. "Palaeozoic regional metamorphism in New England and adjacent areas," in *Studies of Appalachian Geology: Northern and Maritime,* ed. Zen, E-an, White, W. S., Hadley, J. B., and Thompson, J. B., Jr. New York, Wiley-Interscience, 319–27.

—— Robinson, P., Clifford, T. N., and Trask, N. J. J., 1968. "Nappes and gneiss domes in west-central New England," in *Studies of Appalachian Geology: Northern and Maritime,* ed. Zen, E-an, White, W. S., Hadley, J. B., and Thompson, J. B., Jr. New York, Wiley-Interscience, 203–18.

Thompson, R. N., 1968. "A calcic marginal facies of the Panticose granodiorite, Spanish Pyrenees," *Proc. Geol. Assoc.* **79**, 219–26.

Thomson, J., 1877. "On the geology of the island of Islay," *Trans. Geol. Soc. Glasg.* **5**, 200–22.

Tobisch, O. T., 1967. "The influence of early structure on the orientation of late-phase folds in an area of repeated deformation," *J. Geol.* **75**, 554–64.

Tozer, C. F., 1955. "The mode of occurrence of sillimanite in the Glen district, County Donegal, Ireland," *Geol. Mag.* **92**, 310–20.

Townend, R., 1966. "The geology of some granite plutons from western Connemara, County Galway," *Proc. R. Irish Acad.* **65b**, 157–202.

Tuominen, H. V., 1961. "The structural position of the Orijarvi granodiorite and the problem of synkinematic granites," *Bull. Comm. Geol. Fin.* **196**, 499–515.

—— 1964. "The trends of differentiation in percentage diagrams," *J. Geol.* **72**, 855–60.

—— 1966. "Structural control of composition in the Orijarvi granodiorite," *Bull. Comm. Geol. Fin.* **222**, 311–29.

Turner, D. C., 1963. "Ring structures in the Sara–Fier Younger Granite complex, northern Nigeria," *Q. J. Geol. Soc. Lond.* **119**, 345–66.

Turner, F. J., 1968. *Metamorphic Petrology: Mineralogical and Field Aspects.* New York, McGraw-Hill. 403 p.

—— 1970. "Uniqueness versus conformity to pattern in petrogenesis," *Am. Miner.* **55**, 339–48.

—— and Verhoogen, J., 1960. *Igneous and Metamorphic Petrology.* 2nd ed. New York, McGraw-Hill, 694 p.

Tuttle, O. F. and Bowen, N. L., 1958. "The origin of granite in the light of experimental studies in the system $NaAlSi_3O_8$-$KAlSi_3O_8$-SiO_2-H_2O," *Mem. Geol. Soc. Am.* **74**, 153p.

Tyrrell, G. W., 1928. "The geology of Arran," *Mem. Geol. Surv. Scot.*, 292 p.

Vance, J. A., 1961. "Zoned granitic intrusions—an alternative hypothesis of origin," *Bull. Geol. Soc. Am.* **72**, 1723–7.

Vance, J. A., 1965. "Zoning in igneous plagioclase: patchy zoning," *J. Geol.* **73**, 636–51.

────── and Gilreath, J. P., 1967. "The effect of synneusis on phenocryst distribution patterns in some porphyritic igneous rocks," *Am. Miner.* **52**, 529–35.

Van Diver, B. B., 1968. "Origin of Jove Peak orbiculite in Wenatchee Ridge area, northern Cascades, Washington," *Am. J. Sci.* **266**, 110–23.

────── 1969. "A cummingtonite case history, and the influence of Ca and Al on optical properties of cummingtonite-grünerite minerals," *Spec. Paper Geol. Soc. Am.* **121** (*Abstracts* for 1968), 302.

────── 1970. "Origin of biotite orbicules in 'bullseye granite' of Craftsbury, Vermont," *Am. J. Sci.* **268**, 322–40.

Veniale, F., Pigorini, B., and Soggetti, F., 1968. "Petrological significance of the accessory zircon in the granites from Baveno, M. Orfano, and Alzo (north Italy)," *Report 23rd. Int. Geol. Cong.* (*Prague*) **13**, 243–68.

Verhoogen, J., 1951. "The chemical potential of a stressed solid," *Trans. Am. Geophys. Union* **32**, 251.

Vernon, R. H., 1965. "Plagioclase twins in some mafic gneisses from Broken Hill, Australia," *Min. Mag.* **35**, 488–507.

────── 1968. "Microstructures of high-grade metamorphic rocks at Broken Hill, Australia," *J. Petrol.* **9**, 1–22.

Vincent, E. A., 1953. "Hornblende-lamprophyre dykes of basaltic parentage from the Skaergaard area, East Greenland," *Q. J. Geol. Soc. Lond.* **109**, 21–49.

Vistelius, A. B., 1966. "A stochastic model for the crystallization of alaskite and its corresponding transition probabilities," *Doklady Acad. Sci. USSR*, Earth Sci. Sect. **170**, 82–85.

────── 1967. "Courses of crystallization and secondary minerals in certain granites of the Khanka district (Maritime region)," *Doklady Acad. Sci. USSR*, Earth Sci. Sect. **177**, 114–16.

Vogel, T. A., 1970. "The origin of some antiperthites—a model based on nucleation," *Am. Miner.* **55**, 1390–95.

────── and Spence, W. H., 1969. "Relict plagioclase phenocrysts from amphibolite-grade metamorphic rocks," *Am. Miner.* **54**, 522–28.

Voll, G., 1960. "New work on petrofabrics," *Lpool Manchr. Geol. J.* **2**, 503–67.

────── 1964. "Deckenbau und Fazies im Schottischen Dalradian," *Geol. Rdsch.* **53**, 590–612.

Von Platen, H., 1965. "Experimental anatexis and genesis of migmatites," in *Controls of Metamorphism*, ed. Pitcher, W. S. and Flinn. G. W. London, Oliver and Boyd, 203–18.

Waard, D. de, 1949a. "Tectonics of the Mt. Aigoual pluton in the southeastern Cevennes, France," *Kon. Nederl. Akad. Wetens.* **52**, 388–402; 539–550.

────── 1949b. "Diapiric structures," *Kon. Nederl. Akad. Wetens.* **52**, 1027–38.

Wager, L. R. and Brown, G. M., 1951. "A note on rhythmic layering in the ultrabasic rocks of Rhum," *Geol. Mag.* **88**, 166–68.

────── 1968. *Layered Igneous Rocks*. London, Oliver and Boyd, 588p.

Walker, G. P. L. and Leedal, G. P., 1954. "The Barnesmore granite complex, County Donegal," *Sci. Proc. R. Dubl. Soc.* **26**, 207–43.

Walton, M., 1960. "Granite problems," *Science* **131**, 635–45.

Ward, R. F., 1959. "Petrology and metamorphism of Wilmington complex, Delaware, Pennsylvania, and Maryland," *Bull. Geol. Soc. Am.* **70**, 1425–58.

Waters, A. C., 1955. "Volcanic rocks and the tectonic cycle," in *The Crust of the Earth*, ed. Poldervaart, A. *Spec. Paper Geol. Soc. Am.* **62**, 703–22.

Watson, J., 1949. "Late sillimanite in the migmatites of Kildonan," *Geol. Mag.* **85**, 149–62.

────── 1963. "Some problems concerning the evolution of the Caledonides of the Scottish Highlands," *Proc. Geol. Assoc.* **74**, 213–58.

Watson, J., 1964. "Conditions in the metamorphic Caledonides during the period of late-orogenic cooling," *Geol. Mag.* **101**, 457–65.

────── 1967. "Evidence of mobility in reactivated basement complexes," *Proc. Geol. Assoc.* **78**, 211–36.

Watt, W. S., 1965. "Textural and field relationships of basement granitic rocks, Qaersuarssuk, south Greenland," *Meddr. Grønland* **179**, Nr. 8, 34p.

Watterson, J. 1965. "Plutonic development of the Ilordleq area, south Greenland: Part 1. Chronology and the occurrence and recognition of metamorphosed basic rocks," *Meddr. Grønland* **172**, Nr. 7, 145p.

────── 1968a. "Plutonic development of the Ilordleq area, south Greenland: Part 2. Late-kinematic basic dykes," *Meddr. Grønland* **185**, Nr. 3, 104p.

────── 1968b. "Homogeneous deformation of the gneisses of Vesterland, southwest Greenland," *Meddr. Grønland* **175**, Nr. 6, 72p.

Weaver, J. D., 1958. "Utuado pluton, Puerto Rico," *Bull. Geol. Soc. Am.* **69**, 1125–42.

Wegmann, C. E., 1930. "Über Diapirismus (Besonders im Grundgebirge)," *Bull. Comm. Geol. Fin.* **92**, 58–76.

────── 1963. "Tectonic patterns at different levels," *Trans. Geol. Soc. S. Africa*, Annex to vol. 66. 78p.

────── and Schaer, J. F., 1962. "Chronologie et déformations des filons basiques dans les formations Precambriennes du sud de la Norvege," *Norsk. Geol. Tiddskr.* **42**, 371–87.

Weidmann, M., 1964. "Géologie de la région située entre Tigssaluk Fjord et Sermiligârssuk Fjord (partie médiane), S. W. Groenland," *Meddr. Grønland* **169**, Nr. 5, 146p.

Weiss, L. E., 1969. "Flexural slip folding of foliated model materials," in *Proc. Conf. Res. Tectonics*, ed. Baer, A. J. and Norris, D. K. *Paper Geol. Surv. Can.* 68–52, 294–357.

Wells, A. K. and Bishop, A. C., 1954. "The origin of aplites," *Proc. Geol. Assoc.* **65**, 95–114.

────────────── 1955. "An appinitic facies associated with certain granites in Jersey, Channel Islands," *Q. J. Geol. Soc. Lond.* **111**, 143–63.

Wells, M. K., 1954. "The structure of the granophyric quartz dolerite intrusion of Centre 2, Ardnamurchan and the problem of net-veining," *Geol. Mag.* **91**, 293–307.

Westropp, W. H. S., 1867. "On the origin of granite," *Geol. Mag.* **4**, 522–25.

Weymouth, J. H. and Williamson, W. O., 1953. "The effects of extrusion and some other processes on the microstructure of clay," *Am. J. Sci.* **251**, 89–108.

White, A. J. R., Compston, W., and Kleeman, A. W., 1967. "The Palmer granite—a study of a granite within a regional metamorphic environment," *J. Petrol.* **8**, 29–50.

Whitfield, J. M., Rogers, J. J. W., and McEwan, M. C., 1959. "Relationships among textural properties and modal compositions of some granitic rocks," *Geochim. Cosmochim. Acta* **17**, 272–85.

Whitten, E. H. T., 1951. "Cataclastic pegmatites and calc-silicate skarns near Bunbeg, County Donegal," *Min. Mag.* **29**, 737–56.

────── 1953. "Modal and chemical analyses in regional studies," *Geol. Mag.* **90**, 337–44.

────── 1955. "Metasediments of Bunbeg (County Donegal) and their relationship to the surrounding granite," *Proc. Geol. Assoc.* **66**, 51–67.

────── 1957a. "The petrogenetic significance of the contact relationships of the Donegal granite in Gweedore and Cloghaneely," *Geol. Mag.* **94**, 25–39.

────── 1957b. "The Gola granite (County Donegal) and its regional setting," *Proc. R. Irish Acad.* **58B**, 245–92.

────── 1959a. "Tuffisites and magnetite tuffisites from Tory Island, Ireland, and related products of gas action," *Am. J. Sci.* **257**, 113–37.

────── 1959b. "Compositional trends in a granite: modal variation and ghost-stratigraphy in part of the Donegal granite, Eire," *J. Geophys. Res.* **64**, 835–48.

Whitten, E. H. T., 1960a. "Quantitative evidence of palimpsestic ghost stratigraphy from modal analysis of a granitic complex," *Report 21st Int. Geol. Cong.* (Norden) **14**, 182–93.

——— 1960b. "Systematic quantitative areal variation of six granitic massifs," *Bull. Geol. Soc. Am.* **71**, 2002–3.

——— 1961a. "Quantitative areal modal analysis of granitic complexes," *Bull. Geol. Soc. Am.* **72**, 1331–60.

——— 1961b. "Systematic quantitative areal variation in five granitic massifs from India, Canada, and Great Britain," *J. Geol.* **69**, 619–46.

——— 1961c. "Quantitative distribution of major and trace components in rock masses," *Trans. Am. Inst. Min. Metal. Engrs.* **220**, 239–46.

——— 1962. "A new method for determination of the average composition of a granite massif," *Geochim. Cosmochim. Acta* **26**, 545–60.

——— 1963. "Application of quantitative methods in the geochemical study of granite massifs," in *Studies in Analytical Geochemistry*, ed. Shaw, D. M. *Spec. Pub. R. Soc. Can.* **6**, 76–123.

——— 1966a. *Structural Geology of Folded Rocks.* Chicago, Rand McNally, 663p.

——— 1966b. "Quantitative models in the economic evaluation of rock units: illustrated with the Donegal granite and gold-bearing Witwatersrand conglomerates," *Trans. Inst. Min. Metal.* **75B**, 181–98.

Widenfalk, L., 1969. "Electron microprobe analyses of myrmekite plagioclases and coexisting feldspars," *Lithos* **2**, 295–309.

Wiebe, R. A., 1968. "Plagioclase stratigraphy: a record of magmatic conditions and events in a granite stock," *Am. J. Sci.* **266**, 690–703.

Wilkinson, S. B. et al., 1907. "The geology of Islay," *Mem. Geol. Surv. Scot.* 88p.

Williams, Harold, 1965. "The Appalachians in northeastern Newfoundland—a two-sided symmetrical system," *Am. J. Sci.* **262**, 1137–58.

——— 1968. "Wesleyville, Newfoundland," *Map Geol. Surv. Can.* 1227A.

Williams, Howell, 1954. "Problems and progress in volcanology," *Q. J. Geol. Soc. Lond.* **109**, 311–32.

Wilshire, H. G., 1961. "Layered diatremes near Sydney, New South Wales," *J. Geol.* **69**, 473–84.

——— 1967. "The Prospect alkaline diabase-picrite intrusion, New South Wales, Australia," *J. Petrol.* **8**, 97–162.

Wilson, G., 1952. "Ptygmatic structures and their formation," *Geol. Mag.* **89**, 1–21.

Wilson, H. E., 1953. "The petrography of the Old Red Sandstone rocks of the north of Ireland," *Proc. R. Irish Acad.* **55B**, 283–320.

——— and Robbie, J. A., 1966. "Geology of the country around Ballycastle," *Mem. Geol. Surv. Northern Ireland*, 370p.

Wilson, J. T., 1962. "Cabot Fault: an Appalachian equivalent of the San Andreas and Great Glen Faults and some implications for continental displacement," *Nature* **195**, 135–38.

Wilson, R. L., 1964. "The Tertiary dykes of Magho mountain, County Fermanagh," *Irish Nat. J.* **14**, 254–257.

Windley, B., 1965. "The role of cooling cracks formed at high temperatures and of released gas in the formation of chilled basic margins in net-veined intrusions," *Geol. Mag.* **102**, 521–30.

Winkler, H. G. F., 1966. "Der Prozess der Anatexis: Seine Bedeutung für die Genese der Migmatite," *Tschm. Min. Petr. Mitt.* **11**, 266–87.

——— 1967. *Petrogenesis of Metamorphic Rocks.* 2nd ed. Berlin, Springer-Verlag, 237p.

——— and Von Platen, H., 1958. "Experimentelle Gesteinsmetamorphose II. Bildung von anatektischen Schmelzen bei der Metamorphose von NaCl-führenden kalkfreien Tonen," *Geochim. Cosmochim. Acta* **15**, 91–112.

——————— 1961a. "Experimentelle Gesteinsmetamorphose IV. Bildung anatektischen Schmelzen ans metamorphosierten Grauwacken," *Geochim. Cosmochim. Acta* **24**, 48–69.

Winkler, H. G. F., and Von Platen, H., 1961*b*. "Experimentelle Gesteinsmetamorphose V. Experimentelle anatektischen Schmelzen und ihre petrogenetische Bedeutung," *Geochim. Cosmochim. Acta* **24**, 250–59.

Wolfe, M. E., 1969. "A trace fossil from the Lower Dalradian, County Donegal, Eire," *Geol. Mag.* **106**, 274–76.

Wright, A. E. and Bowes, D. R., 1963. "Classification of volcanic breccias: a discussion," *Bull. Geol. Soc. Am.* **74**, 79–86.

——————————— 1968. "Formation of explosion breccias," *Bull. Volc.* **32**, 15–32.

Wright, P. C., 1964. "The petrology, chemistry and structure of the Galway granite of the Carna area, County Galway," *Proc. R. Irish Acad.* **63B**, 239–64.

Wyllie, B. K. N. and Scott, A., 1913. "The plutonic rocks of Garabal Hill," *Geol. Mag.* **10**, 499–508.

Wyllie, P. J., ed. 1967. *Ultramafic and Related Rocks*. New York, Wiley-Interscience, 464p.

———, Cox, K. G. and Biggar, G. M., 1963. "The habit of apatite in synthetic systems of igneous rocks," *J. Petrol.* **3**, 238–43.

——— and Tuttle, O. F., 1959. "Effect of carbon dioxide on the melting of granite and feldspars," *Am. J. Sci.* **257**, 648–55.

Wynne-Edwards, H. R., 1957. "Structure of the Westport concordant pluton in the Grenville, Ontario," *J. Geol.* **65**, 639–49.

——— 1967. "Westport map-area, with special emphasis on the Precambrian rocks," *Mem. Geol. Surv. Can.* **346**, 142p.

Yoder, H. S. and Tilley, C. E., 1962. "Origin of basalt magmas: an experimental study of natural and synthetic rock systems," *J. Petrol.* **3**, 342–532.

Young, D. G. G., 1969. "The gravity anomaly map of County Donegal," *Geophys. Bull. Dubl. Inst. Adv. Stud.* No. **26**, 6p.

Zen, E-an, 1969. "The stability relations of the polymorphs of aluminium silicate: a survey and some comments," *Am. J. Sci.* **167**, 297–309.

Ziegler, A. M., 1970. "Geosynclinal development of the British Isles during the Silurian Period," *J. Geol.* **78**, 445–79.

Zwart, H. J., 1958. "Regional metamorphism and related granitization in the Valle de Aran (central Pyrenees)," *Geol. en Mijnb.* **20**, 18–30.

——— 1964. "The structural evolution of the Palaeozoic of the Pyrenees," *Geol. Rdsch.* **53**, 170–205.

BIBLIOGRAPHY II: THESES ON DONEGAL

Akaad, M. K., 1954. "The geology of the Narin area, County Donegal," *Ph.D. thesis,* Imperial College, London.

*Berger, A. R., 1967. "The Main Donegal Granite and its regional setting: a study of its fabric and structural relationships," *Ph.D. thesis,* University of Liverpool.

Cambray, F. W., 1964. "The Dalradian rocks east of Glenties, County Donegal," *Ph.D. thesis,* King's College, London.

*Cheesman, R. L., 1952. "The geology of the granitic and metamorphic rocks of the Loughros Peninsula, County Donegal," *M.Sc. thesis,* Imperial College, London.

*——— 1956. "The plutonic geology of the area between Glenties and Fintown, County Donegal," *Ph.D. thesis,* Imperial College, London.

Church, W. R., 1962. "The structural and metamorphic history of the Ballyshannon District, County Donegal," *Ph.D. thesis,* University of Wales, Cardiff.

*Curtis, P. J., 1959. "The petrology of the appinitic diorite complex of Meenalargan, County Donegal," *M.Sc. thesis,* Imperial College, London.

*Edmunds, W. M., 1969. "A chemical and mineralogical study of pelitic hornfelses associated with certain granites of County Donegal with particular reference to the paragenesis of garnet and biotite," *Ph.D. thesis,* University of Liverpool.

*Fernandes-Davila, M., 1969. "The petrology and mode of emplacement of the Rosguill Pluton, County Donegal, Eire," *M.Sc. thesis,* University of Liverpool.

French, W. J., 1960. "Appinitic intrusions associated with the granodioritic pluton of Ardara, County Donegal," *Ph.D. thesis,* King's College, London.

Ghobrial, M. G., 1955. "The geology of the Convoy district, County Donegal," *Ph.D. thesis,* University of Liverpool.

Gindy, A. R., 1951. "The country rocks and associated granites north east of Gweebarra Bay, County Donegal," *Ph.D. thesis.* Imperial College, London.

Hall, A., 1964. "The crystallization history of the Rosses and Ardara granite complexes, Donegal, Ireland," *Ph.D. thesis,* King's College, London.

Harvey, J. J. T., 1969. "Postmagmatic alteration in the Rosses granite complex, Donegal," *Ph.D. thesis,* King's College, London.

Howarth, R. J., 1967. "The Boulder Bed group (Dalradian) of County Donegal, Eire," *Ph.D. thesis,* University of Bristol.

Hubbard, J. A. E. B., 1966. "The Ballyshannon Limestone and basal beds of the Carboniferous of northwest Ireland, with particular reference to the conditions of deposition," *Ph.D. thesis,* Bedford College, London.

*Theses referred to in text.

Iyengar, S. V. P., 1948. "The geology of the Maas–Derryloaghan area, County Donegal," *Ph.D. thesis,* Imperial College, London.

Judge, D. C. (later Knill, D. C.), 1957. "The metamorphic geology of Rosguill, County Donegal," *Ph.D. thesis,* Imperial College, London.

*Kemp, J., 1966. "The stratigraphy, structure and metamorphism of the Dalradian rocks of Slieve League, County Donegal, Eire," *M.Sc. thesis,* Leeds University.

*Kirwan, J. L., 1965. "Metamorphic alteration in the dolerites of the Creeslough area, County Donegal, Ireland," *M.Sc. thesis,* King's College, London.

Lemon, G. G., 1966. "The metamorphic rocks of the Sligo District, northwest Ireland," *Ph.D. thesis,* University of Wales, Cardiff.

McCall, G. J. H., 1951. "The geology of the metamorphic and plutonic rocks of the Creeslough area of Donegal, Eire," *Ph.D. thesis,* Imperial College, London.

Mercy, E. L. P., 1956. "The geochemistry of part of the Donegal granite, County Donegal, Eire," *Ph.D. thesis,* Imperial College, London.

*Mithal, R. S., 1952. "The geology of the Portnoo district, County Donegal," *Ph.D. thesis,* Imperial College, London.

Naggar, M. H. E., 1968. "The petrology, geochemistry and mineralogy of some aluminosilicate-bearing rocks from County Donegal, Eire," *Ph.D. thesis,* University of Liverpool.

*Obaid, T. M. S., 1967. "Certain minor intrusions in County Donegal, Eire," *Ph.D. thesis,* University of Liverpool.

Oswald, D. H., 1952. "The Carboniferous rocks between the Ox Mountains and Donegal Bay," *Ph.D. thesis,* University of Glasgow.

Pande, I. C., 1954. "The geology of the Kilmacrenan district, County Donegal," *Ph.D. thesis,* Imperial College, London.

Pitcher, W. S., 1950. "The igneous and metamorphic geology of the Thorr district, County Donegal," *Ph.D. thesis,* Imperial College, London.

Rickard, M. J., 1957. "The structure and stratigraphy of the Dalradian rocks of the Errigal area, Donegal," *Ph.D. thesis,* Imperial College, London.

Simpson, I. M., 1951. "The stratigraphy and structure of the Carboniferous rocks of the Omagh Syncline," *Ph.D. thesis,* University of Glasgow.

Sinha, R. C., 1955. "The geochemistry of the aureoles of the Ardara pluton and Cleengort migmatites, County Donegal, Ireland," *Ph.D. thesis,* Imperial College, London.

Smart, T. B., 1961. "A study of some contact aureoles in County Donegal with special reference to the composition of andalusite-bearing rocks," *M.Sc. thesis,* Kings College, London.

Srivastava, K. K., 1955. "The geology of the Lifford area, Ireland," *M.Sc. thesis,* University of Liverpool.

*Tozer, C. F., 1955. "The geology of the Glen district, County Donegal, with special reference to the emplacement of the granitic rocks," *Ph.D. thesis,* University of London (External).

Whitten, E. H. T., 1953. "The geology of the metamorphic and plutonic rocks of the Gweedore area, County Donegal, Eire," *Ph.D. thesis,* Queen Mary College, London.

*Wood, R. S. R., 1970. "The Dalradian of westernmost Tyrone and adjacent parts of Donegal, Ireland," *Ph.D. thesis,* University of London (External).

Author Index

Ager, D. V., 387
Agrell, S. O., 48, 377
Akaad, M. K., 8, 17, 20, 24, 25, 94, 170, 174-180, 182, 282, 308, 377, 411
Albee, A. L., 314, 315, 325, 377, 412
Allaart, J. H., 222, 223, 337, 377
Allen, R., 342, 397
Alper, A. M., 219, 377
Althaus, E., 320, 325, 326, 377
Ambrose, J. W., 335, 377
Anderson, E. M., 200, 209, 377, 398
Anderson, G. M., 122, 349, 388
Anderson, J. G. C., 10, 17, 27, 28, 33, 36, 38, 53, 70, 77, 79, 144, 377
Andrew, G., 8, 138, 215, 363, 377
Armstrong, J. E., 222, 336, 402
Atherton, M. P., 69, 134, 180, 259, 268, 270, 274, 279, 293, 303, 304, 308, 309, 311, 314-318, 320, 321, 324, 326, 378, 384, 397
Aucott, J. W., 218, 378
Augustithis, S. S., 334, 378
Autran, A., 324, 378

Badgley, P. C., 56, 337, 378
Baer, A. J., 408
Bailey, D. K., 349, 378
Bailey, E. B., 56, 76, 77, 79, 144, 163, 168, 184, 378, 383
Bailey, R. A., 404
Baker, G., 124, 378
Balk, R., 93, 94, 199, 219, 328, 329, 334, 378
Ball, H. W., 387
Barker, D. S., 117, 336, 378
Barr, K. G., 200, 402
Barrell, J., 293, 378
Barrow, G., 294, 378
Barth, T. F. W., 334, 378
Bateman, P. C., 92, 332, 336, 378, 391, 393
Bathurst, R. G. C., 106, 379

Becke, F., 93
Bell, K., 11, 348, 349, 379
Bell, P. M., 320, 401
Bennington, K. O., 219, 379
Bennison, G. M., 77, 379
Berger, A. R., 9, 94, 184, 192, 198, 199, 215-218, 220-222, 225, 230, 239, 247-249, 255, 256, 259, 291, 292, 294, 328-331, 333, 348, 379, 411
Berger, J. F., 6, 379
Bhattacharji, S., 219, 379
Biggar, G. M., 410
Billings, M. P., 16, 200, 290, 293, 329, 379
Binns, R. A., 293, 334, 379
Bishop, A. C., 146, 161, 165, 220, 379, 408
Bishopp, D. W., 366, 379
Black, R., 391
Blake, E. H., 6, 7, 215, 379
Blatt, H., 334, 379
Blow, W. H., 387
Bluck, B. J., 366, 379
Blyth, F. G. H., 223, 379
Boer, J. H. de, 383
Boone, G. M., 105, 334, 379
Booth, B., 332, 333, 379
Boschma, D., 283, 379
Bott, M. H. P., 339, 379
Bowen, N. L., 91, 117, 118, 123, 172, 219, 347, 379, 406
Bowes, D. R., 16, 144, 146, 162, 163, 164, 167, 168, 219, 347, 380, 410
Bowler, C. M. L., 312, 380
Brace, W. F., 335, 380
Bradley, J., 50, 380, 398
Brannock. W. W., 120, 403
Braunstein, J., 397
Brett, R., 93, 380
Brindley, J. C., 144, 324, 336, 380
Brooks, C., 349, 380
Brothers, R. N., 335, 380
Brotherton, M., 47, 50, 346
Brousse, R., 92, 392

Brown, G. M., 106, 218, 407
Brown, P. E., 11, 91, 293, 380
Bryner, L., 162, 380
Buddington, A. F., 90, 91, 129, 200, 335, 336, 337, 380
Bullerwell, W., 362, 363, 380
Burch, S. H., 219, 335, 380
Burke, K. C., 191, 194, 381
Burns, C. A., 335, 377
Burns, K. L., 94, 381
Burwash, R. A., 334, 381
Butler, B. C. M., 314, 381

Callaghan, E., 336, 381
Callaway, C., 6, 7, 17, 18, 26, 32, 381
Cambray, F. W., 1, 8, 9, 17, 28, 31, 55, 61, 69, 70-73, 291, 292, 348, 379, 381, 398, 411
Cannon, R. T., 219, 230, 333, 381
Cannon, W. F., 331, 337, 381
Carey, S. W., 61, 381
Carmichael, I. S. E., 172, 195, 381
Carpenter, J. R., 219, 381
Carstens, H., 219, 233, 334, 381
Carter, N. L., 334, 381
Casey, R., 387
Castle, R. O., 334, 381
Chadwick, B. A., 63, 381
Chadwick, R. A., 220, 381
Charlesworth, J. K., 1, 3, 5, 77, 356, 375, 381
Chayes, F., 92, 93, 110, 120, 123, 334, 347, 381, 382
Cheeseman, R. L., 1, 8, 9, 17, 29, 211, 215, 256, 398, 411
Cheng, Y., 93, 233, 382
Chinner, G. A., 321, 324, 325, 382
Chodos, A. A., 325
Christie, J. M., 334, 379, 381
Church, W. R., 10, 36, 79, 382, 411
Clark, L. D., 378
Clarke, S. P., 325, 382
Claxton, C. W., 218, 382
Cleaves, E. T., 382
Clifford, T. N., 406
Cloos, E., 93, 184, 195, 220, 283, 290, 328, 333, 382

AUTHOR INDEX

Cloos, H., 184, 382
Clough, C. T., 398
Cobbing, E. S., 336, 382
Cole, A. J., 8, 97, 106, 108, 144, 170, 215, 331, 371, 382
Colhoun, E. A., 3, 382
Collette, B. J., 247, 382
Compston, W., 349, 380, 394, 408
Compton, R. R., 184, 323, 324, 335, 382
Cook, A. H., 204, 338, 339, 382, 404
Cooray, P. G., 220, 382
Corvalan, J., 334, 383
Cox, F. C., 66, 383
Cox, K. G., 410
Craig, G. Y., 383, 392, 396
Craig, E. H. Cunningham, 163, 383
Crampton, C. B., 398
Crimes, T. P., 11, 293, 348, 359, 400
Crowder, D. F., 93, 335, 383
Cruise, R. J., 391, 393
Cruse, M. J. B., 48, 383
Cummins, W. A., 59, 383
Cunningham, M. A., 381
Cunningham-Craig, E. H., 163, 383
Curtis, P. J., 152, 346, 411
Cushing, H. P., 337, 383

Dachille, F., 326, 383
Daly, R. A., 198, 335, 336, 383
Dalziel, I. W. D., 223, 292, 392
Daniels, J. L., 336, 383
Davies, R. G., 48, 383
Davila, M. Fernandes, 132, 140, 411
Davis, G. A., 336, 383
Dawson, J. B., 163, 383
Dawson, K. R., 109, 383
Dearman, W. R., 53, 383
Dearnley, R., 292, 383
de Boer, J. H., 383
Deer, W. A., 146, 165, 166, 383
DeHills, S. M., 334, 383
de la Roche, H., 335, 402
Den Tex, E., 53, 336, 383
Deutsch, S., 342, 404
De Waard, D., 184, 290, 383, 407
Dewey, H., 48, 383
Dewey, J. F., 3, 10, 11, 53, 63, 66, 82, 276, 293, 349, 356, 363, 384
Dickson, F. W., 332, 333, 384
Didier, J., 119, 124, 384
Dietrich, R. V., 219, 384
Digman, R. E., 184, 219, 396
Dimroth, E., 50, 384
Diver, B. B. Van, 106, 332, 407
Doe, B. R., 349, 384
Donath, F. A., 56, 384
Donnelly, T. W., 333, 384

Donner, J. J., 387
Dowling, J. W. F., 384
Drescher-Kaden, F. K., 93, 332, 384
Duffield, W. A., 184, 384
Dunham, A. C., 384
Dunham, K. C., 375, 384
Dunlop, G. M., 398
Durrance, E. M., 200, 384
Dury, G. H., 3, 4, 384

Edelman, N., 225, 384
Edmunds, W. M., 48, 69, 132, 134, 135, 181, 256, 279, 291, 303, 306, 314, 315, 321, 326, 378, 384, 411
Edwards, H. R. Wynne, 335, 337, 405, 410
Egan, F. W., 6, 201, 384, 391
Eggler, D. H., 198, 384
Elliot, D., 244, 384
Elsasser, W. M., 351, 384
Elwell, R. W. D., 33, 398
Emeleus, C. H., 218, 385, 389
Emmons, R. C., 219, 332, 385
Engel, A. E. J., 312, 337, 385
Engel, G. C., 312, 337, 385
Erdmannsdörffer, O. H., 93, 385
Erikson, E. H., 336, 385
Escher, A., 336, 385
Eskola, P., 144, 222, 290, 333, 385
Evans, B. W., 319, 320, 325, 385, 397
Exley, C. S., 124, 385
Exner, C., 194, 385

Fairbairn, H. W., 391
Faure, G., 333, 349, 385
Felici, M., 333, 385
Ferguson, J., 218, 385
Fernandes-Davila, M., 132, 140, 411
Fisher, G. W., 382
Fitch, F. J., 11, 342, 385
Fleming, P. D., 53, 397
Fleuty, M. J., 56, 385
Flinn, D., 81, 94, 95, 243, 244, 249, 312, 336, 385, 386, 396
Flinn, G. W., 378, 381, 398, 400, 402, 405, 407
Floyd, P. A., 312, 386
Forbes, D., 7, 386
Foslie, S., 219, 386
Fourmarier, P., 184, 386
Fox, F. G., 56, 58, 386
Fox, P. E., 184, 386
Fraser, A. G., 334, 386
French, W. J., 9, 144-146, 151-155, 157, 161, 163, 166, 386, 411
Fyfe, W. S., 326, 349, 351, 386

Gabrielse, H., 293, 386
Gallagher, M. J., 381
Galwey, A. K., 321, 323, 326, 392
Ganguly, J., 315, 386
Ganly, P., 6, 386
Gansser, A., 336, 386
Gates, O., 164, 386
Gates, R. M., 93, 219, 386
Gault, H. R., 336, 386
Gay, N. C., 244, 246, 247, 386
Geikie, A., 6, 366, 386
George, T. N., 3, 16, 356, 366-370, 386, 387
Gevers, T. W., 336, 387
Ghobrial, M. G., 9, 411
Giesecke, C. L., 6, 387
Gilbert, M. C., 320, 401
Gilluly, J., 8, 349, 387
Gilreath, J. P., 333, 407
Gindy, A. R., 8, 17, 19, 21, 45, 97, 101, 105, 111, 114, 115, 117, 124, 152, 153, 159, 161, 211, 235, 296-299, 387, 411
Ginzburg, I. V., 92, 387
Glen, J. W., 244, 387
Glikson, A. Y., 50, 387
Goodspeed, G. E., 220, 222, 332, 336, 337, 387
Gore, D. J., 334, 387
Graham, R. H., 246, 330, 400
Grasty, R. L., 91, 380
Green, A. H., 7, 387
Green, J. C., 293, 315, 387
Gresens, R. L., 219, 388
Griffin, W. L., 334, 388
Griffith, R. J., 6
Griggs, D. T., 381
Grout, F. F., 119, 183, 336, 351, 388
Guidotti, C. V., 293, 388
Guitard, G., 324, 378
Gunn, W., 398

Hadley, J. B., 377, 398, 402, 406
Hall, A., 9, 117, 144, 146, 149, 165, 166, 168, 170, 172-174, 186, 193-198, 225, 282, 298, 333, 346, 347, 349, 350, 388, 411
Hamilton, D. L., 122, 347, 349, 388, 391
Hamilton, W., 183, 293, 349, 388
Hardie, W. G., 163, 164, 388
Harker, A., 162, 219, 388
Harkness, R., 6, 388
Harland, W. B., 30, 387, 388
Harme, M., 312, 389
Harper, C. T., 11, 389
Harper, J. C., 355, 389
Harpum, J. R., 91, 337, 389
Harris, A. L., 10, 48, 389, 405

AUTHOR INDEX

Harris, P. G., 163, 345, 389
Harry, W. T., 200, 218, 223, 233, 294, 336, 389
Hart, S. R., 293, 389
Harte, B., 53, 54, 66, 79, 81, 270, 274, 277, 389
Harte, W., 7, 389
Hartley, J. J., 36, 77, 355, 389
Harvey, J. J. T., 411
Hatch, F. H., 92, 106, 389
Haughton, S., 7, 201, 389
Hawkes, J. R., 381
Heitanen, A., 92, 281, 293, 389, 390
Helm, D. G., 53, 390
Hess, H. H., 388
Hess, P. C., 279, 314, 390
Hewitt, D. F., 337, 390
Hibbard, M. J., 219, 332, 390
Hickling, N. L., 195, 390
Hill, A. R., 3, 390
Hill, J. B., 144, 390
Hill, P. J., 132
Hills, E. S., 198, 200, 390
Hinxman, L. W., 398
Hirschberg, A., 319-321, 390
Hobbs, B. E., 66, 405
Hodge, B. L., 384
Holland, C. H., 387
Holland, J. G., 335, 390
Holtropp, J. F., 335, 390
Hooper, P. R., 332, 390
Hopgood, A. M., 380
Hopson, C. A., 333, 382
Hoschek, G., 319, 320, 321, 390
Hossak, J. R., 94, 390
Howarth, R. J., 27, 28, 36, 40, 47, 55, 70-72, 81, 283, 291, 390, 411
Hsü, J. K., 335, 390
Hsü, L. C., 319, 320, 390
Hubbard, F. H., 334, 391
Hubbard, J. A. E. B., 366, 369, 370, 391, 411
Hubbard, W. F., 391
Huber, N. K., 378
Hughes, N. F., 387
Hull, E., 6-8, 30, 144, 170, 283, 291, 366, 373, 391
Hunt, T. S., 6, 391
Hurley, P. M., 333, 349, 385, 391
Hutchinson, R. M., 92, 93, 219, 230, 391
Hutchinson, W. W., 332, 391
Hyland, J. S., 373

Ianello, P., 335, 391
Ishi, K., 334, 391
Iyengar, S. V. P., 8, 17, 19, 20, 59, 97, 99, 151, 152, 154, 182, 211, 229, 256, 266, 282, 391, 412

Jacobson, R. R. E., 198, 391
Jaeger, J. C., 183, 317, 318, 391
Jahns, R. H., 219, 225, 337, 391, 395
James, R. S., 347, 391
Jenness, S. E., 293, 391
Johannsen, A., 92, 93, 392
Johannsen, C. E., 244, 392
Johnson, A. L., 384
Johnson, M. R. W., 37, 53, 54, 66, 73, 77, 79, 81, 82, 223, 270, 274, 277, 292, 294, 389, 392, 393, 396, 400
Johnstone, G. S., 18, 38, 77, 392
Jones, K. A., 16, 321, 323, 326, 380, 392
Jones, P. B., 56, 392
Joplin, G. A., 50, 92, 164, 165, 168, 349, 392
Joyce, A. S., 312, 392
Judge, D. C., 412
Jung, J., 92, 392
Juurinen, A., 315, 392

Kaden, F. K. Drescher-, 93, 332, 384,
Kaitaro, S., 198, 200, 223, 392
Kardoss, E. Szadeczky-, 91, 405
Karl, F., 93, 392
Katz, M. B., 335, 392, 398
Kay, M., 363, 379, 380-382, 384, 392, 394, 395, 398
Kays, M. A., 281, 293, 392
Kellaway, G. A., 387
Kelly, J., 7, 392
Kemp, J., 8, 9, 27, 28, 31, 81, 412
Kennedy, G. C., 123, 392
Kennedy, M. J., 279, 392, 398
Kennedy, W. Q., 166, 389, 392
Kent, P. E., 380, 385-387
Kerrick, D. M., 332, 392
Ketskhoveli, D. N., 332, 393
Keyes, C. R., 194, 393
Khitarov, N. I., 326, 399
Kilburn, C., 10, 27, 29, 36, 390, 393
Kilroe, J. R., 6, 8, 291, 384, 391
Kinahan, G. H., 6, 7, 30, 223, 391, 393
King, B. C., 93, 105, 220, 393
King, R. F., 174, 182, 220, 299, 393
Kinloch, E. D., 380
Kirwan, J. L., 48, 50, 268, 269, 314, 412
Kistler, R. W., 336, 393
Kleeman, A. W., 347, 393, 408
Knill, D. C., 8, 16, 19, 21, 24, 25, 40, 45, 48, 55, 56, 58, 59, 63, 77, 79, 89, 132-135, 137, 139-142, 294, 303, 393, 412
Knill, J. L., 8, 16, 19, 21, 24, 25, 29, 36, 37, 40, 45, 48, 54-56, 58, 59, 63, 77, 79, 89, 132-134, 137, 139-142, 294, 303, 393

Knowles, C. R., 382
Knowles, R. R., 61, 393
Koch, G. S., 109, 110, 394
Kock, H., 397
Kolbe, P., 333, 393
Koschmann, A. H., 93, 105, 393
Krauskopf, K. B., 330, 342, 393
Kretz, R., 220, 321, 330, 335, 393, 394
Krupicka, J., 334, 381
Kuenen, P. H., 239, 243, 394
Kulish, Y. A., 106, 394
Kuno, H., 49, 394
Kynaston, H., 144, 390

Lacroix, A., 120
Lambert, R. St. J., 91, 344, 347, 355, 390
Lappin, M. A., 335
Larsen, E. S., 317, 375, 394
Larsen, L. H., 131, 394
Lauerma, R., 290, 394
Laurence, R. D., 334, 394
Lazarenkov, V. G., 119, 394
Leake, B. E., 48, 383, 390, 394
Lee, J. S., 109, 394
Leedal, G. P., 158, 201-206, 208, 336, 346, 359, 362, 373, 374, 394, 407
Leggo, R. J., 10, 11, 90, 341, 349, 394
Lemon, G. G., 412
Leonard, H., 393
Leveson, D. J., 106, 108, 394
Lindsay, J. F., 244, 394
Link, R. F., 109, 110, 394
Lipman, P. W., 92, 219, 336, 394
Litherland, M., 38, 39, 77, 400
Lobjoit, W. M., 325, 394
Lovering, T. S., 129, 317, 394, 395
Lunn, J., 178, 179
Luth, W. C., 320, 344, 348, 395

M'Adam, J., 6, 395
McBirney, A., 163, 395
McCall, G. J. H., 8, 16, 19, 21, 23, 24, 45, 47, 48, 56, 59, 63, 82, 142, 211, 215, 256, 261, 280, 283, 286, 395, 412
McCallien, W. J., 17, 18, 25, 26, 29, 32, 33, 36, 37, 74, 75, 76, 77, 279, 363, 378, 395
MacCulloch, J., 27, 395
MacDonald, G. J. F., 326, 395
McEwan, M. C., 408
MacGregor, A. M., 290, 395
MacGregor, M., 123, 395
M'Henry, A., 30, 391
McKenzie, D. P., 335, 395

Mackenzie, W. S., 207, 395
McKerrow, W. S., 3, 355, 356, 384, 395
MacLeod, W. N., 391
M'Parlan, J., 6, 395
Mahan, S. M., 333, 395
Mallick, D. I. J., 54, 290, 395
Marfunin, A. S., 333, 334, 395
Marmo, V., 91, 219, 334, 395
Martin, N. R., 178, 182, 334, 396
Maufe, H. B., 144, 378
Mayo, E. B., 329, 336, 396
Mead, W. J., 219, 396
Means, W. D., 249, 396
Mehnert, K. R., 93, 105, 119, 120, 123, 222, 239, 331-335, 345, 347, 396
Melamed, V. G., 401
Mercy, E. L. P., 9, 114, 120-124, 128, 130, 186, 194, 196, 197, 198, 293, 313, 314, 342, 345, 346, 396, 412
Merrill, R. B., 347, 396
Mikami, H. M., 184, 219, 396
Miller, A. A., 3, 396
Miller, J. A., 11, 91, 336, 380, 385, 396
Miller, W. J., 222, 223, 336, 396
Misch, P., 270, 335, 396
Mitchell, J. G., 385
Mitchell, R. L., 342, 397
Mitchell, W. F., 6, 8, 291, 384, 391
Mithal, R. S., 8, 9, 17, 25, 412
Miyashiro, A., 312, 396
Money, M., 144, 163
Moorbath, S., 372, 396
Moore, J. G., 378
Moore, R., 390
Morteani, G., 230, 399
Mountjoy, E. W., 56, 399
Mrazec, M. L., 184, 396
Mueller, R. F., 177, 334, 396
Muir, I., 117, 396
Munro, M., 184, 396
Murthy, M. V. N., 82, 397
Murphy, T., 204, 338, 339, 382
Murthy, V. R., 293, 397
Myers, W. B., 183, 293, 349, 388

Naggar, M. H., 9, 134, 180, 256, 259, 268, 270, 274, 279, 293, 304, 308, 309, 311, 314-318, 320, 324, 397, 412
Nairn, A. E. M., 388
Neale, E. R. W., 399
Nevin, C. M., 329
Newall, G., 378, 386, 389, 398, 400, 402
Newton, R. C., 326, 397
Nicholson, R., 290, 397

Nickel, E., 397
Niyogi, D., 335, 397
Nockolds, S. R., 92, 118, 123, 146, 165, 166, 342, 397
Nolan, G. H., 283
Nolan, J., 391
Norris, D. K., 408
Norton, S. A., 293, 294, 325, 406
Nungässer, W., 397

Obaid, T. M. S., 44, 47, 48, 50, 82, 412
O'Brien, G. D., 184, 397
O'Brien, M. V., 371, 372, 397
Oen, I. S., 198, 200, 223, 332-334, 336, 389, 397
Offler, R., 53, 397
Ohta, Y., 332, 397
Okamura, R., 404
Okrusch, M., 325, 397
Opdyke, N. D., 61, 393
Orville, P. M., 334, 397
Oswald, D. H., 366, 367, 369, 370, 371, 387, 412
Owen, T. R., 53, 77, 377

Pabst, A., 119, 397
Page, L. R., 292, 293, 397
Palmer, D. F., 106, 398
Pande, I. C., 8, 17, 211, 215, 256, 412
Pankhurst, R. J., 349, 398
Park, R. G., 53, 219, 380, 398
Parker, R. B., 56, 384
Paterson, M. S., 249, 265, 396, 398
Peach, B. N., 292, 398
Peck, D. L., 404
Peng, C. C. J., 334, 398
Phair, G., 390
Phemister, J., 400
Phemister, T. C., 336, 398
Phillips, W. E. A., 39, 48, 363, 398
Phillips, W. J., 233, 334, 398
Phillips, W. S., 83, 177, 398
Phinney, R. A., 395
Pidgeon, R. T., 10, 11, 90, 341, 394
Pigorini, B., 407
Pinson, W. H., 397, 391
Pitcher, W. S., 1, 8-11, 17, 18, 20, 21, 23-25, 27-29, 32-36, 55, 58-61, 69, 74, 75, 79, 89, 94, 97, 99, 100, 101, 103, 105, 111, 112, 114, 116, 120, 123-129, 134, 139, 151, 152, 157, 159, 163, 165, 180, 184, 186-188, 190, 191, 193, 196-199, 211-216, 219-222, 228-230, 234, 235, 239, 244, 246, 247, 249, 251, 256, 258-261, 266, 268, 269, 272, 274, 276, 277, 279-283, 290-292, 294, 296-299, 300, 302,

Pitcher, W. S. (*continued*) 303, 307, 308, 310-314, 317, 318, 321, 322, 325, 328-331, 335-337, 340, 344, 357, 359-364, 378, 379, 381, 382, 386, 391, 393, 397-402, 405, 407, 412
Piwinski, A. J., 119, 347, 349, 351, 399
Platen, H. von, 123, 197, 347, 348, 407, 409
Platten, I. M., 144, 163, 292, 399
Pockley, R. P. C., 372, 399
Poldervaart, A., 108, 131, 219, 377, 388, 394, 399, 406
Polin, Y. K., 106, 394
Poole, W. H., 61, 399
Powell, D., 10, 36, 66, 279, 399
Prebble, W. M., 398
Preston, J., 290, 334, 373-375, 399
Price, N. J., 199, 331, 399
Price, R. A., 56, 399
Pringle, J., 10, 399
Prinz, M., 219, 243, 399
Prior, D. B., 3, 390
Pugin, V. A., 326, 399
Pulvertaft, T. C. R., 8, 17, 19, 20, 24, 29, 32, 45, 47, 60, 72, 218, 266, 276, 280, 336, 385, 389, 399

Raase, P., 230, 399
Ragan, D. M., 219, 335, 339
Raguin, E., 91, 106, 119, 120, 128, 335, 336, 400
Raleigh, C. B., 334, 335, 381
Ramberg, H., 184, 219, 239, 243, 244, 249, 329, 331, 334, 351, 400
Ramsay, J. G., 56, 63, 65, 66, 94, 234, 239, 243, 244, 246, 247, 249, 251, 283, 330, 400
Ramsbottom, W. H. C., 387
Rao, M. S., 82, 400
Rast, N., 11, 18, 32, 38, 39, 48, 53, 73, 77, 81, 93, 293, 307, 333, 348, 359, 378, 386, 389, 398, 400, 402
Read, H. H., 1, 7, 8, 11, 15, 17, 20, 29, 50, 55, 59, 60, 89, 90, 91, 94, 99-101, 111, 119, 128, 134, 135, 139, 144, 149, 151, 152, 159, 163, 165, 169, 180, 211-216, 219-222, 228-230, 234, 235, 243, 246, 247, 249, 251, 256, 258-261, 266, 268, 269, 272, 274, 276, 277, 279, 280, 281, 283, 290-293, 296-300, 302, 303, 307, 308, 310, 311, 321, 322, 332, 333, 335-337, 340, 345, 391, 398, 400, 401

AUTHOR INDEX

Redden, J. A., 219, 225, 400
Rees, A. I., 244, 401
Reesor, J. E., 118, 293, 386, 401
Reffay, A., 3, 401
Reitan, P., 219, 229, 401
Reverdatto, V. V., 268, 317, 318, 326, 401
Reynolds, D. L., 82, 115, 118, 119, 139, 141, 159, 162-164, 200, 401
Richardson, S. W., 320, 326, 401
Richey, J. E., 139, 142, 184, 200, 389, 401
Rickard, M. J., 8, 16, 18, 19, 21, 23-25, 45, 48, 54-56, 58, 59, 61, 63, 65, 66, 99, 100, 213, 256, 259, 261, 277, 283, 401, 412
Rickards, R. B., 384
Riddihough, R. P., 338, 339, 345, 357, 361, 401
Rinehart, C. D., 378
Ringwood, A. E., 349, 402
Rittman, A., 184, 382
Robbie, J. A., 77, 409
Robert, B., 53, 390
Roberts, J. L., 16, 37, 40, 77, 79, 80, 200, 402
Robertson, J. K., 396
Robinson, P., 406
Robson, G. R., 200, 402
Roche, H. de la, 335, 402
Roddick, J. A., 92, 93, 222, 336, 402
Rodgers, K. A., 335, 380
Rogers, J. J. W., 333-335, 395, 402, 408
Ronner, F., 92, 402
Rose, H. J., 390
Rosenbusch, H., 93, 312, 324, 402
Rosenfeld, J. L., 66, 402
Ross, C. S., 404
Ross, G., 400
Roubault, M., 335, 402
Roy, R., 326, 383
Runner, J. J., 332, 336, 402
Rusnak, G., 244, 402

Sharapov, V. N., 401
Shaw, D. M., 409
Shaw, H. R., 128, 219, 229, 329, 404
Shelley, D., 243, 334, 404
Shengelia, D. M., 332, 393
Sheridan, D. J. R., 366, 391
Silver, L. T., 342, 404
Simkin. T., 219
Simonen, A., 106, 404
Simpson, A., 53, 404
Simpson, I. M., 404, 412
Sinha, R. C., 9, 58, 114, 312-314, 399, 412

Skevington, D., 384
Skiba, W. J., 383
Slobodskoy, R. M., 243, 335, 404
Smart, J., 380
Smart, T. B., 203, 303, 305, 314, 404, 412
Smith, C. H., 219, 404
Smith, H. G., 83, 144, 161, 404
Smith, J. V., 333, 334, 382, 404
Smith, R. L., 200, 404
Smithson, S. B., 332-334, 339, 379, 404
Sorensen, H., 218, 404
Soggetti, F., 407
Sollas, W. J., 374, 404
Solomon, M., 110, 404
Soper, N. J., 184, 293, 380, 404
Spence, W. H., 333, 334, 407
Spencer, A. M., 10, 27, 30, 36, 380, 385, 386, 404
Spjeldnaes, N., 31, 404
Spry, A., 16, 53, 66, 93, 94, 219, 233, 321, 332-335, 381, 404, 405
Srivastava, K. K., 9, 412
Stanton, R. L., 93, 405
Stanton, W. J., 30, 403
Stauffer, M. R., 81, 405
Stephens, N., 4, 405
Stephenson, P. J., 184, 198, 405
Stevens, N. C., 184, 405
Steveson, B. G., 312, 405
Stewart, D., 6, 405
Stewart, F. H., 393, 396, 400
Stone, M., 10, 124, 219, 385, 405
Streckeisen, A. L., 92, 405
Strens, R. G., 325, 326, 405
Stubblefield, J., 387
Sturt, B. A., 48, 405
Sutton, J., 53, 222, 294, 342, 383, 405
Suzuki, Y., 110, 382
Sweatman, T. R., 405
Switzer, G., 375, 394
Sylvester, A. G., 184, 332, 405
Synge, F. M., 4, 405
Szadeczky-Kardoss, E., 91, 405

Talbot, C. J., 243, 251, 405
Talbot, J. L., 66, 335, 405
Tanner, P. W. G., 394
Taubeneck, W. H., 93, 108, 128, 131, 200, 218, 230, 233, 329, 405, 406
Taylor, S. R., 333, 393
Taylor, W. E. G., 48, 406
Tchalenko, J. S., 223, 406
Termier, G., 335, 406
Termier, H., 335, 406

Tex, E. den, 53, 336, 383
Theime, J. G. 106, 406
Thomas, H. H., 184, 401
Thomas, P. K., 48, 77, 384, 406
Thompson, J. B., Jr., 117, 290, 293, 294, 325, 377, 398, 402, 406
Thompson, J. E., 377
Thompson, R. N., 117, 406
Thomson, J., 29, 406
Tilley, C. E., 167, 410
Tobisch, O. T., 65, 406
Townend, R., 117, 406
Tozer, C. F., 8, 17, 33, 211, 214, 226, 256, 276, 324, 398, 406, 412
Trask, N. J. J., 406
Treagus, J. E., 66, 77, 79, 80, 399, 402, 406
Tuominen, H. V., 120, 123, 335, 337, 406
Turner, D. C., 200, 406
Turner, F. J., 50, 120, 219, 268, 279, 294, 317, 319, 326, 406
Tuttle, O. F., 91, 123, 219, 225, 347, 391, 395, 406, 410
Tyrrell, G. W., 162, 406

Ueda, Y., 391

Vance, J. A., 230, 333, 406, 407
Van Diver, B. B., 106, 332, 407
Veniale, F., 333, 407
Verhoogen, J., 50, 120, 219, 326, 406, 407
Vernon, R. H., 333-335, 407
Vincent, E. A., 161, 168, 407
Visteluis, A. B., 333, 407
Vogel, T. A., 333, 334, 407
Voll, G., 38, 63, 77, 82, 93, 333, 334, 407
Volovikova, I. M., 387
Von Platen, H., 123, 197, 347, 348, 407, 409

Waard, D. de, 184, 290, 383, 407
Wager, L. R., 106, 218, 407
Walker, G. P. L., 158, 201-206, 208, 336, 346, 359, 362, 373, 374, 394, 407
Walton, M., 123, 290, 335, 383, 407
Ward, R. F., 219, 407
Waters, A. C., 50, 407
Watson, J., 11, 38, 40, 53, 77, 222, 294, 324, 342, 402, 405, 407, 408
Watt, W. S., 333, 334, 408
Watterson, J., 53, 81, 94, 219, 222, 223, 246, 247, 284, 330, 337, 408

Weaver, J. D., 336, 408
Wegmann, C. E., 53, 184, 222, 337, 408
Weidmann, M., 223, 408
Weiss, L. E., 265, 279, 398, 408
Wells, A. K., 92, 146, 165, 220, 389, 408
Wells, M. K., 92, 150, 389, 408
West, R. G., 387
Westropp, W. H. S., 7, 408
Weymouth, J. H., 223, 408
Wheeler, J. E., 399
White, A. J. R., 333, 408
White, W. S., 293, 377, 398, 402, 406
Whitfield, J. M., 333, 408
Whitten, E. H. T., 8, 56, 58, 97, 99, 102-105, 108-112, 115-117, 120, 122, 125-128, 130, 149, 159, 283, 335, 345, 383, 408, 409, 412

Widenfalk, L., 334, 409
Wiebe, R. A., 332, 409
Wilkinson, S. B., 77, 409
Willgallis, A., 105, 396
Williams, H., 200, 293, 299, 409
Williamson, W. O., 223, 408
Wilshire, H. G., 219, 405, 409
Wilson, G., 243, 409
Wilson, H. E., 77, 366, 409
Wilson, J. T., 363, 409
Wilson, R. L., 373, 375, 409
Windley, B., 150, 409
Winkler, H. G. F., 123, 197, 319-321, 345, 347, 390, 409, 410
Wolfe, M. E., 24, 410
Wood, R. S. R., 8, 9, 33-36, 75, 399, 412
Woodland, A. W., 387
Wright, A. E., 77, 144, 162-164, 167, 168, 379, 380, 410

Wright, P. C., 93, 410
Wright, T. L., 404
Wright, W. B., 383
Wyllie, B. K. N., 166, 410
Wyllie, P. J., 146, 219, 335, 347, 349, 351, 396, 399, 410
Wynne-Edwards, H. R., 335, 337, 405, 410

Yamashita, Y., 391
Yefremova, S. V., 387
Yeliseyeva, O. P., 387
Yoder, H. S., 167, 410
York, D., 380
Young, D. G. G., 338-340, 357, 361, 401, 410

Zen, E-an, 326, 377, 398, 402, 406, 410
Ziegler, A. M., 355, 410
Zwart, H. J., 53, 324, 410

Subject Index

Actinolite, 117, 146, 153, 154
Active stoping, *see* Intrusion mechanisms
Adamellites, 10, 108, 203
 composition, 192, 193
Adinole, 44, 47, 48
Adinolization, 83
Ages, isotopic, K-Ar, 11, 90, 91, 293, 342, 348, 356, 362
 Rb-Sr, 11, 90, 342, 348, 349
 U-Pb, 90
Aghla, 31, 266
Aghla Anticline, 57, 59, 63
Aghla Fault, 358
Aghla Mt., 31, 59
Aghla Syncline, 69-71
Aghyaran Formation, 33, 35, 36
Agmatites, 112, 113, 135, 137, 138
Albite, 194, 197
 rims, 105, 218, 333, 334
 veins, 231, 232
Almandine-amphibolite facies, 53, 54
Almandine garnet, 314, 315, 319-321
Altan Anticline, 57
Altan Limestone, 19, 21
Altan Lough, 19
Altcrin, 191, 193
Aluminosilicates, coexistence of, 259, 305, 307, 309, 311, 319
 porphyroblasts, 272-277
 shimmerized, 280
 stability fields, 313, 319, 320, 324-326
 'sweat-outs', 274
 see also Andalusite, Kyanite, Sillimanite
Amphiboles, 144-147, 151, 153, 154, 159, 165
 crystallization, 349
 porphyroblasts of, 149, 153
 pseudomorphs after carbonate, 69
 replacing pyroxene, 118
Amphibolite facies, 268, 274
Amphibolites, appinitic, 152, 154
 garnet, 44, 47, 50
 schistose, 268
Amphibolite xenoliths, 171, 173, 176

Amphibolitization, 150, 161
Amygdales, 83, 374
Anatexis, 294, 337, 349, 350
 experimental, 347, 350
Anatexites, 351
Andalusite, 100, 101, 133, 134, 203, 268, 279, 293, 302-307, 309, 311, 318, 322
 augen, 302, 309, 310
 dendritic and layeritic growth of, 307
 hornfelses, 302
 host-rock control of, 317, 325
 inclusions in, 276
 stability field, 319, 324, 325
 zoning, 268
 see also Porphyroblasts
Annagary microgranite sheet, 187-188
Antiperthite, 105, 218, 225, 231-234
Antrim, 77, 356, 366, 375
Apatite, 117, 119
Aphort, 185
Aplite-pegmatite, 103, 172, 176, 190, 212, 287
Aplites, 189-191, 204, 207, 221, 225-226
 desilication, 205, 207, 208
 folding, 252
 isotopic age, 90
Aplitic texture, 298
Aplogranite, garnetiferous, 298, 300
 porphyritic, 204
Apophyses, 97, 112, 135, 141, 205, 212, 236, 240
 deformation, 330, 331
Appalachians, 56, 292, 293, 363
Appinites, 89, 90, 143-168, 359
 age, 148-149
 associated lamprophyres, 161
 associated with granites, 349-350
 carbonate porphyroblasts in, 291
 chemical composition, 165-166, 342
 differentiation, 342, 349
 intrusive breccias, 143, 144, 148-149, 156-165

Appinites (*continued*)
 magmas, 145-146, 149-150, 164-168, 349-350
 nomenclature, 144, 145
 origin of, 162-168
 pegmatite-, 150, 151
 Scottish, 144
 structural control, 164
 structures, 150, 151, 159
 textures, 145-147, 150
 see also Xenoliths
Appin Limestone, 21, 25
Appin Phyllites, 21
Appin Quartzite, 21
Appin Striped Transition Series, 21
Aranmore Island, 19, 45, 59, 61, 99, 184, 185
Aranmore-Dunglow fold, 61
Aranmore Syncline, 57, 61
Ardara, 11, 18, 19, 178, 184, 185, 262, 292, 309, 318, 350
Ardara Pluton, 61, 169-184
 age, 90, 91, 170
 appinitic intrusions, 143-168
 aureole, 89, 90, 170, 178-180, 183, 303, 310
 deformation, 177-184, 307, 309, 319, 321, 348
 formation depth, 325
 kyanite and staurolite in, 315-317
 map, 308
 metamorphic zonation, 309
 porphyroblasts in, 309, 310
 structure, 178-180, 182-184
 thermal gradient, 317-318
 width, 318
 carapace deformation, 182-184
 composition, 344, 346
 contact metamorphism, 180-183, 312
 contamination, 172-174, 182
 crystallization differentiation, 342
 emplacement, 11, 89, 169-170, 182-184
 faults, 178
 feldspar transformations in, 334
 folds, 178

419

SUBJECT INDEX

Ardara Pluton (*continued*)
  foliation in, 170, 171, 174-177, 179, 180
  geochemistry, 342
  joint systems, 177-178
  magma, 168, 170-172, 182-184
  rock analysis, 346
  shape, 182, 340
  structural map, 175
  structures, 282
  textures, 170-172, 175
  thrusts, 178
  xenoliths in, 170-176
  xenoliths of, 213, 235, 246
Ardrishaig-Craignish Phyllites, 29, 37
Ards, 19, 23, 261
Ardsbeg, 23
Ards Pelites, 18-20, 23, 39
  crenulation cleavage in, 258, 259, 262
  folds, 261, 262
  porphyroblasts in, 270-275
  structure, 62, 65
  xenoliths in, 262
Ards Quartzite, 18-21, 23, 24, 39, 57, 58
  folds, 60, 65, 99, 100
  mullion structures in, 284, 285
Argyllshire, 17, 18, 25, 27, 29, 36-40, 43, 77, 79, 142-144, 146, 150, 163, 164, 166
Arkoses, 366, 367, 370
Asthenoliths, 350
Atolls, garnet, 279
Augen, 175, 180, 274
  andalusite, 302, 309, 310
  oligoclase, 202
Augite, 45, 48, 269
Aureoles, 10, 45, 47, 134, 135, 139, 140, 144, 149, 302-312, 325, 326
  chronology, 303
  emplacement, 303, 318, 319, 323, 324
  Mg-Fe ratios, 315-317
  mineral zoning in, 303-305, 317
  physico-chemical controls, 317-326
  recrystallization in, 303-312
  see also Ardara Pluton, Barnesmore Pluton, Fanad Pluton, Main Donegal Granite, Rosses Centered Complex, Thorr Pluton and Toories Pluton
Autochthonous granites, 337
Autometamorphism, 10
Autometasomatism, 49, 83, 167

Back Settlement Breccia, 163

Back Strand, 59
Ballachulish, 21, 25, 356
Ballachulish Succession, 10, 17, 21, 38, 39, 77, 347, 356
  relation to Islay Succession, 38
Ballaghdoo Fault, 358
Ballard Antiform, 74, 75
Ballina Syncline, 366
Ballybofey, 18, 27, 34, 54, 357
Ballybofey Antiform, 74, 75, 80
Ballybofey Fold, 34, 35, 75, 79, 82, 357
Ballykillowen Hill Slide, 36, 79
Ballymastocker, 345, 356, 359, 362, 363, 365
Ballymastocker Hill, 363, 365, 366
Ballymore Syncline, 57
Ballyness, 59
Ballyshannon Limestone Group, 367, 369
Banded hornfels, 116
Banded semipelites, 100
Banding, in basic and ultrabasic rocks, 219
  in Main Donegal Granite, 212, 215-219, 222, 226, 227, 233, 251, 252, 255, 290
  in metamorphic rocks, 219
  in pegmatites, 219
  in Upper Falcarragh Pelites, 265, 270
  rhythmic, 106
Barnesbeg, 228, 270, 278, 339
Barnesbeg Gap, 7, 91, 287
Barneslough Fault, 358, 362
Barnesmore, 3, 8, 10, 159, 164, 209, 317
Barnesmore Fault, 358, 359
Barnesmore Pluton, 201-209
  age, 90, 201-202
  aureole, 201-203, 303, 305, 318
  breccia pipes, 204, 208
  composition, 203, 204, 344, 346
  contacts, 203-205, 207-209, 305
  desilication, 205-208
  dykes, 207-208, 374
  emplacement, 11, 89, 90, 204, 205, 207-209
  faults, 359
  fold complex, 53
  igneous arch, 204-205
  map, 202
  rock analyses, 346
  roof pendants, 335-336
  shape, 338-339
  structural relations, 204-207
  thickness, 339
Basaltic dykes, 3, 12, 207-208
Basaltic magma, 50, 375

Basalts, chemical composition of, 165
  flood, 50
  olivine, 374
  see also Tholeiites
Basic rocks, 10, 143-168, 219, 222, 373-375
Basic sills, metamorphism of, 48-49
Basins, deposition of, 366
Batholiths, 222, 236, 349
Beidellite-pitchstone-felsite, 374
Bein an Dubhaich Granite, 336
Belshade Fault, 202, 205, 358, 359, 362, 363
  gravity anomaly, 362
Beryl, 193, 194
Binaniller, 213, 246, 288
Binmore, 264, 266, 270
Binnadoo, 262, 265, 274, 278
Binnadoo Shear Belt, 268
Biotite, 100, 102, 103, 105, 106, 108, 114, 115, 117, 118, 128, 129, 134, 138, 180, 298, 303, 305-307, 311, 312
  alignment, 277
  chloritized, 217, 270, 279, 280, 333, 335
  epidote inclusions in, 194
  foliation, 223
  genesis, 194-195
  host-rock control, 314-317
  inclusions, 278-280
  K-Ar age, 91
  -muscovite knot, 322
  preferred orientation, 220, 229, 232, 277
  replacement by chlorite, 69, 76
  replacement of garnet, 69
  schlieren, 191, 192, 195, 298
  skarns, 117
  stability fields, 319-321, 324
  texture, 103, 232-234
  see also Porphyroblasts
Biotite granites, 115, 187, 195, 211, 212, 296, 298, 300
  composition, 193, 196-197, 298
  petrography, 193
Biotite microgranite, 188, 189, 196
Birroge intrusion pipe, 148, 161
Bishop's Island, 262
Black Slate Association, 19, 27, 39
Blind Rock Dyke, 374, 375
Bloody Foreland, 4, 19, 97, 100, 102, 115, 125
Blue Stack Mts., 3, 17, 76, 201, 373, 374
Boheolan, 26
Boheolan Quartzite, 28, 29, 32
Bonagher Hill, 367

SUBJECT INDEX

Boudins, 135, 175, 179, 185, 212, 229, 236, 238, 243, 244, 249, 251, 252, 262, 265-268, 281, 286, 330
Bouger gravity map, 339
Boulder beds, 27-29, 33, 34
Boultypatrick, 76
Boultypatrick Grit, 28, 33-35
Boundary Fault, 358
Boyeeghter Quartzite, 21, 25
Breaghy Head, 20, 82
Breccia pipes, 148, 149, 157-164, 204, 208
Breccias, intrusion, 157, 161, 162
  marginal, 154-156, 162
  slide, 37, 40
  see also, Explosion breccias and Intrusive breccias
Brecciation, 135, 150, 159, 162-164, 359, 361, 363, 371
Brockagh, 212, 228, 229, 237, 266, 282
Brockagh Limestone, 19
Brownhall Fault, 358
Bruckless, 360
Bruckless Grit, 367
Bunbeg, 120
Bundoran Shales, 367
Burns Mt. Fault, 358, 359

Calabber River, 3
Caladaghlahan Bay, 185
Calcarenites, 370, 371
Calcareous rocks, 18, 19, 24, 31, 36, 40, 111, 115, 125, 128, 367, 370
  metamorphism, 149-150
  reaction with magma, 116-117, 119
Calcareous xenoliths, 115, 117, 119
Calc-pelites, 19, 21, 25, 27, 31, 269
  metasomatism, 154
  porphyroblasts, 148
  xenoliths, 128, 161
Calderas, 200
Caledonian I, 11, 341
Caledonian II, 11
Caledonian dykes, 82-83, 141-142, 292
Caledonian fault system, 202
Caledonian fold belt, 1, 10, 350, 351
Caledonian granites, 87-95, 347
  geochemistry, 342-344, 350
  Sr isotope ratios, 349
Caledonian magmatism, 205, 342, 349
Caledonides, British, 2
  deformation, 348
  structure and metamorphism, 11, 51-83

Caledonides (*continued*)
  tectonic models, 82
Cambrian, 10, 348
Carapace deformation, 182-184
Carbane Fault, 73, 358
Carbane Hill, 97
Carbat Gap, 265
Carbonate porphyroblasts, 54, 69, 148, 291
Carboniferous, 12, 356, 359
  topography, 3, 366
Carboniferous Limestone, 2, 370
Carnaween Fault, 358, 361
Carndonagh, 6
Carrick, 27
Carrick Fault, 358
Carrickfad Point, 266
Carrigan Head Formation, 28, 31
Carrowkeel, 23
Cashel Belt, 55
Cashel Syncline, 57, 65, 99
Castlederg, 79
Castro Daire Complex, 198
Cataclasis, 96, 177, 180, 212, 227-230, 233, 249, 252, 260, 270, 286, 288, 290, 332, 348
Cataclastic rocks, banding, 218
  textures, 222, 229
Cauldron subsidence, *see* Intrusion mechanisms
Cauldrons, 301
Centered complexes, 198, 200
  see also Barnesmore Pluton and Rosses Centered Complex
Central Donegal Facies, 27, 33-35
Central Donegal Succession, 10, 17, 18
Central Highland Quartzite, 40
Chevron folds, 261, 265, 284, 288
Chilled margins, 45, 130, 161, 190, 192, 194, 200, 205, 208, 374
Chlorite, 134, 135, 270, 277-280, 303, 306, 312, 322, 366
  alignment, 277, 280
  -biotite knots, 321
  dehydration reactions, 319-321
  -garnet knots, 100
  inclusions, 279
  magnesium-rich, 69
  replacement of biotite and garnet, 66, 69, 76
  see also Porphyroblasts
Chloritoid, 272, 279
Church Hill, 21, 47, 63, 64, 72
Clady, 3, 367
Claggan Lough, 24
Claggan Lough Limestone, 20, 24
Claggan Lough Pelites, 20
Clasts, dolomite, 28-29, 34

Clasts (*continued*)
  extrabasinal, 30, 40
  granite, 23, 28, 29, 34
  quartzite, 28-29
Claudy Anticline, 74, 77, 80
Claudy Fold, 79
Cleavage, 7, 54, 55, 62, 63, 71-74, 76, 79, 81, 82, 99, 178, 179, 246, 247, 258, 260-262, 268, 272, 274-277, 279, 286, 288, 289, 293, 294, 361
  defined, 54
  see also Crenulation cleavage, Cross-cleavage, Strain-slip
Cleavage-banding, 265, 270
Cleengort Hill, 99, 114, 266, 313
Cleengort Pelites, 20, 25
Clew Bay, 363
Cloghan Green Beds, 26, 33, 37
Cloghboy, 148
Clogher Hill, 159
Clonmass, 45
Clonmass Banded Quartzite, 20, 21
Clonmass Dolomitic Limestone, 20
Clonmass Head, 45, 62
Clonmass Limestone Member, 20, 21, 24
Clonmass Peninsula, 291
Clonmass-Sessiagh Formation *see* Sessiagh-Clonmass Formation
Clonmass Sill, 24, 45-47
Clooney, 91, 181, 310
Clooney Pelites, 20, 25
Clooney Pelitic Schists, *see* Upper Falcarragh Pelites
Cock's Heath Hill, 259, 283
Colonsay explosion breccias, 163
Color indices of Thorr rocks, 110, 112, 117, 127, 128
Conglomerates, 31, 355, 356, 359, 363, 365-367
Connemara, 11, 36, 37, 341
Contact facies, 110-115, 117, 119, 120, 123, 124, 128-130, 138, 139, 141
Contact metamorphism, 99, 100-102, 129, 134-136, 138, 139, 141, 180-183, 213, 294, 312, 375
  controls, 320-327
  related to magma type, 319
  types of, 326-327
Contact metasomatism, 312-314
Contact minerals, deformation and growth of, 303-312
Contact schists, 180, 302, 309, 310
  cleavages, 260
  kyanite-andalusite relations, 325
Contamination, 97, 109, 115, 117, 119, 120-124, 130, 131, 235, 337, 342-344, 351, 375

SUBJECT INDEX

Contamination (*continued*)
  in Ardara Pluton, 172, 173-174, 182
  in Fanad Pluton, 136, 138, 141
  in Rosses Centered Complex, 196, 197
Convergence phenomena, 332
Convoy, 35
Convoy Formation, 33, 35, 36
Cor, 19, 148, 153, 283
Coral associations, 369, 370
Cordierite, 133-135, 180, 279, 293, 303, 305-307, 309, 311, 312, 318
  -microcline-mica, 134
  porphyroblasts, 47, 272, 311
  PT controls, 321, 323
  stability field, 319-321, 323, 324
Cor Quartzite, 20, 21
Corries, 3, 4
Cortlandtite, 147, 150-152
Corundum, 135, 318, 320
Cowal Antiform, 77
Crana Grits, 40
Crana Quartzite, 26, 28, 32, 35-37, 40-42, 70, 71, 74-76
Cranford, 31, 32, 375
Cranford Dolomitic Limestone, 29
Cranford Limestone, 26, 28, 29, 31, 35, 37, 70
Crannogeboy, 148
Creeslough, 10, 19, 21, 24, 25, 39, 58, 60, 72, 142
Creeslough Downwarp, 57, 59, 63, 261, 291
Creeslough Formation, 18-20
  folds, 261, 265
  schistosity, 261
Creeslough Group, 21
Creeslough Succession, 10, 16-25, 39, 40, 356
  cleavage, 55, 61, 63
  correlation with Ballachulish, 17, 21, 25, 76-82
  deformation, 52-65, 71-72
  folds, 18, 19, 52-61, 73
  metadolerite sheets in, 44-45
  metamorphic history, 53-55, 66-69, 76
  relation to Kilmacrenan, 38-39, 72-74, 77
  schistosity, 53, 54, 58, 61, 63, 64
  slides, 18, 23, 24, 48, 55-58, 73, 81, 371
  thickness of members, 16, 18, 19, 21, 25
  xenoliths, 124, 125, 130
Creevagh, 19, 262

Creevagh Limestone, 19
Crenulation cleavage, 53-55, 61-68, 71, 72, 76, 80-82, 99-101, 134, 178, 181, 213, 258-262, 264-267, 270-278, 280, 284, 287, 288, 291, 293, 305, 307, 363
  defined, 54
Cretaceous land surface, 3
Crinan Grits, 29, 37
Croagh, 262
Croaghconnellagh Arch, 205
Croaghgarrow Formation, 33
Croaghpatrick, 187
Croaghubbrid Dark Schists, 28
Croaghubbrid Grits, 41
Croaghubbrid, Pelites, 33, 34
Crockard Hill, 148
Crockastoller, 286, 287
Crockatee, 3, 59, 62, 63, 65, 261, 262, 264, 266, 270, 274
Crockator, 61, 229, 259, 283
Crockmore, 212, 213, 228, 244, 246, 256, 262, 264-266, 268, 272, 287, 288, 340
Crocknawama, 228, 262, 264, 283, 287
Crockunna Schist, 28, 31
Crohy Hills, 19, 21, 58, 99, 102, 299
Crohy Hills Structure, 57, 58, 60
Crohy Pelites and Semipelites, 21
Crohy Quartzite, 21
Crohy Semipelitic Group, 19
Cronamuck Arch, 205
Cronamuck Granodiorite, 203
Cross bands, 216-219, 227, 251
Cross-cleavage, 228-230, 232-234, 249, 252, 260, 261, 264, 268, 279-282, 286, 288-290, 299
Cross-folds, 55, 63, 81
Cross-lamination, 6, 16, 19, 28, 31, 32, 37, 39, 40, 370
Croveenananta Anticline, 75
Crovehy microgranite sheet, 187-188, 189
Crovehy Mt., 128
Crovehy Townland, 189, 190
Crystallization order, 145, 146
Crystalloblastic textures, 170, 171, 176
Culdaff Limestone, 26, 32, 33, 36, 37, 78
Curran Hill, 107
Cymrian Event, 293, 348

Dalradian, 10, 345, 347, 356, 357
  age of folding, 11, 81, 82
  age of metamorphism, 11

Dalradian (*continued*)
  correlation of Donegal and Scottish, 17, 18, 21, 25, 27, 29, 33, 36-38, 73, 76-82
  defined, 6
  deformation chronology, 52, 55, 79-82
  dolerite intrusions, 48
  fauna, 10
  igneous rocks, 1, 2
  metabasites, 365
  sedimentary facies, 39-42
  stratigraphy, 6, 16-40
  see also, Metasediments
Dalriada, 6
Dawros Head Anticline, 64
Dawros Peninsula, 82
Deeside, 77
Deformation, effect on reaction rates, 326
  in Thorr Pluton, 99, 101, 102, 129, 131
  irrotational, 244-247
  of dolerite sheets, 45, 48
  of envelope rocks, 177-184, 258-268, 281-295, 307, 309, 319, 321, 348
  of granitic rocks, 94, 95, 329-332, 348, 350, 351
  of Main Donegal Granite, 258-295
  of marginal sheets, 286-288
  of metasediments, 33, 53-56, 60, 61-65, 71-72, 79-82
  of sedimentary structures, 16
  stretching, 284
  structures, 133, 134, 258-268, 270, 274, 276, 277, 279, 280-288, 329-331
  superimposed, 291-293
  synplutonic, 210-256
  twinning, 334, 335
  see also Cataclasis
Deformation aureole, 294
Deformation ellipsoid, 94, 95, 175, 234, 244-246, 249, 251, 252
Deformation plot, 94, 95
Dehydration reactions, 312-314, 317, 319-321
  experimental, 319, 320
Delta deposits, 40, 367, 370
Dendritic growth, 307
Denudation series, 184
Derg Fault, 358
Derkberg, 212
Derry, 74, 356, 371, 375
Derryhassan Slide, 58
Derryhassan Syncline, 57
Derryloaghan, 213, 217, 236, 238, 243, 244, 246

Derryreel Syncline, 57, 59, 63
Derryveagh, 3, 213, 373
Derryveagh Mt., 212, 221
Derryveagh raft train, 237
Desilication, 205-207, 208
Deuteric reactions, 93, 333, 335
Devlin Fold Belt, 55, 261
Devonian, 11
  erosion surface, 4, 356
  see also Old Red Sandstone
Diachroneity, 53, 79, 81, 369
Diapirism, see Intrusion mechanisms
Diatremes, 162, 164
Differentiation, 150, 152-154, 157,
  165-168, 196-198, 229, 342,
  344, 345, 349, 351
  filter-pressing model, 219
Diorite, apophyses, 139, 141
  gneiss, 135
  xenoliths, 171, 173, 174
Diorites, 118, 119, 130
  orbicular, 106-108
  quartz-, 97, 102, 111, 115, 123,
    131, 259, 318
    abundance in Donegal
      plutons, 345
    classification, 93
    migmatitic, 89
  see also Appinites
Dioritic magma, contamination, 138
  origin, 138
Discontinuous reaction series, 145
Discordant fabric, 330, 331, 334
Doagh, 139, 359
Dolerites, 10, 373-375
  olivine, 374, 375
  quartz-, composition, 314
  see also Metadolerites
Dolomite, 21, 28, 30-32, 35,
  36, 39, 40
Dolomite breccia, 31
Dolomitic Beds, 28, 29
Dolomitic limestone, 19, 20, 26
Domes, granitic, 8, 71, 73
Donegal, aeromagnetic maps, 357
  Bouger gravity map, 339
  early geological mapping, 6
  structural map, 52
Donegal basement rocks, faulting, 356
Donegal basin, 368
Donegal Bay, 1, 375
Donegal Fault, 358
Donegal Granite Complex, 6-8
  age, K-Ar, 90-91, 293, 342, 348,
    Rb-Sr, 11, 341, 348
  associated dykes, 89
  composition, 7, 344-346
  deformation, 348, 350, 351
  emplacement, 8, 87, 89, 338,
    340-342, 347, 350-351

Donegal Granite
  Complex (continued)
  folds, 60-61
  foliation in, 8
  form, 338-341
  geochemistry, 342-344, 347-348
  ghost structures, 57, 60-61
  gravity investigations, 338-340
  regional metamorphism, 341-342
  roof, 357
  unroofing, 355-356
  see also, Ardara Pluton, Barnesmore
    Pluton, Fanad Pluton, Main
    Donegal Pluton, Rosses
    Centered Complex, Thorr
    Pluton, Toories Pluton,
    Trawenagh Bay Pluton
Donegal Highlands, 355, 356, 363,
  366, 370, 371
Donegal-Kingscourt dyke swarm,
  373, 374
Donegal Syncline, 355, 366
Donegal Town, 375
Doochary, 216, 217, 220, 224,
  236-238, 248-250, 255, 265
Doochary Arch, 216
Doon Fault, 361
Doorin-Coolmore Shales, 367
Dooros Anticline, 57, 59
Dooros Point, 45, 46
Drainage, 3, 356
Ductility, 83, 223, 230, 246-248,
  251, 252, 254, 255, 292-294,
  329-332
Dunfanaghy, 23, 47, 56
Dunglow, 8, 61, 91, 118, 124,
  224, 299
Dunlewy, 18-20, 23, 24, 48, 56,
  59, 250
Dunlewy cross-folds, 55, 63
Dunlewy Fault, 264, 358
Dunmore, 59, 62, 148
Dunmore breccia-pipe, 157, 162
Dunmore Head Anticline, 57, 60, 64
Dunmore Syncline, 57
Duntally Limestone, 19
Dykes, acid, 202, 374
  ages of, 341
  aplite, 189-191, 221, 225-226
  basaltic, 3, 12, 201, 207-208
  basic, 222, 373-375
  breccia, 157, 161, 162
  boudinage, 330
  deformation, 220-230
  deformed, geometry of, 230, 249,
    251, 252, 336
  dilational, 222
  dolerite, 373-375
  ductility, 252, 254, 255

Dykes (continued)
  emplacement, 207-208, 221-226
  folding of, 225, 226, 229,
    236-238, 244, 247, 249-252,
    255, 330
  foliation in, 223-228, 237-239, 243,
    244, 247, 249-252
  granitic, 220-227, 239, 241,
    282, 284, 286
  in Main Donegal Granite, 220-230,
    236-244, 247-252, 255
  intrusion, 359, 373, 375
  lamprophyre, 11, 82, 89, 144, 148,
    161, 266, 267, 281-282
  late-stage, 221, 226, 227, 341
  non-dilational, 220
  porphyritic, 187-199, 296
  post tectonic, 292
  ptygmatic folds in, 239
  relict structures in, 336
  ring, 188, 196, 198, 205, 209
  sigmoidal foliation in, 223
  synplutonic, 222, 330, 336
  syntectonic, 251
  Tertiary, 210, 203, 207-208
  see also Felsite dykes,
    Lamprophyre dykes,
    Microgranite dykes,
    Pegmatite dykes
Dyke swarms, 133, 141-142, 296
  acid, 11, 89
  appinitic, 89
  basaltic, 207-208
  felsite, 89
  map, 374
  microgranite, 89, 133
  pegmatite, 89
  Tertiary, 3, 4, 12, 373-375

Easdale Slates, 29, 37
Edenacarnan Syncline, 74, 75
Eglish Fault, 358, 359
Enclaves, see Xenoliths
En echelon structure, 246, 247, 255
Envelopes, see Aureoles
Environment, eugeosynclinal, 50
  high energy, 337
  low energy, 337
  miogeosynclinal, 40
Eocambrian stratigraphy, 26
Epidote, 112, 212, 231-233, 298
  -allanite, 103, 231
  -clinozoisite, 103, 117
  -inclusions in biotite, 194
  porphyroblasts, 72
  veins, 226, 227
Epizonal plutons, layering, 218
Erosion, 355-357
  penecontemporaneous, 370, 371

Erosion (*continued*)
 subaerial, 363, 366
Erosion surfaces, 3
 Carboniferous, 375
 Devonian, 356
 Tertiary, 3, 356, 375
Errigal, 3, 19, 21, 24, 45, 59, 261
Errigal Quartzite, 3, 19, 21
Errigal Syncline, 55-57, 59, 63, 81
Errig Fault, 358
Eugeosynclinal environment, 50
Explosion breccias, 162, 163
Explosion mechanisms, 89
Exsolution, 332-335

Fabric, 80, 81, 187, 191, 261, 262, 268, 284, 288, 291
 granitic, 94-95, 328-337
 L, 281, 282
 L-S, 82, 94-95, 220, 229, 234, 249, 251, 281, 282, 284, 286, 329
 magnetic, 174, 182
 metamorphic, 332-335
 S, 63, 66-68, 73, 82, 185, 220, 234, 262, 265, 266, 270, 274, 276
Fahan Slate-Grit Group, 26, 33, 37
Falcarragh, 23-25, 65, 139, 371
Falcarragh Limestone, 20, 21, 24, 60
Falcarragh Pelites, 23, 25, 135, 228, 372
Falcarragh Pelitic Schists, 21
Fanad, 1, 2, 4, 6, 11, 16-19, 23, 24, 26, 27, 29, 39, 59, 70, 71, 132-134, 138, 139, 141, 302, 303, 305, 306, 318, 350, 356, 365
 map, 133
Fanad Dolomitic Limestone, 29
Fanad Dome, 71
Fanad Pluton, 132-142
 age, 90, 133-134
 aureole, 89, 134-136, 138-141
  formation temperature, 135, 318
  garnet growth in, 315
  metamorphic zonation, 303-306
  metamorphism, 314
  mineral growth, 305, 312
  width, 318
 composition, 136, 340, 344
 contact metamorphism, 303, 305, 306
 contacts, 134-136, 138-141
 deformation structures in, 133-135, 139
 dyke swarm, 133, 141-142
 emplacement, 11, 89, 133, 139-141
 fabric, 139

Fanad Pluton (*continued*)
 folds in, 135, 137
 foliation in, 134
 ghost stratigraphy, 139, 141
 ghost structure, 336
 magma, 136, 138
 magmatic origin, 343
 outcrop, 134, 135, 138
 rock analysis, 346
 texture, 136-138, 140
 xenoliths in, 133, 135, 138-141
Fanad Quartzite, 29
Fault-breccia, 371
Faulting, age of, 359, 360
Faults, 3, 11, 27, 32, 39, 56, 69-73, 75, 134, 138, 202, 205, 264, 356-358, 361-363, 371-373
 dip-slip, 357, 359, 361, 362
 in Ardara Pluton, 178
 in Barnesmore Pluton, 202, 205
 map, 358
 strike-slip, 52, 79, 359
 transcurrent, 11
 wrench, 357-363
 see also Leannan Fault, Strain-slip
Fault systems, correlation of Irish and Scottish, 363, 364
Faymore, 245, 371, 372
Feldspar augen, 233
Feldspar porphyroclasts, 234, 332
Feldspars, 19, 32, 114, 117, 118, 215, 228, 229
 potash, 103, 105, 106, 112, 115, 117, 120, 123, 134, 135, 138, 217, 219, 226, 306
 crystallization, 251
 growth, 333-335
 origin in granite, 105
 porphyroblasts of, 332
 stability field, 319, 320
 zonal distribution of inclusions, 332
 quartz, crystallization, 195-196
 relicts, 193
 textures, 298
 twinning, 93, 105, 333
 see also Albite, Microcline, Microperthite, Oligoclase, Perthite, Plagioclase
Feldspathization, 114-116, 218, 222, 225, 226, 233, 235, 289, 290
Felsite dykes, 11, 82, 220, 221-226, 229, 230
 deformation, 247-248, 281-282, 290
 ductility, 223, 255

Felsite dykes (*continued*)
 emplacement, 223, 227
 folding, 251-252
 foliation in, 223-225
 structures, 248-249
Felsite sills, 82
Felsite xenoliths, 99
Felsites, 144, 157, 359
 appinitic, 89
 beidellite-pitchstone, 374
Ferromagnesian material, rhythmic precipitation, 106
Fibrolite, *see* Sillimanite
Finn Fault, 358
Finn River, 3
Fintown, 21, 24, 25, 28, 29, 39, 58, 60, 71-73, 212, 220, 265, 266, 268, 277, 279, 281, 286, 288, 291, 371
Fintown Dark Schists, 28
Fintown Siliceous Flags, 20
Fintown Succession, 10, 17, 20
Flash injection, 375
Flood basalts, 50
Flow, of magma, 329, 332
 of solid rock, 329-332, 334, 335
Flow mechanisms, 94, 123, 333-335
Flow structures, 212, 214, 215, 218, 219
Fluidal structures, 150
Fluidization hypothesis, 162-164
Fold belts, 1, 10, 43, 342, 349-351
Folding, age of Dalradian, 11, 81, 82
 of dykes, 225, 226, 228, 229, 236-238, 240, 242-244, 247-248, 250-252, 255
 of sills, 238
 of veins, 224, 242-244, 252, 281, 282, 285
Folds, 10, 24, 25, 27, 34, 39, 351, 357
 anticlinal, 18, 19, 35, 57, 59-64, 70, 71, 73-77, 79-81, 178, 368
 chevron, 261, 265, 284, 288
 cross-, 55, 63, 81
 fir-tree, 73
 flexural-slip, 261, 266, 291
 in Ardara Pluton, 178
 in Barnesmore Pluton, 53
 in Creeslough Succession, 18, 19, 52-61, 73, 261, 265
 in granite, 60, 61, 225, 242-244, 246, 249, 252, 287
 in Kilmacrenan Succession, 69-72, 74-76
 in microgranite, 229, 236-238, 240, 247-248, 250-252
 in pegmatites, 224-226, 228, 229, 236-237, 250-252, 255, 265, 285, 287

SUBJECT INDEX

Folds (*continued*)
  in Thorr Pluton, 99, 282
  isoclinal, 62, 74, 76, 80, 185, 240, 261, 266
  lobate, 63
  micro-, 66, 262, 265, 266, 274, 277, 278, 288
  ptygmatic, 239-244
  recumbent, 54, 56, 59, 60, 63, 69, 70, 72, 74, 77, 79-82, 260
  shear, 277
  small-scale, 58-63
  synclinal, 35, 55-57, 59-61, 63, 64, 69-71, 74-77, 80, 81, 99, 178, 355, 366
  upright, 54, 59-64, 70, 72, 73, 75, 76
  see also Main Donegal Granite
Foliation, 7, 8, 89, 150, 174-177, 179, 180, 192, 282
  in Ardara Pluton, 170, 171, 174-177, 179, 180
  in Fanad Pluton, 134
  in porphyry dykes, 192
  in Thorr Pluton, 97, 98
  in Toories Pluton, 185
  see also Dykes, Main Donegal Granite, Mineral alignment
Forceful emplacement, *see* Intrusion mechanisms
Fossils, 10, 12, 24, 367, 369, 370
Fractures, 329-331
  see also Joints

Gabbro, Rb-Sr age, 349
Galena, 371
Galway, 1, 39
Galwolie Hill, 229
Garnet, growth kinetics, 321, 323
  host-rock control, 317
  inclusions in, 278-280
  inclusions of, 276
  nucleation, 322
  replacement by chlorite, 66, 69, 76, 280, 314, 319
  stability field, 319-321
  see also Porphyroblasts
Garnets, 134, 191, 194, 213, 225, 260, 272, 274, 275, 279, 293, 294, 298, 300, 302, 303, 306, 309, 311, 317, 326
  almandine, 314, 315, 319-321
  spessartite, 315
Garnet-amphibolites, 44, 47
Garnetiferous microgranite, K-Ar age, 91
Gartan Bridge Limestone, 29
Gartan Lough, 24, 39, 60
Gartan Lough Group, 47

Gaugin Mountain, 33, 34
Gaugin Quartzite, 28, 33-35
Gas-drilled veins, 350
Geological Survey of Ireland, 1, 6, 8, 132, 170, 201, 371
Geomorphology, 45, 358, 375
Ghost folds, 214, 246, 259, 266, 267, 291
Ghost stratigraphy, 7, 17, 60-61, 89, 102, 110, 124-130, 139, 141, 213, 235, 256, 259, 291, 335-337
Ghost structure, 7, 60-61, 67, 105, 106, 118, 124-126, 128-129, 139, 141, 207, 309, 311, 331, 335-337
  cryptic, 126-128, 131, 337
  overt, 124-126, 335
Ghost synclines, 61
Glacial pavements, 189
Glaciation, 3-5, 30, 40
Glen, 31, 39, 212
Glenaddragh Fault, 358
Glenaboghill, 371
Glenaboghill Limestone, 20
Glencolumbkille, 27, 28, 31, 36, 39, 40, 54, 69, 71, 72, 363
Glencolumbkille Dolomite, 28
Glencolumbkille Formation, 73
Glencolumbkille Limestone, 27-29, 33, 35, 37
Glencolumbkille Pelites, 27, 28, 33, 37, 39, 266, 267
Glendowan, 371, 372
Glendowan Mts., 3, 212
Glendowan Pelites, 20
Glengad Schists, 26, 29
Glengesh, 3, 148
Glengesh Fault, 358
Glen Head Schist, 28
Glenieraragh, 19
Glenieraragh Siliceous Flags, 20
Glenkeo, 232, 270, 272, 274, 276
Glenleheen, 213, 216, 221, 236-239, 242, 244, 246-249, 255
Glen Lough, 212, 244, 246, 255
Glenties, 25, 28, 31, 32, 71, 97, 151, 152, 282, 294
Glenties Series, 20
Glenveagh, 3, 8, 213
Glinsk, 134, 291, 306, 314, 322
Gneiss-dome hypothesis, 290-291
Gneisses, 258
  granitic, 7, 8
  migmatic, 135
  orbicular diorite, 135
  psammitic, 10, 347
  semipelitic, 347
Gneissosity, 218

Gola Island, 108, 110, 120, 124
Gola type granite, 124, 130
  composition, 344
Gorey Hill, 74
Gortahork, 23, 56, 99, 100, 371
Gorteen Lough Striped Flags and Pelites, 29, 32
Gortnatraw Septum, 134, 138
Graded bedding, 16, 19, 24, 25, 31, 32, 37, 39, 42, 45
Graffy Hill Fold, 69, 71
Grampian Event, 348
Grampian folding, 11
Granite, banding in, 212, 215-219, 222, 226, 227, 233, 251, 252, 255
  cooling age, 356, 362
  cross-cleavages in, 229-230, 232-234
  deformation, 210-256, 268, 282, 286, 290
  desilication, 205, 207, 208
  ductility, 246-248, 251, 252, 254, 255, 292, 329-332
  emplacement, 11, 55, 61, 184, 198-200
  erosion, 355-356
  folds, 60, 61, 225, 242-244, 246, 249, 252, 287
  origin, 6-8, 93, 196-198
  -quartzite contacts, 115-116
  structures, 139
  tectonics, 329-331
  textures, 231-232, 246
  xenoliths, *see* Xenoliths
Granites, epidote-bearing, 194
  experimental anatexitic, 347, 350
  field relations, 189-191
  geochemistry, 342-345, 347-348
  granitization, 7
  granophyric, 134
  greisenized, 187, 191, 193, 194, 199, 200
  intrusive, 7-8
  layered, 218
  magmatic, 120, 122
  metamorphic, 7-8
  modal composition, 173, 193, 203
  muscovite, 187, 191, 193-194
  porphyritic, 204-205
  rock analyses, 196, 346
  Tertiary, 374
  tonalitic, 222
  see also Aplites, Biotite granite, Donegal Granite Complex, Pegmatites
Granite sheets, 128, 204, 256, 265, 298
  cataclastic deformation, 227-230

Granite sheets (*continued*)
 deformation, 259, 284-288
 forceful emplacement, 288
 structures, 287
Granite veins, 228, 239, 242-244
Granitic dykes, deformation of, 283, 284, 286
 in Main Donegal Granite, 220-227, 239, 241, 282, 284, 286
Granitic fabric, 94-95, 139, 328-335
Granitic magma, 298-299
 anatexitic, 347, 351
 crystallization, 93, 329, 330, 333
 crystal mushes, 351
 ductility, 329-332
 emplacement, 300-301, 340, 350-351
 generation depth, 348
 linked-anion liquids, 351
 phase diagram, 195
 source, 342, 345-351
 temperature, 348, 351
 volatile content, 348, 350, 351
 water vapour pressures, 347-351
 see also Contamination and Magma
Granitic plutons, map of, 88
 metamorphic zones, 293, 294
 post-kinematic, 339
 see also Intrusion mechanisms
Granitic rocks, classification of, 91-93
 color index, 92, 93, 109, 110, 112, 117, 127
 deformation, 94, 332
 flow structures, 332-335
 mineral content, 91-92
 mineral fabric, 332-335
 preferred orientation in, 93-95
 structure, 93-95
 textures, 93, 332-335
Granitization, 7, 114-116, 128-131, 135, 138, 222, 335, 337
Granodiorite, 89, 99, 102, 111, 115, 117, 119, 128, 134, 135, 137-139, 144, 151-154, 164, 170-172, 203, 205, 211, 212, 318, 339, 350
 abundance in Donegal plutons, 345
 analyses, 346
 apophyses, 112
 inclusions of, 299, 300
 zoning, 219
 biotite-, 138
 classification, 93
 contact facies, 138
Granodioritic dykes, *see* Felsite dykes
Granodioritic magma, contamination, 174, 182

Granodioritization, 138
Granophyric intergrowths, 223
Graphite inclusions, 274
Gravity investigations, 338-340, 362
Gravity slumping, 73
Great Glen Fault, 11, 27, 357, 359, 361, 362
Green Beds, 36
Greencastle Green Beds, 26, 33, 37
Greenland, 10, 223, 336
Greenockite, 371
Greenschist facies, 53, 54
Greisen, K-Ar age, 91
Greisenization, 191, 193, 194, 199, 200
Greywacke, 32, 36, 39
Grossular, 116
Gubbin Hill, 236, 237, 241, 242, 244
Gweebarra Bay, 17, 60
Gweebarra Bridge, 24, 25, 91
Gweebarra Estuary, 19, 24, 25, 59, 112, 114, 153, 211
Gweebarra Fault, 357, 358, 362, 373
Gweedore, 3, 102

Harker variation diagram, 120-122
Harristic texture, 106
Helicitic trails, 47, 58, 66
Hercynian mineralization, 371-372
High-energy environment, 337
Highland Boundary Fault, 77, 80, 357-359, 363
Hornblende, 102, 103, 106, 112, 117, 118, 128, 136, 138, 144-147, 150, 151, 154, 159, 161, 166
 colour changes, 144, 146
 K-Ar age, 91
 -lamprophyres, 144, 149
 preferred orientation, 281, 283, 286
 stability, 167
 textures, 103, 146, 147
Hornblendite, 135, 147, 150-153
 -appinites, 153, 161
Hornblendization, 166
Hornfelses, 16, 24, 45, 48, 87, 100, 105, 112, 114-117, 123, 124, 129
 andalusite, 302
 biotite-muscovite-quartz, 134
 dehydration, 312
 folds, 47
 hornblende, 268
 K-Ar age, 91
 mobilized, 99-102
 pelitic, 305
  kyanite content, 315
Horn Head, 4, 19, 45, 59, 64

Horn Head Slide, 18, 23, 24, 48, 56, 58, 371
Host rock, compositional control of, 314-317
'Hot Spot' model, 293-295
Hybridization, 165-168
Hydrous phase, 305

Idocrase, 116
Igneous activity, Caledonian, 87-95
Igneous arches, 204-205, 216
Illancrone, 184
Ilmenite, 272
 porphyroblasts, 66, 67
Iltay Boundary Slide, 38, 39, 77, 79
Inch Island Limestone Group, 26, 33, 37
Inclusions, 58, 66, 67, 68
 epidote, 194
 in microcline, 105
 in porphyroblasts, 311
 nomenclature and origin, 119
 plagioclase, 105
 zonal distribution in feldspar, 332
 see also Rafts, Xenoliths
Inclusion trails, 58-66, 274, 277-279
Inishboffin, 25, 99, 307
Inishboffin Banded Semipelites, 21
Inishdooey, 23
Inishowen, 17, 18, 26, 29, 31-33, 35, 36, 40, 53, 74-77, 132
Inishowen Grits and Phyllites, 26
Inishowen Head Grits and Slates, 33, 37
Inishtrahull, 347, 361
Iniskeeragh, 184, 185
Intermontane molasse, 356, 363-366
Intrusion breccias, 157, 161, 162
Intrusion lineation, 256
Intrusion mechanisms, 302-303
 active stoping, 87, 89, 96-131
 cauldron subsidence, 11, 89, 90, 186-200, 201-209, 318, 325, 327
 diapirism, 11, 89, 154, 164, 169-185, 290, 292, 295, 325, 331
 explosion, 89
 forceful emplacement, 87, 89, 150, 213, 214, 259, 284, 288, 290-294, 327, 331, 340, 351
 fluidization, 162-164
 gas-drilling, 149
 magmatic wedging, 11, 89
 of dykes, 375
 passive sheet intrusion, 293
 passive stoping, 87, 89
 permissive emplacement, 87, 89, 208, 327, 340

SUBJECT INDEX

Intrusion mechanisms (*continued*)
 piecemeal stoping, 11, 89, 139-141
 rock bursting, 164
 stoping, 301, 318, 336, 351
Intrusive breccias, appinite suite of, 143, 144, 148, 149, 156-165
 nomenclature, 162
 origin, 162-164
Intrusive sheets, 187-189, 193, 196-198, 200
Inver, 375
Island Roy Syncline, 57, 59
Islay, 27, 29, 36, 40, 77
Islay Anticline, 77
Islay Limestone, 29, 37
Islay Quartzite, 29, 36, 37, 40
Islay Succession, 10, 18, 38-40, 77
 correlation with Kilmacrenan, 36, 76-82
 relation to Ballachulish Succession, 38
Isograd, garnet-chlorite, 79
Isotopic ages, *see* Ages
Istotope ratios, 333

Joints, 94, 202, 208, 373, 374
 in Ardara Pluton, 177-178
 in Rosses Centered Complex, 187, 189, 194, 199
Joint systems, 363
Jura Slates, 29, 37

Keadew Strand, 188
Keeldrum, 99, 371
Kildoney Sandstones, 367
Kilkenny, 148
Kilkenny breccia-pipe, 159, 160, 163
Killeter Quartzite, 33, 35, 36
Killin Hill, 367
Killybegs, 357, 359
Killybegs Fault, 358
Killybegs Group, 27, 28
Killygarvan Flags, 26, 29, 32, 33
Killygarvan Limestone, 26
Kilmacrenan, 26, 29, 35, 74, 76, 359, 361
Kilmacrenan Grits, 28, 29, 32
Kilmacrenan Succession, 10, 17, 18, 26-39, 351
 correlation with Islay-Loch Awe Successions, 27, 29, 36-38, 76-82
 deformation, 69-76
 folds, 69-72, 74-76
 inversion, 361
 metadolerite sills, 44-45
 metamorphism, 72, 76
 relation to Creeslough Succession, 38-39, 72-74, 77

Kilmacrenan Succession (*continued*)
 structure, 69-72, 74
 thickness of strata, 29, 31, 32
Kilrean, 147, 148, 151, 152, 166
Kilrean boss, 148, 151
Kiltooris Lough, 148
Kiltyfanned Schist, 28
Kindrum, 23
Kindrum Fault, 134
Kingarrow, 212, 246, 247, 255, 266, 287
Kink-bands, 65, 71, 261, 266, 267, 281
Knockalla, 363, 365
Knockalla Fault, 358
Knockalla Quartzite, 26, 29
Knockateen Slide, 17, 20, 39, 73, 77
Knockateen Thrust, 28, 29, 71
Knockfadda Anticline, 57, 59, 63
Knockletteragh Grits, 28, 32
Knocknafaugher Intrusion, 45
Kyanite, 180, 268, 293, 302, 309, 311, 318, 326
 alteration to muscovite, 279
 host-rock control, 315-317, 325
 inclusions, 311
 in andalusite, 276
 replacement by micaceous pseudomorphs, 76
 stability fields, 319, 320, 325, 326
 see also Porphyroblasts

Labradorite, 269
Lackagh Bridge, 18, 19, 249, 266, 278, 283-287
Lackagh Bridge Quartzite, 21
Lackagh River, 262, 333
Lag, 31
Lag Limestone, 26, 29, 33
Laggan Fault, 363
Laghy Fault, 358
Lakelandian Event, 348
Lamination, 24
Lamprophyres, 11, 82, 83, 89, 144, 148, 161, 266, 267, 359
 deformation, 179, 281-282
 hornblende-, 144, 149, 161
 xenoliths in, 162
 xenoliths of, 99
Laxfordian rocks, composition of, 344
Layeritic growth, 307
Lead isotopes, 349
Lead ores, 371
Leahanmore, 256
Leannan Fault, 3, 10-12, 17, 26, 27, 32, 33, 48, 52, 53, 61, 69, 72, 76, 79, 80, 134, 356-359, 361-363, 365

Leannan Fault (*continued*)
 magnetic survey, 361
 sections, 362
Leannan River, 361
Leckenagh, 189
Lesbellow Inlier conglomerates, 355
Letterkenny, 35, 53, 76, 79
Lettermacaward, 21, 97, 99, 100-102, 111, 307
 geological map, 101
Lettermacaward Alternating Group, 21
Lettermacaward Limestone, 21
Lettermacaward Semipelite-Pelite Group, 21
Leuco-diorites, 166
Leven Schists, 21, 25
Lewisian, 347, 348
Lewisian rocks, composition of, 344, 345, 347
 Sr isotope ratios, 349
Limestone, in intrusive breccia, 157, 161
 silicified, 371
Limestones, 6, 10, 12, 19, 24, 25, 27, 31, 32, 35-37, 39, 367, 370, 371
Limestone xenoliths, 116, 117, 235-237, 239, 241-243, 249, 262
Lineation, *see* Foliation, Mineral alignment, Stretching lineation
Linsfort Black Schists, 26, 29
Liskeeraghan, 147, 148, 151, 154
Liskeeraghan Intrusion, 147, 148, 151, 154, 160
Lismore Limestone, 21
Lit-par-lit injection, 8, 215
Load casts, 19, 23, 24
Loch Avich Grits, 37
Loch Awe, 29, 77, 123
Loch Awe-Lough Foyle Belt, 82
Loch Awe Succession, 18
Loch Awe Synform, 77, 79
Loch Skerrols Thrust, 77
Loch Tay Inversion, 77
Losset, 91, 228
Losset-Lough Greenan Pelites and Semipelites, 20
Lough Agher, 62, 91
Lough Agher Fault, 358
Lough Aleane, 262, 265
Lough Anna, 148
Lough Anoon, 125
Lough Anure, 191
Lough Ardmeen, 282
Lough Belshade, 205, 208
Lough Derg, 10, 27, 35, 344, 347

427

SUBJECT INDEX

Lough Derg Antiform, 79
Lough Derg Psammitic Group, 29, 36
Lough Derg Slide, 36, 79
Lough Ea Fault, 69, 71, 358
Lough Ea Fold, 55, 71
Lough Eske, 367
Lough Eske Formation, 28, 33, 35
Lough Eske Psammites, 40, 41
Lough Finn, 279
Lough Finn Dark Schists and Limestones, 29
Lough Foyle, 356, 362
Lough Foyle Fault, 358, 362, 363
Lough Foyle Succession, 18, 26, 32, 35-37
Lough Foyle Synclinal Complex, 74, 77, 366
Lough Gartan Anticline, 57, 60, 61
Lough Greenan, 25, 60, 61, 260, 262, 270, 274, 311
Lough Greenan Belt, 73
Lough Greenan Syncline, 57, 60, 61, 64
Lough Illion, 111, 124
Lough Keel, 26, 100
Lough Laragh, 152, 191
Lough Nabrack, 91
Lough Nacollum, 204
Lough Nageeragh, 191
Lough Nagreany, 134
Lough Pollrory Alternating Group, 21
Lough Reelan, 262
Loughros, 20, 25, 39, 59, 73, 139, 148, 159, 178, 291
Loughros Anticline, 57, 60, 178
Loughros Formation, 20, 23, 25, 39
Loughros Quartzite, 20
Loughros Semipelite, 20
Lough Salt, 24, 99, 262
Lough Salt Fault, 75, 358, 362
Lough Salt Limestone, 20
Lough Salt Mountain, 73
Lough Swilly, 4, 79
Low-energy environment, 337
Lower Crana Quartzite, 26, 29, 32, 33, 36
Lower Falcarragh Pelites, 18, 20, 21, 23-25
  cleavages, 266
  deformation, 67
  porphyroblasts in, 67
Lower Palaeozoic, 1, 342
Lower Termon Schists, 28

Maas, 19, 21, 23, 24, 45, 59, 61, 63, 64, 90, 99, 100, 211, 212, 259, 266, 270, 282
Maas-Ardlougher Fault, 178
Maas Semipelites, 20

Maas Succession, 10, 17, 20
Maas Syncline, 57, 59, 63, 266
Mafic clots, 103, 106, 136, 138, 174
  exogenic origin, 115, 117-119
Mafic minerals, 120, 136, 138, 139
Maghera Road, 148
Maghery, 60
Magma, acidic, 342
  appinitic, crystallization of, 145, 146, 150, 161, 163, 165, 167, 168
  gas-charged, 157, 159, 161-164, 168
  P-T conditions, 167
  volatile content, 146, 161, 167, 168
  water content, 167, 168, 349-351
basaltic, 50, 351, 375
basic, 10, 202, 350, 351
composition, 196-198, 318
crystallization, 83, 105, 106, 108, 119, 120, 122, 124, 129, 136, 138, 172, 173, 177, 183, 192, 195, 213, 218, 222, 223, 226, 233, 234, 250, 290, 292, 325, 341, 342, 348, 349, 351
deformation, 212, 213, 214, 218, 219, 226, 244
emplacement, 190, 198-200, 259, 284, 288, 290-295
gas-charged, 207
granodioritic, 174, 182
ionic diffusion in, 194
low temperature, 203
monzodioritic, 141
reaction with calcareous rocks, 116-117, 119
reaction with metadolerites, 117
reaction with pelite, 112-115, 119
reaction with quartzite, 115-116
stress fields, 200
temperature control, 194, 197
tholeiitic, 42, 43, 45, 49, 50
vapour pressure, 348, 349
volatile fraction, 189, 190, 348
volatile-rich, 298
wet, 83, 194, 196, 207
see also Contamination, Differentiation, Granitic magma, Intrusion mechanisms
Magmatic differentiation, 120-124
Magmatic flow, 212-214
Magmatic indicators, 332, 333
Magmatic sedimentation, 218
Magmatic textures, 93
Magmatic wedging, 11, 89
Magmatism, 87, 89, 93, 106, 108
Magnesium-iron ratios, 314-317

Magnetic anisotropy, 220
Magnetic fabric, 174, 182, 251
Magnetic survey, 361
Magnetic susceptibility, 299
Magnetite, 48, 106, 118
Main Donegal Granite, 11, 12, 19, 24
  age, K-Ar, 90, 91
  aureole, 211-214, 227-230, 233, 252, 256, 268, 281, 291
  chronology, 257-295
  contact metamorphism, 258, 259, 268-295, 310-311
  deformation, 249-251, 307, 309, 319, 321, 348
  folds, 212-214, 227, 229
  formation depth, 325
  kyanite and staurolite in, 315
  mineralogy, 258, 259, 268-270, 293
  origin, 259, 288-295
  photomicrographs, 271, 275
  porphyroblasts in, 272, 292, 309, 311, 312
  structures, 246, 256-268, 282, 283, 286-288, 291, 294
  tectonics, 259-268, 280-295
  temperature, 318
  width, 318
banding in, 212, 215-219, 222, 226, 227, 233, 251, 252, 255, 329, 333
boundary with Trawenagh Bay Pluton, 211, 259, 298-299
bulk deformation, 244-247, 252, 255
composition, 193, 217, 300, 344, 346, 347
deformation, 52, 64, 65, 89, 210-256, 258-295
deformation chronology, 251-256
deformed dykes, 249, 251, 252
ductility, 223, 230, 246-248, 251, 252, 254, 255
dykes in, 220-230, 236-244, 247-252, 255
emplacement, 11, 55, 89, 98, 259, 284, 288, 290-295, 340
folds in, 55, 211-214, 223, 225-231, 246, 247, 249, 258-270, 277, 281, 282, 284, 285, 287, 288, 291, 294
geological history, 260, 280-281, 283, 286-290
ghost folds, 214, 246
ghost stratigraphy, 17, 60, 61, 128, 213, 235, 256, 336
ghost structures, 60, 61, 128, 336
isochemical metamorphism, 314
magma, 211-214, 218, 219, 222, 223, 226, 233, 234, 244, 255

SUBJECT INDEX

Main Donegal Granite (*continued*)
  marginal sheets, 286-288
  metamorphic grade, 53
  metamorphism, 258, 259,
    268-283, 286-289, 294
  mineral alignment, 211, 212,
    219-239, 244, 246-250,
    252, 255, 259, 262, 264,
    268, 272, 277, 278, 280, 281,
    284, 285, 286, 287, 289, 290,
    291, 292, 293
    geometry of, 244-245, 252
  mineralogy, 211, 212, 218, 226,
    230-234, 256
  origin of, 218-219, 288-295
  petrography, 230-234
  porphyroblasts in, 259, 260,
    269-283, 288-293
  raft trains in, 212-214, 235,
    237, 244, 246, 247, 256
  regional stress pattern, 247-249
  rock analysis, 346
  rock types, 344-345
  shape, 339-340
  shear zones, 330
  structures, 212-214
  synplutonic deformation, 210-256
  tectonics, 259, 260-268, 280-295
  textures, 231, 232
  thermal metamorphism, 281, 291
  thickness, 339-340
  zoning of feldspar inclusions in, 332
  xenoliths in, 211-213, 215, 216,
    221, 222, 223, 224, 227,
    234-247, 249, 252, 255
Main Donegal Pluton, *see* Main
  Donegal Granite
Main Granite, *see* Main Donegal
  Granite
Main Granite (Barnesmore Pluton),
  203, 204
Malchites, 83
Malin, 41, 70, 361
Malin Head Quartzite, 26, 29, 33
Malinmore, 4
Mam Sill, 45-48
Maps, 2, 4, 5, 9, 44, 52, 75, 98,
  101, 127, 133, 175, 188, 202,
  206, 258, 297, 308, 313, 339,
  367, 369, 374
Marble, 24
Marble Hill Dolomite Limestone, 20
Marble Hill Limestone, 21
Marginal sheets, deformation of,
  286-288, 292
Matrix coarsening, 272, 274, 276,
  283, 289, 290
Matrix flattening, 272, 274, 276,
  277, 281, 288

Mayo, 1, 39, 363, 369, 371
Meenacross, 299
Meenalargan, 19, 148, 166, 229, 282
Meenalargan Appinitic Complex,
  152, 166, 182, 282
  xenoliths of, 213
Meenamarragh, 125
Meenatotan, 103, 296, 299, 308
Meenbannad, 193
Meencoolagh Banded Group, 21
Meencorwick, 91, 97
Meenderryherk, 125
Meenlaragh, 60
Melmore, 4, 134-138, 140
Melmore migmatites, 134-137,
  139, 141
Melmore Mixed Group, 21
Melmore Septum, 134, 135, 138
Metabasites, 52, 101, 281, 289, 365
  see also Metadolerites
Metadolerite xenoliths, 141, 235,
  237, 246, 255
Metadolerites, 16, 43-46, 72, 112,
  113, 120, 125, 135, 138, 139
  age, 44, 48
  agmatized, 137
  chemical analyses, 49-50, 117
  composition, 312
  deformation, 45, 48, 179
  distribution, 44
  folds in, 60, 281, 283
  metamorphism, 48-50
  mineralogy, 45, 269, 283, 288, 289
  quartz, 43-50
  reaction with magma, 117-119, 124
  Rosbeg type, 43-44, 47-68
  textures, 45, 48
Metadolerite sheets, 45, 228
Metadolerite sills, 16, 32, 43-50
  aureoles, 16, 47
  boudinage, 266, 267, 281
  graded bedding in, 45
  intrusion, 44, 48
  thickness, 45
  xenoliths, 119, 128
Meta-greywacke, 202
Metamorphic aureoles, *see* Aureoles
Metamorphic events, correlation
  with tectonics, 260,
  280-283, 289
Metamorphic fabric, 332-335
Metamorphic grade, 76, 213,
  222, 280
Metamorphic mineralogy, 49-50,
  268-270, 319, 320
Metamorphic rocks, banding of, 219
Metamorphic textures, correlation
  of, 53
Metamorphic zonation, 101

Metamorphism, 7, 8, 10, 81
  chronology, 11, 53-56, 66-69,
    72, 76
  in Main Donegal Granite
    envelope, 258-260, 268-295
  isochemical, 312-314
  of quartz dolerites, 48, 49
  polyphase, 307, 312, 315
  regional, 54, 55, 79, 89, 260,
    268, 281, 293, 294, 303, 307,
    312, 341-342
    date of, 341, 342
  retrogressive, 54, 68-69, 72, 76,
    260, 270, 277, 279-280,
    289, 290, 303, 314
  thermal, 203, 281, 291, 303,
    318, 319, 321
  see also Contact metamorphism
Metasediments, 1, 3, 6, 11, 15-42,
  114, 116, 134-136, 139,
  140, 366
  age, 10, 11
  composition, 312
  deformation, 11, 33, 52-65, 69-72
  ductility, 294
  folds, 18, 19, 52-61, 73, 79-81
  mapping, 8, 9
  metamorphism, 53-55, 66-69,
    76, 259, 268-281
  sedimentary assemblages, 15-42
  stratigraphy, 10, 16
  structure, 18, 19, 52-65, 73,
    260-268
  thickness, 357
  xenoliths of, 235, 246
  see also Schistosity
Metasomatism, 48, 49, 83, 87, 89,
  93, 101, 102, 106, 115, 116,
  119, 128, 129, 131, 135, 141,
  152, 154, 157, 159, 164,
  167, 191, 200, 207, 243,
  294, 305, 307, 312-314,
  324, 327, 337
Mevagh Mixed Group, 21, 24
Mica, 230, 232, 233, 309, 324
  crenulations, 270, 271, 277
  inclusions, 274
Mickey's Hole, 45
Microcline, 114, 117, 119, 220,
  225, 226, 282
  genesis, 194
  inclusions in, 105
  microperthite, 104, 175, 298
  -plagioclase relations, 104, 105,
    110, 127, 128
  porphyroblasts, 118
  preferred orientation, 102
  -quartz relations, 104, 105
  textures, 216-219, 225, 230-232

Microdiorites, 208
Microfolds, 66, 262, 265, 266, 274, 277, 278, 288
Microgranite dykes, 193, 220-222, 226-230, 236, 237, 247-252
　deformation of, 290, 292
　folding of, 229, 236-238, 240, 247-248, 250-252
Microgranite dyke swarms, 133
Microgranites, 11, 89, 221, 289
　chemical variation, 196-198
　garnetiferous, K-Ar age of, 91
　ductility, 255
　modal composition, 193
　porphyritic, 115
Microgranite sheets, 187-189, 193, 195-198
　intrusion, 198-200
　petrography, 193
Microgranite sills, 215, 229
　deformation, 285
　folding of, 238
Microgranodiorite, 115
Microgranodioritic sheets, 172
Microlithons, 180, 276
　relict, 307, 309
Microstructures, 262, 268, 270, 271, 274-277
　see also Microfolds
Microveins, 233
Migmatite gneiss, 135
Migmatites, 97, 111, 114, 134, 135, 137, 139, 141, 333
　contact, 135, 136
Migmatitization, 8, 89, 114
Milford, 69
Milford Schists, 26, 29, 33
Mimetic growth, 303, 305, 310
Mineral ages, K-Ar, 91
Mineral alignment, 58, 74, 81, 94, 100-103, 105, 125, 139, 141, 183, 191, 195, 284, 292, 299, 302, 303, 329, 330, 334
　see also Foliation, Main Donegal Granite
Mineral assemblages, 83, 230, 259, 302, 303, 307, 309, 314-326, 366
Mineral fabric, 129
Mineral growth, compositional control of, 303
Mineral paragenesis, 194-196, 319-326
Mineral textures, 170, 172, 268, 332-334
Mineral zoning, 134, 303-305, 317, 324, 325
Mineralization, Hercynian, 371-372
　lead isotope ages of, 372

Minerals, 6
　deformation and growth chronology, 303-312
　metamorphic, 268-270, 304
　　experimental equilibria, 320
　preferred orientation, 94
Miogeosynclinal environment, 40
Miogeosynclinal sediments, 50
Mobilization, 99-102, 114, 119, 124, 129, 135, 141, 154, 156, 166, 168, 178, 185, 282, 292, 314, 340, 347, 348, 350, 351
Moine, 2, 27, 36, 38, 79, 344, 345, 347
Moine sediments, 10
Molasse, intermontane, 356, 363-366
Monargan, 148
Monzodiorite, 89, 103, 347
　analyses, 346
　K-Ar age, 91
　mafic aggregates in, 282
　magma, 138, 141
　modal composition, 170, 171
　quartz-, 134, 136, 138, 170-172, 174, 182, 185
　texture, 136-138
　xenoliths in, 118, 170-174, 185
Monzonitic rocks, 205
Monzotonalite, 170, 171, 185, 318
　analyses, 346
　composition, 170, 171, 344, 346
　definition, 93
Moorlagh, 188, 189, 198
Moraines, 3, 4
Mossfield, 39
Mossfield-Lough Salt Fault, 72, 358, 362
Mountcharles Fault, 358, 359
Mountcharles Sandstones, 367
Mourne Mountains granite, 7
Muckish, 19, 24, 59, 64, 261
Muckish Anticline, 57, 59, 63
Muckish Mountain, 59, 62
Mud cracks, 24, 39, 40
Mullaghderg, 106-108, 188
Mullaghduff dyke, 194
Mullions, 258, 268, 282
Mull of Oa Phyllites, 29, 37
Mullyfa Formation, 33, 36
Mulnamin, 59, 166
Mulnamin Anticline, 57, 59, 64, 266
Mulnamin Calc-Silicate Group, 20
Mulnamin fold, 60, 61
Mulnamin Intrusion, 148, 149, 151-153, 166
Mulnamin More, 148, 152, 153
Mulnamin Siliceous Flags, 20
Mulroy Bay, 4, 18, 26, 211

Muscovite, 101, 114, 115, 117, 134, 135, 180, 211, 232, 233, 265, 270, 276, 298, 305-307, 309, 312, 327
　alignment, 277
　dehydration reactions, 319, 320
　gneiss, 194
　host-rock control, 317
　inclusions, 278, 279
　K-Ar age, 91
　see also Porphyroblasts and Xenoliths
Muscovite-biotite schists, 202
Muscovite granites, 187, 191
　modal composition, 193
　petrography, 193-196
Mylonite, 229, 361
Mylonitic flags, 56
Mylonitization, 23
Myrmekite, 105, 172, 217, 231-233, 298, 333, 334

Naran Hill, 161
Naran Hill appinite body, 148, 149, 154-156, 166
Narin, 147
Net-veining, 150, 151, 159, 164
New Bridge, 91, 363
New England, 292, 293, 295
"Newer Granites", 90, 120, 296
Newfoundland, 293, 363
New Red Sandstone, 356
Newry Granite, 164, 165
Non-dilational dykes, 220
North Channel, 363
Northwest Donegal Facies, 27-32
　comparison with Central Donegal Facies, 351
Nucleation, 159, 289, 307, 321, 323-326
Nunnarssuit Complex, 336

Old Red Sandstone, 2, 11, 345, 356, 361, 363-366
Oligoclase, 101, 115
Oligoclase augen, 202
Oligoclase porphyroblasts, see Porphyroblasts
Olivine dolerites, 374, 375
Omagh Syncline, 366
Oolites, 371
Ophitic textures, 45, 48
Orbicular diorite, 106-108, 129, 135
Ordovician, 6, 348
Ore bodies, 371
Orocline, 61
Oughtradeen Group, 33
Owenamarve Fault, 358
Owenator River, 91

SUBJECT INDEX

Owenea Lough, 148
Owengarve Formation, 28, 33, 35
Ox Mts., 368

Palaeocurrents, 40
Palaeogeography, 40, 369-371
Paragenetic sequences, 93
Parautochthonous granites, 337
Pebble beds, 19, 25, 31, 32, 39
Pebble dykes, 162
Pegmatite-aplite, 172, 212
Pegmatite-aplite sheets, 190, 287
Pegmatite-aplite veins, deformation of, 176
Pegmatite dykes, 102, 103, 189-191, 217, 220, 221, 225-230, 285, 287
  deformation of, 290, 292
Pegmatites, 11, 89, 150, 185, 190, 191, 204, 219
  banded, 219
  desilication, 205
  emplacement, 289
  folding of, 224, 236-237, 250-252, 255, 265, 285, 287
  K-Ar age, 91
  zoned, 225
Pegmatite sheets, mineral alignment of, 286
Pegmatite veins, 265-267, 287, 289
Pelites, 19, 21, 24-27, 31, 32, 34-37, 47, 72, 99, 100, 111-116, 119, 120, 123, 134, 135, 154, 159, 160, 262, 318
  chemical composition, 114, 312, 313
  cleavage, 265, 283, 286
  contact facies, 123
  folds in, 262, 265-267
  dehydration, 313, 314
  K-Ar ages, 91
  metamorphism, 66-69, 76, 259, 261, 280, 283
  microstructures, 270-271
  mineral assemblages, 314, 317
  mineralogy, 269, 280-281
  mineral zonation, 303-305
  mobilized, 156
  porphyroblasts in, 283
  reaction with magma, 111-115, 119
Pelitic hornfelses, kyanite content of, 315
Pelitic rocks, 53
  metamorphism, 10
Pelitic xenoliths, 111, 114, 115, 119, 125, 128, 215, 221, 235, 239, 266, 267, 278, 286-288
Peripheral crystallization, 131
Permissive emplacement, see Intrusion mechanisms

Permo-Triassic uplift, 356
Perthitic intergrowths, 105, 334
Perthshire, 10, 40, 351
Perthshire Succession, 357
Petrogenetic indicators, 332-333
Pettigo Fault, 358, 362, 363
Pettigo Limestone Group, 367
Phacoliths, 337
Phosphorus, geochemical culmination of, 119
Phyllites, 32, 36, 72
Phyllitic cleavage, see Crenulation cleavage
Phyllitic schist, 44
Phyllosilicate porphyroblasts, see Porphyroblasts
Phyllosilicates, 76, 270, 280, 293, 386
  growth, 54, 66, 72, 260, 277, 279
  orientation, 279
  textures, 278
Physico-chemical controls, 317-326
Piecemeal stoping, see Intrusion mechanisms
Pillow lavas, 36, 42
Pinch and swell structures, 23, 25, 56, 236, 238, 240, 241, 249, 285, 287
Plagioclase, 104-106, 108, 112, 117, 118, 170, 172, 173, 298, 311, 312
  albite-anorthite ratio, 120
  aligned, 270
  crystallization, 194-196, 334
  inclusions, 104, 105, 231
  -microcline relations, 104, 105, 110
  sericitized, 195, 233, 279-280
  textures, 217-218, 225, 230-233, 333, 334
  twinning, 334
  zoning, 93, 104, 105, 129, 136, 172, 223, 232, 332
Plagioclase porphyroblasts, see Porphyroblasts
Plagiophyre, 83, 144, 148, 161
Plateau basalts, 356, 375
Pollakeeran breccia-pipe, 158, 164
Plutons, emplacement, 303, 329
  see also Intrusion mechanisms, Main Donegal Granite
Pollakeeran Hill, 204
Polnaguill Syncline, 56, 57, 59
Polymorphic inversions, 326
Porphyrites, composition of, 83
Porphyritic granites, 204-205
  modal composition, 193
Porphyritic texture, 103, 115
Porphyroblastic minerals, 180

Porphyroblasts, 58, 63, 315, 319, 322, 326
  chronology, 272, 292
  growth and deformation, 66-69, 76, 180, 181, 288-290, 293
  in Ardara Pluton, 180, 181, 309, 310
  in Main Donegal Granite, 259, 260, 269-280, 281, 283, 288-293, 309, 311, 312
  in Thorr Pluton, 100-106, 114, 118
  of alumino silicates, 272-277
  of amphibole, 149, 153
  of andalusite, 180, 181, 259, 269, 270, 272, 274, 276, 280, 281, 283, 293, 307, 309-311
  of biotite, 54, 68, 69, 72, 76, 134, 269, 270, 272, 277, 279, 291
  of calc-silicate, 148
  of carbonate, 54, 69, 148, 291
  of chlorite, 54, 66-69, 76, 134, 269, 270, 272, 274, 275, 277, 279, 291
  of chloritoid, 272
  of cordierite, 47, 272, 311
  of epidote, 72
  of garnet, 47, 66-68, 72, 259, 269, 270, 272, 273, 277, 280, 281, 289, 292, 322
  of ilmenite, 66, 67
  of kyanite, 259, 270, 272-276, 280, 281, 289, 292
  of microcline, 118
  of muscovite, 101, 269
  of oligoclase, 114, 203, 259, 270, 272, 274, 276, 281, 292
  of phyllosilicates, 69, 277, 279-281, 289, 290, 311, 312
  of plagioclase, 66, 68, 105, 260, 269-273, 279, 280, 289
  of pyrite, 54, 69
  of sillimanite, 259, 269, 270, 272, 274, 280, 281, 289, 311
  of staurolite, 259, 269
  textures, 213, 272-274
  xenoliths in, 270
Porphyroclasts, feldspar, 234
Porphyry, 365, 366
Porphyry dykes, 187-199, 296
  chemical variation, 196-198
  foliation in, 192
  petrography, 194
Porphyry microgranites, 187, 188
Portabrabane, 47
Portacurry, 188
Portacurry ring dyke, 196
Port Askaig Limestone and Black Pelite, 37
Port Askaig Tillite, 23, 27, 28, 29, 31, 33, 34, 37, 73

SUBJECT INDEX

Port Dolomite Limestone, 20
Port Limestone, 21
Port Lough, 21
Portnablagh, 63, 82
Portnoo, 24, 25, 147, 148, 161, 266, 321
Portnoo Intrusion, 147, 154, 157
Portnoo Limestone, 20, 24
Portsalon, 365
Precambrian, 10, 30, 40, 223
Preferred orientation, in granitic rocks, 335
 of inclusions, 211
 of minerals, 94
Priesttown, 34
Psammites, 24, 25, 28, 31, 32, 34-38, 40, 62, 202
Psammitic gneiss, composition of, 347
Psammitic xenoliths, 119, 125, 128, 262
Pseudo-aplite veins, 159
Pseudodykes, 222, 336
Pseudomorphs, 69, 76, 101, 117, 147, 151, 161
Ptygmatic folds, 239-244
Pyrite, 24, 134
Pyrite porphyroblasts, 54, 69
Pyroxene, 144, 145, 147, 151, 166
 replacement by amphibole, 118
Pyrrhotite, 134

Quartz, 19, 32, 105, 106, 110, 112, 115, 117, 118, 120, 135, 136, 145, 266, 272, 273, 305, 307, 311
 alignment, 270, 286
 -fibrolite, 99
 growth, 260, 262
 inclusions, 274
 intergrowths, 272
 -microcline relations, 104, 105
 preferred orientation, 220, 229
 recrystallization, 333-335
 segregations, 100, 101
 stability fields, 319, 320
 "sweat-outs", 274
 textures, 231, 233
Quartz diorite, see Diorite
Quartz-dolerite, composition of, 314
Quartzite, 19-25, 27, 31, 34-36, 39, 40, 97, 99, 100, 102, 111, 116, 131, 135, 137, 318, 361, 363, 365, 370
 -granite contact, 115-116
 gravity measurements, 340
 mullions, 268, 283, 284, 289
 reaction with magma, 115-116
 xenoliths, 102, 116, 125, 161, 213, 235, 236, 285, 286

Quartzite (continued)
 see also Ards Quartzite
Quartzite rocks, sedimentary structures of, 6
Quartz metadolerites, 43-56
Quartzofeldspathic rocks, 17, 114, 115
Quartzofeldspathic schists, composition of, 344, 347
Quartzofeldspathic veins, 219, 228, 229, 237, 239, 240
 folding of, 281, 282, 285
Quartz veins, 220, 226-229, 231, 234, 252, 289
 folding of, 224, 252

Raft trains, 212-214, 235, 237, 244, 246, 247, 256, 299, 300
 deformation of, 247
Raised beach deposits, 363
Raised beaches, 4
Ranney Point intrusion, 45
Raphoe, 35, 36
Raphoe Anticline, 74, 79, 80
Raphoe Syncline, 35, 77
Rathmullen, 26, 29, 33, 74-76, 363
Rathmullen Syncline, 74, 76
Recrystallization, 47, 48, 58, 66, 68, 82, 93, 99, 100-102, 105, 106, 114, 116, 119, 129, 134, 159, 166, 167, 170, 177, 180, 183, 185, 191, 194, 196, 219, 229, 230, 233, 234, 259, 272, 274, 276, 280, 281, 283, 288, 290, 297, 302-312, 326, 333-335
 chronology, 55-56, 66
Red beds, 363, 365
Relict dykes, 222, 336
Relict pseudodykes, 222
Relicts, 104, 105, 193, 332, 337
 see also Xenoliths
Relict structure, see Ghost structure
Remelting, 115, 119, 123, 124, 131
Remobilization, 222, 226, 251, 289, 290
"Resisters", 119, 130, 222
Rhythmic banding, 25, 39, 106
Ring dykes, 188, 196, 198, 205, 209
Ring granite, 193, 194
Ripidolite, 320
Ripple marks, 6, 16, 19, 24, 31, 32
River systems, 3, 4
Roanish, 185
Rock analyses, 49, 50, 58, 120, 196, 346
 sampling, 312-313
Roof breccia, 162
Roof pendants, 109, 128, 134, 139, 335-336

Rosbeg, 20, 43, 44, 47, 48, 50, 60, 82, 148
Rosbeg amphibolites, 44, 47
 chemical analysis, 58
Rosbeg metadolerites, 43-44, 47
 chemical analysis, 49, 50
 deformation, 48
 metamorphic mineralogy, 47, 49, 50
Rosbeg Semipelites, 20, 25
Rosguill, 4, 20, 21, 24, 25, 40, 45, 56, 58, 59, 63, 132-135, 138-141, 294, 303, 312, 359
 map, 133
Rosguill fold, 55, 63
Roskin, 148
Rosnakill Pelitic Schists, 29
Rosses, the, 3, 186, 198, 199, 210, 373
Rosses aplites, age of, 90, 341
Rosses Centered Complex, 11, 186-200
 age, 187
  Rb-Sr, 90, 91, 341
 aureole, 305, 318
 chemical composition, 193, 300, 343, 344, 346, 347
 contacts, 186, 194, 198-200, 333
 crystallization history, 192, 195-198
 dykes, 187-200, 224
 emplacement, 89, 90, 187-190, 198-200
 field relations, 187-192
 gravity investigations, 340
 magma, 190, 194-198
 magmatic origin, 343
 mineral paragenesis, 194-196
 rock analyses, 196, 346
 shape, 340
 Sr isotope ratios, 349
 structural map, 188
 trace elements, 197
 xenoliths in, 187, 188, 191, 196, 199
Rosses Pluton, see Rosses Centered Complex
Rosses dyke swarm, 90, 296
Rosses porphyry swarm, 223

Salt diapirs, 184
Sand Lough, 91
Sandstones, 19, 36, 39, 40, 367, 370
Saussurite-uralite, 268
Scarba Conglomerate Group, 29, 37
Scarba Transition Group, 29, 37
Schistosity, 8, 47, 50, 54, 55, 58, 61, 63, 65, 66, 68, 71, 72, 74, 75, 77, 79, 81, 82, 100, 179-181, 213, 227, 229, 235, 237, 238, 246, 249, 259-262, 265, 270, 272-274, 276, 277, 279-282, 286, 288, 290, 293, 294, 309, 310

SUBJECT INDEX

Schistosity (*continued*)
  definition, 54
Schists, 6, 72, 87, 89, 258, 262, 264-267, 286
  albite-muscovite, 36
  biotite, 268
  biotite-muscovite-chlorite, 10
  biotite-quartz, 302, 310
  calc-silicate, 152, 159
  chlorite, 268
  chlorite-garnet, 213
  cleavage, 283, 288
  contact, 180, 260, 302, 309, 310, 325
  cross-cleavages, 230
  deformed, 32
  garnet-biotite, 76
  inclusions in, 215, 216
  marginal, 213
  metamorphism of, 149
  mineral alignment, 281
  muscovite-biotite, 202
  phyllitic, 44
  psammitic, 10
  quartzo-feldspathic, 344, 347
  semipelitic, 27
  staurolite-garnet-kyanite, 302-303
Schlieren, 102, 139, 174, 218, 235, 244, 350
  biotite, 191, 192, 195, 298
  exogenic origin, 117-119
Scotland, 1, 3, 6, 18, 25, 142
  faults, correlation with Irish, 77, 80, 357, 359, 363, 364
  ghost structure, 336
  'Newer Granites', 90, 296
  see also Argyllshire, Ballachulish, Islay, Loch Awe, Moine
Scottish ice-cap, 3
Scraigs, 266
Sederholm effect, 221, 222, 336
Sedimentary assemblages, 15-42
  black slate association, 19, 27, 39
Sedimentary structures, 6, 16, 19, 23-25, 27, 28, 31, 32, 37, 39, 40, 42, 45, 81, 106, 329, 370
Sedimentation, 357, 359, 367, 370
  turbiditic, 40
Sediments, deltaic, 366, 367, 370
  marine, 366
  perideltaic, 370
  shallow-water, 28, 30, 35, 37, 39, 40, 356, 366-369, 371
Semipelite xenoliths, 112, 113, 119, 125, 262, 282
Semipelites, 18, 19, 24, 25, 31, 32, 36, 100, 123, 134, 137, 259
  quartzofeldspathic, 36
  schistosity, 270

Semipelitic gneiss, 347
Semipelitic rocks, 53
Sericite, 180, 193, 195, 312
Sessiagh Banded Quartzites, 20, 21
Sessiagh Clonmass Formation, 18, 19, 21, 23, 24, 58, 73
  folds, 262, 265
Sessiagh Quartzites, 21
Shales, 367, 370
  black, 19, 27, 39
Shanaghy Green Beds, 33
Shear folds, 277
Shear-fracture filling, 200
Shear zones, 8, 65, 229, 249, 260, 261, 268, 272, 289, 290, 330
Sheephaven, 261
Sheet Complex, 204-205
Sheskinarone, 91, 191, 193, 194
Shimmer aggregates, 312
Shoreline, Dalradian, 40, 42
  Viséan, 369-370
Shuna Limestone, 29, 37
Silica in contact facies, 123
Silicate analysis, 312
Sillimanite, 99-101, 114, 115, 117, 133, 134, 180, 268, 272, 279, 280, 293, 302, 303, 305-309, 311, 312, 314, 318, 322, 327
  alignment, 278
  host-rock control, 317
  mats, 305
  preferred orientation, 276
  stability field, 319, 324
Sillimanite-muscovite-tourmaline, 307
Sillimanite porphyroblasts, *see* Porphyroblasts
Sills, 10
  appinitic, 161
  dolerite, 32, 43-50
  felsite, 82
  granitic, 282, 286
  lamprophyre, 82, 179
  metadolerite, 179, 281, 283
  microgranite, 215, 229
  quartz-dolerite, 314
Siltstones, 32
  granite clasts in, 23
Silurian, 6, 11, 348
  palaeogeography, 355-356
  uplift, 356
Silver Hill, 76
Silver Hill Quartzite, 28, 33
Sir Albert's Bridge, 69
Skarns, 116, 153, 235
Skelpoonagh Bay Limestone, 28
Skialiths, 337
Slickensides, 212, 230, 249, 261, 361, 363

Slide, definition of, 56
Slide breccias, 37, 40
Slides, 17, 18, 20, 23, 24, 26, 36, 38, 39, 48, 54-58, 73, 77, 79, 81, 371
Slieve League, 1, 3, 4, 17, 27, 28, 31, 54, 69-72, 79
Slieve League Anticline, 70, 79
Slieve League Fault, 358
Slieve League Formation, 28, 31, 33, 37
Slieve Tooey, 32
Slieve Tooey Quartzite, 27-29, 31-33, 35-37, 39, 40, 70, 73
Slieve Tooey Syncline, 70
Sligo, 369
Slump structures, 19, 370
Sodic mantle, 232
Solid-state mechanisms, 93
Southern Loughros Group, 148
Sperrin Mountains, 53, 74, 77
Spessartite garnet, 315
Spessartites, 144
Sphalerite, 371
Sphene, 105, 106, 112, 115, 117
Sruhanavarnis, 221
Staghall Group, 20
Staurolite, 134, 180, 268, 293, 302, 309, 311, 312, 318, 326
  alteration to muscovite, 279
  host-rock control, 315, 317
  inclusions, 276, 311
  predating andalusite, 304
  PT controls, 321, 323
  stability fields, 319-324
Staurolite porphyroblasts, *see* Porphyroblasts
Stoping, *see* Intrusion mechanisms
Strabane, 53, 77
Strabane Syncline, 74, 75, 80
Stragar River, 71, 79
Stragar River Syncline, 71
Stragill Group, 26, 29
Strain-slip, *see* Crenulation cleavage
Stranaglogh Fault, 358
Stratigraphic nomenclature, 16, 19
Stretching lineation, 58, 61, 81
Strike-slip, 178, 357, 359, 361-364, 371
Strike swing, 61, 65, 79, 82
Stromatolite reefs, 40
Strontium isotope ratios, 349
Structural map, 52
Structures, in granitic rocks, 93-95
  in Main Donegal Granite envelope, 258-268, 282, 283, 286-288, 291, 294
  in polymetamorphosed sediments, 329

SUBJECT INDEX

Structures (*continued*)
  primary and secondary, 328
Submarine mudflows, 30
Sulphide mineralization, 371
Sulphides, precipitation, 371, 372
Summy Lough, 148
Sun cracks, 23
"Sweat-outs", 101, 272, 274
Swilly Chloritic Schist, 28
Swilly River, 3, 76
Synclines, *see* Folds, synclinal
Synneusis, 333
Synplutonic deformation, 210-256, 348, 351
Synplutonic dykes, 222, 330, 336
Syntectonic crystallization, 270
Syntectonic dykes, 251
Syntectonic growth, 259, 270, 271, 274
Syntectonic minerals, 283

Tallabrista, 161
Tay Nappe, 77, 79, 82
Tayvallich Limestone, 37, 78
Tectonic events, correlation with metamorphism, 269, 280-283, 289
Tectonic thinning, 18, 23
Teelin Point Formation, 28, 31
Teelin Schist, 28, 31
Termon Formation, 26, 28, 31, 32-37, 47, 70, 76
Termon Green Schists, 29
Termon Group, 29
Termon Pelites, 35
Tertiary basaltic lavas, 77, 375
Tertiary dykes, 201, 202, 207-208, 359
Tertiary dyke swarms, 3, 6, 12, 373-375
  map, 374
Tertiary erosion surfaces, 3, 356
Tertiary granite, 374
Textures, aplitic, 298
  appinitic, 145-147, 150
  crystalloblastic, 170, 171, 176
  granitic, 93, 193, 332-335
  harristic, 106
  in Thorr Pluton, 102-106
  metamorphic, 53, 321, 326
  ophitic, 45, 48
  porphyritic, 103
Thermal conductivity, 318
Thermal domes, 348, 351
Thermal metamorphism, 203, 281, 291, 303, 318, 319, 321
Thermal events, 341-342
Thermal spalling, 139, 164
Tholeiites, 10, 43-50, 374

Tholeiites (*continued*)
  chemical composition, 49-50
Thorr, 60, 61, 89, 97, 100, 112, 118-120, 123-128, 131, 141, 211, 213, 255, 282, 302, 307
Thorr Anticline, 57, 60, 61
Thorr Granodiorite, *see* Thorr Pluton
Thorr Pluton, 64, 65
  aureole, 89, 90, 97-102, 109, 111-113, 123, 124, 131
    porphyroblasts in, 307, 308
    temperature, 318
    width, 318
  color indices, 109, 110, 112, 117, 127
  composition, 108-112, 116, 119, 120-124, 127, 128, 342-344, 346
  contact metamorphism, 99-102, 110-124, 128, 129, 305, 307, 308
  contacts, 91-99, 109, 111-117, 119, 123, 124, 129, 130
  contamination, 97, 109, 115, 117, 119-124, 130, 131
  crystallization differentiation, 342
  deformation, 99, 101, 102, 129, 131
  emplacement, 11, 89, 99, 102, 114, 128-131
  folds, 99
  foliation in, 97, 98
  ghost stratigraphy, 17, 60, 102, 110, 124-130, 336, 337
  gravity investigations, 340
  K-Ar age, 90, 91
  magma, 102-120, 123, 129
  map, 98
  origin, 129-131
  outcrop, 97, 111-120
  porphyroblasts in, 100-106, 114, 118
  rock analyses, 120, 346
  shape, 98, 340
  synneusis features, 333
  textures, 102-105, 278
  trend surface analysis, 108-112, 127
  xenoliths in, 60, 97-131
  zoning of feldspar inclusions, 332
Thorr quartz-diorite, 259
Thorr rocks, xenoliths of, 213, 235, 237, 246, 255
Thorr tonalite, folds in, 282
  xenoliths in, 282
Thrusts, 28, 29, 54, 56, 71, 77, 79, 178
Tievealehid, 99
Tillite, 10, 23, 27-30, 34-36, 38-40, 78

Tonalite, 97, 102, 108, 111, 115, 117, 123, 124, 128, 130, 131, 134, 139, 140
  abundance in Donegal Pluton, 345
  classification, 93
  folds in, 282
  -quartz-diorite, 123, 318
  xenoliths, 237, 246
Tonalitic facies, 120, 124
Toome Lough Semipelitic-Pelitic Group, 21
Toories, 184, 185
Toories Pluton, 184, 185
  age, 90
  aureole, 89
  emplacement, 11, 89
  foliation, 185
  shape, 340
Torbenite, 194
Tory Island, 125, 149, 159
Tourmaline, 101, 270, 305
Tourmaline-quartz, 172
Trace elements, 197
Trace fossils, 24, 370
Trawenagh Bay, 8, 11, 373
Trawenagh Bay Pluton, 296-301, 318
  age, 297
    Rb-Sr, 90, 341
  boundary with Main Donegal Granite, 259
  chemical composition, 298, 300, 344, 346, 347
  contacts, 296, 297, 299
  emplacement, 11, 89, 299-301
  envelope, 303
  fabric, 299
  gravity investigations, 340
  magma, 298-299
  map, 297
  petrofabrics, 299
  relation to Main Donegal Pluton, 296, 299-301
  rock analyses, 346
  shape, 340
  Sr isotope ratios, 149
  xenoliths, 298, 299, 301
Tremolite, 269
Trend surface analysis, 108-112, 127
Trondhjemite, 215, 218, 233
Tubberkeen, 107
Tuffisite veins, 149, 159
Turbidites, 32, 39, 41, 42
Twinning in feldspars, 93, 105, 333
Tyrone, 33, 35, 36, 74

Ultrabasic rocks, banding in, 219
  emplacement and deformation, 335
  hornblende-rich, 152, 165, 166
Ultramafic appinites, 168

SUBJECT INDEX

Ultramafic rocks, 109
Unroofing, 355-357
Uplift, 348, 356, 357
Upper Caninian fossils, 369
Upper Crana-Quartzite, 29, 32, 33, 69
Upper Falcarragh Pelites, 20, 21, 25, 47
  chemical analyses, 114
  contact metamorphism, 312
  folds in, 262, 265-267
  porphyroblasts in, 270-275
  textures, 278
  variation in Mg^{2+}, 313, 314
  xenoliths of, 270, 271, 287
Upper Quartzite, 26
Upper Termon Schists, 28
Uralite, 269
  saussurite-, 268

Varangian glaciation, 30, 40
Veins, albite, 231, 232
  epidote, 226, 227
  folding of, 224, 239
  granite, 228, 239, 242, 243
  pegmatite, 103, 224, 265-267, 287, 289
  pseudo-aplite, 159
  quartz, 220, 224, 226, 228, 229, 231, 234, 252, 289
  quartzofeldspathic, 219, 228, 229, 237, 239, 240, 281-284, 285
Viséan, 12
  environment, 370-371
  palaeogeography, 369-371
  shoreline, 369-370
  stratigraphy, 366-368
Viséan Marine Transgression, 356, 366-371
Viséan rocks, thickness of, 366-368
Viséan sediments, deposition of, 360
Vogesites, 144
Von Wolff diagram, 122

Washouts, 19
Watersheds, 3, 4
Way-up, 6, 16, 27, 60
Weardale Granite, 374

Wollastonite, 116, 269
Worm casts, 40
Wrench faults, 357-363

Xenocrysts, 83
Xenoliths, 205, 335-337
  calcareous, 115, 117, 119, 239
  deformation of, 45, 99, 100, 133, 216, 224, 243, 246, 249, 252, 330
  deformation ellipsoids for, 244
  deformed veins, in, 239
  ductility, 252
  feldspathized, 222, 235
  in Blind Rock Dyke, 375
  in dolerite sills, 45, 46
  in Fanad Pluton, 133, 135, 138-141
  in granite, 82, 119, 270, 271, 287
  in granodiorite, 111, 116, 139
  in intrusive breccias, 149, 154, 159, 161, 162
  in lamprophyres, 83, 162
  in Main Donegal Granite, 211-213, 215, 221-224, 227, 234-247, 249, 252, 269, 276
  in monzodiorite, 131, 170-174, 185
  in monzotonalite, 171, 185
  in pelites, 111
  in porphyroblasts, 270
  in Rosses Centered Complex, 187, 189, 191, 196, 199
  internal structures, 235-239, 244-247
  in Thorr Pluton, 60, 97-131, 282
  in tonalite, 124, 282
  in tonalite-quartz diorite, 111
  in Trawenagh Bay Pluton, 298, 299, 301
  irrotational deformation, 244-247
  lineation, 244, 246, 247
  meta-igneous, 115
  mineralogy, 173
  of amphibolite, 171, 173, 176
  of appinites, 149, 171, 173, 213, 235, 246, 255
  of Ardara Pluton, 213, 235, 246
  of calc-silicate, 128, 161
  of Creeslough Succession, 124, 125, 130

Xenoliths (continued)
  of diorite, 171, 173, 174
  of felsite, 99
  of granite, 216, 223, 224, 235, 238, 246, 252
  of granodiorite, 299
  of igneous rocks, 246
  of lamprophyres, 82, 99
  of limestone, 116, 117, 235, 236, 239, 241-243, 249
  of metabasites, 101
  of metadolerites, 119, 125, 139, 171, 235, 237, 246, 255
  of metasediments, 101, 139, 235, 246
  of mobilized hornfels, 45, 46
  of mullioned quartzite, 285, 286
  of pelite, 111, 114, 115, 119, 125, 128, 215, 221, 235, 239, 266, 267, 278, 282, 287, 288, 299
  of psammite, 119, 125, 128, 262
  of quartzite, 102, 116, 125, 213, 235, 236
  of semipelite, 112, 113, 119, 125, 235, 262, 282, 299
  of Thorr rocks, 213, 235, 237, 246, 255, 299
  of tonalite, 189
  of Upper Falcarragh Pelites, 270, 271, 287
  orientation of, 235-238, 246, 247, 299
  plagioclase zoning in, 172
  sedimentary, 7, 8
  shape, 235, 246
  structures of, 175, 176, 185, 244-247
  see also Rafts and Schlieren

Zinc ores, 371
Zircon ages, Th-Pb, 342
  U-Pb, 90
Zircon morphology, 93, 217, 233
Zone of Veins, 114
Zoning, mineral, 134, 303-305, 317, 324, 325
  oscillatory in plagioclase, 172, 332

The authors are indebted to Mrs. G. Flinn for the preparation of this index.

Date Due

UML 735